FINITE-TEMPERATURE FIELD THEORY
Principles and Applications

This book develops the basic formalism and theoretical techniques for studying relativistic quantum field theory at high temperature and density. Specific physical theories treated include QED, QCD, electroweak theory, and effective nuclear field theories of hadronic and nuclear matter. Topics include functional integral representation of the partition function, diagrammatic expansions, linear response theory, screening and plasma oscillations, spontaneous symmetry breaking, the Goldstone theorem, resummation and hard thermal loops, lattice gauge theory, phase transitions, nucleation theory, quark–gluon plasma, and color superconductivity. Applications to astrophysics and cosmology include white dwarf and neutron stars, neutrino emissivity, baryon number violation in the early universe, and cosmological phase transitions. Applications to relativistic nucleus–nucleus collisions are also included. This title, first published in 2006, has been reissued as an Open Access publication on Cambridge Core.

JOSEPH I. KAPUSTA is Professor of Physics at the School of Physics and Astronomy, University of Minnesota, Minneapolis. He received his Ph.D. from the University of California, Berkeley, in 1978 and has been a faculty member at the University of Minnesota since 1982. He has authored over 150 articles in refereed journals and conference proceedings. Since 1997 he has been an associate editor for *Physical Review C*. He is a Fellow of the American Physical Society and of the American Association for the Advancement of Science. The first edition of *Finite-Temperature Field Theory* was published by Cambridge University Press in 1989; a paperback edition followed in 1994.

CHARLES GALE is James McGill Professor at the Department of Physics, McGill University, Montreal. He received his Ph.D. from McGill University in 1986 and joined the faculty there in 1989. He has authored over 100 articles in refereed journals and conference proceedings. Since 2005 he has been the Chair of the Department of Physics at McGill University. He is a Fellow of the American Physical Society.

CAMBRIDGE MONOGRAPHS ON MATHEMATICAL PHYSICS

General editors: P. V. Landshoff, D. R. Nelson, S. Weinberg

[1] Issued as a paperback

Finite-Temperature Field Theory
Principles and Applications

JOSEPH I. KAPUSTA

School of Physics and Astronomy, University of Minnesota

CHARLES GALE

Department of Physics, McGill University

Shaftesbury Road, Cambridge CB2 8EA, United Kingdom

One Liberty Plaza, 20th Floor, New York, NY 10006, USA

477 Williamstown Road, Port Melbourne, VIC 3207, Australia

314–321, 3rd Floor, Plot 3, Splendor Forum, Jasola District Centre, New Delhi – 110025, India

103 Penang Road, #05-06/07, Visioncrest Commercial, Singapore 238467

Cambridge University Press is part of Cambridge University Press & Assessment, a department of the University of Cambridge.

We share the University's mission to contribute to society through the pursuit of education, learning and research at the highest international levels of excellence.

www.cambridge.org
Information on this title: www.cambridge.org/9781009401951

DOI: 10.1017/9781009401968

First published 2006
Reissued as OA 2023

A catalogue record for this publication is available from the British Library.

ISBN 978-1-009-40195-1 Hardback
ISBN 978-1-009-40198-2 Paperback

Contents

v

Preface

What happens when ordinary matter is so greatly compressed that the electrons form a relativistic degenerate gas, as in a white dwarf star? What happens when the matter is compressed even further so that atomic nuclei overlap to form superdense nuclear matter, as in a neutron star? What happens when nuclear matter is heated to such great temperatures that the nucleons and pions melt into quarks and gluons, as in high-energy nuclear collisions? What happened in the spontaneous symmetry breaking of the unified theory of the weak and electromagnetic interactions during the big bang? Questions like these have fascinated us for a long time. The purpose of this book is to develop the fundamental principles and mathematical techniques that enable the formulation of answers to these mind-boggling questions. The study of matter under extreme conditions has blossomed into a field of intense interdisciplinary activity and global extent. The analysis of the collective behavior of interacting relativistic systems spans a rich palette of physical phenomena. One of the ultimate goals of the whole program is to map out the phase diagram of the standard model and its extensions.

This text assumes that the reader has completed graduate level courses in thermal and statistical physics and in relativistic quantum field theory. Our aims are to convey a coherent picture of the field and to prepare the reader to read and understand the original and current literature. The book is not, however, a compendium of all known results; this would have made it prohibitively long. We start from the basic principles of quantum field theory, thermodynamics, and statistical mechanics. This development is most elegantly accomplished by means of Feynman's functional integral formalism. Having a functional integral expression for the partition function allows a straightforward derivation of diagrammatic rules for interacting field theories. It also provides a framework for defining gauge theories on finite lattices, which then enables integration by Monte Carlo techniques. The formal aspects are illustrated with applications drawn from fields of research that are close to the authors' own experience. Each chapter carries its own exercises, reference list, and select bibliography.

The book is based on *Finite-Temperature Field Theory*, written by one of us (JK) and published in 1989. Although the fundamental principles have not changed, there have been many important developments since then, necessitating a new book.

We would like to acknowledge the assistance of Frithjof Karsch and Steven Gottlieb in transmitting some of their results of lattice computations, presented in Chapter 10, and Andrew Steiner for performing the numerical calculations used to prepare many of the figures in Chapter 11. We are grateful to a number of friends, colleagues, and students for their helpful comments and suggestions and for their careful reading of the manuscript, especially Peter Arnold, Eric Braaten, Paul Ellis, Philippe de Forcrand, Bengt Friman, Edmond Iancu, Sangyong Jeon, Keijo Kajantie, Frithjof Karsch, Mikko Laine, Stefan Leupold, Guy Moore, Ulrich Mosel, Robert Pisarski, Brian Serot, Andrew Steiner, and Laurence Yaffe.

1

Review of quantum statistical mechanics

Thermodynamics is used to describe the bulk properties of matter in or near equilibrium. Many scientists, notably Boyle, Carnot, Clausius, Gay-Lussac, Gibbs, Joule, Kelvin, and Rumford, contributed to the development of the field over three centuries. Quantities such as mass, pressure, energy, and so on are readily defined and measured. Classical statistical mechanics attempts to understand thermodynamics by the application of classical mechanics to the microscopic particles making up the system. Great progress in this field was made by physicists like Boltzmann and Maxwell. Temperature, entropy, particle number, and chemical potential are thus understandable in terms of the microscopic nature of matter. Classical mechanics is inadequate in many circumstances however, and ultimately must be replaced by quantum mechanics. In fact, the ultraviolet catastrophe encountered by the application of classical mechanics and electromagnetism to blackbody radiation was one of the problems that led to the development of quantum theory. The development of quantum statistical mechanics was achieved by a number of twentieth century physicists, most notably Planck, Einstein, Fermi, and Bose. The purpose of this chapter is to give a mini-review of the basic concepts of quantum statistical mechanics as applied to noninteracting systems of particles. This will set the stage for the functional integral representation of the partition function, which is a cornerstone of modern relativistic quantum field theory and the quantum statistical mechanics of interacting particles and fields.

1.1 Ensembles

One normally encounters three types of ensemble in equilibrium statistical mechanics. The *microcanonical* ensemble is used to describe an isolated system that has a fixed energy E, a fixed particle number N, and a fixed volume V. The *canonical* ensemble is used to describe a system in contact with a heat reservoir at temperature T. The system can freely exchange energy with the reservoir, but the particle number and volume are fixed. In the *grand canonical* ensemble the system can exchange particles as well as energy with a reservoir. In this ensemble the temperature, volume, and chemical potential μ are fixed quantities. The standard thermodynamic relations are summarized in appendix section A1.1.

In the canonical and grand canonical ensembles, $T^{-1} = \beta$ may be thought of as a Lagrange multiplier that determines the mean energy of the system. Similarly, μ may be thought of as a Lagrange multiplier that determines the mean number of particles in the system. In a relativistic quantum system, where particles can be created and destroyed, it is most straightforward to compute observables in the grand canonical ensemble. For that reason we use the grand canonical ensemble throughout this book. There is no loss of generality in doing so because one may pass over to either of the other ensembles by performing an inverse Laplace transform on the variable μ and/or the variable β. See appendix section A1.2.

Consider a system described by a Hamiltonian H and a set of conserved number operators $\hat{N}_i$. (A hat or caret is used to denote an operator for emphasis or whenever there is the possibility of an ambiguity.) In QED, for example, the number of electrons minus the number of positrons is a conserved quantity, not the number of electrons or positrons separately, because of reactions like $e^+e^- \rightarrow e^+e^+e^-e^-$. These number operators must be Hermitian and must commute with H as well as with each other. They must also be extensive (scale with the volume of the system) in order that the usual macroscopic thermodynamic limit can be taken. The statistical density matrix $\hat{\rho}$ is the fundamental object in equilibrium statistical mechanics:

$$\hat{\rho} = \exp\left[-\beta\left(H - \mu_i \hat{N}_i\right)\right] \tag{1.1}$$

Here and throughout the book a repeated index is assumed to be summed over. In QED the sum would run over two conserved number operators if one allowed for both electrons and muons. The statistical density matrix is used to compute the ensemble average of any desired observable, represented by the operator $\hat{A}$, via

$$A = \langle \hat{A} \rangle = \frac{\text{Tr}\, \hat{A}\hat{\rho}}{\text{Tr}\, \hat{\rho}} \tag{1.2}$$

where Tr denotes the trace operation.

The grand canonical partition function

$$Z = Z(V, T, \mu_1, \mu_2, \ldots) = \text{Tr}\, \hat{\rho} \tag{1.3}$$

is the single most important function in thermodynamics. From it all the thermodynamic properties may be determined. For example, the pressure, particle number, entropy, and energy are, in the infinite-volume limit, given by

$$P = \frac{\partial(T \ln Z)}{\partial V}$$
$$N_i = \frac{\partial(T \ln Z)}{\partial \mu_i} \tag{1.4}$$
$$S = \frac{\partial(T \ln Z)}{\partial T}$$
$$E = -PV + TS + \mu_i N_i$$

1.2 One bosonic degree of freedom

As a simple example consider a time-independent single-particle quantum mechanical mode that may be occupied by bosons. Each boson in that mode has the same energy ω. There may be 0, 1, 2, or any number of bosons occupying that mode. There are no interactions between the particles. This system may be thought of as a set of noninteracting quantized simple harmonic oscillators. It will serve as a prototype of the relativistic quantum field theory systems to be introduced in later chapters. We are interested in computing the mean particle number, energy, and entropy. Since the system has no volume there is no physical pressure.

Denote the state of the system by $|n\rangle$, which means that there are n bosons in the system. The state $|0\rangle$ is called the vacuum. The properties of these states are

$$\langle n|n'\rangle = \delta_{nn'} \quad \text{orthogonality} \tag{1.5}$$

$$\sum_{n=0}^{\infty} |n\rangle\langle n| = 1 \quad \text{completeness} \tag{1.6}$$

One may think of the bras $\langle n|$ and kets $|n\rangle$ as row and column vectors, respectively, in an infinite-dimensional vector space. These vectors form a complete set. The operation in (1.5) is an inner product and the number 1 in (1.6) stands for the infinite-dimensional unit matrix.

It is convenient to introduce creation and annihilation operators, $a^\dagger$ and a, respectively. The creation operator creates one boson and puts it in the mode under consideration. Its action on a number eigenstate is

$$a^\dagger|n\rangle = \sqrt{n+1}|n+1\rangle \tag{1.7}$$

Similarly, the annihilation operator annihilates or removes one boson,

$$a|n\rangle = \sqrt{n}|n-1\rangle \tag{1.8}$$

unless $n = 0$, in which case it annihilates the vacuum,

$$a|0\rangle = 0 \qquad (1.9)$$

Apart from an irrelevant phase, the coefficients appearing in (1.7) and (1.8) follow from the requirements that $a^\dagger$ and a be Hermitian conjugates and that $a^\dagger a$ be the number operator $\hat{N}$. That is,

$$\hat{N}|n\rangle = a^\dagger a|n\rangle = n|n\rangle \qquad (1.10)$$

As a consequence the commutator of a with $a^\dagger$ is

$$[a, a^\dagger] = aa^\dagger - a^\dagger a = 1 \qquad (1.11)$$

We can build all states from the vacuum by repeated application of the creation operator:

$$|n\rangle = \frac{1}{\sqrt{n!}}(a^\dagger)^n|0\rangle \qquad (1.12)$$

Next we need a Hamiltonian. Up to an additive constant, it must be ω times the number operator. Starting with a wave equation in nonrelativistic or relativistic quantum mechanics the additive constant emerges naturally. One finds that

$$H = \tfrac{1}{2}\omega\left(aa^\dagger + a^\dagger a\right) = \omega\left(a^\dagger a + \tfrac{1}{2}\right) = \omega\left(\hat{N} + \tfrac{1}{2}\right) \qquad (1.13)$$

The additive term $\tfrac{1}{2}\omega$ is the zero-point energy. Usually this term can be ignored. Exceptions arise when the vacuum changes owing to a background field, such as the gravitational field or an electric field, as in the Casimir effect. We shall drop this term in the rest of the chapter and leave it as an exercise to repeat the following analysis with the inclusion of the zero-point energy.

The states $|n\rangle$ are simultaneous eigenstates of energy and particle number. We can assign a chemical potential to the particles. This is possible because there are no interactions to change the particle number. The partition function is easily computed:

$$\begin{aligned}
Z &= \mathrm{Tr}\,\mathrm{e}^{-\beta(H-\mu\hat{N})} = \mathrm{Tr}\,\mathrm{e}^{-\beta(\omega-\mu)\hat{N}} \\
&= \sum_{n=0}^{\infty}\langle n|\mathrm{e}^{-\beta(\omega-\mu)\hat{N}}|n\rangle = \sum_{n=0}^{\infty}\mathrm{e}^{-\beta(\omega-\mu)n} \qquad (1.14) \\
&= \frac{1}{1-\mathrm{e}^{-\beta(\omega-\mu)}}
\end{aligned}$$

The mean number of particles is found from (1.4) to be

$$N = \frac{1}{\mathrm{e}^{\beta(\omega-\mu)} - 1} \qquad (1.15)$$

and the mean energy E is ωN. Note that N ranges continuously from zero to infinity as μ ranges from $-\infty$ to ω. Values of the chemical potential, in this system, are restricted to be less than ω on account of the positivity of the particle number or, equivalently, the Hermiticity of the number operator.

There are two interesting limits. One is the classical limit, where the occupancy is small, $N \ll 1$. This occurs when $T \ll \omega - \mu$. In this limit the exponential in (1.15) is large and so

$$N = \mathrm{e}^{-\beta(\omega-\mu)} \qquad \text{classical limit} \qquad (1.16)$$

The other is the quantum limit, where the occupancy is large, $N \gg 1$. This occurs when $T \gg \omega - \mu$.

1.3 One fermionic degree of freedom

Now consider the same problem as in the previous section but for fermions instead of bosons. This is a prototype for a Fermi gas, and later on will help us to formulate the functional integral expression for the partition function involving fermions. These could be electrons and positrons in QED, neutrons and protons in nuclei and nuclear matter, or quarks in QCD.

The Pauli exclusion principle forbids the occupation of a single-particle mode by more than one fermion. Thus there are only two states of the system, $|0\rangle$ and $|1\rangle$. The action of the fermion creation and annihilation operators on these states is as follows:

$$\begin{aligned} \alpha^\dagger|0\rangle &= |1\rangle \\ \alpha|1\rangle &= |0\rangle \\ \alpha^\dagger|1\rangle &= 0 \\ \alpha|0\rangle &= 0 \end{aligned} \qquad (1.17)$$

Therefore, these operators have the property that their square is zero when acting on any of the states,

$$\alpha\alpha = \alpha^\dagger\alpha^\dagger = 0 \qquad (1.18)$$

Up to an arbitrary phase factor, the coefficients in (1.17) are chosen so that α and $\alpha^\dagger$ are Hermitian conjugates and $\alpha^\dagger\alpha$ is the number operator $\hat{N}$:

$$\hat{N}|n\rangle = n|n\rangle \qquad (1.19)$$

It follows that the creation and annihilation operators satisfy the anti-commutation relation

$$\{\alpha, \alpha^\dagger\} = \alpha\alpha^\dagger + \alpha^\dagger\alpha = 1 \tag{1.20}$$

The Hamiltonian is taken to be

$$H = \tfrac{1}{2}\omega(\alpha^\dagger\alpha - \alpha\alpha^\dagger) = \omega\left(\hat{N} - \tfrac{1}{2}\right) \tag{1.21}$$

This form follows from the Dirac equation. Notice that the zero-point energy is equal in magnitude but opposite in sign to the bosonic zero-point energy. In this chapter we drop this term for fermions, as we have for bosons.

The partition function is computed as in (1.14) except that the sum terminates at $n = 1$ on account of the Pauli exclusion principle:

$$\begin{aligned} Z &= \mathrm{Tr}\,\mathrm{e}^{-\beta(H-\mu\hat{N})} = \mathrm{Tr}\,\mathrm{e}^{-\beta(\omega-\mu)\hat{N}} \\ &= \sum_{n=0}^{1}\langle n|\mathrm{e}^{-\beta(\omega-\mu)\hat{N}}|n\rangle = \sum_{n=0}^{1}\mathrm{e}^{-\beta(\omega-\mu)n} \\ &= 1 + \mathrm{e}^{-\beta(\omega-\mu)} \end{aligned} \tag{1.22}$$

The mean number of particles is found from (1.4) to be

$$N = \frac{1}{\mathrm{e}^{\beta(\omega-\mu)} + 1} \tag{1.23}$$

and the mean energy E is ωN. Note that N ranges continuously from zero to unity as μ ranges from $-\infty$ to ∞. Unlike bosons, for fermions there is no restriction on the chemical potential.

As with bosons, there are two interesting limits. One is the classical limit, where the occupancy is small, $N \ll 1$. This occurs when $T \ll \omega - \mu$:

$$N = \mathrm{e}^{-\beta(\omega-\mu)} \quad \text{classical limit} \tag{1.24}$$

which is the same limit as for bosons. The other is the quantum limit. When $T \to 0$ one obtains $N \to 0$ if $\omega > \mu$ and $N \to 1$ if $\omega < \mu$.

1.4 Noninteracting gases

Now let us put particles, either bosons or fermions, into a box with sides of length L. We neglect their mutual interactions, although in principle they must interact in order to come to thermal equilibrium. One can imagine including interactions, waiting until the particles come to equilibrium, and then slowly turning off the interactions. Such a noninteracting gas is often a good description of the atmosphere around us, electrons in a metal or white dwarf star, blackbody photons in a heated cavity or in the cosmic microwave background radiation, phonons in low-temperature materials, neutrons in a neutron star, and many other situations.

In the macroscopic limit the boundary condition imposed on the surface of the box is unimportant. For definiteness we impose the condition that the wave function vanishes at the surface of the box. (Also frequently used are periodic boundary conditions.) The vanishing of the wave function on the surface means that an integral number of half-wavelengths must fit in the distance L:

$$\lambda_x = 2L/j_x \quad \lambda_y = 2L/j_y \quad \lambda_z = 2L/j_z \tag{1.25}$$

where j_x, j_y, j_z are all positive integers. The magnitude of the x component of the momentum is $|p_x| = 2\pi/\lambda_x = \pi j_x/L$, and similarly for the y and z components. Amazingly, quantum mechanics tells us that these relations hold for both nonrelativistic and relativistic motion, for both bosons and fermions.

The full Hamiltonian is the sum of the Hamiltonians for each mode on account of the assumption that the particles do not interact. We use a shorthand notation in which $\mathbf{j}$ represents the triplet of numbers (j_x, j_y, j_z) that uniquely specifies each mode. Thus the Hamiltonian and number operator are

$$H = \sum_{\mathbf{j}} H_{\mathbf{j}}$$
$$\hat{N} = \sum_{\mathbf{j}} \hat{N}_{\mathbf{j}} \tag{1.26}$$

Then the partition function is the product of the partition functions for each mode:

$$Z = \mathrm{Tr}\, e^{-\beta(H - \mu\hat{N})} = \prod_{\mathbf{j}} \mathrm{Tr}\, e^{-\beta(H_j - \mu\hat{N}_j)} = \prod_{\mathbf{j}} Z_{\mathbf{j}} \tag{1.27}$$

Each mode corresponds to the single bosonic or fermionic degree of freedom discussed previously.

According to (1.4) it is $\ln Z$ that is of fundamental interest. From (1.27),

$$\ln Z = \sum_{j_x=1}^{\infty} \sum_{j_1=1}^{\infty} \sum_{j_z=1}^{\infty} \ln Z_{j_x, j_y, j_z} \tag{1.28}$$

In the macroscopic limit, $L \to \infty$, it is permissible to replace the sum from $j_x = 1$ to ∞ with an integral from $j_x = 1$ to ∞. (The correction to this approximation is proportional to the surface area L^2 and the relative contribution is therefore of order $1/L$.) We can then use $dj_x = L d|p_x|/\pi$

to write

$$\ln Z = \frac{L^3}{\pi^3} \int_0^\infty d|p_x| \int_0^\infty d|p_y| \int_0^\infty d|p_z| \ln Z(\mathbf{p}) \qquad (1.29)$$

In all cases to be dealt with in this book the mode partition function depends only on the magnitude of the momentum components. Then the integration over p_x may be extended from $-\infty$ to ∞ if we divide by 2:

$$\ln Z = V \int \frac{d^3p}{(2\pi)^3} \ln Z(p) \qquad (1.30)$$

Note the natural appearance of the phase-space integral $\int d^3x d^3p/(2\pi)^3$ in this expression.

Recalling the mode partition function from the previous sections we have

$$\ln Z = V \int \frac{d^3p}{(2\pi)^3} \ln \left(1 \pm e^{-\beta(\omega-\mu)}\right)^{\pm 1} \qquad (1.31)$$

where the upper sign $(+)$ refers to fermions and the lower sign $(-)$ refers to bosons. From (1.4) and (1.31) we obtain the pressure, particle number, and energy:

$$P = \frac{T}{V} \ln Z$$

$$N = V \int \frac{d^3p}{(2\pi)^3} \frac{1}{e^{\beta(\omega-\mu)} \pm 1} \qquad (1.32)$$

$$E = V \int \frac{d^3p}{(2\pi)^3} \frac{\omega}{e^{\beta(\omega-\mu)} \pm 1}$$

These formulæ for N and E have the simple interpretation of phase-space integrals over the mean particle number and energy of each mode, respectively.

The dispersion relation $\omega = \omega(p)$ determines the energy for a given momentum. For relativistic particles $\omega = \sqrt{p^2 + m^2}$, where m is the mass. The nonrelativistic limit is $\omega = m + p^2/2m$. For phonons the dispersion relation is $\omega = c_s p$, where c_s is the speed of sound in the medium.

There are a number of interesting and physically relevant limits. Consider the dispersion relation $\omega = \sqrt{p^2 + m^2}$. The classical limit corresponds to low occupancy of the modes and is the same for bosons (1.16) and fermions (1.24). The momentum integral for the pressure can be performed and written as

$$P = \frac{m^2 T^2}{2\pi^2} e^{\mu/T} K_2 \left(\frac{m}{T}\right) \qquad \text{classical limit} \qquad (1.33)$$

where K_2 is the modified Bessel function. The nonrelativistic limit of this is

$$P = T \left(\frac{mT}{2\pi}\right)^{3/2} e^{(\mu-m)/T} \qquad \text{classical nonrelativistic limit} \qquad (1.34)$$

Knowing the pressure as a function of temperature and chemical potential we can obtain all other thermodynamic functions by differentiation or by using thermodynamic identities.

The zero-temperature limit for fermions requires that $\mu > m$, otherwise the vacuum state is approached. In this limit all states up to the Fermi momentum $p_\mathrm{F} = \sqrt{\mu^2 - m^2}$ and energy $E_\mathrm{F} = \mu$ are occupied and all states above are empty. The pressure, energy density $\epsilon = E/V$, and number density $n = N/V$ are given by

$$P = \frac{1}{16\pi^2}\left[2\mu^3 p_\mathrm{F} - m^2\mu p_\mathrm{F} - m^4\ln\left(\frac{\mu + p_\mathrm{F}}{m}\right)\right]$$

$$\epsilon = \frac{1}{16\pi^2}\left[\frac{2}{3}\mu p_\mathrm{F}^3 - m^2\mu p_\mathrm{F} + m^4\ln\left(\frac{\mu + p_\mathrm{F}}{m}\right)\right] \qquad (1.35)$$

$$n = \frac{p_\mathrm{F}^3}{6\pi^2}$$

In the nonrelativistic limit,

$$P = \frac{p_\mathrm{F}^5}{30\pi^2 m} \qquad (1.36)$$

$$\epsilon = mn + \frac{3}{2}P \qquad \text{nonrelativistic limit}$$

Electrons and nucleons have spin $1/2$ and these expressions need to be multiplied by 2 to take account of that! The low-temperature limit for bosons will be discussed in the next chapter.

Massless bosons with zero chemical potential have pressure

$$P = \frac{\pi^2}{90}T^4 \qquad (1.37)$$

This is one of the most famous formulae in the thermodynamics of radiation fields.

If time reversal is a good symmetry, a detailed balance must occur among all possible reactions in equilibrium. For example, if the reaction $A + B \rightarrow C + D$ can occur then not only must the reverse reaction, $C + D \rightarrow A + B$, occur but it must happen at the same rate. Detailed balance implies relationships between the chemical potentials. It is shown in standard textbooks that, for the reactions just mentioned, the chemical potentials obey $\mu_A + \mu_B = \mu_C + \mu_D$. For a long-lived resonance that decays according to $X \to A + B$, the formation process $A + B \to X$ must happen at the same rate. The chemical potentials are related by $\mu_X = \mu_A + \mu_B$. Generally any reactions that are allowed by the conservation laws can and will occur. These conservation laws restrict the number of linearly independent chemical potentials. Consider, for example, a system whose only relevant conservation laws are for baryon number and electric charge. There are only two independent chemical potentials, one for baryon number (μ_B) and one for electric charge (μ_Q). Any particle in the system has a chemical potential which is a linear combination of these:

$$\mu_i = b_i \mu_B + q_i \mu_Q \qquad (1.38)$$

Here b_i is the baryon number and q_i the electric charge of the particle of type i. These chemical potentials are all measured with respect to the total particle energy including mass. (The chemical potential μ_i^{NR}, as customarily defined in nonrelativistic many-body theory, is related to ours by $\mu_i^{\mathrm{NR}} = \mu_i - m_i$.) Bosons that carry no conserved quantum number, such as photons and π^0 mesons, have zero chemical potential. Antiparticles have a chemical potential opposite in sign to particles.

The electrically charged mesons π^+ and π^- have electric charges of $+1$ and -1 and therefore equal and opposite chemical potentials, μ_Q and $-\mu_Q$, respectively. The total conserved charge is the number of π^+ mesons minus the number of π^- mesons:

$$Q = V \int \frac{d^3 p}{(2\pi)^3} \left(\frac{1}{e^{\beta(\omega - \mu_Q)} - 1} - \frac{1}{e^{\beta(\omega + \mu_Q)} - 1} \right) \qquad (1.39)$$

and the total energy is

$$E = V \int \frac{d^3 p}{(2\pi)^3} \left(\frac{\omega}{e^{\beta(\omega - \mu_Q)} - 1} + \frac{\omega}{e^{\beta(\omega + \mu_Q)} - 1} \right) \qquad (1.40)$$

If the bosons have nonzero spin s, then the phase-space integrals must be multiplied by the spin degeneracy factor $2s + 1$. An analogous discussion can be given for fermions.

1.5 Exercises

1.1 Prove that the state $|n\rangle$ given in (1.12) is normalized to unity.

1.2 Referring to (1.17), let $|0\rangle$ and $|1\rangle$ be represented by the basis vectors in a two-dimensional vector space. Find an explicit 2×2 matrix representation of the abstract operators α and $\alpha^\dagger$ in this vector space.

1.3 Calculate the partition function for noninteracting bosons, including the zero-point energy. From it calculate the mean energy, particle number, and entropy. Repeat the calculation for fermions.

1.4 Calculate the average energy per particle of a noninteracting gas of massless bosons with no chemical potential. Repeat the calculation for massless fermions.

1.5 Derive an expression like (1.39) or (1.40) for the entropy. Repeat the calculation for fermions.

Bibliography

Thermal and statistical physics

Reif, F. (1965). *Fundamentals of Statistical and Thermal Physics* (McGraw-Hill, New York).

Landau, L. D., and Lifshitz, E. M. (1959). *Statistical Physics* (Pergamon Press, Oxford).

Many-body theory

Abrikosov, A. A., Gorkov, L. P., and Dzyaloshinskii, I. E. (1963). *Methods of Quantum Field Theory in Statistical Physics* (Prentice-Hall, Englewood Cliffs).

Fetter, A. L. and Walecka, J. D. (1971). *Quantum Theory of Many-Particle Systems* (McGraw-Hill, New York).

Negele, J. W. and Orland, H. (1988). *Quantum Many-Particle Systems* (Addison-Wesley, Redwood City).

Numerical evaluation of thermodynamic integrals

Johns, S. D., Ellis, P. J., and Lattimer, J. M., *Astrophys. J.* **473**, 1020 (1996).

2

Functional integral representation of the partition function

The customary approach to nonrelativistic many-body theory is to proceed with the method of second quantization begun in the first chapter. There is another approach, the method of functional integrals, which we shall follow here. Of course, what can be done in one formalism can always be done in another. Nevertheless, functional integrals seem to be the method of choice for most elementary particle theorists these days, and they seem to lend themselves more readily to nonperturbative phenomena such as tunneling, instantons, lattice gauge theory, etc. For gauge theories they are practically indispensable. However, there is a certain amount of formalism that must be developed before we can start to discuss physical applications. In this chapter, we shall derive the functional integral representation of the partition function for interacting relativistic non-gauge field theories. As a check on the formalism, as well as to obtain some feeling for how functional integrals work, we shall then rederive some well-known results on relativistic ideal gases for bosons and fermions.

2.1 Transition amplitude for bosons

Let $\hat{\phi}(\mathbf{x}, 0)$ be a Schrödinger-picture field operator at time $t = 0$ and let $\hat{\pi}(\mathbf{x}, 0)$ be its conjugate momentum operator. The eigenstates of the field operator are labeled $|\phi\rangle$ and satisfy

$$\hat{\phi}(\mathbf{x}, 0)|\phi\rangle = \phi(\mathbf{x})|\phi\rangle \qquad (2.1)$$

where $\phi(\mathbf{x})$ is the eigenvalue, as indicated, a function of $\mathbf{x}$. We also have the usual completeness and orthogonality conditions,

$$\int d\phi(\mathbf{x})|\phi\rangle\langle\phi| = 1 \qquad (2.2)$$

$$\langle\phi_a|\phi_b\rangle = \prod_{\mathbf{x}} \delta(\phi_a(\mathbf{x}) - \phi_b(\mathbf{x})) \tag{2.3}$$

Similarly, the eigenstates of the conjugate momentum field operator satisfy

$$\hat{\pi}(\mathbf{x}, 0)|\pi\rangle = \pi(\mathbf{x})|\pi\rangle \tag{2.4}$$

The completeness and orthogonality conditions are

$$\int \frac{d\pi(\mathbf{x})}{2\pi} |\pi\rangle\langle\pi| = 1 \tag{2.5}$$

$$\langle\pi_a|\pi_b\rangle = \prod_{\mathbf{x}} \delta(\pi_a(\mathbf{x}) - \pi_b(\mathbf{x})) \tag{2.6}$$

The practical meaning of the formal expressions (2.2), (2.3), (2.5), and (2.6) is elucidated in Section 2.6.

Just as in quantum mechanics one may work in coordinate space or in momentum space, one may work here in the field space or in the conjugate momentum space. In quantum mechanics, one goes from one to the other by using

$$\langle x|p\rangle = \mathrm{e}^{ipx} \tag{2.7}$$

In field theory one has the overlap

$$\langle\phi|\pi\rangle = \exp\left(i \int d^3x\, \pi(\mathbf{x})\phi(\mathbf{x})\right) \tag{2.8}$$

In a natural generalization one goes from a denumerably finite number of degrees of freedom N in quantum mechanics to a continuously infinite number of degrees of freedom in quantum field theory: $\sum_{i=1}^{N} p_i x_i \to \int d^3x\, \pi(\mathbf{x})\phi(\mathbf{x})$.

For the dynamics one requires a Hamiltonian, which is now a functional of the field and of its conjugate momentum:

$$H = \int d^3x\, \mathcal{H}(\hat{\pi}, \hat{\phi}) \tag{2.9}$$

Now suppose that a system is in a state $|\phi_a\rangle$ at a time $t = 0$. After a time t_f it evolves into $\mathrm{e}^{-iHt_\mathrm{f}}|\phi_a\rangle$, assuming that the Hamiltonian has no explicit time dependence. The transition amplitude for going from a state $|\phi_a\rangle$ to a state $|\phi_b\rangle$ after a time t_f is thus $\langle\phi_b|\mathrm{e}^{-iHt_\mathrm{f}}|\phi_a\rangle$.

For statistical mechanical purposes we will be interested in cases where the system returns to its original state after the time t_f. To obtain a practical definition of the transition amplitude we use the following prescription: we divide the time interval $(0, t_\mathrm{f})$ into N equal steps of duration

$\Delta t = t_{\mathrm{f}}/N$. Then, at each time interval we insert a complete set of states, alternating between (2.2) and (2.5):

$$
\begin{aligned}
\langle \phi_a | \mathrm{e}^{-iHt_{\mathrm{f}}} | \phi_a \rangle = \lim_{N \to \infty} \int \left(\prod_{i=1}^{N} d\pi_i \, d\phi_i / 2\pi \right) \\
\times \langle \phi_a | \pi_N \rangle \langle \pi_N | \mathrm{e}^{-iH\Delta t} | \phi_N \rangle \langle \phi_N | \pi_{N-1} \rangle \\
\times \langle \pi_{N-1} | \mathrm{e}^{-iH\Delta t} | \phi_{N-1} \rangle \cdots \\
\times \langle \phi_2 | \pi_1 \rangle \langle \pi_1 | \mathrm{e}^{-iH\Delta t} | \phi_1 \rangle \langle \phi_1 | \phi_a \rangle
\end{aligned}
\tag{2.10}
$$

We know that

$$
\langle \phi_1 | \phi_a \rangle = \delta(\phi_1 - \phi_a)
\tag{2.11}
$$

(as a shorthand for (2.3)) and that

$$
\langle \phi_{i+1} | \pi_i \rangle = \exp \left(i \int d^3x \, \pi_i(\mathbf{x}) \phi_{i+1}(\mathbf{x}) \right)
\tag{2.12}
$$

Since $\Delta t \to 0$, we can expand as follows, keeping terms up to first order:

$$
\begin{aligned}
\langle \pi_i | \mathrm{e}^{-iH_i \Delta t} | \phi_i \rangle &\approx \langle \pi_i | (1 - iH_i \Delta t) | \phi_i \rangle \\
&= \langle \pi_i | \phi_i \rangle (1 - iH_i \Delta t) \\
&= (1 - iH_i \Delta t) \exp \left(-i \int d^3x \, \pi_i(\mathbf{x}) \phi_i(\mathbf{x}) \right)
\end{aligned}
\tag{2.13}
$$

where

$$
H_i = \int d^3x \, \mathcal{H} \left(\pi_i(\mathbf{x}), \phi_i(\mathbf{x}) \right)
\tag{2.14}
$$

Putting it all together we get

$$
\begin{aligned}
\langle \phi_a | \mathrm{e}^{-iHt_{\mathrm{f}}} | \phi_a \rangle = \lim_{N \to \infty} \int \left(\prod_{i=1}^{N} d\pi_i \, d\phi_i / 2\pi \right) \delta(\phi_1 - \phi_a) \\
\times \exp \left\{ -i\Delta t \sum_{j=1}^{N} \int d^3x \, [\mathcal{H}(\pi_j, \phi_j) - \pi_j(\phi_{j+1} - \phi_j)/\Delta t] \right\}
\end{aligned}
\tag{2.15}
$$

where $\phi_{N+1} = \phi_a = \phi_1$. The advantage of alternating between π and ϕ for the insertion of a complete set of states is that the Hamiltonian in (2.13) and (2.15) is evaluated at a single point in time.

Taking the continuum limit of (2.15), we finally arrive at the important result

$$\langle \phi_a | e^{-iHt_f} | \phi_a \rangle$$

$$= \int [d\pi] \int_{\phi(\mathbf{x},0)=\phi_a(\mathbf{x})}^{\phi(\mathbf{x},t_f)=\phi_a(\mathbf{x})} [d\phi]$$

$$\times \exp \left[i \int_0^{t_f} dt \int d^3x \left(\pi(\mathbf{x},t) \frac{\partial \phi(\mathbf{x},t)}{\partial t} - \mathcal{H}\left(\pi(\mathbf{x},t), \phi(\mathbf{x},t)\right) \right) \right] \quad (2.16)$$

The symbols $[d\pi]$ and $[d\phi]$ denote functional integration as defined in (2.15). The integration over $\pi(\mathbf{x},t)$ is unrestricted, but the integration over $\phi(\mathbf{x},t)$ is such that the field starts at $\phi_a(\mathbf{x})$ at $t = 0$ and ends at $\phi_a(\mathbf{x})$ at $t = t_f$. Note that all references to operators have gone.

2.2 Partition function for bosons

Recall that

$$Z = \text{Tr}\, e^{-\beta(H - \mu_i \hat{N}_i)} = \sum_a \int d\phi_a \langle \phi_a | e^{-\beta(H - \mu_i \hat{N}_i)} | \phi_a \rangle \quad (2.17)$$

where the sum runs over all states. This expression is very similar to that for the transition amplitude defined in the previous section. In fact we can express Z as an integral over fields and their conjugate momenta by making use of (2.16). In order to make that connection, we switch to an imaginary time variable $\tau = it$. The trace operator in (2.17) simply means that we must integrate over all ϕ_a. Finally, if the system admits a conserved charge then we must make the replacement

$$\mathcal{H}(\pi, \phi) \to \mathcal{H}(\pi, \phi) - \mu\mathcal{N}(\pi, \phi) \quad (2.18)$$

where $\mathcal{N}(\pi, \phi)$ is the conserved charge density. We finally arrive at the fundamental formula

$$Z = \int [d\pi] \int_{\text{periodic}} [d\phi]$$

$$\times \exp \left[\int_0^\beta d\tau \int d^3x \left(i\pi \frac{\partial \phi}{\partial \tau} - \mathcal{H}(\pi, \phi) + \mu\mathcal{N}(\pi, \phi) \right) \right] \quad (2.19)$$

The term "periodic" means that the integration over the field is constrained in such a way that $\phi(\mathbf{x}, 0) = \phi(\mathbf{x}, \beta)$. This follows from the trace operation, setting $\phi_a(\mathbf{x}) = \phi(\mathbf{x}, 0) = \phi(\mathbf{x}, \beta)$. There is no restriction over the π integration. The expression for the partition function (2.19) can readily be generalized to an arbitrary number of fields and conserved charges.

2.3 Neutral scalar field

The most general renormalizable Lagrangian for a neutral scalar field ϕ is

$$\mathcal{L} = \tfrac{1}{2}\partial_\mu\phi\,\partial^\mu\phi - \tfrac{1}{2}m^2\phi^2 - \mathrm{U}(\phi) \tag{2.20}$$

where the potential is

$$\mathrm{U}(\phi) = g\phi^3 + \lambda\phi^4 \tag{2.21}$$

and $\lambda \geq 0$ for the stability of the vacuum. The momentum conjugate to this field is

$$\pi = \frac{\partial\mathcal{L}}{\partial(\partial_0\phi)} = \frac{\partial\phi}{\partial t} \tag{2.22}$$

and the Hamiltonian is obtained through the usual Legendre transformation

$$\mathcal{H} = \pi\frac{\partial\phi}{\partial t} - \mathcal{L} = \tfrac{1}{2}\pi^2 + \tfrac{1}{2}(\nabla\phi)^2 + \tfrac{1}{2}m^2\phi^2 + \mathrm{U}(\phi) \tag{2.23}$$

There is no conserved charge.

We shall evaluate the partition function by returning to the discretized version:

$$Z = \lim_{N\to\infty} \left(\prod_{i=1}^{N} \int_{-\infty}^{\infty} \frac{d\pi_i}{2\pi} \int_{\text{periodic}} d\phi_i \right)$$

$$\times \exp\left(\sum_{j=1}^{N} \int d^3x \Big\{ i\pi_j(\phi_{j+1} - \phi_j) \right.$$

$$\left. - \Delta\tau \left[\tfrac{1}{2}\pi_j^2 + \tfrac{1}{2}(\nabla\phi_j)^2 + \tfrac{1}{2}m^2\phi_j^2 + \mathrm{U}(\phi) \right] \Big\} \right) \tag{2.24}$$

The momentum integrals can be evaluated immediately since they are simply products of Gaussian integrals. We divide position space into M^3 small cubes with $V = L^3$, $L = aM$, $a \to 0$, $M \to \infty$, M being an integer. For convenience and to make sure that Z remains explicitly dimensionless at each stage of the calculation, we write $\pi_j = A_j/(a^3\Delta\tau)^{1/2}$ and integrate A_j from $-\infty$ to ∞. We get

$$\int_{-\infty}^{\infty} \frac{dA_j}{2\pi} \exp\left[-\tfrac{1}{2}A_j^2 + i\left(\frac{a^3}{\Delta\tau}\right)^{1/2}(\phi_{j+1} - \phi_j)A_j \right]$$

$$= (2\pi)^{-1/2} \exp\left(\frac{-a^3(\phi_{j+1} - \phi_j)^2}{2\Delta\tau} \right) \tag{2.25}$$

for each cube. Thus far we have

$$
Z = \lim_{M,N\to\infty} (2\pi)^{-M^3 N/2} \int \left(\prod_{i=1}^{N} d\phi_i \right)
$$

$$
\times \exp \left\{ \Delta\tau \sum_{j=1}^{N} \int d^3x \left[-\frac{1}{2} \left(\frac{\phi_{j+1} - \phi_j}{\Delta\tau} \right)^2 \right. \right.
$$

$$
\left. \left. -\frac{1}{2}(\nabla\phi_j)^2 - \frac{1}{2}m^2\phi_j^2 - U(\phi_j) \right] \right\}
$$

$$(2.26)$$

Taking the continuum limit, we obtain

$$
Z = N' \int_{\text{periodic}} [d\phi] \exp \left(\int_0^\beta d\tau \int d^3 \mathcal{L} \right) \tag{2.27}
$$

The Lagrangian is expressed as a functional of ϕ and of its first derivatives. The formula (2.27) expresses the partition function Z as a functional integral over ϕ of the exponential of the action in imaginary time. The overall normalization constant N' is irrelevant, since multiplication of Z by any constant will not change the thermodynamics.

Next, we turn to the case of noninteracting fields by letting $U(\phi) = 0$. Interactions will be discussed in a later chapter. We define

$$
S = \int_0^\beta d\tau \int d^3x\, \mathcal{L} = -\frac{1}{2} \int_0^\beta d\tau \int d^3x \left[\left(\frac{\partial\phi}{\partial\tau} \right)^2 + (\nabla\phi)^2 + m^2\phi^2 \right] \tag{2.28}
$$

Integrating by parts, and using the periodicity of ϕ, we obtain

$$
S = -\frac{1}{2} \int_0^\beta d\tau \int d^3x\, \phi \left(-\frac{\partial^2}{\partial\tau^2} - \nabla^2 + m^2 \right) \phi \tag{2.29}
$$

The field admits a Fourier expansion:

$$
\phi(\mathbf{x}, \tau) = \sqrt{\frac{\beta}{V}} \sum_{n=-\infty}^{\infty} \sum_{\mathbf{p}} e^{i(\mathbf{p}\cdot\mathbf{x} + \omega_n \tau)} \phi_n(\mathbf{p}) \tag{2.30}
$$

where $\omega_n = 2\pi nT$, owing to the constraint of periodicity that $\phi(\mathbf{x}, \beta) = \phi(\mathbf{x}, 0)$ for all $\mathbf{x}$. The normalization in (2.30) is chosen such that each Fourier amplitude is dimensionless. Substituting (2.30) into (2.29) and recalling that the field is real, we find that

$$
S = -\tfrac{1}{2}\beta^2 \sum_n \sum_{\mathbf{p}} (\omega_n^2 + \omega^2)\phi_n(\mathbf{p})\phi_n^*(\mathbf{p}) \tag{2.31}
$$

with $\omega = \sqrt{\mathbf{p}^2 + m^2}$. The integrand depends only on the magnitude of the field, $A_n(\mathbf{p}) = |\phi_n(\mathbf{p})|$. Integrating out the phases, we get

$$Z = N' \prod_n \prod_{\mathbf{p}} \left\{ \int_{-\infty}^{\infty} dA_n(\mathbf{p}) \exp\left[-\tfrac{1}{2}\beta^2(\omega_n^2 + \omega^2)A_n^2(\mathbf{p})\right] \right\}$$

$$= N' \prod_n \prod_{\mathbf{p}} (2\pi)^{1/2} \left[\beta^2(\omega_n^2 + \omega^2)\right]^{-1/2} \tag{2.32}$$

From the treatment above, we know that a factor of $(2\pi)^{-1/2}$ appears for each momentum integration. Thus, ignoring an overall factor that is independent of volume and temperature,

$$Z = \prod_n \prod_{\mathbf{p}} \left[\beta^2(\omega_n^2 + \omega^2)\right]^{-1/2} \tag{2.33}$$

The partition function can be formally written as

$$Z = N' \int [d\phi] \exp\left[-\tfrac{1}{2}(\phi, D\phi)\right] = N''(\det D)^{-1/2} \tag{2.34}$$

where N'' is a constant. Here D equals $\beta^2(-\partial^2/\partial\tau^2 - \nabla^2 + m^2)$ in $(\mathbf{x}, \tau)$ space and $\beta^2(\omega_n^2 + \omega^2)$ in $(\mathbf{p}, \omega_n)$ space, and $(\phi, D\phi)$ is the inner product on the function space. The expression (2.34) follows from the formula for Riemann integrals with a constant matrix D:

$$\int_{-\infty}^{\infty} dx_1 \cdots dx_n \, e^{-x_i D_{ij} x_j} = \pi^{n/2} (\det D)^{-1/2} \tag{2.35}$$

One may also derive (2.33) using (2.34).

We now have

$$\ln Z = -\tfrac{1}{2} \sum_n \sum_{\mathbf{p}} \ln\left[\beta^2(\omega_n^2 + \omega^2)\right] \tag{2.36}$$

Using the following identities,

$$\ln\left[(2\pi n)^2 + \beta^2\omega^2\right] = \int_1^{\beta^2\omega^2} \frac{d\theta^2}{\theta^2 + (2\pi n)^2} + \ln\left[1 + (2\pi n)^2\right] \tag{2.37}$$

and

$$\sum_{n=-\infty}^{\infty} \frac{1}{n^2 + (\theta/2\pi)^2} = \frac{2\pi^2}{\theta} \left(1 + \frac{2}{e^\theta - 1}\right) \tag{2.38}$$

and dropping a temperature-independent term, we can write

$$\ln Z = -\sum_{\mathbf{p}} \int_1^{\beta\omega} d\theta \left(\frac{1}{2} + \frac{1}{e^\theta - 1}\right) \tag{2.39}$$

Carrying out the integral and dropping terms that are independent of temperature and volume, we finally get

$$\ln Z = V \int \frac{d^3p}{(2\pi)^3} \left[-\tfrac{1}{2}\beta\omega - \ln(1 - e^{-\beta\omega}) \right] \qquad (2.40)$$

This expression is identical to the bosonic version of (1.31) with $\mu = 0$, except that (2.40) includes the zero-point energy. Both

$$E_0 = -\frac{\partial}{\partial\beta} \ln Z_0 = \frac{1}{2} V \int \frac{d^3p}{(2\pi)^3} \, \omega \qquad (2.41)$$

and

$$P_0 = T \frac{\partial}{\partial V} \ln Z_0 = -\frac{E_0}{V} \qquad (2.42)$$

should be subtracted, since the vacuum is a state with zero energy and pressure.

2.4 Bose–Einstein condensation

An interesting system is obtained by considering a theory with a charged scalar field $\mathbf{\Phi}$. The field $\mathbf{\Phi}$ is then complex and describes bosons of positive and negative charge, i.e., they are each other's antiparticle. The Lagrangian density in this case is

$$\mathcal{L} = \partial_\mu \mathbf{\Phi}^* \partial^\mu \mathbf{\Phi} - m^2 \mathbf{\Phi}^* \mathbf{\Phi} - \lambda(\mathbf{\Phi}^* \mathbf{\Phi})^2 \qquad (2.43)$$

This expression has an obvious U(1) symmetry:

$$\mathbf{\Phi} \to \mathbf{\Phi}' = \mathbf{\Phi} e^{-i\alpha} \qquad (2.44)$$

where α is a real constant. This is a global symmetry since the multiplying phase factor is independent of spacetime location.

By Noether's theorem, there is a conserved current associated with each continuous symmetry of the Lagrangian. We can find this current by letting the phase factor α depend on the spacetime coordinate for a moment. In this case the U(1) transformation is

$$\begin{aligned} \mathcal{L} \to \mathcal{L}' &= \partial_\mu(\mathbf{\Phi}^* e^{i\alpha(x)})\partial^\mu(\mathbf{\Phi} e^{-i\alpha(x)}) - m^2 \mathbf{\Phi}^* \mathbf{\Phi} - \lambda(\mathbf{\Phi}^* \mathbf{\Phi})^2 \\ &= \mathcal{L} + \mathbf{\Phi}^* \mathbf{\Phi} \partial_\mu \alpha \, \partial^\mu \alpha + i\partial_\mu \alpha (\mathbf{\Phi}^* \partial^\mu \mathbf{\Phi} - \mathbf{\Phi} \partial^\mu \mathbf{\Phi}^*) \end{aligned} \qquad (2.45)$$

The equation of motion for the "field" $\alpha(x)$ is

$$\partial^\mu \frac{\partial \mathcal{L}'}{\partial(\partial^\mu \alpha)} = \frac{\partial \mathcal{L}'}{\partial \alpha} \qquad (2.46)$$

Since $\partial \mathcal{L}'/\partial \alpha = 0$, it follows that the "current" $\partial \mathcal{L}'/\partial(\partial^\mu \alpha) = \mathbf{\Phi}^* \mathbf{\Phi} \partial_\mu \alpha + i\mathbf{\Phi}^* \partial_\mu \mathbf{\Phi} - i\mathbf{\Phi} \partial_\mu \mathbf{\Phi}^*$ is conserved. We recover our original theory by letting

$\alpha(x) = $ constant. The conserved current density is then

$$j_\mu = i(\boldsymbol{\Phi}^* \partial_\mu \boldsymbol{\Phi} - \boldsymbol{\Phi} \partial_\mu \boldsymbol{\Phi}^*) \tag{2.47}$$

with $\partial^\mu j_\mu = 0$. The conservation law may be verified independently using the equation of motion for $\boldsymbol{\Phi}$. The full current and density are $J_\mu = \int d^3x \, j_\mu(x)$ and $Q = \int d^3x \, j_0(x)$.

It is convenient to decompose $\boldsymbol{\Phi}$ into real and imaginary parts using the real fields ϕ_1 and ϕ_2, $\boldsymbol{\Phi} = (\phi_1 + i\phi_2)/\sqrt{2}$. In terms of the conjugate momenta $\pi_1 = \partial\phi_1/\partial t$, $\pi_2 = \partial\phi_2/\partial t$, the Hamiltonian density and charge are

$$\mathcal{H} = \tfrac{1}{2}\left[\pi_1^2 + \pi_2^2 + (\nabla\phi_1)^2 + (\nabla\phi_2)^2 + m^2\phi_1^2 + m^2\phi_2^2\right] + \tfrac{1}{4}\lambda(\phi_1^2 + \phi_2^2)^2 \tag{2.48}$$

and

$$Q = \int d^3x (\phi_2\pi_1 - \phi_1\pi_2) \tag{2.49}$$

The partition function is

$$Z = \int [d\pi_1][d\pi_2] \int_{\text{periodic}} [d\phi_1][d\phi_2] \times \exp\left[\int_0^\beta d\tau \int d^3x \right.$$

$$\left. \times \left(i\pi_1\frac{\partial\phi}{\partial\tau} + i\pi_2\frac{\partial\phi_2}{\partial\tau} - \mathcal{H}(\pi_1, \pi_2, \phi_1, \phi_2) + \mu(\phi_2\pi_1 - \phi_1\pi_2)\right)\right] \tag{2.50}$$

where we have used a chemical potential associated with the conserved charge Q. Integrating out the conjugate momenta, we get

$$Z = (N')^2 \int_{\text{periodic}} [d\phi_1][d\phi_2]$$

$$\times \exp\left\{\int_0^\beta d\tau \int d^3x \left[-\tfrac{1}{2}\left(\frac{\partial\phi_1}{\partial\tau} - i\mu\phi_2\right)^2 - \tfrac{1}{2}\left(\frac{\partial\phi_2}{\partial\tau} + i\mu\phi_1\right)^2 \right.\right.$$

$$\left.\left. -\tfrac{1}{2}(\nabla\phi_1)^2 - \tfrac{1}{2}(\nabla\phi_2)^2 - \tfrac{1}{2}m^2\phi_1^2 - \tfrac{1}{2}m^2\phi_2^2 - \tfrac{1}{4}\lambda(\phi_1^2 + \phi_2^2)^2\right]\right\} \tag{2.51}$$

where N' is the same divergent normalization factor as before. Notice that the argument of the exponential in (2.51) differs from one's naive expectation of

$$\mathcal{L}(\phi_1, \phi_2, \partial_\mu\phi_1, \partial_\mu\phi_2; \mu = 0) + \mu j_0(\phi_1, \phi_2, i\partial\phi_1/\partial\tau, i\partial\phi_2/\partial\tau)$$

by an amount $\mu^2\boldsymbol{\Phi}^*\boldsymbol{\Phi}$, owing to the momentum dependence of j_0.

The expression (2.51) cannot be evaluated in closed form unless $\lambda = 0$. In this case, the functional integral becomes Gaussian and can then be worked out analogously to that for the free scalar field.

The components of $\boldsymbol{\Phi}$ can be Fourier-expanded:

$$\phi_1 = \sqrt{2}\zeta\cos\theta + \sqrt{\frac{\beta}{V}}\sum_n\sum_{\mathbf{p}} e^{i(\mathbf{p}\cdot\mathbf{x}+\omega_n\tau)}\phi_{1;n}(\mathbf{p})$$

$$\phi_2 = \sqrt{2}\zeta\sin\theta + \sqrt{\frac{\beta}{V}}\sum_n\sum_{\mathbf{p}} e^{i(\mathbf{p}\cdot\mathbf{x}+\omega_n\tau)}\phi_{2;n}(\mathbf{p})$$

(2.52)

Here ζ and θ are independent of $(\mathbf{x},\tau)$ and determine the full infrared behavior of the field; that is, $\phi_{1;0}(\mathbf{p}=\mathbf{0}) = \phi_{2;0}(\mathbf{p}=\mathbf{0}) = 0$. This allows for the possibility of condensation of the bosons into the zero-momentum state. Condensation means that in the infinite-volume limit a finite fraction of the particles resides in the $n=0$, $\mathbf{p}=\mathbf{0}$ state.

Setting $\lambda = 0$ and substituting (2.52) into (2.51) after an integration by parts, see (2.29), we find

$$Z = (N')^2\left(\prod_n\prod_{\mathbf{p}}\int d\phi_{1;n}(\mathbf{p})\,d\phi_{2;n}(\mathbf{p})\right)e^S \qquad (2.53)$$

where

$$S = \beta V(\mu^2 - m^2)\zeta^2 - \tfrac{1}{2}\sum_n\sum_{\mathbf{p}}\Big(\phi_{1;-n}(-\mathbf{p}),\,\phi_{2;-n}(-\mathbf{p})\Big)D\begin{pmatrix}\phi_{1;n}(\mathbf{p})\\\phi_{2;n}(\mathbf{p})\end{pmatrix}$$

and

$$D = \beta^2\begin{pmatrix}\omega_n^2+\omega^2-\mu^2 & -2\mu\omega_n\\ 2\mu\omega_n & \omega_n^2+\omega^2-\mu^2\end{pmatrix}$$

Carrying out the integrations,

$$\ln Z = \beta V(\mu^2 - m^2)\zeta^2 + \ln(\det D)^{-1/2} \qquad (2.54)$$

The second term can be handled as follows:

$$\ln\det D = \ln\left\{\prod_n\prod_{\mathbf{p}}\beta^4\big[(\omega_n^2+\omega^2-\mu^2)^2 + 4\mu^2\omega_n^2\big]\right\}$$

$$= \ln\left\{\prod_n\prod_{\mathbf{p}}\beta^2\big[\omega_n^2+(\omega-\mu)^2\big]\right\} + \ln\left\{\prod_n\prod_{\mathbf{p}}\beta^2\big[\omega_n^2+(\omega+\mu)^2\big]\right\}$$

(2.55)

Putting all this together,

$$\ln Z = \beta V (\mu^2 - m^2)\zeta^2 - \tfrac{1}{2} \sum_n \sum_{\mathbf{p}} \ln \left\{ \beta^2 \left[\omega_n^2 + (\omega - \mu)^2 \right] \right\}$$

$$- \tfrac{1}{2} \sum_n \sum_{\mathbf{p}} \ln \left\{ \beta^2 \left[\omega_n^2 + (\omega + \mu)^2 \right] \right\} \tag{2.56}$$

The last two terms in (2.56) are precisely of the form (2.36). All we need to do is recall (2.40) and make the substitutions $\omega \to \omega - \mu$ and $\omega \to \omega + \mu$, respectively, for the two terms in (2.56). We obtain

$$\ln Z = \beta V (\mu^2 - m^2)\zeta^2 - V \int \frac{d^3 p}{(2\pi)^3}$$

$$\times \left[\beta\omega + \ln \left(1 - e^{-\beta(\omega-\mu)} \right) + \ln \left(1 - e^{-\beta(\omega+\mu)} \right) \right] \tag{2.57}$$

There are several observations we can make about (2.57). The momentum integral is convergent only if $|\mu| \le m$. The parameter ζ appears in the final expression but θ does not, as expected from the U(1) symmetry of the Lagrangian. In this context, since the parameter ζ is not determined *a priori*, it should be treated as a variational parameter that is related to the charge carried by the condensed particles. At fixed β and μ, $\ln Z$ is an extremum with respect to variations of such a free parameter. Thus

$$\frac{\partial \ln Z}{\partial \zeta} = 2\beta V (\mu^2 - m^2)\zeta = 0 \tag{2.58}$$

which implies that $\zeta = 0$ unless $|\mu| = m$, in which case ζ is undetermined by this variational condition. When $|\mu| < m$ we simply recover the results obtained in Chapter 1, namely (1.31).

To determine ζ when $|\mu| = m$, note that the charge density $\rho = Q/V$ is given by

$$\rho = \frac{T}{V} \left(\frac{\partial \ln Z}{\partial \mu} \right)_{\mu=m} = 2m\zeta^2 + \rho^*(\beta, \mu = m) \tag{2.59}$$

where

$$\rho^* = \int \frac{d^3 p}{(2\pi)^3} \left(\frac{1}{e^{\beta(\omega-m)} - 1} - \frac{1}{e^{\beta(\omega+m)} - 1} \right)$$

(The case $\mu = -m$ is handled analogously.) Here the separate contributions from the condensate (the zero-momentum mode) and the thermal excitations are manifest. If the density ρ is kept fixed and the temperature is lowered, μ will decrease until the point $\mu = m$ is reached. If the temperature is lowered even further then $\rho^*(\beta, \mu = m)$ will be less than

ρ. Therefore ζ is given by

$$\zeta^2 = \frac{\rho - \rho^*(\beta, \mu = m)}{2m} \tag{2.60}$$

when $\mu = m$ and $T < T_{\text{c}}$. The critical temperature is determined implicitly by the equation

$$\rho = \rho^*(\beta_{\text{c}}, \mu = m) \tag{2.61}$$

In the nonrelativistic limit, one obtains

$$T_{\text{c}} = \frac{2\pi}{m} \left(\frac{\rho}{\zeta(3/2)}\right)^{2/3} \qquad \rho \ll m^3 \tag{2.62}$$

In the ultrarelativistic limit, one finds

$$T_{\text{c}} = \left(\frac{3\rho}{m}\right)^{1/2} \qquad \rho \gg m^3 \tag{2.63}$$

In the limit $m \to 0$, we have $|\mu| \to 0$ and $T_{\text{c}} \to \infty$. When $m = 0$, all the charge resides in the condensate, at all temperatures, and none is carried by the thermal excitations.

There is a second-order phase transition at T_{c}. This can be shown rigorously by a careful examination of the behavior of the chemical potential $\mu(\rho, T)$ as a function of T near T_{c} with ρ fixed. This analysis is left as an exercise. A more intuitive way to see this involves the general Landau theory of phase transitions [1]. The order parameter ζ drops continuously to zero as T_{c} is approached from below and remains zero above T_{c}. Physically, the reason for a phase transition is the following. At $T = 0$, all the conserved charge can reside in the zero-momentum mode on account of the bosonic character of the particles. (This would not be the case for fermions.) As the temperature is raised, some of the charge is excited out of the condensate. Eventually, the temperature becomes great enough to completely melt, or thermally disorder, the condensate. There is no reason for ζ to drop to zero discontinuously; hence the transition is second order.

2.5 Fermions

We now turn our attention to (Dirac) fermions. In relativistic quantum mechanics, we know that electrons or muons are described by a four-component spinor ψ. The components are identified as ψ_α, with α running from 1 to 4. The motion of a free electron is characterized by a wavefunction

$$\psi(\mathbf{x}, t) = \frac{1}{\sqrt{V}} \sum_{\mathbf{p}} \sum_{s} \sqrt{\frac{M}{E}} \left[b(p, s) u(p, s) \, \mathrm{e}^{-ip \cdot x} + d^*(p, s) v(p, s) \, \mathrm{e}^{ip \cdot x} \right]$$

(2.64)

Here u and v are positive- and negative-energy plane-wave spinors, respectively. The sum on s runs over the two possible spin orientations for a spin-1/2 Dirac fermion. The expansion coefficients $b(p, s)$ and $d^*(p, s)$ are complex functions in relativistic quantum mechanics but become operators in a field theory. As usual, $p \cdot x = p^\mu x_\mu = Et - \mathbf{p} \cdot \mathbf{x}$. Equation (2.64) is normalized as

$$\int d^3x \, \psi^\dagger(\mathbf{x}, t) \psi(\mathbf{x}, t) = \sum_{\mathbf{p}} \sum_{s} \left[|b(p, s)|^2 + |d(p, s)|^2 \right] = 1 \qquad (2.65)$$

In the absence of interactions, the Lagrangian density is

$$\mathcal{L} = \bar{\psi}(i\partial\!\!\!/ - m)\psi \qquad (2.66)$$

The Dirac matrices γ^μ, which are defined by the anticommutators $\{\gamma^\mu, \gamma^\nu\} = 2g^{\mu\nu}$, are in the standard convention

$$\gamma^0 = \begin{pmatrix} 1 & 0 \\ 0 & -1 \end{pmatrix}$$

$$\gamma = \begin{pmatrix} 0 & \boldsymbol{\sigma} \\ -\boldsymbol{\sigma} & 0 \end{pmatrix}$$

(2.67)

Each of these is a 4×4 matrix: "1" denotes the unit 2×2 matrix and $\boldsymbol{\sigma}$ denotes the triplet of Pauli matrices. In (2.66), $\bar{\psi} = \psi^\dagger \gamma^0$ and $\partial\!\!\!/ \equiv \gamma^\mu \partial_\mu = \gamma^\mu \partial/\partial x^\mu$. Written out explicitly,

$$\mathcal{L} = \psi^\dagger \gamma^0 \left(i\gamma^0 \frac{\partial}{\partial t} + i\boldsymbol{\gamma} \cdot \boldsymbol{\nabla} - m \right) \psi \qquad (2.68)$$

The Lagrangian has a global U(1) symmetry, so that $\psi \to \psi \mathrm{e}^{-i\alpha}$ and $\psi^\dagger \to \psi^\dagger \mathrm{e}^{i\alpha}$. Following Noether's theorem, there is a conserved current associated with this symmetry. To find it, we proceed in the same way as we did for the charged scalar field theory. We allow α to depend on x, treating it as an independent field. Under the above phase transformation, $\mathcal{L} \to \mathcal{L} + \bar{\psi} \left[\partial\!\!\!/ \alpha(x) \right] \psi$. Using the equation of motion for $\alpha(x)$, namely $\partial_\mu(\partial \mathcal{L}/\partial[\partial_\mu \alpha(x)]) - \partial \mathcal{L}/\partial \alpha(x) = 0$, we find the conservation law

$$\partial_\mu j^\mu = 0$$

$$j^\mu = \bar{\psi} \gamma^\mu \psi \qquad (2.69)$$

Now we set α = constant to recover our original theory. The total conserved charge is

$$Q = \int d^3x \, j^0 = \int d^3x \, \psi^\dagger \psi \qquad (2.70)$$

For relativistic quantum mechanics in the absence of interactions this is a trivial result because of (2.65).

In the field theory we treat ψ as a basic field. The momentum conjugate to this field is, from (2.68),

$$\Pi = \frac{\partial \mathcal{L}}{\partial(\partial\psi/\partial t)} = i\psi^\dagger \qquad (2.71)$$

because $\gamma^0 \gamma^0 = 1$. Thus, somewhat paradoxically, ψ and $\psi^\dagger$ must be treated independently in the Hamiltonian formalism. The Hamiltonian density is found using the standard procedure:

$$\mathcal{H} = \Pi \frac{\partial\psi}{\partial t} - \mathcal{L} = \psi^\dagger \left(i\frac{\partial}{\partial t} \right)\psi - \mathcal{L} = \bar{\psi}(-i\boldsymbol{\gamma} \cdot \boldsymbol{\nabla} + m)\psi \qquad (2.72)$$

The partition function is

$$Z = \mathrm{Tr}^\dagger e^{-\beta(H-\mu\hat{Q})} \qquad (2.73)$$

Apart from two differences, which could be lost in the formalism if we are not careful, we can follow the steps leading up to (2.19) and write

$$Z = \int [id\psi^\dagger][d\psi] \exp\left[\int_0^\beta d\tau \int d^3x \, \bar{\psi} \left(-\gamma^0 \frac{\partial}{\partial\tau} + i\boldsymbol{\gamma} \cdot \boldsymbol{\nabla} - m + \mu\gamma^0 \right)\psi \right] \qquad (2.74)$$

Recall that ψ and $\psi^\dagger$ are independent fields, which must be integrated independently. In contrast with boson fields, there is no advantage in attempting to integrate the conjugate momentum separately from the field. The two differences mentioned above have to do with the periodicity of the field in imaginary time τ and with the nature of the "classical" (in the path-integral formulation) fields $\psi(\mathbf{x}, \tau)$ and $\psi^\dagger(\mathbf{x}, \tau)$ over which we integrate.

The canonical commutation relations for bosons are

$$\left[\hat{\phi}(\mathbf{x}, t), \, \hat{\pi}(\mathbf{y}, t) \right] = i\hbar\delta(\mathbf{x} - \mathbf{y})$$
$$\left[\hat{\phi}(\mathbf{x}, t), \, \hat{\phi}(\mathbf{y}, t) \right] = [\hat{\pi}(\mathbf{x}, t), \, \hat{\pi}(\mathbf{y}, t)] = 0 \qquad (2.75)$$

and for fermions

$$\left\{ \hat{\psi}_\alpha(\mathbf{x}, t), \, \hat{\psi}_\beta^\dagger(\mathbf{y}, t) \right\} = \hbar\delta_{\alpha\beta}\delta(\mathbf{x} - \mathbf{y}) \qquad (2.76)$$
$$\left\{ \hat{\psi}_\alpha(\mathbf{x}, t), \, \hat{\psi}_\beta(\mathbf{y}, t) \right\} = \left\{ \hat{\psi}_\alpha^\dagger(\mathbf{x}, t), \, \hat{\psi}_\beta^\dagger(\mathbf{y}, t) \right\} = 0$$

These commutation relations are the only ones allowed by the fundamental spin-statistics theorem in relativistic quantum field theory. In the limit $\hbar \to 0$ the field operators are replaced by their eigenvalues. For the case of bosons, those eigenvalues are actually c-number functions, as illustrated in (2.1). We have expressed the partition function as a functional integral over these c-number functions, or "classical fields". For the case of fermions, the $\hbar \to 0$ limit is rather peculiar since the eigenvalues replacing the field operators anticommute with each other! This is of course connected with the Pauli exclusion principle and with the famous spin-statistics theorem. Note that (2.74) instructs us to integrate over these "classical" but anticommuting functions. The mathematics necessary to handle this situation was studied by Grassmann. There are Grassmann variables, Grassmann algebra, and Grassmann calculus.

For a single Grassmann variable η, there is only one anticommutator to define the algebra,

$$\{\eta, \eta\} = 0 \tag{2.77}$$

Because of this, the most general function of η is (using a Taylor series expansion) $f(\eta) = a + b\eta$, where a and b are c-numbers. Integration is defined by

$$\int d\eta = 0$$
$$\int d\eta \, \eta = 1 \tag{2.78}$$

The first of these says that the integral is invariant under the shift $\eta \to \eta + a$, and the second is just a convenient normalization.

In a more general setting, we may have a set of Grassmann variables $\eta_i, i = 1, 2, \ldots N$, and a paired set $\eta_i^\dagger$. The algebra is defined by

$$\{\eta_i, \eta_j\} = \{\eta_i, \eta_j^\dagger\} = \{\eta_i^\dagger, \eta_j^\dagger\} = 0 \tag{2.79}$$

The most general function of these variables may be written as

$$f = a + \sum_i a_i \eta_i + \sum_i b_i \eta_i^\dagger + \sum_{i,j} a_{ij} \eta_i \eta_j + \sum_{i,j} b_{ij} \eta_i^\dagger \eta_j^\dagger$$
$$+ \sum_{i,j} c_{ij} \eta_i^\dagger \eta_j + \cdots + d \eta_1^\dagger \eta_1 \eta_2^\dagger \eta_2 \cdots \eta_N^\dagger \eta_N \tag{2.80}$$

Integration over all variables of (2.80) is defined by

$$\int d\eta_1^\dagger d\eta_1 \cdots d\eta_N^\dagger d\eta_N \, f = d \tag{2.81}$$

Integrals over Grassmann variables were introduced for the explicit purpose of dealing with path integrals over fermionic coordinates. The bibliography at the end of this chapter refers the interested reader to more detailed treatments.

For our purposes, the only integral we need is

$$\int d\eta_1^\dagger d\eta_1 \cdots d\eta_N^\dagger d\eta_N \, e^{\eta^\dagger D\eta} = \det D \qquad (2.82)$$

where D is an $N \times N$ matrix. This formula is simple to prove if $N = 1$ or 2. The general case is left as an exercise for the reader.

As with bosons, it is most convenient to work in $(\mathbf{p}, \omega_n)$ space instead of $(\mathbf{x}, \tau)$ space. In imaginary time we can write

$$\psi_\alpha(\mathbf{x}, \tau) = \frac{1}{\sqrt{V}} \sum_n \sum_{\mathbf{p}} e^{i(\mathbf{p}\cdot\mathbf{x}+\omega_n\tau)} \tilde{\psi}_{\alpha;n}(\mathbf{p}) \qquad (2.83)$$

where both n and $\mathbf{p}$ run over negative and positive values. For an arbitrary function defined over the interval $0 \le \tau \le \beta$, the discrete frequency ω_n can take on the values $n\pi T$. For bosons we argued that we must take $\omega_n = 2\pi n T$ in order that $\phi(\mathbf{x}, \tau)$ be periodic, which followed from the trace operation in the partition function. This can be verified by examining the properties of the thermal Green's function for bosons defined by

$$G_B(\mathbf{x}, \mathbf{y}; \tau_1, \tau_2) = Z^{-1} \operatorname{Tr} \left\{ \hat{\rho} T_\tau \left[\hat{\phi}(\mathbf{x}, \tau_1) \hat{\phi}(\mathbf{y}, \tau_2) \right] \right\} \qquad (2.84)$$

Here T_τ is the imaginary time ordering operator, which for bosons acts as follows:

$$T_\tau \left[\hat{\phi}(\mathbf{x}, \tau_1) \hat{\phi}(\mathbf{y}, \tau_2) \right] = \hat{\phi}(\tau_1) \hat{\phi}(\tau_2) \theta(\tau_1 - \tau_2) + \hat{\phi}(\tau_2) \hat{\phi}(\tau_1) \theta(\tau_2 - \tau_1) \quad (2.85)$$

where θ is the step-function. Using the fact that T_τ commutes with $\hat{\rho} = e^{-\beta K}$, where $K \equiv H - \mu \hat{Q}$, and the cyclic property of the trace we find that

$$\begin{aligned}
G_B(\mathbf{x}, \mathbf{y}; \tau, 0) &= Z^{-1} \operatorname{Tr} \left[e^{-\beta K} \hat{\phi}(\mathbf{x}, \tau) \hat{\phi}(\mathbf{y}, 0) \right] \\
&= Z^{-1} \operatorname{Tr} \left[\hat{\phi}(\mathbf{y}, 0) e^{-\beta K} \hat{\phi}(\mathbf{x}, \tau) \right] \\
&= Z^{-1} \operatorname{Tr} \left[e^{-\beta K} e^{\beta K} \hat{\phi}(\mathbf{y}, 0) e^{-\beta K} \hat{\phi}(\mathbf{x}, \tau) \right] \\
&= Z^{-1} \operatorname{Tr} \left[e^{-\beta K} \hat{\phi}(\mathbf{y}, \beta) \hat{\phi}(\mathbf{x}, \tau) \right] \\
&= Z^{-1} \operatorname{Tr} \left\{ \hat{\rho} T_\tau \left[\hat{\phi}(\mathbf{x}, \tau) \hat{\phi}(\mathbf{y}, \beta) \right] \right\} \\
&= G_B(\mathbf{x}, \mathbf{y}; \tau, \beta) \qquad (2.86)
\end{aligned}$$

(Notice that $\hat{\phi}(\mathbf{y}, \beta) = e^{\beta K} \hat{\phi}(\mathbf{y}, 0) e^{-\beta K}$, in analogy with the real time Heisenberg time-evolution expression $\hat{\phi}(\mathbf{y}, t) = e^{iHt} \hat{\phi}(\mathbf{y}, 0) e^{-iHt}$.) The result (2.86) implies that $\phi(\mathbf{y}, 0) = \phi(\mathbf{y}, \beta)$ and hence $\omega_n = 2\pi n T$.

For fermions, however, instead of (2.85) one has (in direct analogy with the real time Green's functions)

$$T_\tau\left[\hat\psi(\tau_1)\hat\psi(\tau_2)\right] = \hat\psi(\tau_1)\hat\psi(\tau_2)\theta(\tau_1-\tau_2) - \hat\psi(\tau_2)\hat\psi(\tau_1)\theta(\tau_2-\tau_1) \quad (2.87)$$

Following the same steps as in (2.86), one is led to

$$G_{\mathrm{F}}(\mathbf{x},\mathbf{y};\tau,0) = -G_{\mathrm{F}}(\mathbf{x},\mathbf{y};\tau,\beta) \quad (2.88)$$

This implies that

$$\psi(\mathbf{x},0) = -\psi(\mathbf{x},\beta) \quad (2.89)$$

and hence

$$\omega_n = (2n+1)\pi T \quad (2.90)$$

This antiperiodicity required of fermion fields is in no way inconsistent with the trace operation in the partition function. The trace only means that the system returns to its original state after a "time" β. Since the sign of ψ is just an overall phase and hence is not observable, the right-hand side of (2.89) describes the same physical state as the left-hand side.

Now we are ready to evaluate (2.74). Inserting (2.83) and using (2.82) we get

$$Z = \left[\prod_n\prod_{\mathbf{p}}\prod_\alpha \int i d\tilde\psi^\dagger_{\alpha;n}(\mathbf{p}) d\tilde\psi_{\alpha;n}(\mathbf{p})\right] e^S \quad (2.91)$$

where

$$S = \sum_n\sum_{\mathbf{p}} i\tilde\psi^\dagger_{\alpha;n}(\mathbf{p}) D_{\alpha\rho}\tilde\psi_{\rho;n}(\mathbf{p})$$

$$D = -i\beta\left[(-i\omega_n+\mu) - \gamma^0\boldsymbol\gamma\cdot\mathbf{p} - m\gamma^0\right]$$

and so

$$Z = \det D \quad (2.92)$$

In (2.92) the determinantal operation is carried out over both Dirac indices (thus with 4×4 matrices) and in frequency–momentum space. Using

$$\ln\det D = \mathrm{Tr}\ln D \quad (2.93)$$

and (2.67), one finds that

$$\ln Z = 2\sum_n\sum_{\mathbf{p}}\ln\left\{\beta^2\left[(\omega_n+i\mu)^2 + \omega^2\right]\right\} \quad (2.94)$$

Since the summation is over both negative and positive frequencies (2.94) can be put into a form analogous to (2.55),

$$\ln Z = \sum_n \sum_{\mathbf{p}} \left\{ \ln \left[\beta^2 (\omega_n^2 (\omega - \mu)^2) \right] + \ln \left[\beta^2 (\omega_n^2 (\omega + \mu)^2) \right] \right\} \quad (2.95)$$

Following (2.37), we write

$$\ln \left[(2n+1)^2 \pi^2 + \beta^2 (\omega \pm \mu)^2 \right] = \int_1^{\beta^2 (\omega \pm \mu)^2} \frac{d\theta^2}{\theta^2 + (2n+1)^2 \pi^2}$$
$$+ \ln \left[1 + (2n+1)^2 \pi^2 \right] \quad (2.96)$$

The sum over n may be carried out by using the summation formula

$$\sum_{n=-\infty}^{\infty} \frac{1}{(n-x)(n-y)} = \frac{\pi (\cot \pi x - \cot \pi y)}{y - x} \quad (2.97)$$

This gives

$$\sum_{n=-\infty}^{\infty} \frac{1}{(2n+1)^2 \pi^2 + \theta^2} = \frac{1}{\theta} \left(\frac{1}{2} - \frac{1}{e^\theta + 1} \right) \quad (2.98)$$

Integrating over θ and dropping terms that are independent of β and μ, we finally obtain

$$\ln Z = 2V \int \frac{d^3 p}{(2\pi)^3} \left[\beta \omega + \ln \left(1 + e^{-\beta(\omega - \mu)} \right) + \ln \left(1 + e^{-\beta(\omega + \mu)} \right) \right] \quad (2.99)$$

This result agrees with that derived in Chapter 1 using completely different methods.

Notice the factor 2 in (2.99). This factor comes out automatically and owes its existence to the spin-1/2 nature of the fermions. Separate contributions from particles (μ) and antiparticles ($-\mu$) are evident. Finally, this formula also contains a contribution from the zero-point energy.

To recapitulate, the difference between fermions and bosons in the functional integral approach to the partition function is essentially twofold. First, for fermions we must integrate over Grassmann variables instead of c-number variables. Contrast the result (2.92), $Z = \det D$, for fermions with the result (2.34), $Z = (\det D)^{-1/2}$, for bosons. Integration over c-number variables would have led to a factor -1 in (2.99) instead of the factor 2. Second, and this is related to the first, is the fact that the fermion fields are actually antiperiodic in imaginary time, with period β, instead of periodic as is the case for bosons. The consequence is that $\omega_n = (2n+1)\pi T$ for fermions whereas $\omega_n = 2\pi n T$ for bosons. These two points account for the difference between (2.57) (with $\zeta = 0$, of course) and (2.99).

2.6 Remarks on functional integrals

The notation used for functional integration (and differentiation!) is deceptively simple. It must be kept simple, for if we think back on the tremendous progress made in mechanics and electromagnetism in the nineteenth century, it was certainly made easier by the introduction of compact notation for differentiation, integration, and vectors. This also seems to be the case with functional methods in modern quantum physics. However, it is also clear that the mathematical symbols we are using represent rather exotic entities. For example, (2.6) uses a Dirac delta function whose argument is a difference between two functions. A less formal and compact, but more practical, way to view these objects is to start with a complete orthonormal set of real functions for the physical problem of interest. Call this set $w_n(x)$, with n any positive integer. Then any function may be written as

$$a(x) = \sum_{n=1}^{\infty} a_n w_n(x) \tag{2.100}$$

Another function may be expressed as

$$b(x) = \sum_{n=1}^{\infty} b_n w_n(x) \tag{2.101}$$

Then

$$\delta\left(a(x) - b(x)\right) = \prod_{n=1}^{\infty} \delta(a_n - b_n) \tag{2.102}$$

and

$$\int [da(x)] = \prod_{n=1}^{\infty} \int_{-\infty}^{\infty} da_n \tag{2.103}$$

and so on. Most physical problems are defined on the space of a continuous variable, such as position. For such problems it is intuitively obvious that the functional integral ought to be divergent in general since the possible functional configurations form an uncountably infinite set. Indeed, it seems that the extent to which mathematical rigor can be applied to functional integrals is still uncertain. This should be no surprise since they are just a means of phrasing the physical content of relativistic quantum field theory. The extent to which mathematical rigor can be applied in the operator formalism is probably no more certain, because of the highly singular nature of the products of field operators at a point. For physical problems defined on a space of discrete variables, some mathematical rigor can be applied. This is one reason why certain spacetime theories are defined on a spacetime lattice. This will be studied in Chapter 10.

2.7 Exercises

2.1 For the charged scalar field show, by direct application of the equation of motion for $\mathbf{\Phi}$, that $j_\mu = i(\mathbf{\Phi}^* \partial_\mu \mathbf{\Phi} - \mathbf{\Phi} \partial_\mu \mathbf{\Phi}^*)$ is conserved.

2.2 If j_μ is conserved show that

$$\dot{Q} = \frac{d}{dt} \int d^3x \, j_0(\mathbf{x}, t) = 0$$

2.3 Obtain (2.62) and (2.63), starting from (2.59) to (2.61).

2.4 For Bose–Einstein condensation, consider μ as a function of ρ and T. If ρ is held fixed, show that μ and $\partial\mu/\partial T$ are continuous but $\partial^2\mu/\partial T^2$ is discontinuous at T_c.

2.5 Prove (2.82).

2.6 Fill in the steps leading from (2.91)–(2.93) to (2.94).

2.7 When $m = 0$ show that (2.99) can be evaluated in closed form, leading to $P = T \ln Z/V = \mu^4/12\pi^2 + \mu^2 T^2/6 + 7\pi^2 T^4/180$.

Reference

1. Landau, L. D., and Lifshitz, E. M. (1959). *Statistical Physics* (Pergamon Press, Oxford).

Bibliography

Path integrals in quantum mechanics

Feynman, R. P., *Phys. Rev.* **91**, 1291 (1953).
Feynman, R. P. and Hibbs, A. R. (1965). *Quantum Mechanics and Path Integrals* (McGraw-Hill, New York).

Functional integrals in field theory

Lee, T. D. (1982). *Particle Physics and Introduction to Field Theory* (Harwood Academic, London).
Nash, C. (1978). *Relativistic Quantum Fields* (Academic Press, New York).
Ramond, P. (1989). *Field Theory: A Modern Primer* (Addison-Wesley, Redwood City).

Functional integrals in field theory at finite temperature

Bernard, C., *Phys. Rev. D* **9**, 3312 (1974).

Thermal phase transitions

Landau, L. D., and Lifshitz, E. M. (1959). *Statistical Physics* (Pergamon Press, Oxford).

Relativistic Bose–Einstein condensation

Kapusta, J. I., *Phys. Rev. D* **24**, 426 (1981).
Haber, H. E., and Weldon, H. A., *J. Math. Phys.* **23**, 1852. (1982).

Grassmann variables

Chandlin, D. J., *Nuovo Cim.* **4**, 231 (1956).
Berezin, F. A. (1966). *Method of Second Quantization* (Academic Press, New York).
McLerran, L. D., *Rev. Mod. Phys.* **58**, 1021 (1986).

3
Interactions and diagrammatic techniques

Unfortunately it is not possible to carry out the functional integration in closed form when the Lagrangian contains terms that are more than quadratic in the fields. The reader is invited to verify this. Thus, it is important to develop approximation techniques. An approximation that is expected to be useful when the interactions are weak is found by expanding the partition function in powers of the interaction. The convergence properties of these perturbation expansions have not been established with any degree of mathematical rigor, however. An alternative approach is to evaluate the partition function containing a given Lagrangian on a spacetime lattice using numerical Monte Carlo methods. This approach is described in Chapter 10.

3.1 Perturbation expansion

Consider a single scalar field ϕ. Other, more physical, theories such as QED, QCD, the Glashow–Weinberg–Salam model, and effective nuclear models will be considered in later chapters. The reader must be prepared now to learn some basic techniques before tackling more complicated but physically relevant theories.

The partition function is

$$Z = N' \int [d\phi] \mathrm{e}^{S} \tag{3.1}$$

The action can be decomposed as

$$S = S_0 + S_\mathrm{I} \tag{3.2}$$

where S_0 is at most quadratic in the field and S_I, the part due to interactions, is of higher order. We may expand (3.1) in a power series in

33

the part due to interaction, S_I:

$$Z = N' \int [d\phi] e^{S_0} \sum_{l=0}^{\infty} \frac{1}{l!} S_I^l \tag{3.3}$$

Taking the logarithm on both sides we get

$$\ln Z = \ln \left(N' \int [d\phi] e^{S_0} \right) + \ln \left(1 + \sum_{l=1}^{\infty} \frac{1}{l!} \frac{\int [d\phi] e^{S_0} S_I^l}{\int [d\phi] e^{S_0}} \right)$$

$$= \ln Z_0 + \ln Z_I \tag{3.4}$$

This explicitly separates the interaction contributions from the ideal gas contribution, which we have evaluated already. The relevant quantity that we actually need to compute is

$$\langle S_I^l \rangle_0 = \frac{\int [d\phi] e^{S_0} S_I^l}{\int [d\phi] e^{S_0}} \tag{3.5}$$

which is the value of S_I raised to an arbitrary positive integral power and averaged over the unperturbed ensemble, represented by S_0. The normalization of the functional integration is now irrelevant, as it cancels in the expression (3.5).

3.2 Diagrammatic rules for $\lambda\phi^4$ theory

The task of actually evaluating (3.4) and (3.5) is significantly more difficult than our compact notation would suggest. It is in fact useful to associate diagrams with the mathematical expressions in the expansion. Diagrams are a common language in particle physics, nuclear physics, statistical physics and condensed matter physics and allow for the exchange of ideas and concepts among these different disciplines.

Consider the lowest-order correction to $\ln Z_0$ in $\lambda\phi^4$ theory. It is

$$\ln Z_1 = \frac{-\lambda \int d\tau \int d^3x \int [d\phi] e^{S_0} \phi^4(\mathbf{x}, \tau)}{\int [d\phi] e^{S_0}} \tag{3.6}$$

If we express $\phi(\mathbf{x}, \tau)$ as a Fourier series as in (2.30), and insert this into (3.6) we get

$$\ln Z_1 = -\lambda \int d\tau \int d^3x \sum_{n_1, \ldots, n_4} \sum_{\mathbf{p}_1, \ldots, \mathbf{p}_4} \frac{\beta^2}{V^2}$$

$$\times \exp[i(\mathbf{p}_1 + \cdots + \mathbf{p}_4) \cdot \mathbf{x}] \exp\left[i(\omega_{n_1} + \cdots + \omega_{n_4})\tau\right] \frac{A}{B}$$

$$\tag{3.7}$$

where

$$A = \prod_l \prod_{\mathbf{q}} \int d\tilde{\phi}_l(\mathbf{q}) \exp\left[-\tfrac{1}{2}\beta^2(\omega_l{}^2 + \mathbf{q}^2 + m^2)\tilde{\phi}_l(\mathbf{q})\tilde{\phi}_{-l}(-\mathbf{q})\right]$$

$$\times \tilde{\phi}_{n_1}(\mathbf{p}_1)\tilde{\phi}_{n_2}(\mathbf{p}_2)\tilde{\phi}_{n_3}(\mathbf{p}_3)\tilde{\phi}_{n_4}(\mathbf{p}_4)$$

and

$$B = \prod_l \prod_{\mathbf{q}} \int d\tilde{\phi}_l(\mathbf{q}) \exp\left[-\tfrac{1}{2}\beta^2(\omega_l^2 + \mathbf{q}^2 + m^2)\tilde{\phi}_l(\mathbf{q})\tilde{\phi}_{-l}(-\mathbf{q})\right]$$

The integrations over $\mathbf{x}$ and τ yield a factor $\beta V \delta_{n_1+\cdots+n_4,0}\,\delta_{\mathbf{p}_1+\cdots+\mathbf{p}_4,0}$. The numerator of the whole expression for $\ln Z_1$ will be zero by symmetric integration unless $n_3 = -n_1$, $\mathbf{p}_3 = -\mathbf{p}_1$ and $n_4 = -n_2$, $\mathbf{p}_4 = -\mathbf{p}_2$, or the other two permutations thereof. This will satisfy the constraints of the Kronecker deltas and the integrals will factorize. The integrals in the numerator are canceled by those in the denominator except for the two corresponding to $l = n_1$, $\mathbf{q} = \mathbf{p}_1$ and $l = n_2$, $\mathbf{q} = \mathbf{p}_2$, and the other two permutations. Using

$$\frac{\int_{-\infty}^{\infty} dx\, x^2 e^{-ax^2/2}}{\int_{-\infty}^{\infty} dx\, e^{-ax^2/2}} = \frac{1}{a} \tag{3.8}$$

we obtain

$$\ln Z_1 = -3\lambda\beta V \left(T\sum_n \int \frac{d^3 p}{(2\pi)^3} \mathcal{D}_0(\omega_n, \mathbf{p})\right)^2 \tag{3.9}$$

Here we have defined the propagator in frequency–momentum space as

$$\mathcal{D}_0(\omega_n, \mathbf{p}) = \frac{1}{\omega_n^2 + \mathbf{p}^2 + m^2} \tag{3.10}$$

The expression (3.9) can be associated with a diagram in the following way. Remember that we are calculating $\ln Z_1$ to first order in λ. With $\phi^4(\mathbf{x}, \tau)$ we associate a cross with four arms (because of the fourth power of ϕ), with the vertex located at $(\mathbf{x}, \tau)$:

$$\phi^4(\boldsymbol{x}, \tau): \qquad \times\, (\boldsymbol{x}, \tau)$$

After expressing each field $\phi(\mathbf{x}, \tau)$ as a Fourier series we draw the figure

$$\begin{array}{cc}(\boldsymbol{p}_2, \omega_{n_2}) & (\boldsymbol{p}_3, \omega_{n_3}) \\ & \times \\ (\boldsymbol{p}_1, \omega_{n_1}) & (\boldsymbol{p}_4, \omega_{n_4})\end{array}$$

The directions of the arrows reflect the signs of the momenta and frequencies. By convention, we draw them pointing towards the vertex, but we could have chosen a convention in which they all point away. The functional integration vanishes unless $n_3 = -n_1$, $\mathbf{p}_3 = -\mathbf{p}_1$ and $n_4 = -n_2$, $\mathbf{p}_4 = -\mathbf{p}_2$, etc. Thus we connect the ends in pairs. There are three possible pairings. We then have

$$\ln Z_1 = 3 \;\; \text{⟨}\!\bigcirc\!\!\bigcirc\!\text{⟩} \qquad\qquad (3.11)$$
$$(\mathbf{p}_1, \omega_{n_1})(\mathbf{p}_2, \omega_{n_2})$$

With each closed loop we associate a factor

$$T \sum_n \int \frac{d^3 p}{(2\pi)^3} \mathcal{D}_0(\omega_n, \mathbf{p})$$

With the vertex we associate a factor $-\lambda$ (coming from $\mathcal{L}_\mathrm{I} = -\lambda\phi^4$) and a factor

$$\beta \delta_{\omega_\mathrm{in}, \omega_\mathrm{out}} V \delta_{\mathbf{p}_\mathrm{in}, \mathbf{p}_\mathrm{out}} \rightarrow \beta \delta_{\omega_\mathrm{in}, \omega_\mathrm{out}} (2\pi)^3 \delta(\mathbf{p}_\mathrm{in} - \mathbf{p}_\mathrm{out})$$

Since the arguments of the frequency–momentum-conserving deltas are zero we simply get an overall factor βV. The factor V makes $\ln Z_1$ a properly extensive quantity. Pictorially, (3.11) corresponds precisely with (3.9).

Next we look at order λ^2 in $\ln Z_\mathrm{I}$. From (3.4) it is

$$\ln Z_2 = -\frac{1}{2} \left(\frac{\int [d\phi] \mathrm{e}^{S_0} S_\mathrm{I}}{\int [d\phi] \mathrm{e}^{S_0}} \right)^2 + \frac{1}{2} \frac{\int [d\phi] \mathrm{e}^{S_0} S_\mathrm{I}^2}{\int [d\phi] \mathrm{e}^{S_0}} \qquad (3.12)$$

The first term in (3.12) is simply

$$-\frac{1}{2}(\ln Z_1)^2 = -\frac{1}{2} \left(3 \;\; \bigcirc\!\!\bigcirc \;\; \otimes \;\; 3 \;\; \bigcirc\!\!\bigcirc \right) \qquad (3.13)$$

The second term in (3.12) may be analyzed algebraically using functional integrals or it may be analyzed diagrammatically. Choosing the latter approach, we draw two crosses corresponding to the factors $\phi^4(\boldsymbol{x}, \tau)$ and $\phi^4(\boldsymbol{x}', \tau')$ contained in $\frac{1}{2}\langle S_\mathrm{I}^2 \rangle_0$:

We then pair the lines as before. Counting in the factor one-half and all the possible pairings, we obtain

$$\frac{1}{2} \times 3 \; \bigcirc\!\!\bigcirc \; \otimes 3 \; \bigcirc\!\!\bigcirc \; + \; \frac{6 \times 6 \times 2}{2} \; \bigcirc\!\!\bigcirc\!\!\bigcirc$$

$$+ \; \frac{4 \times 3 \times 2}{2} \; \bigoplus \quad\quad\quad (3.14)$$

Combining (3.13) and (3.14), we observe that all the disconnected diagrams cancel. We are thus left with

$$\ln Z_2 = 36 \; \bigcirc\!\!\bigcirc\!\!\bigcirc \; + \; 12 \; \bigoplus \quad\quad\quad (3.15)$$

What is needed at some arbitrary order N in the perturbative expansion of $\ln Z_{\mathrm{I}}$ should now be clear. We formally expand in powers of λ:

$$\ln Z_{\mathrm{I}} = \sum_{N=1}^{\infty} \ln Z_N \quad\quad\quad (3.16)$$

where $\ln Z_N$ is proportional to λ^N. The "finite-temperature Feynman rules" at order N are:

1 Draw all connected diagrams.
2 Determine the combinatoric factor for each diagram.
3 Include a factor $T \sum_n \int [d^3p/(2\pi)^3] \mathcal{D}_0(\omega_n, \mathbf{p})$ for each line.
4 Include a factor $-\lambda$ for each vertex.
5 Include a factor $(2\pi)^3 \delta(\mathbf{p}_{\mathrm{in}} - \mathbf{p}_{\mathrm{out}}) \beta \delta_{\omega_{\mathrm{in}}, \omega_{\mathrm{out}}}$ for each vertex, corresponding to energy(frequency)–momentum conservation. There will be one factor $\beta(2\pi)^3 \delta(\mathbf{0}) = \beta V$ left over.

We now understand why $\mathcal{D}$ is called a propagator: it propagates a particle (or field) from one vertex to the next. We have illustrated the cancellation mechanism only at second order. However, it is clear why disconnected diagrams cancel. If, at some order, there existed a contribution that was the product of K connected diagrams then this contribution would be proportional to V^K. If we have done our job correctly, then $\ln Z_{\mathrm{I}}$ is an extensive quantity proportional to V and thus no such contribution can arise.

The formal proof that in $\ln Z_{\mathrm{I}}$ the disconnected diagrams cancel goes as follows. From (3.3) and (3.5) we have

$$Z_{\mathrm{I}} = \sum_{l=0}^{\infty} \frac{1}{l!} \langle S_{\mathrm{I}}^l \rangle_0 \qu\quad\quad (3.17)$$

In general, $\langle S_{\mathrm{I}}^l \rangle_0$ can be written as a sum of terms, each of which is a product of connected parts (see (3.14)). Denoting a connected part by a subscript c, we may write

$$\langle S_{\mathrm{I}}^l \rangle_0 = \sum_{a_1,a_2,\ldots=0}^{\infty} \frac{l!}{a_1! a_2! (2!)^{a_2} a_3! (3!)^{a_3} \cdots} \langle S_{\mathrm{I}} \rangle_{0c}^{a_1} \langle S_{\mathrm{I}}^2 \rangle_{0c}^{a_2} \cdots \delta_{a_1+2a_2+3a_3+\cdots,l}$$

(3.18)

The combinatoric factor takes into account indistinguishability, and the Kronecker delta picks out the contribution of order λ^l. Substituting (3.18) into (3.17) and summing over l eliminates the Kronecker delta:

$$Z_{\mathrm{I}} = \sum_{a_1,a_2,\ldots=0}^{\infty} \frac{\langle S_{\mathrm{I}} \rangle_{0c}^{a_1}}{a_1!} \frac{\langle S_{\mathrm{I}}^2 \rangle_{0c}^{a_2}}{a_2! (2!)^{a_2}} \cdots = \exp\left(\sum_{n=1}^{\infty} \frac{1}{n!} \langle S_{\mathrm{I}}^n \rangle_{0c} \right)$$

(3.19)

Hence $\ln Z_1$ is simply the sum of the connected diagrams.

As an example, let us apply these rules to the second diagram of (3.15). We get

$$\bigcirc\!\!\!\!\ominus = \beta V (-\lambda^2) T \sum_{n_1} \int \frac{d^3 p_1}{(2\pi)^3} \cdots T \sum_{n_4} \int \frac{d^3 p_4}{(2\pi)^3}$$

$$\times \mathcal{D}_0(\omega_{n_1}, \mathbf{p}_1) \cdots \mathcal{D}_0(\omega_{n_4}, \mathbf{p}_4)(2\pi)^3 \delta(\mathbf{p}_1 + \cdots + \mathbf{p}_4) \beta \delta_{n_1+\cdots+n_4,0}$$

(3.20)

The evaluation of expressions such as (3.20) is not simple and will be discussed in detail in Section 3.4. The diagrammatic technique is a convenient means for keeping track of the combinatoric factors and the order of the coupling constant in a perturbative expansion of the partition function. It circumvents much of the tedious algebra associated with the direct evaluation of functional integrals.

3.3 Propagators

We shall define a finite-temperature propagator in position space by

$$\mathcal{D}(\mathbf{x}_1, \tau_1; \mathbf{x}_2, \tau_2) = \langle \phi(\mathbf{x}_1, \tau_1) \phi(\mathbf{x}_2, \tau_2) \rangle$$

(3.21)

where the angle brackets denote an ensemble average. Owing to translation invariance, $\mathcal{D}$ depends only on $\mathbf{x}_1 - \mathbf{x}_2$ and $\tau_1 - \tau_2$. The Fourier transform is, with $\mathbf{x}_1 = \mathbf{x}$, $\mathbf{x}_2 = 0$, $\tau_1 = \tau$, $\tau_2 = 0$,

$$\mathcal{D}(\omega_n, \mathbf{p}) = \int_0^\beta d\tau \int d^3 x \, e^{-i(\mathbf{p}\cdot\mathbf{x}+\omega_n\tau)} \mathcal{D}(\mathbf{x}, \tau)$$

$$= \sum_{n_1,n_2} \sum_{\mathbf{p}_1,\mathbf{p}_2} \frac{\beta}{V} \langle \tilde{\phi}_{n_1}(\mathbf{p}_1) \tilde{\phi}_{n_2}(\mathbf{p}_2) \rangle \int_0^\beta d\tau \int d^3 x$$

$$\times \exp[i(\mathbf{p}_1 - \mathbf{p}) \cdot \mathbf{x}] \exp[i(\omega_{n_1} - \omega_n)\tau]$$

(3.22)

The ensemble average vanishes by symmetric integration unless $n_1 = -n_2$, $\mathbf{p}_1 = -\mathbf{p}_2$. Then

$$\mathcal{D}(\omega_n, \mathbf{p}) = \beta^2 \langle \tilde{\phi}_n(\mathbf{p}) \tilde{\phi}_{-n}(-\mathbf{p}) \rangle \tag{3.23}$$

We remind the reader at this point of the concept of a functional derivative. Consider the integral

$$I = I[f] = \int dx\, f(x) w(x)$$

where $w(x)$ is some weight function and I is a functional of $f(x)$, i.e., it depends on the function $f(x)$. The functional derivative of I with respect to $f(y)$ is

$$\frac{\delta I}{\delta f(y)} = w(y)$$

The generalization to more complicated functionals of $f(x)$ is immediate.

Recalling (3.3), (2.31), and (3.10), we discover that $\mathcal{D}(\omega_n, \mathbf{p})$ can be expressed as a functional derivative of $\ln Z$ with respect to $\mathcal{D}_0(\omega_n, \mathbf{p})$. Then

$$\mathcal{D}(\omega_n, \mathbf{p}) = \beta^2 \frac{\int [d\phi] e^S \tilde{\phi}_n(\mathbf{p}) \tilde{\phi}_{-n}(-\mathbf{p})}{\int [d\phi] e^S}$$
$$= -2 \frac{\delta \ln Z}{\delta \mathcal{D}_0^{-1}} = 2 \mathcal{D}_0^2 \frac{\delta \ln Z}{\delta \mathcal{D}_0} \tag{3.24}$$

Unless otherwise indicated, the symbol $\mathcal{D}$ will from now on refer to the propagator in frequency–momentum space.

We define the self-energy $\Pi(\omega_n, \mathbf{p})$ by

$$\mathcal{D}(\omega_n, \mathbf{p}) = \left[\omega_n^2 + \mathbf{p}^2 + m^2 + \Pi(\omega_n, \mathbf{p}) \right]^{-1}$$
$$= (1 + \mathcal{D}_0 \Pi)^{-1} \mathcal{D}_0 \tag{3.25}$$

We shall see shortly that, in the absence of interactions, $\Pi = 0$ and $\mathcal{D} = \mathcal{D}_0$, the free-particle propagator. Using (3.25) and (3.24),

$$(1 + \mathcal{D}_0 \Pi)^{-1} = 2 \mathcal{D}_0 \frac{\delta \ln Z}{\delta \mathcal{D}_0} \tag{3.26}$$

Recall from (2.33) that

$$\ln Z_0 = \tfrac{1}{2} \sum_n \sum_{\mathbf{q}} \ln \left[\mathcal{D}_0(\omega_n, \mathbf{q}) \beta^{-2} \right] \tag{3.27}$$

Thus

$$\frac{\delta \ln Z_0}{\delta \mathcal{D}_0(\omega_n, \mathbf{p})} = \tfrac{1}{2} \mathcal{D}_0^{-1}(\omega_n, \mathbf{p}) \tag{3.28}$$

and so (3.26) becomes

$$(1 + \mathcal{D}_0 \Pi)^{-1} = 1 + 2\mathcal{D}_0 \frac{\delta \ln Z_{\mathrm{I}}}{\delta \mathcal{D}_0} \tag{3.29}$$

It is useful to consider the formal expansion of Π in a power series in λ:

$$\Pi = \sum_{l=1}^{\infty} \Pi_l \tag{3.30}$$

Here Π_l is ostensibly of order l in the coupling constant. Let us see how (3.29) works at the first few orders. Expanding to first order, we obtain

$$1 - \mathcal{D}_0 \Pi_1 = 1 + 2\mathcal{D}_0 \frac{\delta \ln Z_1}{\delta \mathcal{D}_0}$$

$$= 1 + 2\mathcal{D}_0 \frac{\delta}{\delta \mathcal{D}_0}\left(3\ \text{⚭}\right)$$

$$= 1 + 12\mathcal{D}_0\ \text{◯} \tag{3.31}$$

Thus, differentiating $\ln Z_1$ with respect to $\mathcal{D}_0$ is equivalent to cutting each line in the diagram, as inspection of (3.9) shows. A factor 2 appears because we can choose either of the two lines in the "figure 8". Thus

$$\Pi_1 = -12\ \text{◯} \tag{3.32}$$

Continuing in this way, we seek the second-order contribution to Π. Again differentiating (3.29) and keeping terms of order λ^2, we obtain

$$-\mathcal{D}_0 \Pi_2 + \mathcal{D}_0 \Pi_1 \mathcal{D}_0 \Pi_1 = 2\mathcal{D}_0 \frac{\delta \ln Z_2}{\delta \mathcal{D}_0}$$

$$= 2\mathcal{D}_0 \frac{\delta}{\delta \mathcal{D}_0}\left(36\ \text{⚭⚭} + 12\ \text{⊖}\right)$$

$$= 144\mathcal{D}_0\ \text{8} + 96\mathcal{D}_0\ \text{⊖}$$

$$+144\mathcal{D}_0\ \text{◯◯} \tag{3.33}$$

The term $\mathcal{D}_0 \Pi_1 \mathcal{D}_0 \Pi_1$ on the left-hand side simply cancels the last diagram on the right-hand side. Thus

$$\Pi_2 = -144\ \text{8} - 96\ \text{⊖} \tag{3.34}$$

The last diagram in (3.33) is one-particle reducible; that is, by cutting one line we can break the diagram into two disconnected parts. The first two diagrams in (3.33) are not of that form, they are one-particle irreducible (1PI). It is apparent that the one-particle reducible diagrams arise from the iteration of $\mathcal{D}_0\Pi$ in the denominator of (3.29),

$$\Pi = -2\left(\frac{\delta \ln Z_\mathrm{I}}{\delta \mathcal{D}_0}\right)_{1PI} \tag{3.35}$$

where by 1PI we mean that only the 1PI diagrams contribute to Π. The procedure is then as follows. First draw all diagrams which contribute to $\ln Z_\mathrm{I}$ up to a given order, then differentiate with respect to $\mathcal{D}_0$, and, lastly, throw away the one-particle reducible diagrams. This yields the diagrammatic expansion of Π.

3.4 First-order corrections to Π and $\ln Z$

Let us evaluate the one-loop diagram in (3.32). It yields the expression

$$\Pi_1 = 12\lambda T \sum_n \int \frac{d^3p}{(2\pi)^3} \frac{1}{\omega_n^2 + \omega^2} \tag{3.36}$$

where $\omega^2 = \mathbf{p}^2 + m^2$. We could do the frequency sum using (2.97), but there is a more elegant method, which we sketch below.

Suppose we want to evaluate a frequency sum of the form

$$T \sum_{n=-\infty}^{\infty} f(p_0 = i\omega_n = 2\pi n T i) \tag{3.37}$$

Here we think of p_0 as the fourth component of a Minkowski four-vector. We may express (3.37) as a contour integral,

$$\frac{T}{2\pi i} \oint_c dp_0 f(p_0) \frac{1}{2}\beta \coth\left(\frac{1}{2}\beta p_0\right) \tag{3.38}$$

where the contour C is as shown in the following figure:

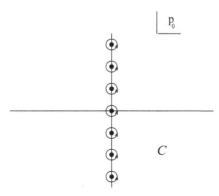

The function $\frac{1}{2}\beta\coth(\frac{1}{2}\beta p_0)$ has poles at $p_0 = 2\pi nTi$ and is everywhere else bounded and analytic. The contour can be deformed into

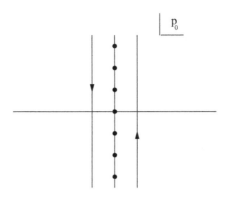

Then, with a suitable rearrangement of the exponentials in the hyperbolic cotangent, we get

$$\frac{1}{2\pi i}\int_{i\infty-\epsilon}^{-i\infty-\epsilon}dp_0 f(p_0)\left(-\frac{1}{2}-\frac{1}{e^{-\beta p_0}-1}\right)$$
$$+\frac{1}{2\pi i}\int_{-i\infty+\epsilon}^{i\infty+\epsilon}dp_0 f(p_0)\left(\frac{1}{2}+\frac{1}{e^{\beta p_0}-1}\right) \tag{3.39}$$

Setting $p_0 \to -p_0$ in the first integral,

$$T\sum_{n=-\infty}^{\infty}f(p_0=i\omega_n) = \frac{1}{2\pi i}\int_{-i\infty}^{i\infty}dp_0\,\frac{1}{2}\left[f(p_0)+f(-p_0)\right]$$
$$+\frac{1}{2\pi i}\int_{-i\infty+\epsilon}^{i\infty+\epsilon}dp_0\,[f(p_0)+f(-p_0)]\,\frac{1}{e^{\beta p_0}-1} \tag{3.40}$$

This expression is correct as long as $f(p_0)$ has no singularities along the imaginary p_0 axis. The frequency sum then naturally separates into a temperature-independent part (the vacuum part) and a part containing the Bose–Einstein distribution (the matter part). In some sense, the replacement of frequency sums by contour integrals, as in (3.40), is equivalent to switching from imaginary time (discrete frequencies in Euclidean space) to real time (continuous energies in Minkowski space). However this is only a matter of mathematical convenience and involves no new physics.

Using (3.40), Π_1 can now be evaluated. With $f(p_0) = -1/(p_0^2 - \omega^2)$ we obtain

$$\Pi_1 = \Pi_1^{\text{vac}} + \Pi_1^{\text{mat}} \tag{3.41}$$

where

$$\Pi_1^{\text{vac}} = 12\lambda \int \frac{d^4 p}{(2\pi)^4} \frac{1}{p_4^2 + \mathbf{p}^2 + m^2}$$

$$\Pi_1^{\text{mat}} = 12\lambda \int \frac{d^3 p}{(2\pi)^3} \frac{1}{\omega} \frac{1}{e^{\beta\omega} - 1}$$

For Π_1^{vac} we have simply defined $p_4 = ip_0$ and $d^4 p = dp_4 d^3 p$, with p_4 integrated from $-\infty$ to ∞. This is the standard result of $T = 0$ field theory in four Euclidean dimensions. For Π_1^{mat}, we have closed the contour about the only pole in the right-hand half-plane, located at $p_0 = \omega$. There is no surface contribution since the integrand falls off sufficiently rapidly as $|p_0| \to \infty$.

The vacuum contribution to Π is actually divergent. This divergence needs to be regulated. The most straightforward way of doing this is to place a high-momentum cutoff, Λ_c, on $p \equiv \sqrt{p_4^2 + \mathbf{p}^2}$. Since the solid angle subtended by a hypersphere in n dimensions is $\Omega_n = 2\pi^{n/2}[\Gamma(n/2)]^{-1}$, we get

$$\Pi_1^{\text{vac}} = \frac{3\lambda}{2\pi^2} \int_0^{\Lambda_c} \frac{p^3 dp}{p^2 + m^2} = \frac{3\lambda}{4\pi^2} \left[\Lambda_c^2 - m^2 \ln \left(\frac{\Lambda_c^2 + m^2}{m^2} \right) \right]$$
$$\to \frac{3\lambda}{4\pi^2} \left[\Lambda_c^2 - m^2 \ln \left(\frac{\Lambda_c^2}{m^2} \right) \right] \tag{3.42}$$

where the arrow indicates that terms that vanish as $\Lambda_c \to \infty$ have been dropped. At $T = 0$, the inverse propagator to first order in λ is

$$\mathcal{D}^{-1}(p_4, \mathbf{p}) = p_4^2 + \mathbf{p}^2 + m^2 + \Pi_1^{\text{vac}} \tag{3.43}$$

In order to avoid a divergent mass we add a counterterm $-\frac{1}{2}\delta m^2 \phi^2$ to the Lagrangian. Treating this as an additional interaction, we see from (3.4) and (3.5) that to lowest order this counterterm contributes to $\ln Z_I$ as

$$-\tfrac{1}{2}\delta m^2 \langle \phi^2 \rangle_0 = -\tfrac{1}{2} \; \text{⬤} \tag{3.44}$$

The cross represents δm^2. The corresponding contribution to the self-energy is, from (3.35),

$$\longrightarrow\!\!\times\!\!\longrightarrow = \delta m^2 \tag{3.45}$$

Adding (3.45) to (3.42) we obtain the renormalized self-energy. We choose the counterterm so that

$$\Pi_1^{\text{vac,ren}} = \frac{3\lambda}{4\pi^2} \left[\Lambda_c^2 - m^2 \ln \left(\frac{\Lambda_c^2}{m^2} \right) \right] + \delta m^2 = 0 \tag{3.46}$$

Then m remains as the physical mass of the particle. More generally, we expand δm^2 in a power series in λ:

$$\delta m^2 = \sum_{n=1}^{\infty} c_n(\Lambda_c)\lambda^n \qquad (3.47)$$

and determine coefficients such that $\Pi^{\text{vac,ren}}(p^2 = M^2) = 0$ at each order in perturbation theory, at some arbitrary subtraction point M. This is part of the renormalization program, which is outside the scope of this book. The relevance of the renormalization group at finite temperature and chemical potential is discussed briefly in Chapter 4.

The complete renormalized self-energy at $T > 0$ at first order in λ is then

$$\Pi_1^{\text{ren}} = 12\lambda \int \frac{d^3 p}{(2\pi)^3} \frac{1}{\omega} \frac{1}{e^{\beta\omega} - 1} \rightarrow \lambda T^2 \qquad (3.48)$$

where the arrow indicates its value as $m \rightarrow 0$. Notice that Π_1^{ren} is finite and vanishes when $T = 0$. It is also momentum independent, but this is not generally true for higher-order diagrams.

Next we calculate the lowest-order correction to $\ln Z$. It is

$$3 \bigcirc\bigcirc - \tfrac{1}{2} \bigodot = -3\lambda\beta V \left(T \sum_n \int \frac{d^3 p}{(2\pi)^3} \mathcal{D}_0(\omega_n, \mathbf{p}) \right)^2$$

$$- \tfrac{1}{2}\beta V \delta m^2 T \sum_n \int \frac{d^3 p}{(2\pi)^3} \mathcal{D}_0(\omega_n, \mathbf{p})$$

$$= -3\lambda\beta V \left(\int \frac{d^3 p}{(2\pi)^3} \frac{1}{\omega} \frac{1}{e^{\beta\omega} - 1} \right)^2$$

$$+ \frac{3\lambda\beta V}{256\pi^4} \Lambda_c^4 \left[1 - \left(\frac{m^2}{\Lambda_c^2} \right) \ln\left(\frac{\Lambda_c^2}{m^2} \right) \right]^2 \qquad (3.49)$$

The last term is a (divergent) contribution to the zero-point energy and pressure of the vacuum (at $T = 0$, $P = \ln Z/\beta V = -E/V$). Since only pressure and energy differences are physically measurable, this term does not contribute to the finite-temperature pressure. If we agree to normalize the vacuum pressure and energy density to zero then the physical pressure contribution at order λ is

$$P_1 = -3\lambda \left(\int \frac{d^3 p}{(2\pi)^3} \frac{1}{\omega} \frac{1}{e^{\beta\omega} - 1} \right)^2 \qquad (3.50)$$

When $m = 0$ and $\lambda \ll 1$ then, from (2.40) and (3.50),

$$P = T^4 \left(\frac{\pi^2}{90} - \frac{\lambda}{48} + \cdots \right) \qquad (3.51)$$

The pressure should be proportional to T^4 by dimensional analysis.

In these calculations, no new ultraviolet (high-momentum or short-distance) divergences appear at finite temperature. All such divergences are already present at $T = 0$; whatever regulation and renormalization is necessary at $T = 0$ is sufficient at $T > 0$ as well. This can be understood in three alternative ways. (i) We construct a complete set of states that are eigenstates of the Hamiltonian with energies E_S. In practice, for an interacting field theory this is usually impossible, but let us imagine it has been done. Then the partition function is obtained directly as

$$Z = \sum_S e^{-\beta E_S} \qquad (3.52)$$

and no new $T > 0$ divergences can arise. (ii) We go back to the transition amplitude (2.16) and compute this quantity as a function of t_f. To obtain the thermodynamic partition function we simply analytically continue from real to imaginary time. (iii) We recall that in the diagrammatic expansions each internal loop involves a frequency sum. The frequency sum can be expressed as a sum of contour integrals, one corresponding to $T = 0$ and the other to $T > 0$; see (3.40). The vacuum integral can give rise to quadratic or logarithmic ultraviolet divergences. The finite-temperature integral is cut off exponentially in the ultraviolet region by the Bose–Einstein distribution. That is to say, the very-short-distance behavior of the theory is unaffected by finite temperature.

3.5 Summation of infrared divergences

The next-order contribution to $\ln Z$ when $m = 0$ is actually of order $\lambda^{3/2}$ and not λ^2 because of a finite-temperature infrared divergence in the perturbative expansion. To see this, consider the second diagram in (3.15). To study its infrared structure, let $n_1 = n_2 = n_3 = n_4 = 0$ and $\mathbf{p}_1 \sim \mathbf{p}_2 \sim \mathbf{p}_3 \sim \mathbf{p}_4 \sim \mathbf{p}$. In the limit $\mathbf{p} \to 0$ this diagram behaves like $\beta V \lambda^2 T^3 dp$, which is infrared convergent. The first diagram in (3.15) has an entirely different structure. Each of the two end loops is proportional to Π_1. Setting $n = 0$ in the middle loop and letting $\mathbf{p}$ denote the three-momentum flowing in that loop, we find that the behavior is $\beta V \Pi_1^2 T dp \, p^{-2}$. This is infrared (small-p) divergent. This divergence is unrelated to the ultraviolet divergences of the field theory at $T = 0$. This new divergence at $T > 0$, when $m = 0$, is due to the fact that $\Pi_1 \neq 0$. The boson develops a dynamically generated mass-squared, $m_{\text{eff}}^2 = \Pi_1^{\text{ren}} = \lambda T^2$. However, we are expanding perturbatively with a propagator that has zero mass. The dynamically generated mass should then damp the infrared divergence.

At order λ^N it is easy to see that the dominant infrared divergent diagram is

$$\frac{(2 \times 3!)^N}{2N} \quad \text{⬡} \quad (N \text{ loops}) \sim \beta V \Pi_1^N T dp \; p^{-2(N-1)} \tag{3.53}$$

The combinatoric factor arises as follows: a factor 3! for two connecting lines at each vertex; a factor 2 for the connection of the remaining two lines at each vertex to the adjacent vertices; a factor $(N-1)!/2$ giving the number of ways to order the vertices in a circle; and a factor $1/N!$ coming from $S_1^N/N!$. We see that the divergence becomes more and more severe at each successive order. Because of the similarity in structure, it is possible to sum this infinite series of diagrams. Summing from $N = 2$ to ∞ we get

$$\frac{1}{2}\beta V T \sum_n \int \frac{d^3 p}{(2\pi)^3} \sum_{N=2}^{\infty} \frac{1}{N} \left[-\Pi_1(\omega_n, \mathbf{p}) \mathcal{D}_0(\omega_n, \mathbf{p}) \right]^N$$

$$= \frac{1}{2} \left[\frac{1}{2} \; \text{⬡} \; - \frac{1}{3} \; \text{△} \; + \cdots \right]$$

$$= -\frac{1}{2}\beta V T \sum_n \int \frac{d^3 p}{(2\pi)^3} \left[\ln(1 + \Pi_1 \mathcal{D}_0) - \Pi_1 \mathcal{D}_0 \right] \tag{3.54}$$

This set of diagrams is sometimes called the set of ring, correlation, or plasmon diagrams in the literature (Gell-Mann and Brueckner) [1]. A more complete summation of the sub-dominant infrared divergent diagrams actually yields the full self-energy in (3.54) instead of the self-energy calculated to first order.

In obtaining (3.54) we summed only the diagrams from (3.53). In addition, there will be diagrams like (3.53) except that any number of the exterior loops are replaced by crosses corresponding to the mass counterterm δm^2. Including those as well (this is left as an exercise), the factor Π_1 in (3.54) is replaced by $\Pi_1^{\text{ren}} = \lambda T^2$:

$$-\frac{1}{2}\beta V T \sum_n \int \frac{d^3 p}{(2\pi)^3} \left[\ln\left(1 + \frac{\lambda T^2}{\omega_n^2 + \mathbf{p}^2}\right) - \frac{\lambda T^2}{\omega_n^2 + \mathbf{p}^2} \right] = \frac{\beta V}{12\pi} \lambda^{3/2} T^4 + \cdots$$

$$\tag{3.55}$$

The $\lambda^{3/2}$ term arises solely from the $n = 0$ mode, which yields the dominant infrared divergence. The $n \neq 0$ modes produce higher-order corrections in λ. The origin of this nonanalyticity in λ is the fact that the boson acquires a mass proportional to $\lambda^{1/2} T$. The weak coupling expansion for the pressure is thus

$$P = \frac{\pi^2}{90}T^4 \left[1 - \frac{15}{8}\left(\frac{\lambda}{\pi^2}\right) + \frac{15}{2}\left(\frac{\lambda}{\pi^2}\right)^{3/2} + \cdots \right] \tag{3.56}$$

A nonanalyticity in the couplings of this type is also found in QED and QCD. It is good to see how this happens in a simpler theory such as $\lambda\phi^4$ first.

The same nonanalyticity should also be expected in Π because of the close relationship between Π and $Z_{\rm I}$, as expressed in (3.35). The dominant infrared contribution at order λ^N comes from the diagram

$$-(2 \times 3!)^N \underset{}{\bigcirc} \quad (N-1 \text{ loops})$$

$$= 12\lambda T \sum_n \int \frac{d^3p}{(2\pi)^3} \left[-\Pi_1(\omega_n, \mathbf{p}) \right]^{N-1} \mathcal{D}_0^N(\omega_n, \mathbf{p}) \tag{3.57}$$

Summing (3.57) from $N = 1$ to ∞, we obtain

$$\Pi = 12\lambda T \sum_n \int \frac{d^3p}{(2\pi)^3} \frac{1}{\omega_n^2 + \mathbf{p}^2 + \Pi_1}$$

$$= -12 \underset{}{\bigcirc} \tag{3.58}$$

which has the nice interpretation that the free propagator is replaced by the first-order-corrected propagator $\mathcal{D}_1$ in the one-loop self-energy diagrams. Actually, we are suppressing all other similar diagrams involving the replacement of Π_1 by δm^2, just as in (3.54). Taking into account the mass counterterms simply replaces Π_1 by $\Pi_1 + \delta m^2 = \Pi_1^{\rm ren}$ in (3.58). Recalling (3.41) leads to

$$\Pi_1^{\rm ren} = \lambda T^2 \left[1 - 3\left(\frac{\lambda}{\pi^2}\right)^{1/2} + \cdots \right] \tag{3.59}$$

Thus there is a term in the self-energy of order $\lambda^{3/2}$, just as in $\ln Z_{\rm I}$.

3.6 Yukawa theory

The simplest theory involving interacting fermions is one in which fermions are coupled to a neutral field by the Yukawa interaction

$$\mathcal{L}_{\rm I} = g\bar{\psi}\psi\phi \tag{3.60}$$

The perturbation expansion in this case proceeds as in the previous sections with only a few changes. Since

$$S_I = \int_0^\beta d\tau \int d^3x \, \mathcal{L}_I(\mathbf{x}, \tau) \tag{3.61}$$

is linear in ϕ, it follows that $\langle S_I^l \rangle = 0$ if l is odd. Here the expansion of $\ln Z$ is formally an expansion in g^2. The lowest-order correction to $\ln Z_0$ is

$$\ln Z_2 = \tfrac{1}{2}\langle S_I^2 \rangle_0 = \tfrac{1}{2}g^2 \int d\tau_1 d\tau_2 \int d^3x_1 d^3x_2 \sum_{n_1,\dots,n_4} \sum_{l_1,l_2} \sum_{\mathbf{p}_1,\dots,\mathbf{p}_4} \sum_{\mathbf{q}_1,\mathbf{q}_2} \frac{\beta}{V^3}$$

$$\times \exp[i(\mathbf{q}_1 + \mathbf{p}_3 - \mathbf{p}_1) \cdot \mathbf{x}_1] \exp[i(\mathbf{q}_2 + \mathbf{p}_4 - \mathbf{p}_2) \cdot \mathbf{x}_2]$$

$$\times \exp[i(\omega_{l_1} + \omega_{n_3} - \omega_{n_1})\tau_1] \exp[i(\omega_{l_2} + \omega_{n_4} - \omega_{n_2})\tau_2] \frac{A}{B} \tag{3.62}$$

where

$$A = \prod_{\alpha,n,l} \prod_{\mathbf{p},\mathbf{q}} \int d\tilde{\bar{\psi}}_{\alpha;n}(\mathbf{p}) d\tilde{\psi}_{\alpha;n}(\mathbf{p}) d\tilde{\phi}_l(\mathbf{q})$$

$$\times e^{S_0} \tilde{\bar{\psi}}_{\rho;n_1}(\mathbf{p}_1) \tilde{\psi}_{\rho;n_3}(\mathbf{p}_3) \tilde{\bar{\psi}}_{\gamma;n_2}(\mathbf{p}_2) \tilde{\psi}_{\gamma;n_4}(\mathbf{p}_4) \tilde{\phi}_{l_1}(\mathbf{q}_1) \tilde{\phi}_{l_2}(\mathbf{q}_2)$$

and

$$B = \prod_{\alpha,n,l} \prod_{\mathbf{p},\mathbf{q}} \int d\tilde{\bar{\psi}}_{\alpha;n}(\mathbf{p}) d\tilde{\psi}_{\alpha;n}(\mathbf{p}) d\tilde{\phi}_l(\mathbf{q}) e^{S_0}$$

Furthermore,

$$S_0 = \beta \sum_n \sum_{\mathbf{p}} \tilde{\bar{\psi}}_{\alpha;n}(\mathbf{p}) \mathcal{G}_0^{-1}(\omega_n, \mathbf{p})_{\alpha\rho} \tilde{\psi}_{\rho;n}(\mathbf{p})$$

$$- \tfrac{1}{2}\beta^2 \sum_n \sum_{\mathbf{p}} \tilde{\phi}_n(\mathbf{p}) \mathcal{D}_0^{-1}(\omega_n, \mathbf{p}) \tilde{\phi}_{-n}(-\mathbf{p}) \tag{3.63}$$

The free-particle fermion propagator $\mathcal{G}_0$ is defined, in analogy to the free-particle boson propagator, as

$$\mathcal{G}_0^{-1}(\omega_n, \mathbf{p}) = \not{p} - M \tag{3.64}$$

Here $p_0 \equiv i\omega_n + \mu$, M is the fermion mass, and m is the boson mass (see (2.91)). We have changed our variable of integration from $i\psi^\dagger$ to $\bar{\psi}$, which is conventional.

The integration over the spatial and temporal coordinates in (3.62) can be done immediately. It leads to an overall factor of $\beta^2 V^2$ and to the constraints

$$\mathbf{p}_1 = \mathbf{p}_3 + \mathbf{q}_1 \quad \mathbf{p}_2 = \mathbf{p}_4 + \mathbf{q}_2 \quad n_1 = n_3 + l_1 \quad n_2 = n_4 + l_2 \tag{3.65}$$

The integration over the scalar field leads to the constraints

$$\mathbf{q}_2 = -\mathbf{q}_1 \quad l_2 = -l_1 \tag{3.66}$$

The integration over the spinor field leads to either of the following constraints:

$$\mathbf{p}_1 = \mathbf{p}_3 \quad \mathbf{p}_2 = \mathbf{p}_4 \quad n_1 = n_3 \quad n_2 = n_4$$

or

$$\mathbf{p}_1 = \mathbf{p}_4 \quad \mathbf{p}_2 = \mathbf{p}_3 \quad n_1 = n_4 \quad n_2 = n_3 \tag{3.67}$$

These two possibilities lead to two topologically distinct diagrams:

$$\ln Z_2 = \tfrac{1}{2} \; \bigcirc \!\!-\!\!-\!\!-\!\! \bigcirc \; - \tfrac{1}{2} \; \bigcirc \tag{3.68}$$

The first of these represents

$$\tfrac{1}{2}\beta V \frac{g^2}{m^2} \left(T \sum_n \int \frac{d^3p}{(2\pi)^3} \operatorname{Tr} \mathcal{G}_0(\omega_n, \mathbf{p}) \right)^2 \tag{3.69}$$

and the second represents

$$-\tfrac{1}{2}\beta V g^2 T^2 \sum_{n_1 n_2} \int \frac{d^3p_1 d^3p_2}{(2\pi)^6}$$
$$\times \operatorname{Tr}[\mathcal{G}_0(\omega_{n_1}, \mathbf{p}_1)\mathcal{D}_0(\omega_{n_2} - \omega_{n_1}, \mathbf{p}_2 - \mathbf{p}_1)\mathcal{G}_0(\omega_{n_2}, \mathbf{p}_2)] \tag{3.70}$$

The solid lines represent fermions and the broken lines represent bosons. The arrows on the fermion lines indicate the flow of fermion number and follow from the fact that in (3.62) a $\bar{\psi}$ must always be matched to a ψ to get a nonzero contribution. The trace operation in (3.69) and (3.70) is over the Dirac indices. The minus sign in (3.70) comes from anticommuting the fermion fields (which are Grassmann variables) to put them into the canonical ordering of (2.80) and (2.81). The boson line in the first diagram carries zero frequency and momentum and gives rise to the factor $\mathcal{D}_0(0, \mathbf{0}) = m^{-2}$. The reader is encouraged to verify that indeed (3.68)–(3.70) follow directly from the functional integral of (3.62).

The "finite-temperature Feynman rules" are similar to those listed in Section 3.2. The new aspects are:

1 There is a factor $T \sum_n \int [d^3p/(2\pi)^3]\mathcal{G}_0(\omega_n, \mathbf{p})$ for each fermion line.
2 There is a factor g at each vertex.
3 There is a trace over Dirac indices for each closed fermion loop as well as a minus sign coming from the Grassmann nature of the fermion field.
4 All connected diagrams are constructed from the following elementary vertex:

It turns out that the one-particle reducible diagrams (one of which is seen in (3.68)) arise from the fact that the scalar field ϕ develops a nonzero thermal average. It is in some sense analogous to a Bose–Einstein condensate. This condensate is however driven by the interaction with the fermions. All such diagrams can be summed by the use of the mean field expansion. This will be illustrated in later chapters.

The frequency sum for fermions can be converted to contour integrals in a manner closely paralleling the procedure for bosons. The fermion propagator depends on the combination $p_0 = i\omega_n + \mu$ with $\omega_n = (2n + 1)\pi T$. A straightforward analysis yields

$$
T \sum_n f(p_0 = i\omega_n + \mu) = -\frac{1}{2\pi i} \int_{-i\infty+\mu+\epsilon}^{i\infty+\mu+\epsilon} dp_0 \ f(p_0) \frac{1}{e^{\beta(p_0-\mu)} + 1}
$$
$$
- \frac{1}{2\pi i} \int_{-i\infty+\mu-\epsilon}^{i\infty+\mu-\epsilon} dp_0 \ f(p_0) \frac{1}{e^{\beta(\mu-p_0)} + 1}
$$
$$
+ \frac{1}{2\pi i} \oint_C dp_0 \ f(p_0) + \frac{1}{2\pi i} \int_{-i\infty}^{i\infty} dp_0 \ f(p_0)
$$
$$
(3.71)
$$

The contour C is as shown in the following figure:

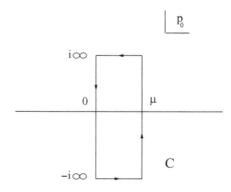

The first two terms in (3.71) correspond to particle and antiparticle contributions and vanish at $T = 0$. The third term is T-independent and gives the $T = 0$ finite-density contribution. The last term is the vacuum contribution.

The boson self-energy Π is still defined by (3.25) and still satisfies (3.35). From (3.68) the lowest-order diagram is

$$\Pi_2 = \text{-----}\bigcirc\text{-----} \tag{3.72}$$

The fermion self-energy Σ is defined by the equation

$$\mathcal{G}^{-1} = \mathcal{G}_0^{-1} + \Sigma = \not{p} - M + \Sigma(\omega_n, \mathbf{p}) \tag{3.73}$$

In position space, the full fermion propagator is defined by

$$\mathcal{G}(\mathbf{x}_1 - \mathbf{x}_2, \tau_1 - \tau_2) = \langle T_\tau [\bar{\psi}(\mathbf{x}_1, \tau_1)\psi(\mathbf{x}_2, \tau_2)] \rangle \tag{3.74}$$

where the angle brackets denote the exact ensemble average. It can be shown that the analog of (3.35) is

$$\Sigma = \frac{\delta \ln Z_{\mathrm{I}}}{\delta \mathcal{G}_0} \tag{3.75}$$

From (3.68) the lowest-order diagrams are

$$\Sigma_2(\omega_n, \mathbf{p}) = \underset{}{\text{------}} - \underset{}{\text{------}} \tag{3.76}$$

Explicit evaluations of loop diagrams involving fermions will be taken up in the later chapters on theories that represent nature.

3.7 Remarks on real time perturbation theory

The perturbative treatment discussed up to now has been in the so-called imaginary time formalism. The functional integral representation of the partition function involves an integration over "imaginary time" from 0 to β. A Fourier decomposition of the fields leads to a discrete frequency sum; for bosons $\omega_n = 2\pi n T$ and for fermions $\omega_n = (2n + 1)\pi T$.

The one-loop expression in (3.41) can be written alternatively as

$$\Pi_1 = 12\lambda \int_{-\infty}^{\infty} \frac{d^3p}{(2\pi)^3} \int_{-\infty}^{\infty} \frac{dp_0}{2\pi} \left(\frac{i}{p^2 - m^2 + i\epsilon} + \frac{2\pi}{e^{\beta|p_0|} - 1}\delta(p^2 - m^2) \right) \tag{3.77}$$

This has the interpretation that the propagator consists of the sum of a vacuum part and a finite-temperature part. Instead of a summation over discrete frequencies there is an integration over a real, continuous, energy p_0. Because of the presence of the Dirac delta function, the finite-temperature contribution is trivial to obtain.

The above suggests the possibility of a "real time" perturbation theory. The rules of Section 3.2 could perhaps be modified according to

$$T \sum_n \to \int_{-\infty}^{\infty} \frac{dp_0}{2\pi}$$

$$\frac{1}{\omega_n^2 + \omega^2} \to \frac{i}{p^2 - m^2 + i\epsilon} + \frac{2\pi}{e^{\beta|p_0|} - 1} \delta(p^2 - m^2) \qquad (3.78)$$

$$\beta \delta_{\omega_{\text{in}}, \omega_{\text{out}}} \to 2\pi \delta(p_{\text{in}}^0 - p_{\text{out}}^0)$$

The advantage would be that no frequency sums need to be done. The finite-temperature contributions are naturally separated. In addition, the Green's functions so obtained are functions of real Minkowski momenta p^μ, which facilitates certain applications such as linear response analyses, discussed in Chapter 6.

Unfortunately, there are cases where the simple substitution (3.78) does not work [2]. As an example, consider a massive boson field with a cubic self-interaction. The one-loop self-energy diagram is

$$\Pi(k) = \underset{k}{\longleftarrow} \bigcirc \underset{k}{\longrightarrow} \qquad (3.79)$$

The (unrenormalized) self-energy evaluated at zero four-momentum ($k = 0$) is

$$T \sum_n \int \frac{d^3 p}{(2\pi)^3} \frac{1}{(\omega_n^2 + \omega^2)^2} \qquad (3.80)$$

This expression is logarithmically divergent in the ultraviolet regime ($|p| \to \infty$), but this divergence is regulated by the usual $T = 0$ counterterm; no new $T > 0$ divergences appear. If we perform the substitution (3.78) we obtain

$$\int_{-\infty}^{\infty} \frac{dp_0}{2\pi} \int \frac{d^3 p}{(2\pi)^3} \left(\frac{i}{p^2 - m^2 + i\epsilon} + \frac{2\pi}{e^{\beta|p_0|} - 1} \delta(p^2 - m^2) \right)^2 \qquad (3.81)$$

There is now a severe mathematical singularity owing to the square of the delta function. The expression (3.81) is ill-defined.

It is possible to formulate a real time perturbation theory reminiscent of (3.78) [3, 4]. Essentially, the number of independent fields doubles. Instead of a single scalar field ϕ, we encounter two scalar fields, ϕ_1 and ϕ_2, called type-1 and type-2 fields. The propagator becomes a 2×2 matrix even though the bosons are neutral and have no spin. The propagator is

$$
D = \begin{pmatrix} \cosh\theta & \sinh\theta \\ \sinh\theta & \cosh\theta \end{pmatrix} \begin{pmatrix} \dfrac{i}{p^2 - m^2 + i\epsilon} & 0 \\ 0 & \dfrac{-i}{p^2 - m^2 - i\epsilon} \end{pmatrix} \begin{pmatrix} \cosh\theta & \sinh\theta \\ \sinh\theta & \cosh\theta \end{pmatrix}
$$

$$
= \begin{pmatrix} \dfrac{i}{p^2 - m^2 + i\epsilon} & 0 \\ 0 & \dfrac{-i}{p^2 - m^2 - i\epsilon} \end{pmatrix}
$$

$$
+ \frac{2\pi}{e^{\beta|p_0|} - 1} \delta(p^2 - m^2) \begin{pmatrix} 1 & -e^{\beta|p_0|/2} \\ -e^{\beta|p_0|/2} & 1 \end{pmatrix} \tag{3.82}
$$

There are two types of vertex. A type-1 vertex has only type-1 fields emerging from it and has its usual value, while a type-2 vertex has only type-2 fields emerging from it and has a value opposite in sign to its type-1 counterpart. For example, for a cubic coupling,

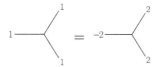

The ultimate reason for this field doubling is to avoid singularities of the type that arise in (3.81). Explicit calculations show the cancellation arising from the two components. In this regard, the delta function appearing in (3.82) is represented as

$$
\delta(p^2 - m^2) = \lim_{\epsilon \to 0} \frac{1}{\pi} \frac{\epsilon}{(p^2 - m^2)^2 + \epsilon^2} \tag{3.83}
$$

Similar field doublings appear for spin-1/2 fermions and for spin-1 vector bosons.

It is interesting that perturbation theory at finite temperature can be formulated directly in real time as well as in imaginary time. Our preference is for the imaginary time formalism and this is the one adopted in this book.

3.8 Exercises

3.1 Derive the diagrammatic rules for the neutral scalar field with a cubic self-interaction $g\phi^3$ in $5 + 1$ dimensions. Derive the lowest-order diagrams for $\ln Z_{\mathrm{I}}$ and Π.

3.2 Derive (3.35) to all orders.

3.3   Show that $\frac{1}{2}\beta \coth\left(\frac{1}{2}\beta p_0\right)$ has simple poles at $p_0 = 2\pi n T i$ with residue 1 and is elsewhere analytic and bounded.

3.4   Show that Π_1 is replaced by $\Pi_1^{\text{ren}} = \Pi_1 + \delta m^2$ in (3.54) when the corresponding diagrams with counterterms are included.

3.5   Show that (3.59) follows from (3.41) when $\lambda \ll 1$.

3.6   Prove (3.71).

3.7   Prove (3.75).

References

1. Gell-Mann, M., and Brueckner, K. A., *Phys. Rev.* **106**, 364 (1957).
2. Dolan, L., and Jackiw, R., *Phys. Rev D* **9**, 3320 (1974).
3. Umezawa, H., Matsumoto, H., and Tachiki, M. (1982). *Thermo Field Dynamics and Condensed States* (North-Holland, Amsterdam).
4. Niemi, A. J. and Semenoff, G. W., *Ann. Phys.* **152**, 105 (1984); *Nucl. Phys.* **B230**, 181 (1984).

Bibliography

Classic papers

Dolan, L., and Jackiw, R., *Phys. Rev. D* **9**, 3320 (1974).
Weinberg, S., *Phys. Rev. D* **9**, 3357 (1974).
Norton, R. E. and Cornwall, J. M., *Ann. Phys.* **91**, 106 (1975).
Lindé, A., *Rep. Prog. Phys.* **42**, 389 (1979).

Real time perturbation theory

Umezawa, H., Matsumoto, H., and Tachiki, M. (1982). *Thermo Field Dynamics and Condensed States* (North-Holland, Amsterdam).
Niemi, A. J. and Semenoff, G. W., *Ann. Phys.* **152**, 105 (1984); *Nucl. Phys.* **B230**, 181 (1984).
Landsman, N. P. and van Weert, Ch. G., *Phys Rep.* **145**, 141 (1987).

Ring diagram contribution

Gell-Mann, M., and Brueckner, K. A., *Phys. Rev.* **106**, 364 (1957).

4

Renormalization

Relativistic quantum field theories generally display infinities at sufficiently high order in a loop expansion. These infinities must first be regulated, meaning that cutoffs are applied to yield finite results that can be manipulated with some mathematical rigor. The results are then renormalized, so that the parameters of the Lagrangian and the cutoffs are eliminated in favor of physical observables such as electric charge and mass. If there are only a finite number of cutoffs as the number of loops increases, the theory is said to be renormalizable and the cutoffs can always be eliminated in favor of a finite number of observables. If the number of required cutoffs increases without bound as the number of loops increases then the theory is said to be nonrenormalizable and one must specify an infinite number of observables to define the theory. The general opinion is that a *fundamental* theory of nature should be renormalizable. This is based on the belief that there are only a finite number of independent parameters in our universe. An *effective* theory only needs to describe nature over a finite range of distances or momenta, and such a theory need not be renormalizable. In this chapter we consider the basic aspects of a renormalizable theory and its implications for finite temperatures. For definiteness we study a scalar field theory; the same principles apply to more complicated theories, such as the gauge theories to be studied in later chapters.

4.1 Renormalizing $\lambda\phi^4$ theory

Recall that the interaction contribution to the partition function is given by

$$\ln Z_{\mathrm{I}} = \ln \left(\frac{\int [d\phi]\mathrm{e}^{S}}{\int [d\phi]\mathrm{e}^{S_0}} \right) \tag{4.1}$$

For the $\lambda\phi^4$ theory the Lagrangian is

$$\mathcal{L} = \tfrac{1}{2}\partial_\mu\phi\,\partial^\mu\phi - \tfrac{1}{2}m^2\phi^2 - \lambda\phi^4 \qquad (4.2)$$

We found in Chapter 3 that we needed to add a counterterm $-\tfrac{1}{2}\delta m^2\phi^2$, which is equivalent to saying that $m^2 = m_R^2 + \delta m^2$, where m_R is the renormalized mass. The cutoff dependence of the self-energy at lowest order could be canceled by a suitable choice of δm^2.

Now we investigate what happens when we scale the field and the coupling constant. Write

$$\phi = \mathcal{Z}_3^{1/2}\phi_R \qquad (4.3)$$

Notice that we can integrate with $[d\phi_R]$ since $\mathcal{Z}_3$ cancels between the numerator and denominator in (4.1). We also write

$$\lambda = \mathcal{Z}_1\mathcal{Z}_3^{-2}\lambda_R \qquad (4.4)$$

The scaling factors $\mathcal{Z}_1$ and $\mathcal{Z}_3$ are known as the coupling constant and the wavefunction renormalization, respectively. Usually in the literature the symbol Z instead of $\mathcal{Z}$ is used for these, but here we do not want to confuse them with the partition function.

The Lagrangian becomes

$$\begin{aligned}
\mathcal{L} &= \tfrac{1}{2}\big[\partial_\mu\phi_R\partial^\mu\phi_R - \left(m_R^2 + \delta m^2\right)\phi_R^2\big]\mathcal{Z}_3 - \lambda_R\phi_R^4\,\mathcal{Z}_1 \\
&= \mathcal{L}_R + \tfrac{1}{2}\big[\partial_\mu\phi_R\partial^\mu\phi_R - m_R^2\phi_R^2\big](\mathcal{Z}_3 - 1) \\
&\quad - \tfrac{1}{2}\mathcal{Z}_3\delta m^2\phi_R^2 - \lambda_R\phi_R^4\left(\mathcal{Z}_1 - 1\right)
\end{aligned} \qquad (4.5)$$

where

$$\mathcal{L}_R = \tfrac{1}{2}\partial_\mu\phi_R\partial^\mu\phi_R - \tfrac{1}{2}m_R^2\phi_R^2 - \lambda_R\phi_R^4 \qquad (4.6)$$

The Lagrangian is thus expressed as a function of the renormalized field and of the renormalized mass and coupling constant. The latter two have numerical values that must be determined by experiment. All cutoff dependence resides in the unobservable quantities $\mathcal{Z}_1$, $\mathcal{Z}_3$, and δm^2. In a perturbative renormalization scheme they should have power series expansions

$$\mathcal{Z}_1 = 1 + \sum_{n=1}^{\infty} a_n\lambda_R^n$$

$$\mathcal{Z}_3 = 1 + \sum_{n=1}^{\infty} b_n\lambda_R^n \qquad (4.7)$$

$$\delta m^2 = \sum_{n=1}^{\infty} c_n\lambda_R^n$$

The coefficients a_n, b_n, c_n will depend in general upon the ultraviolet cutoff Λ_{c}.

All renormalizable field theories can be dealt with in the manner sketched above. The reader is referred to the excellent texts on relativistic quantum field theory listed in the bibliography at the end of the chapter for a full discussion of the renormalization program.

We remark again that whatever regularization and renormalization is necessary and sufficient at zero temperature and chemical potential is also necessary and sufficient at finite temperature and chemical potential. (Recall the discussion in Section 3.4.)

4.2 Renormalization group

For the moment consider the $\lambda\phi^4$ theory at $T = 0$ and with $m_{\mathrm{R}} = 0$. Generalization to $m_{\mathrm{R}} > 0$ and other theories is straightforward. The finite-temperature effects are studied in Section 4.4.

Let $\Gamma^{(n)}$ be a 1PI Green's function of n powers of the field ϕ. The statement that the theory is renormalizable means that

$$\mathcal{Z}_3^{n/2}\left(\lambda, \frac{\Lambda_{\mathrm{c}}}{M}\right)\Gamma^{(n)}(p, \lambda, \Lambda_{\mathrm{c}}) = \Gamma_{\mathrm{R}}^{(n)}(p, \lambda_{\mathrm{R}}, M) \tag{4.8}$$

The unrenormalized Green's function depends on the unrenormalized coupling and on the cutoff Λ_{c}. The symbol p can represent one momentum or a set of momenta $(p_1, p_2, \ldots)$. Since $\mathcal{Z}_3$ is dimensionless it can only depend on λ and on Λ_{c}/M. What is M? Green's functions are typically infinite, so we must specify their value at some particular point, for example, $p^2 = M^2$, using one or other of the following diagrams:

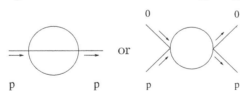

We could require $\Gamma_{\mathrm{R}}^{(n)}$ to have its free-field value at $p^2 = M^2$, that is,

$$\Gamma_{\mathrm{R}}^{(n)}(p^2 = M^2, \lambda_{\mathrm{R}}, M) = \Gamma_{\mathrm{R}}^{(n)}(p^2 = M^2, 0, M) \tag{4.9}$$

as is frequently done, but the choice is arbitrary. Physical results should be independent of the renormalization scheme, in particular, independent of the choice of M.

The requirement of renormalizability has consequences. To see them, take the total derivative of the left- and right-hand sides of (4.8) with respect to M, keeping λ and Λ_c fixed:

$$M\frac{d}{dM}\left(\mathcal{Z}_3^{n/2}\Gamma^{(n)}\right) = M\left(\frac{\partial \mathcal{Z}_3^{n/2}}{\partial M}\right)_{\lambda,\Lambda_c}\Gamma^{(n)}$$

(4.10)

$$M\frac{d}{dM}\Gamma_{\mathrm{R}}^{(n)} = M\left(\frac{\partial \Gamma_{\mathrm{R}}^{(n)}}{\partial M}\right)_{\lambda_{\mathrm{R}}} + \left(\frac{\partial \Gamma_{\mathrm{R}}^{(n)}}{\partial \lambda_{\mathrm{R}}}\right) M\left(\frac{\partial \lambda_{\mathrm{R}}}{\partial M}\right)_{\lambda,\Lambda_c}$$

Now for sake of convenience of notation define

$$\gamma_{(n)} = -\mathcal{Z}_3^{-n/2} M\left(\frac{\partial \mathcal{Z}_3^{n/2}}{\partial M}\right)_{\lambda,\Lambda_c}$$

$$= \mathcal{Z}_3^{-n/2}\Lambda_c\left(\frac{\partial \mathcal{Z}_3^{n/2}}{\partial \Lambda_c}\right)_{\lambda,M}$$

$$= \frac{1}{2}n\Lambda_c \mathcal{Z}_3^{-1}\left(\frac{\partial \mathcal{Z}_3}{\partial \Lambda_c}\right)_{\lambda,M}$$

$$= n\gamma_{(1)}$$

(4.11)

and

$$\beta_\lambda = M\left(\frac{\partial \lambda_{\mathrm{R}}}{\partial M}\right)_{\lambda,\Lambda_c} = -\Lambda_c\left(\frac{\partial \lambda_{\mathrm{R}}}{\partial \Lambda_c}\right)_{\lambda,M}$$

(4.12)

in the conventional notation. The quantity β_λ must not be confused with the inverse temperature. Putting these all together, we arrive at the renormalization-group equation

$$\left(M\frac{\partial}{\partial M} + \beta_\lambda\frac{\partial}{\partial \lambda_{\mathrm{R}}} + \gamma_{(n)}\right)\Gamma_{\mathrm{R}}^{(n)} = 0$$

(4.13)

All the $\Gamma_{\mathrm{R}}^{(n)}$ must satisfy this equation on account of renormalizability. It expresses the invariance of physical observables under changes in M, the renormalization scale.

The renormalized 1PI Green's function has the general functional form

$$\Gamma_{\mathrm{R}}^{(n)} = p^D z\left(\frac{p}{M}, \lambda_{\mathrm{R}}\right)$$

where D is the dimension of $\Gamma^{(n)}$ and z is a dimensionless function of the two displayed dimensionless variables. After substitution into (4.13), factoring out p^D, and then defining $x = M/p$, $y = \lambda_{\mathrm{R}}$, we obtain the linear, homogeneous, first-order partial differential equation

$$\left(x\frac{\partial}{\partial x} + \beta_\lambda(y)\frac{\partial}{\partial y} + \gamma_{(n)}(y)\right)z(x,y) = 0$$

(4.14)

This equation can be solved by the method of characteristics. The solution is

$$z = f(u(x, y)) \exp \left(\int_x^{x_0} \gamma_{(n)}(x') \frac{dx'}{x'} \right) \tag{4.15}$$

Here x_0 is a reference point, f is an arbitrary function, and $u(x, y) = c$ represents the relationship between x and y when they satisfy the differential equation

$$x \frac{dy}{dx} = \beta_\lambda(y) \tag{4.16}$$

The solution to this equation involves one constant of integration, corresponding to c. What is meant by $\gamma_{(n)}(x)$ is $\gamma_{(n)}(y(x))$ where $y(x)$ is determined from $u(x, y) = c$. Translating this back into the original notation we have the solution to (4.13) as

$$\Gamma_{\mathrm{R}}^{(n)} = G \left(p, \bar{\lambda} \left(\frac{M'}{M} \right) \right) \exp \left(\int_{M/p}^{M'/p} \gamma_{(n)}(x) \frac{dx}{x} \right) \tag{4.17}$$

The function G is arbitrary and undetermined by the renormalization-group equation. The renormalization-group running coupling $\bar{\lambda}$ satisfies the differential equation

$$\chi \frac{d\bar{\lambda}}{d\chi} = \beta_\lambda(\bar{\lambda}) \tag{4.18}$$

where $\chi = M'/M$, subject to the condition

$$\bar{\lambda}(\chi = 1) = \lambda_{\mathrm{R}} \tag{4.19}$$

The exponential in (4.17) is referred to as the anomalous dimension of $\Gamma_{\mathrm{R}}^{(n)}$.

To the lowest nontrivial order, β_λ is computed to be (Exercise 4.1)

$$\beta_\lambda(\bar{\lambda}) = \frac{9}{2\pi^2} \bar{\lambda}^2 \tag{4.20}$$

The differential equation to be solved is

$$\chi \frac{d\bar{\lambda}}{d\chi} = \frac{9}{2\pi^2} \bar{\lambda}^2 \tag{4.21}$$

The solution satisfying (4.19) is

$$\bar{\lambda} = \frac{\lambda_{\mathrm{R}}}{1 - (9/4\pi^2)\lambda_{\mathrm{R}} \ln \chi^2} \tag{4.22}$$

The denominator may be expanded in a power series in λ_R:

$$\bar{\lambda} = \lambda_R \sum_{n=0}^{\infty} \left[\frac{9\lambda_R}{4\pi^2} \ln \left(\frac{M'^2}{M^2} \right) \right]^n \tag{4.23}$$

This expansion may be arrived at in a completely independent manner, as follows. At each order in perturbation theory compute the logarithmic contribution of the highest power. This is known as the leading-log approximation. One obtains the same result as a consequence of the renormalization group.

The renormalization-group running coupling $\bar{\lambda}$ does not depend on M and λ_R separately but only on a particular combination of them. In (4.22) define

$$\frac{9}{2\pi^2} \ln \Lambda \equiv \lambda_R^{-1} + \frac{9}{2\pi^2} \ln M \tag{4.24}$$

Furthermore, let us choose $M'^2 = p^2$, the only natural scale in the problem. Then

$$\bar{\lambda} = \frac{4\pi^2}{9 \ln (\Lambda^2/p^2)} \tag{4.25}$$

The effective coupling $\bar{\lambda}$ no longer depends on the coupling λ_R originally appearing in the Lagrangian! This is often referred to as dimensional transmutation. There is no longer an intrinsic coupling constant, but in its place there is an intrinsic energy scale Λ (not to be confused with the cutoff Λ_c). The effective coupling $\bar{\lambda}$ depends on the momentum p. As $p/\Lambda \to 0$, we have $\bar{\lambda} \to 0$, which is infrared freedom. The coupling effectively goes to zero at large distance so that weak coupling expansions should be quite accurate there. Since to lowest order the beta function β_λ is positive, it follows that $\bar{\lambda}$ must be larger at short distances. In fact, from (4.25), $\bar{\lambda} \to \infty$ as $p/\Lambda \to 1$. This is certainly an artifact of the lowest-order perturbation expansion of β_λ, but nevertheless it indicates that the coupling grows as the distance decreases.

4.3 Regularization schemes

We have regulated the divergences in the scalar field theory by placing an upper limit on the integration over four-momentum in Euclidean space. There are alternative regularization procedures, dimensional regularization being the most commonly used by far. Dimensional regularization is almost indispensable in gauge theories. The idea is to work in $n = 4 - \epsilon$ dimensions where integrals converge and then analytically continue to $\epsilon \to 0$.

Consider the self-energy in scalar field theory. The one-loop expression in Minkowski space is

$$\Pi_1^{\text{vac}} = -12\lambda\kappa^\epsilon \int \frac{d^n l}{(2\pi)^n} \frac{i}{l^2 - m^2 + i\epsilon}$$

$$= -\frac{12\lambda\kappa^\epsilon}{(2\pi)^n} \frac{\pi^{n/2}\Gamma(1 - n/2)}{m^{2-n}} \tag{4.26}$$

This scheme requires the introduction of a mass scale κ to compensate for the deviation from four dimensions and so ensuring that λ remains dimensionless. The Γ function has poles at the negative integers. Using

$$\Gamma(-n + \delta) = \frac{(-1)^n}{n!} \left(\frac{1}{\delta} + \psi(n + 1) + \mathcal{O}(\delta) \right), \tag{4.27}$$

with

$$\psi(n + 1) = 1 + \frac{1}{2} + \cdots + \frac{1}{n} - \gamma_E \tag{4.28}$$

where γ_E is Euler's constant, we find that

$$\Pi_1^{\text{vac}} = \frac{3\lambda m^2}{4\pi^2} \left[\frac{2}{\epsilon} + \psi(2) + \ln\left(\frac{\kappa^2}{m^2}\right) + \ln 4\pi + \mathcal{O}(\epsilon) \right] \tag{4.29}$$

This may be compared with the momentum cutoff scheme

$$\Pi_1^{\text{vac}} = \frac{3\lambda}{4\pi^2} \left[\Lambda_c^2 - m^2 \ln\left(\frac{\Lambda_c^2}{m^2}\right) + \mathcal{O}\left(\frac{m^4}{\Lambda_c^2}\right) \right] \tag{4.30}$$

In the "minimal subtraction" scheme (MS) none of the constant terms are absorbed into the mass, only the divergent $1/\epsilon$ term. In the "modified minimal subtraction" scheme ($\overline{\text{MS}}$) the finite constant terms are absorbed too. A similar absorption is made for the renormalized coupling.

The arbitrariness in choosing the counterterms is a reflection of the whole regularization and renormalization program in quantum field theory. After expressing physical observables in terms of them, there should be no difference. However, the intrinsic scale Λ does depend on the scheme; for example, there are Λ_{MOM}, Λ_{MS}, $\Lambda_{\overline{\text{MS}}}$, and so on. Their numerical values will in general be different. This is nowhere more apparent than in QCD.

4.4 Application to the partition function

Now we investigate the implications of the renormalization group for the partition function. Let T replace p. As given in (4.1), $\ln Z_I$ is comparable with a Green's function that is zeroth order in the field. It has dimension exactly four and no anomalous dimension. Thus, (4.17) instructs us to replace λ_R with $\bar{\lambda}$. If we had an exact expression for $\ln Z$ then the choice of renormalization scale M would indeed be arbitrary. Since we only compute a finite number of terms in a weak-coupling expansion, we should choose M in an optimal way so as to minimize the contribution of higher-order terms. For the massless self-interacting scalar field we take $M = bT$, where b is a number of order unity, since this is the only energy scale in the problem. We then have

$$\bar{\lambda} = \frac{2\pi^2}{9\ln(\Lambda/bT)} \tag{4.31}$$

As $T/\Lambda \to 0$ the thermodynamics is well approximated by a gas of non-interacting massless bosons. As $T/\Lambda \to 1$ the system becomes strongly coupled and the weak coupling expansion is no longer a reasonable approximation. What really happens at very high temperatures is unknown.

In Chapter 3 the pressure was calculated up to order $\lambda^{3/2}$. It has been calculated up to order λ^2 by Frenkel, Saa, and Taylor [1], and to order $\lambda^{5/2}$ by Parwani and Singh [2]. Using the minimal subtraction scheme,

$$
\begin{aligned}
P = \frac{\pi^2}{90}T^4 \Bigg\{ &1 - \frac{5}{24}\left(\frac{9\lambda_R}{\pi^2}\right) + \frac{5}{18}\left(\frac{9\lambda_R}{\pi^2}\right)^{3/2} \\
&- \frac{5}{36}\left(\frac{9\lambda_R}{\pi^2}\right)^2 \left[\frac{3}{4}\ln\left(\frac{2\pi T}{M}\right) + c_1\right] \\
&+ \frac{5}{36}\left(\frac{9\lambda_R}{\pi^2}\right)^{5/2}\left[\ln\left(\frac{9\lambda_R}{\pi^2}\right) + \frac{3}{2}\ln\left(\frac{2\pi T}{M}\right) + c_2\right] \Bigg\}
\end{aligned} \tag{4.32}
$$

is obtained. Here the prime has been dropped from the M in accordance with the notation in Section 4.2. The constants are given by

$$c_1 = \frac{3}{8}\ln(4\pi) + \frac{1}{2}\frac{\zeta'(-3)}{\zeta(-3)} - \frac{\zeta'(-1)}{\zeta(-1)} + \frac{\gamma_E}{8} + \frac{59}{60} \approx -0.606\,85$$

$$c_2 = \frac{\zeta'(-1)}{\zeta(-1)} + \frac{\gamma_E}{4} - 2\ln 3 - \frac{5}{4} \approx -1.317\,87 \tag{4.33}$$

If the scale M is held fixed then the perturbative expansion is not reliable at high temperatures on account of the logarithmic terms $\ln(2\pi T/M)$. The renormalization group tells us that we should not choose M constant but proportional to the temperature. If we choose $M = bT$ then the large temperature-dependent logarithms are of order unity. Indeed, if we choose the coefficient b just right then there is no contribution of order λ_R^2 at all! It is compensated by corresponding contributions at higher orders in λ_R. Equivalently, we can eliminate the logarithmic terms $\ln(2\pi T/M)$ by re-expressing the pressure in terms of the renormalization-group running coupling from (4.23),

$$\bar{\lambda} = \lambda_R\left[1 + \frac{9\lambda_R}{2\pi^2}\ln\left(\frac{bT}{M}\right)\right] + \mathcal{O}(\lambda_R^3) \tag{4.34}$$

The result, of course, is the same.

4.5 Exercises

4.1 Derive (4.20) from the definition (4.12). *Hint:* The renormalized coupling λ_R can be determined from the expression $-(1/3!)(\delta^2 \ln Z_1/\delta \mathcal{D}_0^2)_{1PI}$. Use (3.11) and (3.15) to obtain a diagrammatic expansion for λ_R. You will find that the order-λ^2 correction is given by a single one-loop diagram. Note that you only need the cutoff (Λ_c) dependence to determine β_λ.

4.2 Verify the claim surrounding (4.34).

4.3 Make a plot illustrating the convergence of the expansion of the pressure in (4.32) using $M = T$. Repeat the exercise with $M = \pi T$, $2\pi T$, and $2\pi T e^{4c_1/3}$.

4.4 Derive a renormalization-group equation for $\Gamma_R^{(n)}$ at finite temperature as well as finite momentum, and then find the solution. Discuss how you might want to choose the optimal value of M when there are two variables, p and T.

References

1. Frenkel, J., Saa, A. V., and Taylor, J. C., *Phys. Rev. D* **46**, 3670 (1992).
2. Parwani, R. R. and Singh, H., *Phys. Rev. D* **51**, 4518 (1995).

Bibliography

Renormalization in quantum field theory

Bjorken, J. D. and Drell, S. D. (1965). *Relativistic Quantum Fields* (McGraw-Hill, New York).

Nash, C. (1978). *Relativistic Quantum Fields* (Academic Press, New York).

Ramond, P. (1989). *Field Theory: A Modern Primer* (Addison-Wesley, Redwood City).

Peskin, M. E., and Schroeder, D. V. (1995). *An Introduction to Quantum Field Theory* (Addison-Wesley, Reading).

Weinberg, S. (1995). *The Quantum Theory of Fields* (Cambridge University Press, Cambridge).

Renormalization group

Collins, J. C. (1984). *Renormalization* (Cambridge University Press, Cambridge).

Politzer, H. D., *Phys. Rep.* **14**, 129 (1974).

Renormalization group and high temperature

Collins, J. C., and Perry., M. J., *Phys. Rev. Lett.* **34**, 1353 (1975).

5

Quantum electrodynamics

The single most important field theory is electromagnetism. It is responsible for atomic structure and for the great diversity of materials around us: solids, liquids, and gases. The development of nonrelativistic many-body theory was stimulated primarily by solid state and condensed matter physics, where the potentials used all derive from electromagnetism. This compels us to study quantum electrodynamics at high temperatures and densities where the motion of the electrons becomes relativistic. In metals, the density of plasma electrons rarely exceeds a few electrons per cubic angstrom. This means that the Fermi momentum, $k_{\mathrm{F}} = (3\pi^2 n_e)^{1/3}$, is of order 10 keV at most. Unfortunately, it is difficult to test relativistic many-body theory in the basement of the physics building in table-top experiments! Our attention must then be directed toward astrophysical and cosmological applications. Dense astrophysical objects, such as white dwarf stars, will be considered in Chapter 16.

There is another reason for developing the theory of QED at high temperature and density, and that is the extension to a nonabelian gauge theory, quantum chromodynamics (QCD). We may be able to study QCD at high energy density in terrestrial experiments by colliding energetic heavy nuclei (see Chapter 14).

5.1 Quantizing the electromagnetic field

First, let us consider the electromagnetic field in the absence of charged particles. From classical physics we can write down a field strength tensor as

$$F_{\mu\nu} = \partial_\mu A_\nu - \partial_\nu A_\mu \tag{5.1}$$

where A_μ is the vector potential. The electric and magnetic fields are

$$E_i = -F_{0i} = F_{i0}$$
$$B_i = \tfrac{1}{2}\epsilon_{ijk}F_{jk} \quad \text{or} \quad \mathbf{B} = \nabla \times \mathbf{A} \tag{5.2}$$

The Lagrangian density is

$$\mathcal{L} = -\tfrac{1}{4}F_{\mu\nu}F^{\mu\nu} \tag{5.3}$$

Notice that $F_{\mu\nu}$ is invariant under the local (or x-dependent) transformation

$$A_\mu(\mathbf{x}, t) \to A_\mu(\mathbf{x}, t) - \partial_\mu \alpha(\mathbf{x}, t) \tag{5.4}$$

where $\alpha(\mathbf{x}, t)$ is some smooth function of x_μ. Since the field strength tensor is invariant under this transformation, so are the electric and magnetic fields, and so is the Lagrangian. This is called a U(1) gauge symmetry.

To quantize the theory and to compute a partition function, we need a Hamiltonian formulation. In order to do this, we must agree on a gauge to work in. A convenient gauge for this purpose is the axial gauge

$$A_3(\mathbf{x}, t) = 0 \tag{5.5}$$

Actually (5.5) does not entirely fix the gauge, as any gauge function $\alpha(x, y, t)$ that is independent of z leaves (5.5) unchanged. We shall fix this residual gauge freedom later.

The conjugate momenta are defined by

$$\pi_\mu = \frac{\partial \mathcal{L}}{\partial(\partial_0 A_\mu)} = F_{0\mu} \tag{5.6}$$

(This should not viewed be as a tensor equation but as true component by component.) Since $F_{\mu\nu}$ is antisymmetric in its Lorentz indices, it follows that $\pi_0 = 0$. Thus A_0 is not a dynamical field, it is a dependent field. The independent fields are A_1 and A_2 with conjugate momenta

$$\pi_1 = F_{01} = -E_1 = \partial_0 A_1 - \partial_1 A_0$$
$$\pi_2 = F_{02} = -E_2 = \partial_0 A_2 - \partial_2 A_0 \tag{5.7}$$

These two independent fields actually correspond to the two polarization degrees of freedom of free radiation.

The z component of the electric field is

$$E_3 = F_{30} = \partial_3 A_0 \tag{5.8}$$

Since $A_3 = 0$ there is no momentum conjugate to A_3; hence E_3, like A_0, must be a dependent field. We can determine E_3 by an application of

Gauss's law, which, in the absence of charged particles, is

$$\nabla \cdot \mathbf{E} = 0 \tag{5.9}$$

Thus,

$$E_3(x, y, z, t) = \int_{z_0}^{z} dz' [\partial_1 \pi_1(x, y, z', t) + \partial_2 \pi_2(x, y, z', t)] + P(x, y, t) \tag{5.10}$$

and

$$A_0(x, y, z, t) = \int_{z_0}^{z} dz'' E_3(x, y, z'', t) + Q(x, y, t) \tag{5.11}$$

Here, P and Q are smooth functions of x, y, and t. The gauge is not completely fixed until these two functions are specified. They may be determined by specifying the values of A_0 and E_3 at $z = z_0$ for all x, y, and t.

The Hamiltonian may now be found from the Lagrangian in the canonical way (see (2.23)). Dropping surface terms we find the well-known result

$$\mathcal{H} = \tfrac{1}{2}\mathbf{E}^2 + \tfrac{1}{2}\mathbf{B}^2 = \tfrac{1}{2}\pi_1^2 + \tfrac{1}{2}\pi_2^2 + \tfrac{1}{2}E_3^2(\pi_1, \pi_2) + \tfrac{1}{2}\mathbf{B}^2 \tag{5.12}$$

The partition function is

$$Z = \int [d\pi_1][d\pi_2] \int_{\text{periodic}} [dA_1][dA_2]$$
$$\times \exp\left[\int_0^\beta d\tau \int d^3x \left(i\pi_1 \frac{\partial A_1}{\partial \tau} + i\pi_2 \frac{\partial A_2}{\partial \tau} - \mathcal{H} \right) \right] \tag{5.13}$$

Since we have a free-field theory, we should be able to calculate Z exactly. However, in the present form this is not easy since it is a rather complicated function of π_1 and π_2.

To put (5.12) and (5.13) in a more manageable form we insert the identity

$$1 = \int [d\pi_3] \delta(\pi_3 + E_3(\pi_1, \pi_2)) \tag{5.14}$$

and replace E_3 with $-\pi_3$ in the integrand. Note that, despite the suggestive notation, π_3 is not the conjugate momentum of any field; (5.14) is simply the condition on E_3 that ensures Gauss's law. Now

$$\delta(\pi_3 + E_3(\pi_1, \pi_2)) = \delta(\nabla \cdot \boldsymbol{\pi}) \det\left(\frac{\partial(\nabla \cdot \boldsymbol{\pi})}{\partial \pi_3} \right) \tag{5.15}$$

Furthermore,

$$\det\left(\frac{\partial(\nabla \cdot \boldsymbol{\pi})}{\partial \pi_3} \right) = \det(\partial_3) \tag{5.16}$$

Thus far we have

$$
Z = \int [d\pi_1][d\pi_2][d\pi_3] \int_{\text{periodic}} [dA_1][dA_2]\delta(\nabla \cdot \boldsymbol{\pi}) \det(\partial_3)
$$

$$
\times \exp\left[\int_0^\beta d\tau \int d^3x \left(i\pi_1\frac{\partial A_1}{\partial\tau} + i\pi_2\frac{\partial A_2}{\partial\tau} - \tfrac{1}{2}\boldsymbol{\pi}^2 - \tfrac{1}{2}\mathbf{B}^2\right)\right] \quad (5.17)
$$

The constraint of Gauss's law can be implemented alternatively by using the integral representation of the delta function. In vacuum field theory we would write

$$
\delta(\nabla \cdot \boldsymbol{\pi}) = \int [dA_0] \exp\left(i \int d^4x \, A_0 \, \nabla \cdot \boldsymbol{\pi}\right) \quad (5.18)
$$

where A_0 is some auxiliary field, or a Lagrange multiplier field. At finite temperature we make the replacement $t \to -i\tau$ and now also $A_0 \to iA_0$. Thus

$$
\delta(\nabla \cdot \boldsymbol{\pi}) = \int [dA_0] \exp\left(i \int_0^\beta d\tau \int d^3x \, A_0 \, \nabla \cdot \boldsymbol{\pi}\right) \quad (5.19)
$$

Using this representation to implement Gauss's law, we may integrate over $\boldsymbol{\pi}$ directly:

$$
Z = \int [d\pi_1][d\pi_2][d\pi_3] \int [dA_0][dA_1][dA_2] \det(\partial_3)
$$

$$
\times \exp\left[\int_0^\beta d\tau \int d^3x \left(i\pi_1\frac{\partial A_1}{\partial\tau} + i\pi_2\frac{\partial A_2}{\partial\tau} - i\nabla A_0 \cdot \boldsymbol{\pi} - \tfrac{1}{2}\boldsymbol{\pi}^2 - \tfrac{1}{2}\mathbf{B}^2\right)\right]
$$

$$
= \int [dA_0][dA_1][dA_2] \det(\partial_3)
$$

$$
\times \exp\left\{\int_0^\beta d\tau \int d^3x \left[\tfrac{1}{2}\left(i\frac{\partial \mathbf{A}}{\partial\tau} - i\nabla A_0\right)^2 - \tfrac{1}{2}\mathbf{B}^2\right]\right\} \quad (5.20)
$$

where $\mathbf{A} = (A_1, A_2, 0)$ and we have ignored an irrelevant overall normalization constant. Notice that the argument of the exponential is

$$
\tfrac{1}{2}\mathbf{E}^2 - \tfrac{1}{2}\mathbf{B}^2 = \mathcal{L} \quad (5.21)
$$

Making this identification and inserting the factor

$$
1 = \int [dA_3]\delta(A_3) \quad (5.22)
$$

we arrive at

$$
Z = \int_{\text{periodic}} [dA^\mu]\delta(A_3) \det(\partial_3) \exp\left(\int_0^\beta d\tau \int d^3x \, \mathcal{L}\right) \quad (5.23)
$$

The axial gauge is not necessarily a convenient gauge to use for practical computations. Furthermore, it is not immediately apparent that (5.23) is a gauge-invariant expression for Z. Take an arbitrary gauge specified by $F = 0$, where F is some function of A^μ and its derivatives. For the axial gauge above, $F = A_3$. For this gauge, (5.23) becomes

$$Z = \int_{\text{periodic}} [dA^\mu] \delta(F) \det \left(\frac{\partial F}{\partial \alpha} \right) \exp \left(\int_0^\beta d\tau \int d^3x \, \mathcal{L} \right) \quad (5.24)$$

Equation (5.24) is manifestly gauge invariant: $\mathcal{L}$ is invariant, the gauge-fixing factor times the Jacobian of the transformation $\delta(F) \det(\partial F/\partial \alpha)$ is invariant, and the integration is over all four components of the vector potential. Equation (5.24) reduces to (5.23) in the case of the axial gauge $A_3 = 0$. We know this is correct since it was derived from first principles in the Hamiltonian formulation of the gauge theory, $Z = \text{Tr} \, e^{-\beta H}$.

5.2 Blackbody radiation

It is important to verify that (5.24) describes blackbody radiation with two polarization degrees of freedom. We shall do this in two different gauges, the axial gauge $A_3 = 0$ and the covariant Feynman gauge.

In the axial gauge, we rewrite (5.20) as

$$Z = \int [dA_0][dA_1][dA_2] \det(\partial_3) \, e^{S_0} \quad (5.25)$$

where

$$S_0 = \tfrac{1}{2} \int d\tau \int d^3x \, (A_0, A_1, A_2)$$
$$\times \begin{pmatrix} \nabla^2 & -\partial_1 \partial_\tau & -\partial_2 \partial_\tau \\ -\partial_1 \partial_\tau & \partial_2^2 + \partial_3^2 + \partial_\tau^2 & -\partial_1 \partial_2 \\ -\partial_2 \partial_\tau & -\partial_1 \partial_2 & \partial_1^2 + \partial_3^2 + \partial_\tau^2 \end{pmatrix} \begin{pmatrix} A_0 \\ A_1 \\ A_2 \end{pmatrix}$$

We may express the determinant of ∂_3 as a functional integral over a complex ghost field C, which is a Grassmann field with spin 0:

$$\det(\partial_3) = \int [d\bar{C}][dC] \exp \left(\int_0^\beta d\tau \int d^3x \, \bar{C} \partial_3 C \right) \quad (5.26)$$

(This is (2.82) generalized to an infinite number of degrees of freedom.) These ghost fields C and $\bar{C}$ are not physical fields since they do not appear in the Hamiltonian. Furthermore, since they are anticommuting scalar fields they violate the spin-statistics theorem. They are simply a convenient functional integral representation of the determinant of an operator. The great usefulness of these fictitious ghost fields will be in nonabelian gauge theories.

In frequency–momentum space the partition function is expressed as

$$\ln Z = \ln \det(\beta p_3) - \tfrac{1}{2} \ln \det D \tag{5.27}$$

where

$$D = \beta^2 \begin{pmatrix} \mathbf{p}^2 & -\omega_n p_1 & -\omega_n p_2 \\ -\omega_n p_1 & \omega_n^2 + p_2^2 + p_3^2 & -p_1 p_2 \\ -\omega_n p_2 & -p_1 p_2 & \omega_n^2 + p_1^2 + p_3^2 \end{pmatrix}$$

Carrying out the determinantal operation,

$$\ln Z = \tfrac{1}{2} \mathrm{Tr} \ln \left(\beta^2 p_3^2 \right) - \tfrac{1}{2} \mathrm{Tr} \ln \left[\beta^6 p_3^2 \left(\omega_n^2 + \mathbf{p}^2 \right)^2 \right]$$

$$= \ln \left\{ \prod_n \prod_{\mathbf{p}} \left[\beta^2 (\omega_n^2 + \mathbf{p}^2) \right]^{-1} \right\}$$

$$= 2V \int \frac{d^3 p}{(2\pi)^3} \left[-\tfrac{1}{2} \beta \omega - \ln(1 - \mathrm{e}^{-\beta\omega}) \right] \tag{5.28}$$

Here, $\omega = |\mathbf{p}|$. Comparison with (2.40) shows that (5.28) describes massless bosons with two spin degrees of freedom in thermal equilibrium; in other words, blackbody radiation.

A family of covariant gauges is given by the condition

$$F = \partial^\mu A_\mu - f(\mathbf{x}, \tau) = 0 \tag{5.29}$$

where f is an arbitrary function. Under a gauge transformation,

$$F = \partial^\mu (A_\mu - \partial_\mu \alpha) - f = \partial^\mu A_\mu - f - \partial^2 \alpha \tag{5.30}$$

and $\partial F/\partial \alpha = -\partial^2$. Inserting into (5.24) yields

$$Z = \int [dA_\mu] \det(-\partial^2) \, \delta(\partial^\mu A_\mu - f) \exp \left(\int_0^\beta d\tau \int d^3 x \, \mathcal{L} \right) \tag{5.31}$$

The physics contained in Z is unchanged if we first multiply by

$$\exp \left(-\frac{1}{2\rho} \int d\tau \int d^3 x \, f^2 \right)$$

and then do a functional integration over f,

$$Z = \int [dA_\mu] \det(-\partial^2) \exp \left(\int d\tau \int d^3 x \, \mathcal{L}_{\mathrm{eff}} \right) \tag{5.32}$$

where

$$\mathcal{L}_{\text{eff}} = \mathcal{L} - \frac{1}{2\rho}(\partial^\mu A_\mu)^2$$

is the effective Lagrangian, including the gauge-fixing term, and ρ is any real number. The Feynman gauge corresponds to the choice $\rho = 1$ and the Landau gauge to $\rho = 0$.

The partition function should be independent of α and should be the same as in the axial gauge. For simplicity, we examine Z in the Feynman gauge. Then,

$$\int d\tau \int d^3x \, \mathcal{L}_{\text{eff}} = \tfrac{1}{2} \int d\tau \int d^3x \, A_\mu (\partial_\tau^2 + \nabla^2) A_\mu \qquad (5.33)$$

where the summation over μ is Euclidean because in (5.19) we let $A_0 \rightarrow iA_0$. We again employ a ghost field to write

$$\det(-\partial^2) = \int [d\bar{C}][dC] \exp\left(\int d\tau \int d^3x \, (\partial^\mu \bar{C})(\partial_\mu C)\right) \qquad (5.34)$$

Combining (5.32) with (5.34), we get

$$\ln Z = 2\left(\tfrac{1}{2}\right) \text{Tr} \ln\left[\beta^2 \left(\omega_n^2 + \mathbf{p}^2\right)\right] + 4\left(-\tfrac{1}{2}\right) \text{Tr} \ln\left[\beta^2 \left(\omega_n^2 + \mathbf{p}^2\right)\right] \qquad (5.35)$$

The four degrees of freedom of the A_μ field combine with the two degrees of freedom of the C (ghost) field, which contribute with the opposite sign, to produce just the correct number of physical degrees of freedom. The complex ghost field cancels the unphysical degrees of freedom of the longitudinal and timelike photons. Equation (5.35) is the same as (5.28).

5.3 Diagrammatic expansion

Photons interact with fermions (to be specific, we shall consider electrons) with the interaction Lagrangian

$$\mathcal{L}_{\text{I}} = -e\bar{\psi}A\!\!\!/\psi \qquad (5.36)$$

where e is the electronic charge. By far the most frequently used gauges are the covariant gauges. The partition function is

$$Z = \int [d\bar{C}][dC][dA_\mu][d\bar{\psi}][d\psi] \exp\left(\int d\tau \int d^3x \, \mathcal{L}\right) \qquad (5.37)$$

where

$$\mathcal{L} = \mathcal{L}_0 + \mathcal{L}_{\text{I}}$$

and

$$\mathcal{L}_0 = \bar{\psi}(i\,\slashed{\partial} - m + \mu\gamma^0)\psi - \tfrac{1}{4}F^{\mu\nu}F_{\mu\nu}$$
$$- \frac{1}{2\rho}(\partial^\mu A_\mu)^2 + (\partial^\mu \bar{C})(\partial_\mu C)$$

The ghost field does not interact with any other field but serves only to cancel two of the four gauge-field degrees of freedom in the ideal gas term.

The partition function and other physical quantities of interest may be formally expanded in a power series in $\mathcal{L}_I$ or e. The diagrammatic rules closely parallel those discussed in Chapter 3. The bare propagators and vertex are:

fermion
$$\mathcal{G}_0 = \frac{1}{\slashed{p} - m} = \xrightarrow{\qquad p \qquad}$$
$$p_0 = (2n+1)\pi T i + \mu$$

$$(5.38)$$

photon
$$\mathcal{D}_0^{\mu\nu} = \frac{1}{p^2}[g^{\mu\nu} - (1-\rho)p^\mu p^\nu / p^2] = \underset{\mu \qquad\qquad \nu}{\wwwwwwwwww}$$
$$p_0 = 2n\pi T i$$

vertex
$$-e\gamma^\mu = $$

As an example, the lowest-order correction to the ideal gas of photons, electrons, and positrons is

$$\ln Z_2 = -\tfrac{1}{2}\,\bigcirc\!\!\!\sim\!\!\!\bigcirc$$

$$(5.39)$$

The photon self-energy at one loop is

$$\Pi_{\mu\nu} = \mathcal{D}_{\mu\nu}^{-1} - \mathcal{D}_{0\,\mu\nu}^{-1} = \sim\!\!\bigcirc\!\!\sim$$

$$(5.40)$$

5.4 Photon self-energy

The photon self-energy is related to the inverse of the full and bare propagators by

$$\Pi_{\mu\nu} = \mathcal{D}_{\mu\nu}^{-1} - \mathcal{D}_{0\,\mu\nu}^{-1}$$

$$(5.41)$$

The inverse propagator is related to the propagator by

$$\mathcal{D}^{\mu\alpha}\mathcal{D}_{\alpha\nu}^{-1} = g^\mu{}_\nu$$

$$(5.42)$$

The propagator and the self-energy satisfy certain fundamental constraints. To discover them, it is convenient to work with $k_0 = 2n\pi T i$ analytically continued to arbitrary complex values. (Recall our analysis of Section 3.4. This continuation will be taken up again in Chapter 6.) Let

k^μ be the four-momentum of the photon. Current conservation requires that $\Pi_{\mu\nu}$ be transverse,

$$k^\mu \Pi_{\mu\nu} = 0 \tag{5.43}$$

and gauge invariance requires that

$$k^\mu k^\nu \mathcal{D}_{\mu\nu} = \rho \tag{5.44}$$

in a covariant gauge specified by ρ. Both these constraints hold at $T > 0$, $\mu \neq 0$, as well as in the vacuum. The interested reader is referred to Fradkin [1] for a proof of (5.43). The proof of (5.44) will now be outlined.

Consider making the gauge transformation $A_\mu \to A_\mu - \partial_\mu \alpha$, $\psi \to e^{ie\alpha}\psi$ in the partition function as expressed in (5.37). All terms are manifestly independent of α apart from the gauge-fixing term, which becomes

$$-\frac{1}{2\rho}(\partial^\mu A_\mu - f)^2$$

where

$$f = \partial^2 \alpha$$

By construction, the partition function is gauge invariant. Therefore, if we functionally differentiate $\ln Z$ with respect to f any number of times, we must get zero. In particular,

$$\frac{\delta \ln Z}{\delta f(\mathbf{x}, \tau)} = \frac{\langle \partial^\mu A_\mu(\mathbf{x}, \tau)\rangle}{\rho} - \frac{f(\mathbf{x}, \tau)}{\rho} = 0$$

$$\frac{\delta^2 \ln Z}{\delta f(\mathbf{x}, \tau)\delta f(\mathbf{x}', \tau')} = \frac{\langle \partial^\mu A_\mu(\mathbf{x}, \tau)\partial^\nu A_\nu(\mathbf{x}', \tau')\rangle}{\rho^2} - \frac{\langle \partial^\mu A_\mu(\mathbf{x}, \tau)\rangle f(\mathbf{x}', \tau')}{\rho^2}$$

$$-\frac{\delta(\tau - \tau')\delta(\mathbf{x} - \mathbf{x}')}{\rho} = 0 \tag{5.45}$$

Evaluating (5.45) at $f = 0$ and taking the Fourier transform, we obtain (5.44). A constraint on the thermal average of a product of N vector potentials is likewise obtained by differentiating N times $\ln Z$ with respect to f, and then setting $f = 0$.

The propagator, its inverse, and the self-energy, are all symmetric second-rank tensors. Assuming rotational invariance (which would not be correct for a solid) the most general tensor of this type is a linear combination of $g_{\mu\nu}$, $k_\mu k_\nu$, $u_\mu u_\nu$, and $k_\mu u_\nu + k_\nu u_\mu$. Here $u_\mu = (1, 0, 0, 0)$ specifies the rest frame of the many-body system. Taking into account

(5.41) to (5.44) we obtain the general forms

$$\Pi^{\mu\nu} = GP_{\mathrm{T}}^{\mu\nu} + FP_{\mathrm{L}}^{\mu\nu}$$

$$\mathcal{D}^{\mu\nu} = \frac{1}{G - k^2}P_{\mathrm{T}}^{\mu\nu} + \frac{1}{F - k^2}P_{\mathrm{L}}^{\mu\nu} + \frac{\rho}{k^2}\frac{k^\mu k^\nu}{k^2} \qquad (5.46)$$

$$(\mathcal{D}^{-1})^{\mu\nu} = (G - k^2)P_{\mathrm{T}}^{\mu\nu} + (F - k^2)P_{\mathrm{L}}^{\mu\nu} + \frac{k^\mu k^\nu}{\rho}$$

The quantities F and G are scalar functions of k^0 and $|\mathbf{k}|$. The two projection operators are four-dimensionally transverse, but one is also three-dimensionally transverse (P_{T}) while the other is three-dimensionally longitudinal (P_{L}):

$$P_{\mathrm{T}}^{00} = P_{\mathrm{T}}^{0i} = P_{\mathrm{T}}^{i0} = 0$$

$$P_{\mathrm{T}}^{ij} = \delta^{ij} - k^i k^j / \mathbf{k}^2 \qquad (5.47)$$

$$P_{\mathrm{L}}^{\mu\nu} = k^\mu k^\nu / k^2 - g^{\mu\nu} - P_{\mathrm{T}}^{\mu\nu}$$

These have the properties

$$P_{\mathrm{L}}^{\mu\sigma} P_{\mathrm{L}\sigma\nu} = -P_{\mathrm{L}\nu}^{\mu}$$

$$P_{\mathrm{T}}^{\mu\sigma} P_{\mathrm{T}\sigma\nu} = -P_{\mathrm{T}\nu}^{\mu}$$

$$k_\mu P_{\mathrm{T}}^{\mu\nu} = k_\mu P_{\mathrm{L}}^{\mu\nu} = 0 \qquad (5.48)$$

$$P_{\mathrm{L}}^{\mu\sigma} P_{\mathrm{T}\sigma\nu} = 0$$

$$P_{\mathrm{L}\mu}^{\mu} = -1$$

$$P_{\mathrm{T}\mu}^{\mu} = -2$$

In the vacuum there is no preferred rest frame, so the vector u_μ cannot play any role (it is not defined). Also, in the vacuum $\Pi^{\mu\nu}$ must be proportional to $g^{\mu\nu} - k^\mu k^\nu / k^2$; hence $F = G$. Furthermore, G can only depend on k^2. At finite temperature and density, however, F and G can depend on $k^0 = u \cdot k$ and $|\mathbf{k}| = \sqrt{(u \cdot k)^2 - k^2}$ separately, owing to the lack of Lorentz invariance.

Let us evaluate the photon self-energy at the one-loop level. From (5.40),

$$\Pi^{\mu\nu} = e^2 T \sum_l \int \frac{d^3p}{(2\pi)^3} \operatorname{Tr}\left(\gamma^\nu \frac{1}{\not{p} - m} \gamma^\mu \frac{1}{\not{p} + \not{k} - m}\right) \qquad (5.49)$$

Here $p^0 = (2l + 1)\pi T i + \mu$ and $k^0 = 2n\pi T i$. We can always write $\Pi^{\mu\nu} = \Pi_{\mathrm{vac}}^{\mu\nu} + \Pi_{\mathrm{mat}}^{\mu\nu}$, where

$$\Pi_{\mathrm{vac}}^{\mu\nu} = \lim_{\substack{T \to 0 \\ \mu \to 0}} \Pi^{\mu\nu} \qquad (5.50)$$

is the vacuum self-energy and $\Pi^{\mu\nu}_{\text{mat}}$ is the remainder due to the presence of matter. The vacuum part is discussed in many textbooks on field theory, such as Peskin and Schroeder [2]. The matter part is readily evaluated:

$$\Pi^{00}_{\text{mat}} = -\frac{e^2}{\pi^2} \, \text{Re} \int_0^\infty \frac{dp\, p^2}{E_p} N_{\text{F}}(p) \left[1 + \frac{4E_p k^0 - 4E_{\text{P}}^2 - k^2}{4p\omega} \ln\left(\frac{R_+}{R_-}\right) \right]$$

(5.51)

$$\Pi^\mu_{\text{mat}\,\mu} = -2\frac{e^2}{\pi^2} \, \text{Re} \int_0^\infty \frac{dp\, p^2}{E_p} N_{\text{F}}(p) \left[1 - \frac{2m^2 + k^2}{4p\omega} \ln\left(\frac{R_+}{R_-}\right) \right]$$

Here

$$\omega = |\mathbf{k}| \qquad k^2 = k_0^2 - \omega^2 \qquad E_p = \sqrt{\mathbf{p}^2 + m^2}$$

$$N_{\text{F}}(p) = \frac{1}{e^{\beta(E_p - \mu)} + 1} + \frac{1}{e^{\beta(E_p + \mu)} + 1}$$

$$R_\pm = k^2 - 2k_0 E_p \pm 2p\omega$$

Also, the reader should note that here we define the action of the operator Re as follows: $\text{Re}\, f(k^0) = \frac{1}{2}[f(k^0) + f(-k^0)]$.

Various limits of (5.51) are of physical interest. They correspond to the screening of electric and magnetic fields and plasma oscillations. These topics are discussed in Chapter 6 in particular, in the context of linear response theory.

5.5 Loop corrections to $\ln Z$

5.5.1 Two loops

The lowest-order correction to $\ln Z$ due to interactions is the two-loop diagram seen in (5.39). There are two methods of doing explicit calculations with such diagrams. In the traditional method the frequency sums are performed directly. Another method uses analytic continuation and contour integrals, as discussed in Chapter 3. Both methods must of course give the same answer, but usually the contour integral method is much easier.

From (5.39), we have in the Feynman gauge the exchange contribution

$$\frac{\ln Z_{\text{ex}}}{\beta V} = -\frac{1}{2} e^2 \int \frac{d^3 p}{(2\pi)^3} \frac{d^3 q}{(2\pi)^3} \frac{d^3 k}{(2\pi)^3} (2\pi)^3 \delta(\mathbf{p} - \mathbf{q} - \mathbf{k})$$

$$\times T^3 \sum_{n_p, n_q, n_k} \beta \delta_{n_p, n_q + n_k} \frac{\text{Tr}[\gamma^\mu (\not{p} + m)\gamma_\mu (\not{q} + m)]}{k^2 (p^2 - m^2)(q^2 - m^2)}$$

(5.52)

The trace is readily carried out. Apart from integration over three-momenta this becomes

$$-8T^3 \sum_{n_p,n_q,n_k} \beta\delta_{n_p,n_q+n_k} \frac{2m^2 - p\cdot q}{k^2(p^2 - m^2)(q^2 - m^2)} \tag{5.53}$$

The Kronecker delta may be written as

$$\beta\delta_{n_p,n_q+n_k} = \int_0^\beta d\theta \exp[\theta(p^0 - q^0 - k^0)]$$

$$= \frac{\exp\left[\beta(p^0 - q^0 - k^0)\right] - 1}{p^0 - q^0 - k^0} \tag{5.54}$$

where $p^0 = (2n_p + 1)\pi Ti + \mu$, $q^0 = (2n_q + 1)\pi Ti + \mu$, and $k^0 = 2n_k\pi Ti$. Since q^0 and k^0 enter the argument of the exponential with minus signs we multiply by $-\exp[\beta(k^0 + q^0 - \mu)]$, which is unity when evaluated on the integers. This procedure ensures that the integrands of the contour integrals fall off exponentially before the θ integration is performed, so that one never need worry about contributions from contours distorted out to infinity. This procedure also guarantees that the normal vacuum is recovered in the limit of zero temperature and chemical potential (see the discussion in the papers of Norton and Cornwall [3] and Kapusta [4]). With this analytic continuation of the Kronecker delta, (5.53) becomes

$$-8T \sum_{n_k} \frac{1}{k^2} T \sum_{n_p} \frac{1}{p^2 - m^2} T \sum_{n_q} \frac{1}{q^2 - m^2} I(p^0, q^0, k^0) \tag{5.55}$$

where

$$I(p^0, q^0, k^0) = \frac{2m^2 - p\cdot q}{p^0 - q^0 - k^0}\{\exp[\beta(k^0 + q^0 - \mu)] - \exp[\beta(p^0 - \mu)]\}$$

Notice that I has no singularities. Hence, each of the sums may be converted to a contour integral via (3.40) and (3.71), and these contour integrations may be performed simultaneously and independently. For example,

$$T \sum_{n_p} \frac{1}{p^2 - m^2} I(p^0, q^0, k^0)$$

$$= I(E_p, q^0, k^0)\frac{N_F^-(p)}{2E_p} + I(-E_p, q^0, k^0)\frac{N_F^+(p) - 1}{2E_p}$$

$$T \sum_{n_k} \frac{1}{k^2} I(p^0, q^0, k^0) \tag{5.56}$$

$$= -I(p^0, q^0, \omega)\frac{N_B(k)}{2\omega} - I(p^0, q^0, -\omega)\frac{N_B(k) + 1}{2\omega}$$

where the fermion and boson occupation numbers are

$$N_{\mathrm{F}}^{\pm}(p) = \frac{1}{\exp[\beta(E_p \pm \mu)] + 1}$$

$$N_{\mathrm{B}}(k) = \frac{1}{\exp(\beta\omega) - 1} \tag{5.57}$$

and $\omega = |\mathbf{k}|$, $E_p = \sqrt{\mathbf{p}^2 + m^2}$.

As is evident, the contour integration method has two obvious advantages over the direct summation method. First, the contour integrals may be evaluated independently of each other whereas the direct summations must be done in consecutive order. This is a great algebraic simplification, which becomes more pronounced as the complexity of the diagram increases. Second, the contour integration puts each particle on its mass shell automatically.

When (5.56) is used to evaluate (5.55), one finds terms that are quadratic in the occupation numbers, terms that are linear, and terms that are independent of the occupation numbers. Those that are independent represent the energy shift of the vacuum and are not of interest to us. Those that are linear are canceled by the fermion and photon vacuum self-energy renormalizations. These are represented as

the angled parentheses indicating that the $T = \mu = 0$ limit of the subgraph is to be taken (cf. (3.49)). Putting all the above together we find the two-loop result:

$$\frac{\ln Z_{\mathrm{ex}}}{\beta V} = -\frac{1}{6}e^2 T^2 \int \frac{d^3 p}{(2\pi)^3} \frac{N_{\mathrm{F}}(p)}{E_p} - \frac{1}{2}e^2 \int \frac{d^3 p}{(2\pi)^3} \frac{d^3 q}{(2\pi)^3} \frac{1}{E_p E_q}$$

$$\times \left\{ \left(1 + \frac{2m^2}{(E_p - E_q)^2 - (\mathbf{p} - \mathbf{q})^2}\right) [N_{\mathrm{F}}^-(p) N_{\mathrm{F}}^-(q) + N_{\mathrm{F}}^+(p) N_{\mathrm{F}}^+(q)] \right.$$

$$\left. + \left(1 + \frac{2m^2}{(E_p + E_q)^2 - (\mathbf{p} - \mathbf{q})^2}\right) [N_{\mathrm{F}}^-(p) N_{\mathrm{F}}^+(q) + N_{\mathrm{F}}^+(p) N_{\mathrm{F}}^-(q)] \right\} \tag{5.58}$$

where $N_{\mathrm{F}} = N_{\mathrm{F}}^+ + N_{\mathrm{F}}^-$. This is referred to as the exchange term because in the $T = 0$ limit it arises from the exchange of the three-momenta of a pair of fermions in the Fermi sea. Various limits of the exchange term are of interest, and so are listed below; note that the Fermi momentum is

$p_{\mathrm{F}} = \sqrt{\mu^2 - m^2}$ when $|\mu| > m$:

$$\frac{\ln Z_{\mathrm{ex}}}{\beta V} = -\frac{e^2}{(2\pi)^4} \left\{ \frac{3}{2} \left[\mu p_{\mathrm{F}} - m^2 \ln \left(\frac{\mu + p_{\mathrm{F}}}{m} \right) \right]^2 - p_{\mathrm{F}}^4 \right\} \quad (T = 0) \quad (5.59)$$

$$\frac{\ln Z_{\mathrm{ex}}}{\beta V} = -\frac{e^2}{288} \left(5T^4 + \frac{18}{\pi^2} \mu^2 T^2 + \frac{9}{\pi^4} \mu^4 \right) \quad (m = 0) \quad\quad (5.60)$$

$$\frac{\ln Z_{\mathrm{ex}}}{\beta V} = \frac{e^2}{2(2\pi)^3} m^2 T^2 e^{2(\mu - m)/T} \quad (T \ll m - \mu \ll m) \quad\quad (5.61)$$

Equation (5.59) will be useful in our discussion of white dwarf stars. Equation (5.60) will reappear in QCD plasma. Equation (5.61) modifies the classical ideal gas equation of state to $P = nT - e^2 n^2 / 8mT$.

5.5.2 Ring diagrams

The next order to contribute is not e^4 as naively expected but e^3 when $T > 0$ and $e^4 \ln e^2$ when $T = 0$ but $\mu \neq 0$. These arise from the set of ring diagrams shown in (3.54), where the photon self-energy is given to lowest order by (5.40),

$$\frac{\ln Z_{\mathrm{ring}}}{\beta V} = -\tfrac{1}{2} T \sum_n \int \frac{d^3 k}{(2\pi)^3} \mathrm{Tr} \left\{ \ln \left[1 + \mathcal{D}_0(k)\Pi(k) \right] - \mathcal{D}_0(k)\Pi(k) \right\} \quad (5.62)$$

Making use of the explicit forms of $\mathcal{D}_0$ and Π as given by (5.46), we may carry out the trace operation to obtain

$$\frac{\ln Z_{\mathrm{ring}}}{\beta V} = -\tfrac{1}{2} T \sum_n \int \frac{d^3 k}{(2\pi)^3} \left\{ 2 \left[\ln \left(1 - \frac{G(n,\omega)}{k^2} \right) + \frac{G(n,\omega)}{k^2} \right] \right.$$
$$\left. + \ln \left(1 - \frac{F(n,\omega)}{k^2} \right) + \frac{F(n,\omega)}{k^2} \right\} \quad (5.63)$$

Note that F and G are functions of n (since $k_0 = 2\pi n T i$) and $\omega = |\mathbf{k}|$. The terms involving G have a coefficient of 2 relative to the terms involving F. The reason is that there are two transverse degrees of freedom but only one longitudinal degree of freedom ($P_{T\mu}^\mu = -2$, $P_{L\mu}^\mu = -1$). Note that the expressions (5.62) and (5.63) are manifestly gauge invariant since the ρ-dependent part of $\mathcal{D}_0$ vanishes, as a consequence of current conservation, when it multiplies Π.

Since $-k^2 = \omega^2 + 4\pi^2 T^2 n^2$, the logarithms may be expanded to second order in F and G to give an e^4 contribution, as long as $n \neq 0$. If either $F(n = 0, \omega \to 0)$ or $G(n = 0, \omega \to 0)$ does not vanish then expansions of the logarithms do not converge. To isolate this potential infrared divergence, we write

$$-\tfrac{1}{2}T \int \frac{d^3k}{(2\pi)^3} \left[2\ln\left(1 + \frac{G(0,0)}{\omega^2}\right) - \frac{2G(0,0)}{\omega^2} + \ln\left(1 + \frac{F(0,0)}{\omega^2}\right) - \frac{F(0,0}{\omega^2} \right]$$

$$(5.64)$$

The remaining terms, which are explicitly of order e^4 and which have no infrared divergence, are

$$\tfrac{1}{4}T \int \frac{d^3k}{(2\pi)^3} \left\{ \sum_{n\neq0} \left[2\left(\frac{G(n,\omega)}{k^2}\right)^2 + \left(\frac{F(n,\omega)}{k^2}\right)^2 \right] + 2\left(\frac{G(0,\omega)}{\omega^2}\right)^2 \right.$$

$$\left. - 2\left(\frac{G(0,0)}{\omega^2}\right)^2 + \left(\frac{F(0,\omega)}{\omega^2}\right)^2 - \left(\frac{F(0,0)}{\omega^2}\right)^2 \right\} \quad (5.65)$$

Upon examination of (5.46) and (5.51) we find that $G(n=0, \omega \to 0) = 0$ but

$$F(n=0, \omega \to 0) = \frac{e^2}{\pi^2} \int_0^\infty \frac{dp}{E_p} (p^2 + E_p^2) N_F(p) \qquad (5.66)$$

After integrating over k in (5.64), we find the order-e^3 contribution,

$$\frac{\ln Z_{\text{ring}}}{\beta V} = \frac{T}{12\pi} F^{3/2}(0,0) \qquad (5.67)$$

This result, nonanalytic in $\alpha = e^2/4\pi$, is precisely analogous to our result in Chapter 3 for the massless $\lambda\phi^4$ theory. The nonanalyticity here arises because interactions at finite temperature and density generate a static electric screening mass for the photon.

There are several interesting limits of $F(n=0, \omega \to 0)$. In the ultrarelativistic limit ($m = 0$),

$$F(0,0) = e^2 \left(\frac{T^2}{3} + \frac{\mu^2}{\pi^2} \right) \qquad (5.68)$$

In the nonrelativistic limit and with classical statistics,

$$F(0,0) = \frac{2e^2}{T} \left(\frac{mT}{2\pi} \right)^{3/2} e^{(\mu-m)/T} \qquad (5.69)$$

which, when inserted in (5.67), is the well-known Debye–Hückel formula.

At zero temperature, the discrete frequency of the photon becomes continuous and the $n = 0$ mode cannot be isolated. From (3.40),

$$\lim_{T\to 0} T \sum_n = \frac{1}{2\pi i} \int_{-i\infty}^{i\infty} dk_0 = \frac{1}{2\pi} \int_{-\infty}^{\infty} dk_4 \qquad (5.70)$$

At $T = 0$ it is convenient to work in Euclidean space with $k_4 = -ik_0$ and with $\bar{k}^2 = k_4^2 + \mathbf{k}^2 = -k^2 \geq 0$. Both F and G are functions of $|\bar{k}|$ and ϕ, where $\tan \phi = |\mathbf{k}|/k_4$. Then (5.63) becomes

$$
\frac{\ln Z_{\text{ring}}}{\beta V} = -\frac{1}{(2\pi)^3} \int_0^\infty d\bar{k}^2\, \bar{k}^2 \int_0^{\pi/2} d\phi \sin^2 \phi
$$

$$
\times \left\{ 2\left[\ln\left(1 + \frac{G(\bar{k}^2, \phi)}{\bar{k}^2}\right) - \frac{G(\bar{k}^2, \phi)}{\bar{k}^2}\right] \right.
$$

$$
\left. + \ln\left(1 + \frac{F(\bar{k}^2, \phi)}{\bar{k}^2}\right) - \frac{F(\bar{k}^2, \phi)}{\bar{k}^2} \right\} \tag{5.71}
$$

The potential infrared divergence in (5.71) may be isolated by setting $\bar{k} = 0$ whenever possible in the integrand:

$$
-\frac{1}{(2\pi)^3} \int_0^\infty d\bar{k}^2\, \bar{k}^2 \int_0^{\pi/2} d\phi \sin^2 \phi \left\{ 2\ln\left(1 + \frac{G(0, \phi)}{\bar{k}^2}\right) + \ln\left(1 + \frac{F(0, \phi)}{\bar{k}^2}\right) \right.
$$

$$
\left. - \frac{2G(0, \phi) + F(0, \phi)}{\bar{k}^2} + \frac{1}{2\bar{k}^2}\frac{1}{\bar{k}^2 + \mu^2}[F^2(0, \phi) + 2G^2(0, \phi)] \right\} \tag{5.72}
$$

Notice the term $1/(\bar{k}^2 + \mu^2)$. The choice of μ^2 is arbitrary; any choice independent of e^2 will give the same coefficient of $e^4 \ln e^2$. After integrating over $\bar{k}^2$, (5.72) becomes

$$
-\frac{1}{2(2\pi)^3} \int_0^{\pi/2} d\phi \sin^2 \phi \left\{ F^2(0, \phi)\left[\ln\left(\frac{F(0, \phi)}{\mu^2}\right) - \frac{1}{2}\right] \right.
$$

$$
\left. + 2G^2(0, \phi)\left[\ln\left(\frac{G(0, \phi)}{\mu^2}\right) - \frac{1}{2}\right] \right\} \tag{5.73}
$$

The explicit forms of $F(0, \phi)$ and $G(0, \phi)$ may be substituted in (5.73) and the integration performed. A lengthy analysis yields

$$
\frac{\ln Z_{\text{ring}}}{\beta V} = -\frac{e^4 \ln e^2}{128\pi^6}\left[(6 - 4\ln 2)\mu p_{\text{F}}^3 - 5\mu^2 p_{\text{F}}^2 + 4\mu^3 p_{\text{F}} \ln\left(\frac{\mu + p_{\text{F}}}{\mu}\right) \right.
$$

$$
+ 6\mu m^2 p_{\text{F}} \ln\left(\frac{\mu + p_{\text{F}}}{2^{5/3}\mu}\right) - m^2(4\mu^2 + m^2)\ln^2\left(\frac{\mu + p_{\text{F}}}{m}\right)
$$

$$
\left. + m^2\mu(4\mu^2 + m^2)\frac{I(a)}{p_{\text{F}}} \right] \tag{5.74}
$$

where

$$
I(a) = \int_0^\infty \frac{dx}{a^2 x^2 - 1} \ln\left(\frac{x + 1}{x - 1}\right) \qquad a = \frac{\mu}{p_{\text{F}}}
$$

The ultrarelativistic limit is

$$\frac{\ln Z_{\text{ring}}}{\beta V} = -\frac{e^4 \ln e^2}{128\pi^6}\mu^4 \tag{5.75}$$

and the nonrelativistic limit is

$$\frac{\ln Z_{\text{ring}}}{\beta V} = -\frac{e^4 \ln e^2}{48\pi^6}(1 - \ln 2)\mu p_{\text{F}}^3 \tag{5.76}$$

5.5.3 Three loops at finite density

The three-loop diagrams not included in the ring sum are

(5.77)

The evaluation of these diagrams is technically quite involved because of overlapping ultraviolet divergences. For further discussion, see Freedman and McLerran [5] and Baluni [6].

The result of evaluating (5.77) together with the order-e^4 contribution from the sum of ring diagrams is

$$P = \frac{\mu^4}{12\pi^2}\left[1 - \frac{3}{2}\frac{\alpha(M)}{\pi} - \frac{3}{2}\left(\frac{\alpha(M)}{\pi}\right)^2\ln\left(\frac{\alpha(M)}{\pi}\right)\right.$$
$$\left. - \frac{1}{2}\left(\frac{\alpha(M)}{\pi}\right)^2\ln\left(\frac{\mu^2}{M^2}\right) + (2.118\,19)\left(\frac{\alpha(M)}{\pi}\right)^2\right] \tag{5.78}$$

Certain integrals had to be done numerically in producing this result, giving the number in the coefficient of α^2. The photon wavefunction renormalization constant $\mathcal{Z}_3$ was defined at a Euclidean subtraction point $\bar{k}^2 = M^2$. Equivalently, the photon self-energy was renormalized in such a way that $F(\bar{k}^2 = M^2, \mu = 0) = G(\bar{k}^2 = M^2, \mu = 0) = 0$.

The choice of subtraction energy M is completely arbitrary. In (5.78) notice that a logarithm of μ/M appears. At higher orders of α, higher powers of the logarithm will appear. Therefore, to reduce the importance of higher-order terms at high density we are free to choose $M = \mu$. (The optimum choice of the constant of proportionality between M and μ is not known.) Then (5.78) becomes

$$P = \frac{\mu^4}{12\pi^2}\left[1 - \frac{3}{2}\frac{\alpha(\mu)}{\pi} - \frac{3}{2}\left(\frac{\alpha(\mu)}{\pi}\right)^2\ln\left(\frac{\alpha(\mu)}{\pi}\right)\right.$$
$$\left. + (2.118\,19)\left(\frac{\alpha(\mu)}{\pi}\right)^2\right] \tag{5.79}$$

The next question is: What is $\alpha(\mu)$? From our knowledge of the renormalization group we know that $\alpha(\mu)$ must satisfy a renormalization-group equation. To lowest order,

$$M\frac{d\alpha}{dM} = c_0\alpha^2 \tag{5.80}$$

In massless QED the constant c_0 is computed to be $2\pi/3$. Realizing that we have chosen $M = \mu$ to suppress large logarithms at high density, we find that the renormalization-group running coupling is

$$\alpha\left(\frac{\mu}{\mu_0}\right) = \frac{\alpha(1)}{1 - [2\alpha(1)/3\pi]\ln(\mu/\mu_0)} \tag{5.81}$$

Here μ_0 is some reference scale and $\alpha(1)$ is the value of the coupling at that scale. Just as in (4.24), (4.25) for the massless $\lambda\phi^4$ theory, we can combine $\alpha(1)$ and μ_0 into one constant Λ. Then

$$\alpha\left(\frac{\mu}{\Lambda}\right) = \frac{3\pi}{2\ln(\Lambda/\mu)} \tag{5.82}$$

Here Λ is the intrinsic energy scale of massless QED. This theory is not asymptotically free. Therefore, when $\mu \ll \Lambda$ the coupling $\alpha(\mu/\Lambda)$ is very small. The perturbative expansion of the partition function for a cold high-density electron gas converges rapidly until $\mu \simeq \Lambda$ is reached. This limitation is not of practical significance because the intrinsic energy scale $\Lambda \sim m_e e^{137}$ is astronomically large ($m_e = 0.511$ MeV).

It is apparent that the perturbation series for P in (5.79) is rapidly convergent at non-astronomically-large densities because $\alpha/\pi \simeq 2.3 \times 10^{-3}$.

5.5.4 Three loops at finite temperature

The pressure for finite temperature QED has been calculated for $\mu = 0$ up to order e^5. We first show results up to e^4. See Corianò and Parwani [7] for the details (especially on the delicate handling of the singularities). The usual zero-temperature ultraviolet singularities are regularized through dimensional regularization, ensuring that the physical result is gauge invariant. Evaluating the diagrams (5.77) at finite temperature for N_f electron flavors (physical QED corresponds to $N_f = 1$), with the appropriate counterterms, yields

$$\frac{P}{T^4} = \frac{\pi^2}{45}\left(1 + \frac{7}{4}N_f\right) - \frac{5e^2 N_f}{288}$$

$$+ \frac{e^3}{12\pi}\left(\frac{N_f}{3}\right)^{3/2} + \frac{e^4 N_f}{\pi^6}(0.4056 \pm 0.0030)$$

$$- e^4 N_f^2\left[\frac{0.4667 \pm 0.0020}{\pi^6} + \frac{5}{1728\pi^2}\ln\left(\frac{T}{M}\right)\right] \tag{5.83}$$

The above also includes the e^4 contribution for the set of ring diagrams discussed earlier. As before the uncertainties in the quantities are due to the numerical evaluation of some integrals. Note that the coupling in the expression for the pressure is to be evaluated at some renormalization scale M. This scale M may be chosen on physical grounds: for example, setting $M = T$ will eliminate the logarithm at this and higher orders. An alternative procedure is to use renormalization-group arguments to relate the coupling e at some scale M to that at another scale set by the temperature. Doing this, one may write

$$e^2(T) = e^2 \left[1 + \frac{e^2 N_{\rm f}}{6\pi^2} \ln \left(\frac{T}{M} \right) \right] + \mathcal{O}(e^6) \tag{5.84}$$

where e is the coupling in (5.83). Defining $\alpha(T) = e^2(T)/4\pi$, (5.83) can be written as

$$\frac{P}{T^4} = \frac{\pi^2}{45} \left(1 + \frac{7}{4} N_{\rm f} \right) - \frac{5\pi^2}{72} \frac{\alpha(T) N_{\rm f}}{\pi} + \frac{2\pi^2}{9\sqrt{3}} \left(\frac{\alpha(T) N_{\rm f}}{\pi} \right)^{3/2}$$
$$+ \left(\frac{0.658 \pm 0.006}{N_{\rm f}} - 0.757 \pm 0.004 \right) \left(\frac{\alpha(T) N_{\rm f}}{\pi} \right)^2 + \mathcal{O} \left(\alpha(T)^{5/2} \right) \tag{5.85}$$

The logarithm in (5.83) has disappeared and has been absorbed into the renormalization-group redefinition of the coupling constant.

The order-e^5 contribution is then obtained by resumming the boson propagators in (5.77) through a ring insertion, as discussed previously. The details appear in Parwani and Corianò [8]; the result is

$$\frac{P_5}{T^4} = \left(\frac{\alpha(T) N_{\rm f}}{\pi} \right)^{5/2} \left(\frac{\pi^2 [1 - \gamma_{\rm E} - \ln(4/\pi)]}{9\sqrt{3}} - \frac{\pi^2}{2N_{\rm f}\sqrt{3}} \right) \tag{5.86}$$

where $\gamma_{\rm E}$ is Euler's constant.

5.6 Exercises

5.1 Prove (5.12).

5.2 Derive the blackbody radiation formula from (5.32) for arbitrary ρ.

5.3 Discuss what happens when the nonlinear gauge $F = A^\mu A_\mu - f(\mathbf{x}, \tau) = 0$ is chosen.

5.4 Derive the free-photon propagator given by (5.38).

5.5 Obtain the general forms given in (5.46) for the in-medium photon propagator and its inverse.

5.6 Repeat the calculation in the text for $\ln Z_{\rm ex}$ but with an arbitrary covariant gauge parameter ρ. Is the result independent of ρ?

5.7 Using (5.51) and (5.46), find the limits of F and G when $k^2 = k_0^2 - \mathbf{k}^2 = 0$.

5.8 Determine the combinatoric factors for the two diagrams of (5.77).

5.9 Derive (5.63) from (5.62).

5.10 Calculate the relative contributions to the pressure in QED at finite temperature and zero electron mass from orders 0, 2, 3, 4 and 5 in e for arbitrary N_f.

References

1. Fradkin, E. S., *Proc. Lebedev Phys. Inst.* **29**, 6 (1965).
2. Peskin, M. E., and Schroeder, D. V. (1995). *An Introduction to Quantum Field Theory* (Addison-Wesley, Reading).
3. Norton, R. E., and Cornwall, J. M., *Ann. Phys. (NY)* **91**, 106 (1975).
4. Kapusta, J. I., *Nucl. Phys.* **B148**, 461 (1979).
5. Freedman, B. A., and McLerran, L. D., *Phys. Rev. D* **16**, 1130, 1147, 1169 (1977).
6. Baluni, V., *Phys. Rev. D* **17**, 2092 (1978).
7. Corianò, C., and Parwani, R., *Phys. Rev. Lett.*, **73**, 2398 (1994).
8. Parwani, R., and Corianò, C., *Nucl. Phys.* **B434**, 56 (1995).

Bibliography

Quantization of the electromagnetic field in the functional integral formalism

Abers, E. S., and Lee, B.W., *Phys. Rep.* **9**, 1 (1973).
Lee, T. D. (1982). *Particle Physics and Introduction to Field Theory* (Harwood Academic, London).
Ramond, P. (1989). *Field Theory: A Modern Primer* (Addison-Wesley, Redwood City).

Relativistic QED at finite temperature and density

Akhiezer, I. A. and Peletminskii, S. V., *Sov. Phys. JETP* **11**, 1316 (1960).

6

Linear response theory

Suppose that a solid is hit with a hammer. Sound waves will propagate outwards from the point of contact. How is the frequency of the sound wave related to its wave number? How does a light wave propagate through plasma? What happens when a charge impurity is embedded in an electrically neutral medium? Is it screened, and if so how is that screening described quantitatively? If a medium is disturbed by a small amount one might expect its response also to be small. The quantitative formalism for dealing with small disturbances is called linear response theory. The beauty of the theory is that the response of the system can be expressed as a folding of the external source causing the disturbance with a response function that is computed using equilibrium correlation functions not dependent on the strength of the external source. Therefore, details of the internal dynamics of the thermodynamic system can be studied using weak external probes. Other areas of science where linear response theory has proven to be extremely useful are quite extensive, and include x-ray scattering from crystals and molecules, electron scattering from protons and nuclei, and sound waves generated by earthquakes propagating through the earth's interior.

6.1 Linear response to an external field

Suppose we apply some external field to our system, which is initially in equilibrium. The goal of linear response theory is to calculate the change in the ensemble average value of any operator $Y(\mathbf{x}, t)$ caused by the external field, to first order in that external field.

Let

$$H'(t) = H + H_{\text{ext}}(t) \tag{6.1}$$

where H is the unperturbed Hamiltonian (but which still contains inter-actions) and $H_{\text{ext}}(t)$ is the perturbation that couples the external field to the system. We will imagine that $H_{\text{ext}}(t)$ vanishes when $t < t_0$, so that the system has had plenty of time to achieve equilibrium in the past. The exact equation of motion for Y is

$$\frac{\partial Y(\mathbf{x}, t)}{\partial t} = i \left[H'(t), Y(\mathbf{x}, t) \right] \tag{6.2}$$

Let $|j\rangle$ be an eigenstate of H (in the Heisenberg picture). Then it follows that the time rate of change of the expectation value of Y in the state $|j\rangle$ is

$$\begin{aligned}
\frac{\partial \langle j | Y(\mathbf{x}, t) | j \rangle}{\partial t} &= i \langle j | \left[H'(t), Y(\mathbf{x}, t) \right] | j \rangle \\
&= i \langle j | \left[H_{\text{ext}}(t), Y(\mathbf{x}, t) \right] | j \rangle
\end{aligned} \tag{6.3}$$

Equation (6.3) is exact, but it is generally impossible to solve it in closed form. At this point we assume that H_{ext} causes only a small change in the expectation value of Y. Then to first order in H_{ext} we can integrate (6.3) as

$$\begin{aligned}
\delta \langle j | Y(\mathbf{x}, t) | j \rangle &= \langle j | Y(\mathbf{x}, t) | j \rangle - \langle j | Y(\mathbf{x}, t_0) | j \rangle \\
&= i \int_{t_0}^{t} dt' \langle j | \left[H_{\text{ext}}(t'), Y(\mathbf{x}, t) \right] | j \rangle
\end{aligned} \tag{6.4}$$

Now take the (grand canonical) ensemble average,

$$\delta \langle Y(\mathbf{x}, t) \rangle = \frac{\sum_j e^{-\beta K_j} \delta \langle j | Y(\mathbf{x}, t) | j \rangle}{\sum_j e^{-\beta K_j}} \tag{6.5}$$

Here $K = H - \mu_i N_i$, where allowance is made for an arbitrary number of conserved charges. Using (1.1) and (6.4) in (6.5), we obtain

$$\delta \langle Y(\mathbf{x}, t) \rangle = i \int_{t_0}^{t} dt' \, \text{Tr} \left\{ \hat{\rho} \left[H_{\text{ext}}(t'), Y(\mathbf{x}, t) \right] \right\} \tag{6.6}$$

This expresses the change in the ensemble-average value of Y in terms of the commutator of H_{ext} and Y evaluated in the unperturbed ensemble, represented by $\hat{\rho}$. We reiterate that (6.6) is correct to first order in H_{ext}.

As an example, consider a real scalar field ϕ that is coupled to an external source $J(\mathbf{x}, t)$ via

$$H_{\text{ext}}(t) = \int d^3 x \, J(\mathbf{x}, t) \hat{\phi}(\mathbf{x}, t) \tag{6.7}$$

We are interested in the change in the ensemble-average value of $\hat{\phi}$ when the external source is turned on. Putting (6.7) into (6.6) with $Y = \hat{\phi}$

gives

$$\delta\langle\hat{\phi}(\mathbf{x},t)\rangle = -i \int_{t_0}^{t} dt' \int d^3x' J(\mathbf{x}',t') \, \mathrm{Tr}\left\{\hat{\rho}\left[\hat{\phi}(\mathbf{x},t),\hat{\phi}(\mathbf{x}',t')\right]\right\} \quad (6.8)$$

At this point it is useful to introduce the following quantities:
the time-ordered propagator,

$$iD(\mathbf{x},t;\mathbf{x}',t') = \mathrm{Tr}\left\{\hat{\rho}T_t\left(\hat{\phi}(\mathbf{x},t)\hat{\phi}(\mathbf{x}',t')\right)\right\} \quad (6.9)$$

the retarded Green's function,

$$iD^{\mathrm{R}}(\mathbf{x},t;\mathbf{x}',t') = \mathrm{Tr}\left\{\hat{\rho}\left[\hat{\phi}(\mathbf{x},t),\hat{\phi}(\mathbf{x}',t')\right]\right\}\theta(t-t') \quad (6.10)$$

the advanced Green's function,

$$iD^{\mathrm{A}}(\mathbf{x},t;\mathbf{x}',t') = -\mathrm{Tr}\left\{\hat{\rho}\left[\hat{\phi}(\mathbf{x},t),\hat{\phi}(\mathbf{x}',t')\right]\right\}\theta(t'-t) \quad (6.11)$$

In (6.9), T_t is the time-ordering operator. Then (6.8) becomes

$$\delta\langle\hat{\phi}(\mathbf{x},t)\rangle = \int_{-\infty}^{\infty} dt' \int d^3x' J(\mathbf{x}',t')D^{\mathrm{R}}(\mathbf{x},t;\mathbf{x}',t') \quad (6.12)$$

Here we have let $t_0 \to -\infty$ and have set the upper limit of integration over t' to ∞ on account of (6.10).

Since the unperturbed system is in thermal equilibrium, D^{R} must depend only on $\mathbf{x} - \mathbf{x}'$ and $t - t'$ (the former would not be true for a solid or crystal, of course). We insert the Fourier transforms

$$D^{\mathrm{R}}(\mathbf{x}-\mathbf{x}',t-t') = \int \frac{d^3k\,d\omega}{(2\pi)^4} \, \mathrm{e}^{i[\mathbf{k}\cdot(\mathbf{x}-\mathbf{x}')-\omega(t-t')]} D^{\mathrm{R}}(\omega,\mathbf{k}) \quad (6.13)$$

$$J(\mathbf{x}',t') = \int \frac{d^3p\,d\alpha}{(2\pi)^4} \, \mathrm{e}^{i(\mathbf{p}\cdot\mathbf{x}'-\alpha t')} \tilde{J}(\alpha,\mathbf{p}) \quad (6.14)$$

into (6.12) to obtain

$$\delta\langle\hat{\phi}(\mathbf{x},t)\rangle = \int \frac{d^3k\,d\omega}{(2\pi)^4} \, \mathrm{e}^{i(\mathbf{k}\cdot\mathbf{x}-\omega t)} \tilde{J}(\omega,\mathbf{k})D^{\mathrm{R}}(\omega,\mathbf{k}) \quad (6.15)$$

or

$$\delta\langle\tilde{\phi}(\omega,\mathbf{k})\rangle = \tilde{J}(\omega,\mathbf{k})D^{\mathrm{R}}(\omega,\mathbf{k}) \quad (6.16)$$

which is a very aesthetic form. The change in the ensemble average of the field, in frequency–momentum space, is equal to the external source times the retarded Green's function.

6.2 Lehmann representation

The question arises how the real time Green's functions required in the linear response approach to dynamical perturbations are obtained. Are they related to the imaginary time propagators studied in previous chapters? In fact they should be, since all dynamical information in a quantum theory is contained in the matrix elements of operators. If both the real time and imaginary time correlation functions can be expressed in terms of matrix elements then a connection can be made. These expressions are referred to as Lehmann representations. We shall work them out for a real scalar field. It is straightforward to do the same for complex scalar fields and for fields with spin, the main complication being the tensorial structures.

Consider the fully interacting ensemble average of a product of scalar field operators. Suppose that the states $|n\rangle$ form a complete set of eigenstates of the Hamiltonian and of the momentum operator. Starting with

$$iD^+(x, y) = \langle \hat{\phi}(x)\hat{\phi}(y) \rangle = \frac{1}{Z} \sum_n e^{-\beta E_n} \langle n|\hat{\phi}(x)\hat{\phi}(y)|n\rangle \qquad (6.17)$$

we insert a complete set of states between the field operators:

$$\langle \hat{\phi}(x)\hat{\phi}(y) \rangle = \frac{1}{Z} \sum_{m,n} e^{-\beta E_n} \langle n|\hat{\phi}(x)|m\rangle \langle m|\hat{\phi}(y)|n\rangle \qquad (6.18)$$

Under the assumption that the system is translation invariant in both time and space, the matrix elements at x are related to the matrix elements at $x = 0$ as follows:

$$\langle n|\hat{\phi}(x)|m\rangle = e^{i(p_n - p_m)\cdot x} \langle n|\hat{\phi}(0)|m\rangle \qquad (6.19)$$

Thus the explicit representation of the average of the product of fields is

$$iD^+(x - y) = \frac{1}{Z} \sum_{m,n} e^{-\beta E_n} e^{i(p_n - p_m)\cdot(x-y)} \langle n|\hat{\phi}(0)|m\rangle \langle m|\hat{\phi}(0)|n\rangle \qquad (6.20)$$

The Fourier transform (we use the same symbol D in coordinate space and momentum space for ease of notation)

$$D^+(k) = \int d^4z \, e^{ik\cdot z} D^+(z) \qquad (6.21)$$

can be expressed in terms of the spectral density

$$\rho^+(k) = \frac{1}{Z} \sum_{m,n} e^{-\beta E_n} (2\pi)^3 \delta(k - p_m + p_n) |\langle n|\hat{\phi}(0)|m\rangle|^2 \qquad (6.22)$$

as

$$iD^+(k) = 2\pi \rho^+(k) \qquad (6.23)$$

This spectral density is positive definite. The Dirac delta functions do not affect this, since one can always work in a large but finite box for which the energy and momentum modes are discrete, replacing the Dirac delta functions by Kronecker delta functions.

In a similar way we define

$$iD^-(x,y) = -\langle\hat{\phi}(y)\hat{\phi}(x)\rangle \tag{6.24}$$

whose Fourier transform is also expressed in terms of a spectral density:

$$iD^-(k) = 2\pi\rho^-(k) \tag{6.25}$$

where

$$\rho^-(k) = -e^{-\beta k_0}\rho^+(k) \tag{6.26}$$

The minus sign comes from the definition and the Boltzmann factor comes from interchanging the labels m and n in the sum over states and using energy conservation. Obviously this spectral density is negative definite.

The ensemble average of the commutator is

$$D^{\mathrm{n}}(x-y) = -i\langle[\hat{\phi}(x),\hat{\phi}(y)]\rangle = D^+(x-y) + D^-(x-y) \tag{6.27}$$

where the superscript "n" denotes the "normal" commutator-defined Green's function. Its spectral density is

$$\rho^{\mathrm{n}}(k) = \rho^+(k) + \rho^-(k) = \left(1 - e^{-\beta k_0}\right)\rho^+(k) = -\left(e^{\beta k_0} - 1\right)\rho^-(k)$$

$$= \frac{1}{Z}\sum_{m,n}\left(e^{-\beta E_n} - e^{-\beta E_m}\right)(2\pi)^3\delta(k - p_m + p_n)|\langle n|\hat{\phi}(0)|m\rangle|^2 \tag{6.28}$$

For linear response theory the most relevant correlation function is the retarded propagator

$$D^{\mathrm{R}}(z) = \theta(z_0)D^{\mathrm{n}}(z) \tag{6.29}$$

Associated with it is the advanced propagator

$$D^{\mathrm{A}}(z) = -\theta(-z_0)D^{\mathrm{n}}(z) \tag{6.30}$$

Straightforward manipulations show that these can be expressed as integrals over the spectral density ρ^{n}:

$$D^{\mathrm{R}}(k) = -\int_{-\infty}^{\infty}\frac{d\omega}{\omega - k_0 - i\varepsilon}\rho^{\mathrm{n}}(\omega,\mathbf{k}) \tag{6.31}$$

$$D^{\mathrm{A}}(k) = -\int_{-\infty}^{\infty}\frac{d\omega}{\omega - k_0 + i\varepsilon}\rho^{\mathrm{n}}(\omega,\mathbf{k}) \tag{6.32}$$

The imaginary parts of these functions are proportional to the spectral density,

$$\mathrm{Im}\,D^{\mathrm{R}}(k) = -\mathrm{Im}\,D^{\mathrm{A}}(k) = -\pi\rho^{\mathrm{n}}(k) \tag{6.33}$$

and their real parts are equal,

$$\operatorname{Re} D^{\mathrm{R}}(k) = \operatorname{Re} D^{\mathrm{A}}(k) \tag{6.34}$$

under the assumption that k is real.

Now we come to the connection with the imaginary time propagator, for which the finite-temperature perturbation theory was developed. From (3.21) we know that

$$
\begin{aligned}
\mathcal{D}(\mathbf{x}, \tau) &= \langle \hat{\phi}(\mathbf{x}, \tau) \hat{\phi}(0) \rangle \\
&= \frac{1}{Z} \sum_n e^{-\beta E_n} \langle n | \hat{\phi}(\mathbf{x}, \tau) \hat{\phi}(0) | n \rangle \\
&= \frac{1}{Z} \sum_{m,n} e^{-\beta E_n} \langle n | \hat{\phi}(\mathbf{x}, \tau) | m \rangle \langle m | \hat{\phi}(0) | n \rangle
\end{aligned} \tag{6.35}
$$

Just as in (2.86), the field evolves in imaginary time according to

$$\hat{\phi}(\mathbf{x}, \tau) = e^{H\tau} \hat{\phi}(\mathbf{x}, 0) e^{-H\tau} \tag{6.36}$$

which leads to

$$\mathcal{D}(\mathbf{x}, \tau) = \frac{1}{Z} \sum_{m,n} e^{-\beta E_n} e^{\tau(E_n - E_m)} e^{i(\mathbf{p}_m - \mathbf{p}_n) \cdot \mathbf{x}} \langle n | \hat{\phi}(0) | m \rangle \langle m | \hat{\phi}(0) | n \rangle \tag{6.37}$$

Following the conventions of Chapter 3, the Fourier transform is

$$
\begin{aligned}
\mathcal{D}(\omega_n, \mathbf{k}) &= \int_0^\beta d\tau \int d^3 x \, e^{-i(\mathbf{k} \cdot \mathbf{x} + \omega_n \tau)} \mathcal{D}(\mathbf{x}, \tau) \\
&= \frac{1}{Z} \sum_{m,n} (2\pi)^3 \delta(\mathbf{k} - \mathbf{p}_m + \mathbf{p}_n) \langle n | \hat{\phi}(0) | m \rangle \langle m | \hat{\phi}(0) | n \rangle \\
&\quad \times \frac{e^{-\beta E_m} - e^{-\beta E_n}}{E_n - E_m - i\omega_n}
\end{aligned} \tag{6.38}
$$

which can be written in terms of the spectral density as

$$\mathcal{D}(\omega_n, \mathbf{k}) = \int_{-\infty}^\infty \frac{d\omega}{\omega + i\omega_n} \rho^{\mathrm{n}}(\omega, \mathbf{k}) \tag{6.39}$$

Thus the advanced and retarded propagators can be obtained from the finite-temperature propagator by analytic continuation as follows:

$$D^{\mathrm{R}}(k) = -\mathcal{D}(\omega_n \to ik_0 - \varepsilon) \tag{6.40}$$
$$D^{\mathrm{A}}(k) = -\mathcal{D}(\omega_n \to ik_0 + \varepsilon) \tag{6.41}$$

The spectral density ρ^{n} determines both the real time and imaginary time propagators and is therefore a very important function.

A concrete example of these relations is provided by a free field. The imaginary time propagator is $\mathcal{D} = 1/(\omega_n^2 + \mathbf{k}^2 + m^2)$. From this one immediately obtains $\rho^n = \text{sign}(k_0)\,\delta(k_0^2 - \mathbf{k}^2 - m^2)$. This shows quite directly that all the weight is concentrated on the mass shell of the particle. Generally, for interacting particles in a medium, this weight gets spread out over a finite range of energies. The free-particle spectral density has two obvious properties that generalize to interacting systems. One is the symmetry in the sign of the energy and the other is an integral over the energy.

The spectral density ρ^n given in (6.28) has the symmetry

$$\rho^n(-\omega, -\mathbf{k}) = -\rho^n(\omega, \mathbf{k}) \tag{6.42}$$

Here $k_0 = \omega$. In a rotationally invariant system, for every state with energy E_n and momentum $\mathbf{p}_n$ there is a state with the same energy and the opposite momentum. Therefore

$$\rho^n(\omega, -\mathbf{k}) = \rho^n(\omega, \mathbf{k}) \tag{6.43}$$

Combining the above symmetries we conclude that ρ^n is an odd function of the energy:

$$\rho^n(-\omega, \mathbf{k}) = -\rho^n(\omega, \mathbf{k}) \tag{6.44}$$

The canonical commutation relation can be usefully employed to derive a sum rule on the spectral density. Take the spatial Fourier transform of

$$\lim_{x_0 \to 0} \frac{\partial}{\partial x_0} D^n(x) = -i\langle [\hat{\pi}(0, \mathbf{x}), \hat{\phi}(0, \mathbf{0})]\rangle = -\delta(\mathbf{x}) \tag{6.45}$$

and use the Lehmann representation for D^n. One concludes that

$$\int_{-\infty}^{\infty} d\omega\, \omega \rho^n(\omega, \mathbf{k}) = 1 \tag{6.46}$$

This sum rule is naturally obeyed by the free-particle spectral density. It also implies that interactions might modify the shape of the function ρ^n but that the total integrated weight is constant.

6.3 Screening of static electric fields

Let us apply an external static electric field $\mathbf{E}_{cl}$, as might be generated by an imposed charge distribution, to a QED plasma and observe the response. The Hamiltonian density for this interaction is

$$\mathcal{H}_{ext} = \mathbf{E} \cdot \mathbf{E}_{cl} \tag{6.47}$$

The external field $\mathbf{E}_{cl}$ is a classical field, not a quantum operator like $\mathbf{E}$ and $\mathbf{B}$. It depends on position but not on time.

The change in the electric field caused by the introduction of the external field into the plasma can be computed using (6.6):

$$\delta\langle E_i(\mathbf{x}, t)\rangle = -i \int_{-\infty}^{\infty} dt' \int d^3x' E_j^{\mathrm{cl}}(\mathbf{x}') \operatorname{Tr}\{\hat{\rho} \left[E_i(\mathbf{x}, t), E_j(\mathbf{x}', t')\right]\} \theta(t - t')$$

(6.48)

Thus we need to know the commutator of two electric field operators. Using the expression for the electric field in terms of the vector potential and the canonical commutation relations, one readily finds that

$$
\begin{aligned}
\langle\left[E_i(\mathbf{x}, t), E_j(\mathbf{x}', t')\right]\rangle \theta(t - t') = {}& \partial_i \partial_j' \left\{\langle[A_0(\mathbf{x}, t),\, A_0(\mathbf{x}', t')]\rangle\theta(t - t')\right\} \\
& - \partial_i \partial_0' \left\{\langle[A_0(\mathbf{x}, t),\, A_j(\mathbf{x}', t')]\rangle\theta(t - t')\right\} \\
& - \partial_0 \partial_j' \left\{\langle[A_i(\mathbf{x}, t),\, A_0(\mathbf{x}', t')]\rangle\theta(t - t')\right\} \\
& + \partial_0 \partial_0' \left\{\langle[A_i(\mathbf{x}, t),\, A_j(\mathbf{x}', t')]\rangle\theta(t - t')\right\} \\
& - i\delta_{ij}\delta(\mathbf{x} - \mathbf{x}')\delta(t - t')
\end{aligned}
$$

(6.49)

The real time photon propagator is

$$D_{\mu\nu}^{\mathrm{R}}(\mathbf{x} - \mathbf{x}',\, t - t') = i \operatorname{Tr}\{\hat{\rho} \left[A_\mu(\mathbf{x}, t), A_\nu(\mathbf{x}', t')\right]\} \theta(t - t')$$

(6.50)

where the sign is chosen to be compatible with the definition of the imaginary time propagator in Section 5.3. It depends only on the differences $\mathbf{x} - \mathbf{x}'$ and $t - t'$, owing to translation invariance in a plasma and to the assumption of thermal equilibrium. Combining (6.48) to (6.50) we obtain the net electric field in the medium,

$$
\begin{aligned}
E_i^{\mathrm{net}}(\mathbf{x}, t) &= E_i^{\mathrm{cl}}(\mathbf{x}) + \delta\langle E_i(\mathbf{x}, t)\rangle \\
&= \int_{-\infty}^{\infty} dt' \int d^3x' E_j^{\mathrm{cl}}(\mathbf{x}') \\
&\quad \times \left(-\partial_i \partial_j' D_{00}^{\mathrm{R}} + \partial_i \partial_0' D_{0j}^{\mathrm{R}} + \partial_0 \partial_j' D_{i0}^{\mathrm{R}} - \partial_0 \partial_0' D_{ij}^{\mathrm{R}}\right)
\end{aligned}
$$

(6.51)

where the arguments of $D_{\mu\nu}^{\mathrm{R}}$ are $\mathbf{x} - \mathbf{x}'$ and $t - t'$ as in (6.50). The Fourier transforms are

$$D_{\mu\nu}^{\mathrm{R}}(\mathbf{x} - \mathbf{x}',\, t - t') = \int \frac{d^3k\, d\omega}{(2\pi)^4}\, e^{i\mathbf{k}\cdot(\mathbf{x}-\mathbf{x}')} e^{-i\omega(t-t')} D_{\mu\nu}^{\mathrm{R}}(\omega, \mathbf{k})$$

(6.52)

and

$$\mathbf{E}_{\mathrm{cl}}(\mathbf{x}') = \int \frac{d^3p}{(2\pi)^3}\, e^{i\mathbf{p}\cdot\mathbf{x}'} \mathbf{E}_{\mathrm{cl}}(\mathbf{p})$$

(6.53)

Substitution in (6.51) gives

$$
E_i^{\text{net}}(\mathbf{x}) = -\int \frac{d^3k}{(2\pi)^3} \, e^{i\mathbf{k}\cdot\mathbf{x}} E_j^{\text{cl}}(\mathbf{k}) \Big[k_i k_j D_{00}^{\text{R}}(\omega, \mathbf{k}) + \omega k_i D_{0j}^{\text{R}}(\omega, \mathbf{k})
$$
$$
\left. + \omega k_j D_{i0}^{\text{R}}(\omega, \mathbf{k}) + \omega^2 D_{ij}^{\text{R}}(\omega, \mathbf{k}) \Big]\right|_{\omega=0} \tag{6.54}
$$

Note that the $\omega = 0$ limit is a consequence of the static nature of the applied field.

In covariant gauges the propagator is given in (5.46). In such gauges the last three terms in (6.51) vanish. Hence the net electric field in momentum space is

$$
E_i^{\text{net}}(\mathbf{k}) = \frac{k_i k_j E_j^{\text{cl}}(\mathbf{k})}{\mathbf{k}^2 + F(\omega = 0, \mathbf{k})} \tag{6.55}
$$

For a plasma, the net electric field must point in the same direction as the applied external field owing to rotational invariance. The magnitudes can be related by multiplying both sides of the above equation by k_i and summing over i. Thus

$$
\mathbf{E}_{\text{net}}(\mathbf{k}) = \frac{\mathbf{E}_{\text{cl}}(\mathbf{k})}{\epsilon(\mathbf{k})} \tag{6.56}
$$

where $\epsilon(\mathbf{k})$ is the static dielectric constant and is given by

$$
\epsilon(\mathbf{k}) = 1 + \frac{F(0, \mathbf{k})}{\mathbf{k}^2} \tag{6.57}
$$

This result may be obtained in other gauges as well. In the temporal axial gauge, $A_0(\mathbf{x}, t) = 0$, the propagator is given by

$$
\mathcal{D}_{00} = 0 \qquad \mathcal{D}_{0i} = \mathcal{D}_{0i} = 0
$$
$$
\mathcal{D}_{ij} = \frac{1}{G - k^2}\left(\delta_{ij} - \frac{k_i k_j}{\mathbf{k}^2}\right) + \frac{1}{F - k^2}\frac{k^2}{k_0^2}\frac{k_i k_j}{\mathbf{k}^2} \tag{6.58}
$$

Insertion of (6.58) into (6.54) again yields (6.55), although it is interesting that in this gauge the contributing term is $[\omega^2 D_{ij}^{\text{R}}(\omega, \mathbf{k})]_{\omega=0}$. In the Coulomb gauge, $\nabla \cdot \mathbf{A}(\mathbf{x}, t) = 0$, the propagator is given by

$$
\mathcal{D}^{\mu\nu} = \frac{1}{G - k^2} P_{\text{T}}^{\mu\nu} + \frac{1}{F - k^2}\frac{k^2}{\mathbf{k}^2} u^\mu u^\nu \tag{6.59}
$$

where $u^\mu = (1, 0, 0, 0)$ defines the rest frame of the medium. The self-energy $\Pi^{\mu\nu}$ is related to F and G just as in (5.46). This may be verified by returning to the definition of the self-energy in terms of the full and bare propagators, which may be written as

$$
\mathcal{D}^{\mu\nu} = \mathcal{D}_0^{\mu\nu} - \mathcal{D}_0^{\mu\alpha}\Pi_{\alpha\beta}\mathcal{D}^{\beta\nu} \tag{6.60}
$$

Substitution of (6.59) into (6.54) again yields (6.55).

It must be emphasized that (6.55) is an exact result, to be used with the exact expression for $F(0, \mathbf{k})$ or with the best available approximation to it. The only assumption is that the applied field $\mathbf{E}_{\text{cl}}$ is weak enough to justify the linearity approximation.

The dielectric function is the screening factor. In the limit of no interactions, where $e \to 0$ and $F \to 0$, the net electric field equals the applied field. In the absence of matter $T = 0$ and $\mu = 0$, but with interactions turned on, $e \neq 0$, there is still a modification of the applied electric field known as vacuum polarization. When $|\mathbf{k}| \ll m_e$, one finds that

$$F_{\text{vac}}(0, \mathbf{k}) \approx -\frac{\alpha}{15\pi} \frac{|\mathbf{k}|^4}{m_e^2} \tag{6.61}$$

When $|\mathbf{k}| \gg m_e$,

$$F_{\text{vac}}(0, \mathbf{k}) \approx -\frac{\alpha}{3\pi} \mathbf{k}^2 \ln\left(\frac{\mathbf{k}^2}{M^2}\right) \tag{6.62}$$

where M is the renormalization energy scale. One may think of virtual electron–positron pairs continually popping out of and back into the vacuum to produce this modification of the applied field. Since $\epsilon(|\mathbf{k}| > 0) \neq 1$, one may in this sense think of the vacuum as a medium. Furthermore we may think of the dielectric constant as the ratio of the squared net observed charge at momentum transfer $\mathbf{k}$ to the squared ordinary electric charge at zero momentum transfer,

$$\frac{\alpha_{\text{net}}(\mathbf{k})}{\alpha} = \frac{1}{1 + F_{\text{vac}}(0, \mathbf{k})/\mathbf{k}^2} \tag{6.63}$$

Substitution of (6.62) into (6.63) gives exactly the lowest-order renormalization-group result, (5.82) with $\mu \to |\mathbf{k}| \gg m_e$, which is no coincidence.

The one-loop finite-temperature and finite-density contribution to F is in general a complicated function of $\mathbf{k}$. It is given in (5.51) since $F(0, \mathbf{k}) = -\Pi^{00}(0, \mathbf{k})$. At very short distances, $|\mathbf{k}| \gg T$ and μ, the vacuum contribution dominates, $F_{\text{vac}} \gg F_{\text{mat}}$. At very long distances, $|\mathbf{k}| \ll T$ and μ, the matter contribution dominates, $F_{\text{vac}} \ll F_{\text{mat}}$. At distances much less than the average interparticle spacing, many-body effects cannot be important and one recovers the vacuum. At distances much greater than the average interparticle spacing, many-body effects are most important. In fact $F_{\text{vac}}/F_{\text{mat}} \propto \mathbf{k}^2$ as $\mathbf{k} \to 0$, modulo logarithms. Recalling (5.66), (5.68), and (5.69), we define the QED electric mass m_{el} by $m_{\text{el}}^2 = F(0, \mathbf{k} \to 0)$. Then, approximately,

$$\epsilon(\mathbf{k}) \approx 1 + \frac{F_{\text{vac}}(0, \mathbf{k})}{\mathbf{k}^2} + \frac{m_{\text{el}}^2}{\mathbf{k}^2} \tag{6.64}$$

Linear response theory gives both vacuum polarization and plasma screening.

6.4 Screening of a point charge

As a concrete demonstration of a situation commonly encountered, place a static charge Q_1 at $\mathbf{x}_1$ and another static charge Q_2 at $\mathbf{x}_2$. What is the change in free energy of the QED plasma as a function of separation? Analogous problems arise in condensed matter physics when treating an impurity or defect.

From Gauss's law,

$$\nabla \cdot \mathbf{E}_1^{\text{cl}} = Q_1 \, \delta(\mathbf{x} - \mathbf{x}_1) \tag{6.65}$$

we obtain

$$\mathbf{E}_1^{\text{cl}}(\mathbf{x}) = \int \frac{d^3 k}{(2\pi)^3} \, e^{i\mathbf{k}\cdot\mathbf{x}} \mathbf{E}_1^{\text{cl}}(\mathbf{k})$$

where

$$\mathbf{E}_1^{\text{cl}} = -i\frac{\mathbf{k}}{\mathbf{k}^2} \, e^{-i\mathbf{k}\cdot\mathbf{x}_1} \, Q_1 \tag{6.66}$$

Similar equations are obtained for charge 2. The change in free energy is

$$V(\mathbf{x}_1, \mathbf{x}_2) = \frac{1}{2} \int d^3 x \left[\mathbf{E}_1^{\text{cl}}(\mathbf{x}) \cdot \langle \mathbf{E}_2(\mathbf{x}) \rangle + \mathbf{E}_2^{\text{cl}}(\mathbf{x}) \cdot \langle \mathbf{E}_1(\mathbf{x}) \rangle \right]$$

where

$$\langle \mathbf{E}_1(\mathbf{x}) \rangle = \mathbf{E}_1^{\text{net}}(\mathbf{x}) \qquad \langle \mathbf{E}_2(\mathbf{x}) \rangle = \mathbf{E}_2^{\text{net}}(\mathbf{x}) \tag{6.67}$$

After some manipulation, this takes the form

$$V(\mathbf{r} = \mathbf{x}_1 - \mathbf{x}_2) = Q_1 Q_2 \int \frac{d^3 k}{(2\pi)^3} \frac{e^{i\mathbf{k}\cdot\mathbf{r}}}{\mathbf{k}^2 + F(0, \mathbf{k})} \tag{6.68}$$

When r is very large, the dominant contribution to the integral comes from $\mathbf{k} \approx 0$. For this case, we replace $F(0, \mathbf{k})$ by its infrared limit m_{el}^2. Then we get

$$V(r) = \frac{Q_1 Q_2}{4\pi} \frac{e^{-m_{\text{el}}r}}{r} \tag{6.69}$$

which is a screened Coulomb potential with inverse screening length m_{el}.

At $T = \mu = 0$ one may compute the change in the form of Coulomb's law due to vacuum polarization by expanding (6.68) to first order in F and substituting in (6.61). The result is

$$\Delta V_{\text{C}} = \frac{\alpha}{15\pi m_{\text{e}}^2} Q_1 Q_2 \delta(\mathbf{r}) \tag{6.70}$$

This result was first obtained by Uehling [1] and by Serber [2]. See also Bjorken and Drell [3]. Its effect has been measured in the Lamb shift in atomic hydrogen.

At low temperatures, $T \ll |\mu|$, the functional form of (6.69) is not correct even at long distances; it turns out that it is not a good approximation to replace $F(0, \mathbf{k})$ by its infrared limit m_{el}^2 because of the sharp Fermi surface.

The formula (5.51) gives the matter part of F at one-loop order for arbitrary values of external energy, momentum, temperature, and chemical potential. Evaluating it at zero energy (which is the same as at zero Matsubara frequency) and $T = 0$ gives

$$
F_{\mathrm{mat}}(0, k)
$$
$$
= \frac{e^2}{24\pi^2} \left[16\mu k_{\mathrm{F}} - 4k^2 \ln \left(\frac{\mu + k_{\mathrm{F}}}{m} \right) - \frac{\mu(4\mu^2 - 3k^2)}{k} \ln \left(\frac{k - 2k_{\mathrm{F}}}{k + 2k_{\mathrm{F}}} \right)^2 \right.
$$
$$
+ \frac{(2m^2 - k^2)\sqrt{k^2 + 4m^2}}{k}
$$
$$
\left. \times \ln \left(\frac{2\mu^2(k^2 + 2m^2) - 2\mu k k_{\mathrm{F}} \sqrt{k^2 + 4m^2} - m^2(k^2 + 4m^2)}{2\mu^2(k^2 + 2m^2) + 2\mu k k_{\mathrm{F}} \sqrt{k^2 + 4m^2} - m^2(k^2 + 4m^2)} \right) \right]
$$
$$
\tag{6.71}
$$

Here $k_{\mathrm{F}} = \sqrt{\mu^2 - m^2}$ is the Fermi momentum and $k = |\mathbf{k}|$. The vacuum part is derived in many books on QED, such as Berestetskii, Lifshitz, and Pitaevskii [4] and Quigg [5]. It is

$$
F_{\mathrm{vac}}(0, k)
$$
$$
= -\frac{e^2}{4\pi^2} k^2 \left[\frac{4m^2}{3} \frac{M^2 - k^2}{M^2 k^2} \right.
$$
$$
+ \frac{1}{3M} \left(1 - \frac{2m^2}{M^2} \right) \sqrt{M^2 + 4m^2} \ln \left(\frac{\sqrt{M^2 + 4m^2} - M}{\sqrt{M^2 + 4m^2} + M} \right)
$$
$$
\left. - \frac{1}{3k} \left(1 - \frac{2m^2}{k^2} \right) \sqrt{k^2 + 4m^2} \ln \left(\frac{\sqrt{k^2 + 4m^2} - k}{\sqrt{k^2 + 4m^2} + k} \right) \right]
$$
$$
\tag{6.72}
$$

where M is an arbitrary subtraction point such that $F_{\mathrm{vac}}(0, M) = 0$.

The integrand in (6.68) has poles at $k = \pm i m_{\mathrm{el}} \approx \pm i \sqrt{F(0, k \to 0)}$. The contribution from these poles gives a Debye screening function of the form (6.69). The integrand also has a pair of branch points at $k = 2k_{\mathrm{F}} \pm i\varepsilon$ and a mirror pair at $k = -2k_{\mathrm{F}} \pm i\varepsilon$. The branch cuts can be taken to be vertical lines going up from the points $k = \pm 2k_{\mathrm{F}} + i\varepsilon$ and vertical lines going down from the points $k = \pm 2k_{\mathrm{F}} - i\varepsilon$. The contribution to the potential between point charges from these branch cuts is tedious but straightforward to evaluate. The result for asymptotically large r is

$$V(r) = \frac{Q_1 Q_2 e^2}{4\pi^3} \frac{\mu}{(4+a)^2} \left\{ \frac{m^2}{\mu^2} \frac{\cos 2k_{\mathrm{F}} r}{(k_{\mathrm{F}} r)^3} - \frac{\sin 2k_{\mathrm{F}} r}{(k_{\mathrm{F}} r)^4} \right.$$
$$\left. \times \left[\frac{e^2}{\pi^2} \frac{m^4}{\mu^3 k_{\mathrm{F}}} \frac{1}{4+a} \left(\ln(4 k_{\mathrm{F}} r) + \gamma_{\mathrm{E}} - \frac{3}{2} \right) - \frac{16}{4+a} \frac{m^2}{\mu^2} + \frac{m^4}{2\mu^4} - \frac{k_{\mathrm{F}}^2}{\mu^2} \right] \right\}$$
(6.73)

where $a = F(0, 2k_{\mathrm{F}})/k_{\mathrm{F}}^2$. The terms neglected in this expression are one order higher either in $1/r$ or in e^2. The contribution from the branch cuts dominates the Debye contribution at large r because the latter falls exponentially in r whereas the former falls as a power.

There are two especially interesting limits of this potential. Let us write $Q_i = Z_i e$. The nonrelativistic limit, $k_{\mathrm{F}} \ll m$, is

$$V(r) = \frac{Z_1 Z_2 e^2 \xi k_{\mathrm{F}}}{2\pi(4+\xi)^2} \frac{\cos(2 k_{\mathrm{F}} r)}{(k_{\mathrm{F}} r)^3}$$
(6.74)

with

$$\xi = \frac{e^2}{2\pi^2} \frac{m}{k_{\mathrm{F}}}$$

This form of screening is usually referred to as Friedel oscillation in low-temperature physics (Fetter and Walecka [6]) and can be observed in the nuclear magnetic resonance lines in dilute alloys [7].

The relativistic limit, $k_{\mathrm{F}} \gg m$, is

$$V(r) = Z_1 Z_2 \frac{\bar{\alpha}^2}{4\pi} \frac{\sin 2 k_{\mathrm{F}} r}{k_{\mathrm{F}}^3 r^4}$$
(6.75)

This involves the renormalization-group running coupling

$$\bar{\alpha} = \frac{\alpha}{1 - (2\alpha/3\pi) \ln(4\mu/eM)} = \frac{3\pi}{2 \ln(e \Lambda_{\mathrm{MOM}}/4\mu)}$$
(6.76)

where Λ_{MOM} is the QED scale parameter. This is familiar from (5.80)–(5.82). The relativistic results were obtained by Sivak [8] and by Kapusta and Toimela [9]. There may be applications to the dense matter present in white dwarf and neutron stars.

Finally, consider what happens at small but nonzero temperature, $T \ll |\mu|$. The sharp Fermi surface is smeared over a thickness T in the energy. Consequently, the branch cuts do not extend to the real axis, and the branch points are shifted by an amount $2\pi\mu T i/k_{\mathrm{F}}$. Then the asymptotic formula for the potential must be multiplied by the factor $\exp(-2\pi\mu T r/k_{\mathrm{F}})$. When $T^2 > e^2 k_{\mathrm{F}}^3/4\pi^4 \mu$ the contribution from the pole, $k \sim i m_{\mathrm{el}}$, begins to dominate the oscillating terms coming from the branch cuts. For a white dwarf star with $k_{\rm F} = 4m_e$ the crossover would be at 3×10^8 K or 30 keV.

6.5 Exact formula for screening length in QED

It is possible to derive an exact formula for the screening length of static electric fields in QED. This formula connects the screening length to the thermodynamic equation of state and so is a very interesting relation indeed.

An exact expression for the photon self-energy, known as the Schwinger–Dyson equation [10, 11], is

$$\Pi^{\mu\nu}(k) = e^2 T \sum_{n_p} \int \frac{d^3p}{(2\pi)^3} \, {\rm Tr}\left[\gamma^\mu \mathcal{G}(p) \Gamma^\nu(p, p-k) \mathcal{G}(p-k)\right]$$

(6.77)

Here the blobs on the fermion lines represent the exact fermion propagator $\mathcal{G}$, and the blob at the vertex represents the exact photon–fermion vertex function Γ^μ. The latter depends in general on the incoming fermion momentum p and the outgoing fermion momentum $p - k$. To lowest order, the photon–fermion vertex function is the point (or contact) coupling appearing in the Lagrangian,

$$\Gamma_0^\mu = \gamma^\mu \tag{6.78}$$

Corrections due to interactions may be found order by order, by applying the formula

$$-e\Gamma = (\delta \ln Z / \delta \Gamma_0)_{\rm 1PI} \tag{6.79}$$

which may be derived in a way analogous to (3.35). For example, from (5.39), (5.62), and (5.77) we obtain

$$-e\Gamma^\mu(p, p-k) = \quad \mu \; + \; \mu \; + \; \cdots \tag{6.80}$$

It should be clear intuitively that a relation exists between the fermion propagator and the photon–fermion vertex, since the latter represents the propagation of a fermion while emitting a photon of momentum k. To see what this relation might be, consider the free-fermion inverse propagator

$$\mathcal{G}_0^{-1}(p) = \not{p} - m \tag{6.81}$$

We notice that

$$\frac{\partial \mathcal{G}_0^{-1}}{\partial p_\mu} = \gamma^\mu = \Gamma_0^\mu \tag{6.82}$$

It turns out that the exact result is

$$\frac{\partial \mathcal{G}^{-1}}{\partial p_\mu} = \lim_{\delta_\mu \to 0} \Gamma^\mu(p, p - \delta_\mu) \tag{6.83}$$

where only the μ-component of the four-vector δ_μ is nonzero. Equation (6.83) is known as the differential form of Ward's identity. It relates the momentum derivative of the exact inverse fermion propagator to the exact photon–fermion vertex in the limit $k \to 0$.

The only change in the derivations of the Schwinger–Dyson equations and the Ward identity at $T > 0$ and $\mu \neq 0$ is the substitution of the frequency sums for energy integrals (the interested reader should consult Bjorken and Drell [3]; see also Fradkin [12], whose arguments we are following here).

At finite temperature and density, in the imaginary time formalism $p^0 = (2n_p + 1)\pi T i + \mu$. Thus from (6.83)

$$\frac{\partial \mathcal{G}^{-1}}{\partial \mu} = \Gamma^0(p, p) \tag{6.84}$$

The screening length follows from Π^{00} in the static infrared limit. Combining (6.77) and (6.84) yields

$$m_{\text{el}}^2 = -\Pi^{00}(k_0 = 0, \mathbf{k} \to 0)$$

$$= -e^2 T \sum_{n_p} \int \frac{d^3 p}{(2\pi)^3} \, \text{Tr} \left(\gamma^0 \mathcal{G}(p) \frac{\partial \mathcal{G}^{-1}}{\partial \mu}(p) \mathcal{G}(p) \right)$$

$$= e^2 \frac{\partial}{\partial \mu} T \sum_{n_p} \int \frac{d^3 p}{(2\pi)^3} \, \text{Tr} \left[\gamma^0 \mathcal{G}(p) \right]$$

$$= e^2 \left(\frac{\partial n}{\partial \mu} \right)_T = e^2 \frac{\partial^2 P(\mu, T)}{\partial \mu^2} \tag{6.85}$$

The electric screening length is directly related to the equation of state.

To see how remarkable (6.85) is, notice that the static infrared limit of the photon propagator at one-loop order is determined by the pressure of a *noninteracting* fermion gas. To show the power of (6.85) we recall the formula for $P(\mu, T)$ for a massless electron–positron plasma from Exercise 2.7 and from (5.60), (5.67), and (5.68). Since the pressure is known to order e^3 when both T and μ are nonzero, the inverse screening length is known to order e^5. For $\mu = 0$,

$$m_{\text{el}}^2 = \left(\frac{e^2}{3} - \frac{e^4}{8\pi^2} + \frac{e^5}{4\sqrt{3}\pi^3} + \cdots \right) T^2 \qquad (6.86)$$

This expression is phrased in terms of a fixed coupling constant e evaluated at a fixed scale. Let us denote that scale by M_0. At some other scale M the coupling constant changes to

$$e^2(M) = e^2(M_0) \left[1 + \frac{e^2(M_0)}{6\pi^2} \ln \left(\frac{M}{M_0} \right) \right] \qquad (6.87)$$

according to the renormalization group. Then

$$m_{\text{el}}^2 = \left\{ \frac{e^2(M)}{3} - \frac{e^4(M)}{18\pi^2} \left[\ln \left(\frac{M}{M_0} \right) + \frac{9}{4} \right] + \frac{e^5(M)}{4\sqrt{3}\pi^3} + \cdots \right\} T^2 \qquad (6.88)$$

The issue is how best to choose M and M_0 to minimize higher-order contributions. This may be resolved as follows.

Return to (6.68) and expand $F(0, \mathbf{k})$ in powers of $|\mathbf{k}|$, keeping terms up to and including $\mathbf{k}^2$. Including both the vacuum and finite-temperature parts, and using the above expression for $m_{\text{el}}^2(e)$, the integrand of (6.68) becomes

$$\frac{e^2}{m_{\text{el}}^2(e) + \{1 - (e^2/6\pi^2) [\ln(\pi T/M) + 4/3 - \gamma_{\text{E}}]\} \mathbf{k}^2}$$

If we use the electric screening mass to one-loop order only, this becomes

$$\frac{\bar{e}^2}{m_{\text{el}}^2(\bar{e}) + \mathbf{k}^2}$$

where

$$\bar{e}^2(T) = \frac{e^2}{1 - (e^2/6\pi^2) [\ln(\pi T/M) + 4/3 - \gamma_{\text{E}}]} = \frac{6\pi^2}{\ln \left(e^{\gamma_{\text{E}} - 4/3} \Lambda / \pi T \right)} \qquad (6.89)$$

The fixed coupling constant has been replaced by the renormalization-group running coupling with the absolute scale determined naturally. If we use the electric screening mass to order e^5 we get

$$m_{\text{el}}^2 = \left\{ \frac{e^2}{3} + \frac{e^4}{18\pi^2} \left[\ln \left(\frac{\pi T}{M} \right) - \gamma_{\text{E}} - \frac{11}{12} \right] + \frac{e^5}{4\sqrt{3}\pi^3} + \cdots \right\} T^2 \qquad (6.90)$$

Expressing e in terms of $\bar{e}$ gives *exactly* the formula (6.85) with the fixed coupling replaced by the running coupling:

$$m_{\text{el}}^2 = \left(\frac{\bar{e}^2}{3} - \frac{\bar{e}^4}{8\pi^2} + \frac{\bar{e}^5}{4\sqrt{3}\pi^3} + \cdots \right) T^2 \qquad (6.91)$$

This has been verified in an explicit diagrammatic analysis by Blaizot, Iancu, and Parwani [13] (the constant following the logarithm in this work is different, on account of the different renormalization schemes).

The relation (6.85) can be understood very simply. Insert a charge Q at location $\mathbf{x}$ in an electron–positron plasma. If the plasma has a charge density $-en$ then there must be a net background charge density to ensure charge neutrality. Denote this background charge density by en_0, so that in equilibrium $n = n_0$. Owing to the insertion of the charge Q there will be a rearrangement of electrons and positrons in the plasma. The condition of local hydrostatic equilibrium requires a balance of forces:

$$-\nabla P = en\mathbf{E}_{\text{net}} \tag{6.92}$$

Poisson's equation is

$$\nabla \cdot \mathbf{E}_{\text{net}} = [Q\delta(\mathbf{x}) - e(n - n_0)] \tag{6.93}$$

where $\mathbf{E}_{\text{net}}$ is the net electric field due to the external charge Q and the consequent rearrangement of the charged particles in the plasma. In equilibrium, T must be uniform but the charge chemical potential μ may depend on position. Thus we write $P = P(\mu, T)$, $T = \text{constant}$, $\mu = \mu(\mathbf{x})$, and seek to determine $\mu(\mathbf{x})$. Let μ_0 denote the chemical potential in the absence of the charge, and let $\delta\mu(\mathbf{x})$ denote the difference after the introduction of the charge. Then $\nabla P = (\partial P/\partial \mu)\nabla \delta\mu$, and $n - n_0 = (\partial n/\partial \mu)\delta\mu$. Taking the divergence of $\mathbf{E}_{\text{net}}$ in (6.92), identifying it with (6.93), and using the above information we arrive at the expressions

$$\left(\nabla^2 - m_{\text{el}}^2\right)\delta\mu = -eQ\delta(\mathbf{x})$$
$$m_{\text{el}}^2 = e^2\frac{\partial^2 P}{\partial \mu^2} \tag{6.94}$$

which have the solution

$$\delta\mu(r) = \frac{eQ}{4\pi r}\,\mathrm{e}^{-m_{\text{el}}r} \tag{6.95}$$

This is the Thomas–Fermi approximation. Clearly (6.95) is only valid for large r, since the derivation assumes that $|\delta\mu/\mu| \ll 1$. At short distances, the momentum dependence of $F(0, \mathbf{k})$ in (6.68) cannot be neglected and the Thomas–Fermi result is modified. This result is also incorrect for a cold Fermi gas, as already detailed in Section 6.4.

6.6 Collective excitations

Instead of applying a static external field, let us hit the system with an impulsive perturbation. Without loss of generality, we may focus on a single Fourier component. Thus, for the scalar field discussed in Section 6.1, we take

$$J(\mathbf{x}, t) = J_0(\mathbf{k})\, e^{i\mathbf{k}\cdot\mathbf{x}}\delta(t)$$
$$\tilde{J}(\omega, \mathbf{q}) = (2\pi)^3 J_0(\mathbf{k})\delta(\mathbf{q} - \mathbf{k}) \tag{6.96}$$

This leads to the field response

$$\delta\langle\hat{\phi}(\mathbf{x}, t)\rangle = J_0(\mathbf{k})\, e^{i\mathbf{k}\cdot\mathbf{x}} \int_{-\infty}^{\infty} \frac{d\omega}{2\pi} e^{-i\omega t}\, D^{\mathrm{R}}(\omega, \mathbf{k}) \tag{6.97}$$

The retarded Green function is analytic in the upper half-plane. Suppose that it has a simple pole located at $\omega = \omega(\mathbf{k}) - i\gamma(\mathbf{k})$ with $\gamma(\mathbf{k}) \geq 0$. Then

$$D^{\mathrm{R}}(\omega, \mathbf{k}) = \frac{R(\mathbf{k})}{\omega - \omega(\mathbf{k}) + i\gamma(\mathbf{k})} \tag{6.98}$$

where $R(\mathbf{k})$ is the residue. Evaluation of (6.97) leads to

$$\delta\langle\hat{\phi}(\mathbf{x}, t)\rangle = -i J_0(\mathbf{k}) R(\mathbf{k})\, e^{i(\mathbf{k}\cdot\mathbf{x} - \omega(\mathbf{k})t)}\, e^{-\gamma(\mathbf{k})t} \tag{6.99}$$

The field response is a traveling wave with dispersion relation $\omega(\mathbf{k})$ and damping constant $\gamma(\mathbf{k})$.

For a free field with mass m, $\omega(\mathbf{k}) = \sqrt{\mathbf{k}^2 + m^2}$, $\gamma(\mathbf{k}) = 0$, and $R(\mathbf{k}) = \frac{1}{2}\omega(\mathbf{k})$. The amplitude of the wave is proportional to the residue and to the Fourier amplitude of the impulse.

6.7 Photon dispersion relation

Let us consider a QED plasma at such high temperature or density that the electron mass may be neglected. Based on the previous discussion, we would expect that the poles of the photon propagator would give the dispersion relations for traveling electromagnetic waves in the plasma. However, this requires careful consideration owing to the fact that the photon propagator is gauge dependent.

Transverse oscillations have a dispersion relation determined by

$$k_0^2 = \mathbf{k}^2 + G(k_0, \mathbf{k}) \tag{6.100}$$

in the temporal axial gauge (6.58), in the Coulomb gauge (6.59), and in the covariant gauges (5.46). We write $k_0 = \omega - i\gamma$ and assume weak damping, otherwise the oscillations would not propagate. The above equation can be decomposed into real and imaginary parts:

$$\omega^2 = \mathbf{k}^2 + \operatorname{Re} G(\omega, \mathbf{k})$$
$$\gamma = -\frac{\operatorname{Im} G(\omega, \mathbf{k})}{2\omega} \tag{6.101}$$

Even at one-loop order, $G(k_0, \mathbf{k})$ is a complicated function. In general, the solutions can only be found numerically. In the limit of short or long wavelengths, however, analytical results may be found.

For short wavelengths we expect that the modification of the free-photon dispersion relation $\omega = |\mathbf{k}|$ by medium effects will be small. The reason is that if we probe the system at distances considerably less than the average interparticle spacing then medium effects should tend to zero. Thus we look for a solution when $\omega \approx |\mathbf{k}| \gg T, |\mu|$. From Exercise 5.7 we know that

$$G(\omega = |\mathbf{k}|) = \frac{1}{2}e^2 \left(\frac{1}{3}T^2 + \frac{\mu^2}{\pi^2} \right) \equiv m_{\mathrm{P}}^2 \qquad (6.102)$$

which is precisely $\frac{1}{2}m_{\mathrm{el}}^2 = \frac{1}{2}F(k_0 = 0, \mathbf{k} \to 0)$ to order e^2. The short-wavelength dispersion relation is then

$$\omega^2 = \mathbf{k}^2 + m_{\mathrm{P}}^2 + \cdots \qquad (6.103)$$

Clearly this is a gauge invariant result. One may think of the high-momentum photons as having acquired a mass m_{P} due to plasma interactions.

For the long-wavelength transverse oscillations, we expect a substantial modification of the free-photon dispersion relation owing to many-body effects. The oscillatory electric and magnetic fields will cause any nearby electrons and positrons to be accelerated, giving the oscillation inertia. In fact, one might expect that it would take a finite amount of energy to excite an oscillation with vanishing momentum. To look for a solution to (6.100) and (6.101) we calculate G in the limit $|k^2| = |k_0^2 - \mathbf{k}^2| \ll T^2$. The functions F and G may be obtained from a combination of (5.46), (5.48), and (5.51). This is a straightforward calculation leading to [14]

$$G(k_0, \mathbf{k}) = m_{\mathrm{P}}^2 - \frac{1}{2}F(k_0, \mathbf{k}) \qquad (6.104)$$

and

$$F(k_0, \mathbf{k}) = -2m_{\mathrm{P}}^2 \frac{k^2}{|\mathbf{k}|^2} \left[1 - \frac{k_0}{4|\mathbf{k}|} \ln \left(\frac{k_0 + |\mathbf{k}|}{k_0 - |\mathbf{k}|} \right)^2 \right] \qquad (6.105)$$

It must be emphasized that these expressions are valid not just for the case where $|k_0|$ and $|\mathbf{k}|$ are small compared with T but also near the light cone. In fact, note that $G(k^2 = 0) = m_{\mathrm{P}}^2$, the same as the limit obtained from the exact one-loop expression for G.

For small momenta we find that

$$G(|\mathbf{k}| \ll \omega < T, |\mu|) = \omega_{\mathrm{P}}^2 \left(1 + \frac{\mathbf{k}^2}{5\omega^2} + \cdots \right) \qquad (6.106)$$

The plasma frequency ω_P is related to the electric mass and the photon mass via $\omega_P^2 = \frac{1}{3}m_{el}^2 = \frac{2}{3}m_P^2$ at order e^2 when the electron mass is set to zero. The long-wavelength dispersion relation for transverse excitations is

$$\omega^2 = \omega_P^2 + \tfrac{6}{5}\mathbf{k}^2 + \cdots \qquad (6.107)$$

Indeed, it does take a finite energy to excite one of these modes even at zero momentum.

Next we turn to longitudinal oscillations, or compressional charge-density waves. Without doing the full linear response analysis in each gauge, we would expect that the dispersion relation is determined by the poles of the following functions in the specified gauge:

temporal axial

$$\frac{1}{k^2 - F} \frac{k^2}{k_0^2}$$

Coulomb

$$\frac{1}{k^2 - F} \frac{k^2}{\mathbf{k}^2}$$

covariant

$$\frac{1}{k^2 - F} \qquad (6.108)$$

Some of the subtleties involved in gauge invariance now arise.

Consider the limit of no interactions. Then $F = 0$, and the covariant gauges produce the spectrum $\omega = |\mathbf{k}|$, whereas in the temporal axial and Coulomb gauges there is no wave propagation. This could have been anticipated. Free electromagnetic radiation is transversely polarized. The temporal axial and Coulomb gauges are physical gauges in the sense that they have the correct number of polarization degrees of freedom, namely, two. The covariant gauges are unphysical in the same sense since they have four degrees of freedom. The extra two degrees of freedom are canceled by the ghosts in the partition function. There is nothing wrong in all this, but one must be careful to ask only physical questions of the theory. The situation is not altered when interactions are turned on at $T = \mu = 0$. Recall that $F = (k^2/\mathbf{k}^2)\Pi_{00}$. It turns out that Π_{00} is not singular enough at $k^2 = 0$ to cancel the factor of k^2. For example, to order e^2,

$$F_{vac} = \frac{\alpha}{3\pi}k^2 \ln\left(\frac{-k^2}{M^2}\right) \qquad (6.109)$$

The covariant gauges still have a singularity at $k^2 = 0$, a branch point due to pair production, while the other two gauges do not. The conclusion is that short-wavelength longitudinal excitations do not propagate.

The spectrum of long-wavelength longitudinal excitations in the plasma is manifestly gauge invariant and is determined by

$$k_0^2 = \mathbf{k}^2 + F(k_0, \mathbf{k}) \tag{6.110}$$

or equivalently

$$\mathbf{k}^2 = \Pi_{00}(k_0, \mathbf{k}) \tag{6.111}$$

Decomposing into real and imaginary parts gives

$$\mathbf{k}^2 = \operatorname{Re}\Pi_{00}(\omega, \mathbf{k})$$
$$\gamma_{\mathrm{L}} = \frac{\operatorname{Im}\Pi_{00}(\omega, \mathbf{k})}{\partial \operatorname{Re}\Pi_{00}(\omega, \mathbf{k})/\partial\omega} \tag{6.112}$$

Expanding Π_{00} in powers of $\mathbf{k}^2/k_0^2$ leads to

$$\Pi_{00}(\omega, \mathbf{k}) = \omega_{\mathrm{P}}^2\left(1 + \frac{3\mathbf{k}^2}{5\omega^2} + \cdots\right)\frac{\mathbf{k}^2}{\omega^2} \tag{6.113}$$

and finally to the dispersion relation

$$\omega^2 = \omega_{\mathrm{P}}^2 + \tfrac{3}{5}\mathbf{k}^2 + \cdots \tag{6.114}$$

The energy at zero momentum for longitudinal and transverse excitations is the same, which is no surprise since at zero momentum there is no distinction between longitudinal and transverse modes.

At arbitrary momentum, the dispersion relation cannot be obtained by analytic means for either mode. They must be found by numerical methods.

It is interesting that the damping constants as determined by the approximate expressions (6.104) and (6.105) are zero. However, if one returns to the exact one-loop expressions for F and G it turns out that

$$\gamma_{\mathrm{T}} = \gamma_{\mathrm{L}} = \frac{e^2}{24\pi}\omega_{\mathrm{P}} \tag{6.115}$$

at zero momentum. The origins of the various factors in this result are not difficult to find. The factor e^2 comes from the square of the photon–electron or photon–positron vertex and the factor ω_{P} comes from phase space.

Lastly, notice that the propagator in the covariant gauges has a term $\rho k^\mu k^\nu/k^2$. Clearly, no physical significance should be attached to this pole since the residue is proportional to the gauge parameter and in fact vanishes in the Landau gauge $\rho = 0$.

6.8 Electron dispersion relation

The electron propagator is

$$\mathcal{G}(p) = \frac{1}{\not{p} - m_{\mathrm{e}} + \Sigma(p)} \tag{6.116}$$

In the Feynman gauge the one-loop expression for the self-energy is

$$\Sigma(p) = e^2 T^2 \sum_{n_k} \sum_{n_q} \int \frac{d^3k}{(2\pi)^3} \frac{d^3q}{(2\pi)^3} \frac{1}{k^2} \gamma^\mu \frac{1}{\not{q} - m_{\mathrm{e}}} \gamma_\mu \beta \delta_{n_p, n_k + n_q} (2\pi)^3 \delta(\mathbf{p} - \mathbf{k} - \mathbf{q}) \tag{6.117}$$

At very high temperature the electron mass may be neglected. The evaluation of this self-energy is rather tedious but straightforward. The vacuum contribution may be found in numerous textbooks. Here we shall focus on the matter contribution.

The leading contribution at order T^2 and μ^2 is [15]

$$\Sigma_{\mathrm{mat}}^0 = -\frac{m_{\mathrm{F}}^2}{8|\mathbf{p}|} \ln \left(\frac{p_0 + |\mathbf{p}|}{p_0 - |\mathbf{p}|} \right)^2$$

$$\boldsymbol{\Sigma}_{\mathrm{mat}} = \frac{m_{\mathrm{F}}^2}{2|\mathbf{p}|^2} \mathbf{p} \left[1 - \frac{p_0}{4|\mathbf{p}|} \ln \left(\frac{p_0 + |\mathbf{p}|}{p_0 - |\mathbf{p}|} \right)^2 \right] \tag{6.118}$$

where $m_{\mathrm{F}}^2 = \frac{1}{2}(m_{\mathrm{P}}^2 + \frac{1}{3}e^2 T^2) = \frac{1}{4}e^2(T^2 + \mu^2/\pi^2)$. Equations (6.118) may be compared with the corresponding expressions for F and G for the photon self-energy. Although the electron self-energy is in general gauge dependent, the leading contributions (6.118) can be shown to be independent of the gauge. As with the photon self-energy, it must be emphasized that these expressions are valid not only for small electron energy and momentum but also near the light cone at any momentum.

The poles of the propagator are determined by

$$[p_0 + \Sigma_{\mathrm{mat}}^0(p_0, \mathbf{p})]^2 = [\mathbf{p} + \boldsymbol{\Sigma}_{\mathrm{mat}}(p_0, \mathbf{p})]^2 \tag{6.119}$$

There are two undamped solutions to this equation, referred to as $\omega_+(\mathbf{p})$ and $\omega_-(\mathbf{p})$. They can be expressed in parametric form as

$$\omega_\pm^2(\mathbf{p}) = z^2 \mathbf{p}_\pm^2(z)$$

$$\mathbf{p}_\pm^2(z) = \omega_{\mathrm{F}}^2 \left[\frac{\pm 1}{z \mp 1} \mp \frac{1}{2} \ln \left(\frac{z+1}{z-1} \right) \right] \tag{6.120}$$

with $\omega_{\mathrm{F}}^2 = \frac{1}{2}m_{\mathrm{F}}^2$ and $z > 1$. The two solutions are shown in Figure 6.1.

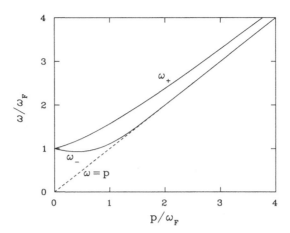

Fig. 6.1. The two branches ($\omega_{\pm}$) of the electron dispersion relation are shown. For comparison, the dispersion relation of a massless particle is also plotted (broken line).

At momenta that are high in comparison with T and μ, the solutions become

$$
\begin{aligned}
\omega_+^2 &= \mathbf{p}^2 + m_{\mathrm{F}}^2 + \cdots \\
\omega_-^2 &= \mathbf{p}^2 + 4\mathbf{p}^2 \exp\left(-\frac{4\mathbf{p}^2}{m_{\mathrm{F}}^2} - 1\right) + \cdots
\end{aligned}
\tag{6.121}
$$

whereas at low momentum the solutions become

$$
\omega_{\pm} = \omega_{\mathrm{F}} \pm \frac{1}{3}|\mathbf{p}| + \frac{1}{3}\frac{\mathbf{p}^2}{\omega_{\mathrm{F}}} + \cdots
\tag{6.122}
$$

The low-momentum spectra have an optical character. The high-momentum spectrum for ω_+ solution may be used to define a finite-temperature and finite-density fermion mass, just as the photon mass was defined at high momentum.

It is interesting to examine the behavior of the propagator in the vicinity of the poles. In the high-momentum limit,

$$
\begin{aligned}
\mathcal{G}(\omega \to \omega_+, p \gg m_{\mathrm{F}}) &\approx \frac{1}{2}\frac{\gamma^0 - \hat{\mathbf{p}}\cdot\boldsymbol{\gamma}}{\omega - \omega_+} \\
\mathcal{G}(\omega \to \omega_-, p \gg m_{\mathrm{F}}) &\approx \frac{2\mathbf{p}^2}{m_{\mathrm{F}}^2}\exp\left(-\frac{4\mathbf{p}^2}{m_{\mathrm{F}}^2} - 1\right)\frac{\gamma^0 + \hat{\mathbf{p}}\cdot\boldsymbol{\gamma}}{\omega - \omega_-}
\end{aligned}
\tag{6.123}
$$

and in the low-momentum limit

$$\mathcal{G}(\omega \to \omega_+, p \ll m_{\mathrm{F}}) \approx \frac{4}{3} \frac{\gamma^0 - \hat{\mathbf{p}} \cdot \boldsymbol{\gamma}}{\omega - \omega_+}$$

$$\mathcal{G}(\omega \to \omega_-, p \ll m_{\mathrm{F}}) \approx \frac{4}{3} \frac{\gamma^0 + \hat{\mathbf{p}} \cdot \boldsymbol{\gamma}}{\omega - \omega_-} \tag{6.124}$$

These display a number of features. The ω_+ solution has the same relation between chirality and helicity as free electrons whereas the ω_- solution has the opposite relation between chirality and helicity. This is true for all momenta, not just in the limits. It suggests that the ω_+ branch represents the modification of the dispersion relation of a real electron in the plasma, whereas the ω_- branch is a true collective excitation. Indeed, the residue of that branch vanishes as the momentum becomes large, which is the vacuum limit. The residue of the ω_+ branch in the high-momentum limit is the same as for free electrons. Finally, notice that the residues are the same as the momentum tends to zero since there is no distinction between different polarizations when the particle is at rest.

6.9 Kubo formulae for viscosities and conductivities

Many physical systems can be described using fluid dynamics. In the context of this book, examples of such systems are stars, the early universe and, to some extent, high-energy nuclear collisions. The state of the fluid can be described in terms of its temperature and chemical potentials, specified as functions of space and time, together with an equation of state. The dynamics of the fluid is described by equations of motion based on the energy–momentum tensor $T^{\mu\nu}(x)$. Here $x^\mu = (t, \mathbf{x})$. The local energy density is T^{00}, the local momentum density is T^{i0}, and the flux of these quantities in the direction j is $T^{\mu j}$. Local conservation of energy and momentum is expressed as

$$\partial_\nu T^{\mu\nu} = 0 \tag{6.125}$$

This conservation law is general and makes no assumption about local equilibrium or the types of interactions. Without loss of generality the energy–momentum tensor can always to taken to be symmetric. As a concrete example, the energy–momentum tensor for a set of N noninteracting particles labeled by index n is

$$T^{\mu\nu}(x) = \sum_{n=1}^{N} \frac{p_n^\mu p_n^\nu}{E_n} \delta(\mathbf{x} - \mathbf{x}_n(t)) \tag{6.126}$$

where $\mathbf{x}_n(t)$ is the trajectory of the nth particle.

In a field theory with independent fields labeled ϕ_n and Lagrangian $\mathcal{L}$ the energy–momentum tensor is found in the usual way to be

$$T^{\mu\nu} = \sum_n \frac{\partial \mathcal{L}}{\partial_\mu \phi_n} \partial^\nu \phi_n - g^{\mu\nu} \mathcal{L} \qquad (6.127)$$

Specific examples include a self-interacting scalar field

$$T^{\mu\nu} = \partial^\mu \phi \, \partial^\nu \phi - g^{\mu\nu} \mathcal{L} \qquad (6.128)$$

and the electromagnetic field

$$T^{\mu\nu} = F^\mu_{\ \rho} F^{\rho\nu} + \tfrac{1}{4} g^{\mu\nu} F^{\alpha\beta} F_{\alpha\beta} \qquad (6.129)$$

To evaluate these in a classical field theory, the solutions to the field equations are inserted into these expressions. In a quantum theory the fields are operators and the expressions are therefore also operators. In a fluid, the expressions may be averaged over spacetime volumes that are large compared with typical thermal wavelengths and correlation lengths but small compared with distances and times over which local energy and momentum densities vary appreciably; this averaging process is referred to as coarse-graining.

Coarse-graining is easy to describe but usually difficult to implement. It can be done in numerical simulations, of course. In a hydrodynamic or perfect-fluid description, the assumption of local thermal equilibrium is made. Then the energy–momentum tensor is

$$T^{\mu\nu} = -Pg^{\mu\nu} + wu^\mu u^\nu \qquad (6.130)$$

where P is the local pressure, $w = \epsilon + P$ is the local enthalpy density, and $u^\mu = (\gamma, \gamma \mathbf{v})$ is the local flow velocity relative to some fixed reference frame. In a frame in which the fluid is locally at rest, $u^\mu = (1, 0, 0, 0)$, $T^{00} = \epsilon$, $T^{ij} = P\delta_{ij}$, and $T^{i0} = 0$. In general the trace of the energy–momentum tensor is $T^\mu_{\ \mu} = \epsilon - 3P$. For a noninteracting gas of massless particles, $\tfrac{1}{3}\epsilon = P$ and the trace vanishes. If there are conserved charges, such as baryon number or electric charge, there is an additional conservation law or equation of motion for each. For example, the baryon current is

$$J^\mu_B = n_B u^\mu \qquad (6.131)$$

where $n_B = J^0_B$ is the local baryon density. The conservation law is

$$\partial_\mu J^\mu_B = 0 \qquad (6.132)$$

Note that the baryon number flows with the same four-velocity as appeared in the energy–momentum tensor. The local pressure, energy, and baryon densities are related through the equation of state. Equivalently they can all be expressed in terms of T and μ_B.

When variations in temperature and chemical potential become appreciable over length scales that are not large compared with thermal wavelengths or correlation lengths then gradients in the thermodynamic variables must be taken into account. In a typical nonrelativistic fluid the massive particles carry the energy and momentum so that energy, momentum, and baryon number all flow together with only very minor departures associated with thermal conductivity. In a relativistic fluid, meaning one in which P is not much less than ϵ, or equivalently in which the temperature and chemical potential are not much less than the mass of the particles, the situation is more complicated. The energy and momentum may flow with a velocity different from that of the baryons if the system has gradients that are not negligibly small. The situation is then described in terms of (first-order) relativistic viscous-fluid dynamics. Dissipative contributions are added to the energy–momentum tensor:

$$T^{\mu\nu} = -Pg^{\mu\nu} + wu^{\mu}u^{\nu} + \Delta T^{\mu\nu}$$
$$J_B^{\mu} = n_B u^{\mu} + \Delta J_B^{\mu}$$

(6.133)

The dissipative terms are proportional to first-order derivatives of the flow velocity, temperature, and chemical potential. There are two common definitions of the flow velocity in relativistic dissipative fluid dynamics.

In the Eckart approach u^{μ} is the velocity of baryon number flow [16]. The dissipative terms must satisfy the conditions $\Delta J_B^{\mu} = 0$ and $u_{\mu}u_{\nu}\Delta T^{\mu\nu} = 0$, the latter following from the requirement that T^{00} be the energy density in the local (baryon) rest frame. The most general form of $\Delta T^{\mu\nu}$ is given by

$$\Delta T^{\mu\nu} = \eta(\Delta^{\mu}u^{\nu} + \Delta^{\nu}u^{\mu}) + \left(\tfrac{2}{3}\eta - \zeta\right)H^{\mu\nu}\partial_{\rho}u^{\rho}$$
$$- \chi(H^{\mu\alpha}u^{\nu} + H^{\nu\alpha}u^{\mu})Q_{\alpha}$$

(6.134)

Here

$$H^{\mu\nu} = u^{\mu}u^{\nu} - g^{\mu\nu}$$

(6.135)

is a projection tensor normal to u^{μ},

$$\Delta_{\mu} = \partial_{\mu} - u_{\mu}u^{\beta}\partial_{\beta}$$

(6.136)

is a derivative normal to u^{μ}, and

$$Q_{\alpha} = \partial_{\alpha}T - Tu^{\rho}\partial_{\rho}u_{\alpha}$$

(6.137)

is the heat flow vector, whose nonrelativistic limit is $\mathbf{Q} = -\nabla T$. Furthermore, η is the shear viscosity, ζ is the bulk viscosity, and χ is the thermal conductivity. The entropy current is

$$s^{\mu} = su^{\mu} + \frac{1}{T}u_{\nu}\Delta T^{\mu\nu}$$

(6.138)

and is defined is such a way that $u_\mu s^\mu = s$, the local entropy density. Its divergence is

$$\partial_\mu s^\mu = \frac{\eta}{2T} \left(\partial_i w^j + \partial_j u^i - \frac{2}{3}\delta^{ij}\nabla\cdot\mathbf{u} \right)^2$$
$$+ \frac{\zeta}{T}\left(\nabla\cdot\mathbf{u}\right)^2 + \frac{\chi}{T^2}\left(\nabla T + T\dot{\mathbf{u}}\right)^2 \qquad (6.139)$$

All three dissipation coefficients must be non-negative to ensure that entropy can never decrease.

In the Landau–Lifshitz approach, u^μ is the velocity of energy transport. The dissipative part of the energy–momentum tensor satisfies $u_\mu\Delta T^{\mu\nu} = 0$, and ΔJ_B^μ is not constrained to be zero. In this case the most general form of the energy–momentum tensor is

$$\Delta T^{\mu\nu} = \eta(\Delta^\mu u^\nu + \Delta^\nu u^\mu) + \left(\tfrac{2}{3}\eta - \zeta\right)H^{\mu\nu}\partial_\rho u^\rho \qquad (6.140)$$

The baryon current is modified to

$$\Delta J_B^\mu = \chi\left(\frac{n_B T}{w}\right)^2 \Delta^\mu\left(\frac{\mu_B}{T}\right) \qquad (6.141)$$

The three coefficients η, ζ, and χ are the same as in the Eckart approach. This can be proven in a variety of ways. For example, even though the entropy current in this approach is different, being

$$s^\mu = su^\mu - \frac{\mu_B}{T}\Delta J_B^\mu \qquad (6.142)$$

its divergence is the same. Physical, observable, results cannot depend on how one defines the frame of reference.

In the above approaches the dissipative coefficients are taken to be phenomenological constants or, rather, functions of temperature and chemical potential. However, it ought to be possible to derive them from the microscopic theory. In particular, it ought to be possible to derive them using linear response theory since departures from local thermal equilibrium are assumed to be small. Indeed this is so, and the resulting formulae are named after Kubo [17].

Consider the problem of pure baryon number diffusion in the absence of energy flow. The most direct approach to use in this case is that of Landau and Lifshitz: the vanishing of the energy flux implies that the flow velocity is zero. The equation of continuity for the baryon current, including dissipation, reduces to a diffusion equation for the baryon chemical potential:

$$\partial\mu_B/\partial t = D\,\nabla^2\mu_B \qquad (6.143)$$

Here

$$D \equiv \frac{\chi T}{dn_B/d\mu_B} \left(\frac{n_B}{w}\right)^2 \tag{6.144}$$

is the diffusion constant. A single Fourier mode $\exp[i(\mathbf{k} \cdot \mathbf{x} - \omega t)]$ will relax towards equilibrium as $\exp(-D k^2 t)$.

A nonuniform baryon distribution can be achieved by the imposition of an external force that is turned on and off, allowing the system to relax back towards equilibrium. It does not matter how this is done. For example, we could take the coupling Hamiltonian to be

$$H_{\text{ext}}(t) = \int d^3x \, \hat{J}_B^\mu(\mathbf{x}, t) J_\mu^{\text{ext}}(\mathbf{x}, t) \tag{6.145}$$

where J_μ^{ext} is an external perturbing current. The response of the baryon current is given in the usual way by

$$\delta \langle \hat{J}_B^\mu(\omega, \mathbf{k}) \rangle = J_\nu^{\text{ext}}(\omega, \mathbf{k}) B_R^{\mu\nu}(\omega, \mathbf{k}) \tag{6.146}$$

where $B_R^{\mu\nu}(\omega, \mathbf{k})$ is the Fourier transform of the retarded current–current correlation or response function:

$$i B_R^{\mu\nu}(\mathbf{x}, t; \mathbf{x}', t') = \left\langle \left[\hat{J}_B^\mu(\mathbf{x}, t), \, \hat{J}_B^\nu(\mathbf{x}', t') \right] \right\rangle \theta(t - t') \tag{6.147}$$

Since baryon number is conserved the most general form of the response function is

$$B_R^{\mu\nu} = B_L P_L^{\mu\nu} + B_T P_T^{\mu\nu} \tag{6.148}$$

where B_L and B_T are longitudinal and transverse response functions. Without loss of generality it is convenient to parametrize the longitudinal response function, or equivalently the time–time component, as

$$B_R^{00}(\omega, \mathbf{k}) = \frac{\mathbf{k}^2}{k^2} B_L(\omega, \mathbf{k}) = \frac{i\mathbf{k}^2 D(\omega, \mathbf{k})}{\omega + i\mathbf{k}^2 D(\omega, \mathbf{k})} B_R^{00}(\omega = 0, \mathbf{k}) \tag{6.149}$$

Here $D(\omega, \mathbf{k})$ is an unknown function. It is expected to be a smooth function of ω and $\mathbf{k}$, whereas the response function itself is expected to have singularities, usually poles. If we define $D \equiv D(\omega \to 0, \mathbf{k} \to 0)$, and if there is a slow perturbing variation in the baryon density, then the density will relax back towards equilibrium with a dispersion relation determined by the pole of the response function, namely, $\omega = -iD\mathbf{k}^2$. Therefore we may identify this D with the diffusion constant in the dissipative fluid dynamics calculation.

The diffusion constant can be extracted directly from the response function. First,

$$D = \lim_{\omega \to 0} \lim_{\mathbf{k} \to 0} \frac{i}{\omega} \frac{B_L(\omega, \mathbf{k})}{B_L(0, \mathbf{k})} \tag{6.150}$$

Now $B_{\mathrm{L}}(\omega = 0, \mathbf{k} \to 0) = -B_{\mathrm{R}}^{00}(\omega = 0, \mathbf{k} \to 0) = \partial^2 P/\partial\mu_B^2 = \partial n_B/\partial\mu_B$. (The reasoning is the same as for the electric screening mass.) Furthermore,

$$B_{\mathrm{L}}(\omega, |\mathbf{k}| \to 0) = \hat{k}^i \hat{k}^j \, B_{\mathrm{R}}^{ij}(\omega, |\mathbf{k}| \to 0) \tag{6.151}$$

where $\hat{k}^i = k^i/|\mathbf{k}|$ is a unit vector in the direction of $\mathbf{k}$. Putting all this together and using the rotational symmetry yields a linear response formula for the thermal conductivity:

$$\chi T = \frac{1}{3}\left(\frac{w}{n_B}\right)^2 \lim_{\omega \to 0} \frac{1}{\omega} \int d^4 x \, e^{i\omega t} \left\langle \left[\hat{J}_B^i(t, \mathbf{x}), \, \hat{J}_B^i(0, \mathbf{0})\right]\right\rangle \theta(t) \tag{6.152}$$

The factor $(w/n_B)^2$ arises in the conversion of baryon current to enthalpy current. Alternatively, (6.152) could be written in terms of the spectral densities for the longitudinal part of the baryon response function as

$$\chi T = \frac{1}{3}\left(\frac{w}{n_B}\right)^2 \lim_{\omega \to 0} \frac{1}{\omega}\rho_{\mathrm{L}}^{\mathrm{n}}(\omega, |\mathbf{k}| = 0)$$

$$= \frac{1}{3T}\left(\frac{w}{n_B}\right)^2 \lim_{\omega \to 0} \rho_{\mathrm{L}}^{+}(\omega, |\mathbf{k}| = 0) \tag{6.153}$$

The latter equality follows from the relation $\rho^{\mathrm{n}} = (1 - e^{-\beta\omega})\rho^+$, as discussed in Section 6.2.

There are Kubo-type linear-response expressions for the viscosities too. These may be derived in a way analogous to that for the thermal conductivity since $T^{\mu\nu}$ may be viewed as representing a set of four conserved currents. One obtains

$$\eta = \frac{1}{20} \lim_{\omega \to 0} \frac{1}{\omega} \int d^4 x \, e^{i\omega t} \left\langle \left[\mathcal{S}^{ij}(t, \mathbf{x}), \, \mathcal{S}^{ij}(0, \mathbf{0})\right]\right\rangle \theta(t) \tag{6.154}$$

$$\zeta = \frac{1}{2} \lim_{\omega \to 0} \frac{1}{\omega} \int d^4 x \, e^{i\omega t} \langle[\mathcal{P}(t, \mathbf{x}), \, \mathcal{P}(0, \mathbf{0})]\rangle \theta(t) \tag{6.155}$$

where $\mathcal{P} = -\frac{1}{3}T^i{}_i$ represents the trace of the momentum tensor (the pressure in equilibrium) and $\mathcal{S}^{ij} = T^{ij} - \delta^{ij}\mathcal{P}$ represents the traceless part. These follow from the dispersion relation for the transverse part of the momentum density,

$$\omega = -iD_{\mathcal{S}}\mathbf{k}^2 \tag{6.156}$$

where $D_{\mathcal{S}} = \eta/w$, and from the dispersion relation for pressure waves,

$$\omega^2 - v_{\mathcal{P}}^2\mathbf{k}^2 + iD_{\mathcal{P}}\omega\mathbf{k}^2 = 0 \tag{6.157}$$

where $D_\mathcal{P} = (\frac{4}{3}\eta + \zeta)/w$ (when the thermal conductivity is neglected). In terms of the spectral densities we have

$$\eta = \frac{1}{20} \lim_{\omega \to 0} \frac{1}{\omega} \rho^{\mathrm{n}}_{\mathcal{S}\mathcal{S}}(\omega, |\mathbf{k}| = 0) = \frac{1}{20T} \lim_{\omega \to 0} \rho^{+}_{\mathcal{S}\mathcal{S}}(\omega, |\mathbf{k}| = 0) \quad (6.158)$$

$$\zeta = \frac{1}{2} \lim_{\omega \to 0} \frac{1}{\omega} \rho^{\mathrm{n}}_{\mathcal{P}P}(\omega, |\mathbf{k}| = 0) = \frac{1}{2T} \lim_{\omega \to 0} \rho^{+}_{\mathcal{P}P}(\omega, |\mathbf{k}| = 0) \quad (6.159)$$

It is worth noting that in all these formulae the relevant transport coefficient is proportional to a diffusion constant with dimensions of length. In a multicomponent fluid those particles or fields with the longest diffusion length tend to dominate the transport coefficient.

In a similar manner one may derive an expression for the electrical conductivity, which is the coefficient in Ohm's law $\mathbf{J}_{\mathrm{EM}} = \sigma_{\mathrm{el}} \mathbf{E}$:

$$\sigma_{\mathrm{el}} = \frac{1}{6} \lim_{\omega \to 0} \frac{1}{\omega} \int d^4 x \, e^{i\omega t} \left\langle \left[\hat{J}^i_{\mathrm{EM}}(t, \mathbf{x}), \, \hat{J}^i_{\mathrm{EM}}(0, \mathbf{0}) \right] \right\rangle \theta(t) \quad (6.160)$$

This may also be expressed in terms of the corresponding spectral density.

The viscosities in $\lambda \phi^4$ theory have been calculated by Jeon [18] and Jeon and Yaffe [19]. In the limit of weak coupling and high temperature, the shear viscosity is

$$\eta = 5.28 T^3 / \lambda^2 \quad (6.161)$$

The parametric dependence of η on T and λ is straightforward. Recall that $\eta = w D_\mathcal{S}$. A diffusion constant may be estimated as $n\langle \sigma v \rangle$, where n is an average density, σ is a cross section, and v is the speed of the particles. For massless, or nearly massless, particles, $v \approx 1$, $n \propto T^3$, and $w \propto T^4$. The thermally averaged elastic cross section in $\lambda \phi^4$ theory is proportional to λ^2 / T^2. Putting this all together yields the estimate $\eta \propto T^3 / \lambda^2$, in agreement with the result quoted above. However, to calculate the overall coefficient is not easy. This may be seen immediately by the inverse dependence of η on λ. An infinite set of ladder diagrams corresponding to elastic scattering must be summed along with finite-temperature self-energy insertions. The calculation is ultimately reduced to a single integral equation that is solved numerically. The bulk viscosity for point particles with no internal degrees of freedom undergoing local interactions is generally much smaller than the shear viscosity. For the $\lambda \phi^4$ theory the bulk viscosity is nonzero at high temperature because of inelastic scatterings. When these are taken into account it is found that

$$\zeta = 0.002\,14\,\lambda \ln^2(1.55\lambda)\, T^3 \quad (6.162)$$

The ratio of the two viscosities $\zeta/\eta = \lambda^3 \ln^2(1.55\lambda)/2470$. For $\lambda = 1/10$ the ratio is 1.4×10^{-6} and for $\lambda = 1$ it is 7.8×10^{-5}. The thermal and electrical conductivities have no meaning in this theory since there is no conserved charge.

The shear viscosity, diffusion constant, and electrical conductivity have been evaluated at high temperature in gauge theories, to lowest order in the gauge coupling but to all orders in the logarithm of the coupling, by Arnold, Moore, and Yaffe [20]. Rather than applying the Kubo formulae directly they found it more expedient to do a numerical calculation based on the Boltzmann transport equation. For one flavor of lepton (electrons) the results are

$$D = \frac{0.596}{\alpha^2 \ln(1.46/\alpha)} \frac{1}{T} \tag{6.163}$$

$$\eta = \frac{2.39}{\alpha^2 \ln(5.99/\alpha)} T^3 \tag{6.164}$$

$$\sigma_{\text{el}} = \frac{2.50}{\alpha \ln(1.46/\alpha)} T \tag{6.165}$$

and for two flavors (electrons and muons) they are

$$D = \frac{0.392}{\alpha^2 \ln(1.08/\alpha)} \frac{1}{T} \tag{6.166}$$

$$\eta = \frac{1.53}{\alpha^2 \ln(2.33/\alpha)} T^3 \tag{6.167}$$

$$\sigma_{\text{el}} = \frac{3.29}{\alpha \ln(1.08/\alpha)} T \tag{6.168}$$

Here D refers to (conserved) lepton number diffusion. These QED expressions have an extra logarithmic factor arising from the screening of the long-range Coulomb force. The corresponding results for QCD will be discussed in later chapters.

6.10 Exercises

6.1 Find the linear response of the fermion number density to an applied neutral scalar field $\phi_{\text{ext}}(\mathbf{x}, t)$ for a Yukawa theory with interaction $\mathcal{L}_{\text{I}} = g\bar{\psi}\psi\phi$.

6.2 Repeat the analysis of Section 6.2 for a charged scalar field with a chemical potential.

6.3 Derive (6.47) for the interaction Lagrangian (5.36). You may choose whichever gauge you prefer.

6.4 Repeat the analysis leading to (6.70) but in the opposite limit, that of vanishing electron mass.

6.5 Derive the low-momentum expansion for $F(0, \mathbf{k})$ at finite temperature and chemical potential.

6.6 Derive the limiting form, (6.104) and (6.105), of the photon self-energy.

6.7 Is there an expression analogous to (6.120) for the photon dispersion relations?

6.8 Find the relationship between the flow velocities in the Eckart and the Landau–Lifshitz approaches.

6.9 Transport coefficients may be expressed in terms of differing correlation functions. As an example of this, express the thermal conductivity in terms of the density–density correlation function instead of the current–current one.

6.10 Derive the Kubo formula for the electrical conductivity.

References

1. Uehling, E. A., *Phys. Rev.* **48**, 55 (1935).
2. Serber, R., *Phys. Rev.* **48**, 49 (1935).
3. Bjorken, J. D., and Drell, S. D. (1964). *Relativistic Quantum Mechanics* (McGraw-Hill, New York).
4. Berestetskii, V. B., Lifshitz, E. M., and Pitaevskii, L. P. (1982). *Quantum Electrodynamics* (Pergamon Press, Oxford).
5. Quigg, C. (1983). *Gauge Theories of the Strong, Electromagnetic and Weak Interactions* (Addison-Wesley, Reading).
6. Fetter, A. L., and Walecka, J. D. (1971). *Quantum Theory of Many-Particle Systems* (McGraw-Hill, New York).
7. Kohn, W., and Vosko, S. H., *Phys. Rev.* **119**, 912 (1960).
8. Sivak, H., *Physica* **A129**, 408 (1985).
9. Kapusta, J. I., and Toimela, T., *Phys. Rev. D* **37**, 3731 (1988).
10. Schwinger, J., *Proc. Natl. Acad. Sci.* **37**, 452 (1951).
11. Dyson, F. J., *Phys. Rev.* **75**, 486 (1949); 1736 (1949).
12. Fradkin, E. S., *Proc. Lebedev Phys. Inst.* **29**, 6 (1965).
13. Blaizot, J.-P., Iancu, E., and Parwani, R. R., *Phys. Rev. D* **52**, 2543 (1995).
14. Weldon, H. A., *Phys. Rev. D* **26**,1394 (1982).
15. Klimov, V. V., *Yad. Fiz.* **33**, 1734 (1981).
16. Eckart, C., *Phys. Rev.* **58**, 919 (1940).
17. Kubo, R., *J. Phys. Soc. Japan* **12**, 570 (1957).
18. Jeon, S., *Phys. Rev. D* **52**, 3591 (1995).
19. Jeon, S. and Yaffe, L. G., *Phys. Rev. D* **53**, 5799 (1996).
20. Arnold, P., Moore, G. D., and Yaffe, L. G., *JHEP* **0305**, 051 (2003); **0011**, 001 (2000).

Bibliography

Linear response theory

Fetter, A. L., and Walecka, J. D. (1971). *Quantum Theory of Many-Particle Systems* (McGraw-Hill, New York).

Forster, D. (1975). *Hydrodynamic Fluctuations, Broken Symmetry, and Correlation Functions* (W. A. Benjamin, Reading).

QED

Berestetskii, V. B., Lifshitz, E. M., and Pitaevskii, L. P. (1982). *Quantum Electrodynamics* (Pergamon Press, Oxford).
Quigg, C. (1983). *Gauge Theories of the Strong, Weak, and Electromagnetic Interactions* (Addison-Wesley, Reading).

Analytic properties of Green's functions

Fradkin, E. S., *Proc. Lebedev Phys. Inst.* **29**, 6 (1965).
Fetter, A. L., and Walecka, J. D. (1971). *Quantum Theory of Many-Particle Systems* (McGraw-Hill, New York).
Doniach, S., and Sondheimer, E. H. (1974). *Green's Functions for Solid State Physicists* (Benjamin-Cummings, Reading).

Relativistic viscous fluid dynamics

Weinberg, S. (1972). *Gravitation and Cosmology* (Wiley, New York).
Landau, L. D., and Lifshitz, E. M. (1987). *Fluid Mechanics* (Pergamon Press, Oxford).

Kubo-type relations and hydrodynamics

Kadanoff, L. P., and Martin, P. C., *Ann. Phys. (NY)* **24**, 419 (1963).

Spontaneous symmetry breaking and restoration

In the standard model of particle physics, which has been thoroughly tested to energies above 100 GeV, a central role is played by scalar fields introduced in the Lagrangian with a negative mass-squared. These fields are introduced to spontaneously break a gauge symmetry and so yield the massive vector mesons W and Z, as observed in nature, in the framework of a renormalizable field theory. This is the Higgs mechanism, to be discussed in Section 7.4, and more specifically in the Glashow–Weinberg–Salam model of electroweak interactions in Chapter 15. Spontaneous symmetry breaking is more general, and arises in the strong interactions too as is elucidated in later chapters. We now turn our attention to a simple model to illustrate the phenomenon. This will be followed by a general statement of Goldstone's theorem, and a consideration of loop corrections and of the Higgs model.

7.1 Charged scalar field with negative mass-squared

Consider a complex scalar field Φ with Lagrangian

$$\mathcal{L} = \partial_\mu \Phi^* \partial^\mu \Phi - m^2 \Phi^* \Phi - \lambda (\Phi^* \Phi)^2 \tag{7.1}$$

This Lagrangian has a global U(1) symmetry $\Phi \to \Phi e^{-i\alpha}$, as discussed in Section 2.4. What happens if $m^2 = -c^2 < 0$? First suppose that $\lambda = 0$. Then in frequency–momentum space the action is

$$S_0 = -\tfrac{1}{2}\beta^2 \sum_n \sum_{\mathbf{p}} \left(\omega_n^2 + \mathbf{p}^2 - c^2 \right)$$

$$\times \left[\phi_{1;n}(\mathbf{p})\phi_{1;-n}(-\mathbf{p}) + \phi_{2;n}(\mathbf{p})\phi_{2;-n}(-\mathbf{p}) \right] \tag{7.2}$$

where $\Phi = \phi_1 + i\phi_2$ in the usual notation. This action is not negative definite and therefore the functional integral is not convergent. Another

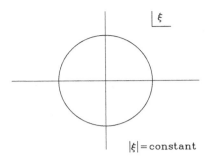

Fig. 7.1.

way to see this is to recall the expression for the partition function, (2.40), and simply replace m^2 with $-c^2$:

$$\ln Z_0 = 2V \int \frac{d^3p}{(2\pi)^3} \left[-\tfrac{1}{2}\beta\omega - \ln\left(1 - e^{-\beta\omega}\right) \right] \tag{7.3}$$
$$\omega = \sqrt{\mathbf{p}^2 - c^2}$$

The dispersion relation indicates an instability when $|\mathbf{p}| < c$.

The basic problem is that the potential is unbounded from below when $\lambda = 0$. To stabilize the system we require $\lambda > 0$, which means a repulsive interaction between the particles. The aforementioned instability occurs at small momenta. This suggests that the bosons condense, or accumulate, in the zero-momentum mode. Therefore, following the discussion of Bose–Einstein condensation, we separate out explicitly the static infrared part of the field as

$$\Phi = \xi + \chi$$
$$\Phi^* = \xi^* + \chi^* \tag{7.4}$$

Here ξ is a constant and $\chi_{n=0}(\mathbf{p} = 0) = 0$; that is, the thermal average $\langle \Phi \rangle = \xi$. Owing to the global U(1) symmetry, $\mathcal{L}$ depends only on the magnitude of ξ and not on its phase, as illustrated in Figure 7.1. For convenience we shall choose ξ real.

In terms of the shifted field, the Lagrangian is given by

$$\mathcal{L} = -U(\xi) + \mathcal{L}_0 + \mathcal{L}_I \tag{7.5}$$

where

$$U(\xi) = -c^2\xi^2 + \lambda\xi^4$$
$$\mathcal{L}_0 = \tfrac{1}{2}\partial_\mu\chi_1\partial^\mu\chi_1 - \tfrac{1}{2}\left(6\lambda\xi^2 - c^2\right)\chi_1^2$$
$$\qquad + \tfrac{1}{2}\partial_\mu\chi_2\partial^\mu\chi_2 - \tfrac{1}{2}\left(2\lambda\xi^2 - c^2\right)\chi_2^2$$
$$\mathcal{L}_I = -\sqrt{2}\lambda\xi\left(\chi_1^2 + \chi_2^2\right)\chi_1 - \tfrac{1}{4}\lambda\left(\chi_1^2 + \chi_2^2\right)^2$$

In addition, $\mathcal{L}$ contains terms linear in χ_1 and χ_2, but these contribute nothing and may be dropped. (Using the Fourier expansion (2.30), we see that these terms contribute to the action an amount proportional to $\int_0^\beta d\tau \int d^3x \, \chi(\mathbf{x}, \tau) \propto \chi_{n=0}(\mathbf{p} = \mathbf{0})$.) The procedure of shifting the field in this way and regarding χ_1 and χ_2 as the elementary excitations instead of ϕ_1 and ϕ_2 is called the mean field expansion. The mean field potential energy density is $\mathrm{U}(\xi)$, as we show below. The mean field masses can be read off from $\mathcal{L}_0$ as

$$\begin{aligned} \bar{m}_1^2 &= 6\lambda\xi^2 - c^2 \\ \bar{m}_2^2 &= 2\lambda\xi^2 - c^2 \end{aligned} \qquad (7.6)$$

Finally, notice that a cubic interaction is induced if $\xi \neq 0$.

The mean field approximation is obtained by calculating $\ln Z$ with the neglect of $\mathcal{L}_\mathrm{I}$. One might expect this to be a good approximation if both λ and $\lambda\xi$ are small. At this point, it is convenient to introduce the thermodynamic potential density Ω. For a uniform infinite volume system we have the relationship

$$\Omega(T, \xi) = -P(T, \xi) = -\frac{T}{V} \ln Z \qquad (7.7)$$

We know from thermodynamical considerations (Landau and Lifshitz [1]; Reif [2]) that in thermal equilibrium Ω is a minimum with respect to variations in ξ, when ξ is treated as a variational parameter. Intuitively, this can be recognized by remembering that in equilibrium the pressure is spatially uniform and that a local fluctuation to a state of lower pressure is obviously unstable. In the mean field approximation,

$$\Omega(T, \xi) = \mathrm{U}(\xi) + \int \frac{d^3p}{(2\pi)^3} \left[\tfrac{1}{2}\omega_1 + \tfrac{1}{2}\omega_2 \right.$$
$$\left. + T \ln\left(1 - \mathrm{e}^{-\beta\omega_1}\right) + T \ln\left(1 - \mathrm{e}^{-\beta\omega_2}\right) \right] \qquad (7.8)$$
$$\omega_i = \sqrt{\mathbf{p}^2 + \bar{m}_i^2}$$

The vacuum energy density is $\Omega(T = 0, \xi)$.

The classical energy density, obtained by neglecting the zero-point energy in the fields, is

$$\Omega_{\mathrm{cl}}(T = 0, \ \xi) = \mathrm{U}(\xi) = -c^2\xi^2 + \lambda\xi^4 \qquad (7.9)$$

This potential has a minimum at $\xi_0^2 = \xi^2(T = 0) = c^2/2\lambda$, as shown in Figure 7.2. The potential energy density has a local maximum at $\xi = 0$. This explains the instability encountered earlier. Instead of expanding about this local maximum, we should expand about the global minimum at ξ_0. The mean field masses, that is, the masses of small excitations about

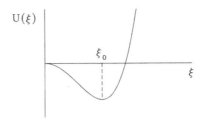

Fig. 7.2

the equilibrium field configuration, are

$$\bar{m}_1^2(T = 0) = 2c^2$$
$$\bar{m}_2^2(T = 0) = 0$$

$$(7.10)$$

These results are rather transparent. If we allow for complex values of ξ then the potential would still have the shape illustrated if we rotated the curve about the vertical axis. So we plot U along the z-axis and take the complex ξ plane to define the x- and y- axes. Since U depends only on $|\xi|$, we obtain the famous "bottom of the wine bottle" shape. Radial excitations of the field have a mass $\sqrt{2}c$, while rotational excitations have zero mass. Since the Lagrangian written in terms of Φ and Φ^* has a global U(1) symmetry, it is clear that if we change the phase of the field everywhere in space at the same time there will be no change in the energy of the system. This static, infinite-wavelength, zero-momentum excitation circles around the bottom of the potential in the complex ξ plane. This excitation is called a Goldstone boson. The U(1) symmetry apparent in (7.1) is not so obvious in (7.5). It is said to be spontaneously broken, since the vacuum exhibits a lesser symmetry than the Lagrangian. The real and imaginary components of the field exhibit different masses. The existence of a Goldstone boson in such a case is guaranteed by Goldstone's theorem, which is discussed in more detail in the next section.

There are a number of analogies with more common systems. In a ferromagnetic metal all the spins line up at $T = 0$. Since there is no preferred direction in which they should point, rotational symmetry is spontaneously broken. Spin waves with vanishing momentum carry no energy; their dispersion relation is $\omega = c_s k$. When the ends of a rod are subjected to sufficient force, the lowest-energy state is achieved when the rod is bowed. Since there is no preferred direction in which the rod should bow, rotational symmetry is spontaneously broken. The energy of a rotating bent rod, $\omega = l^2/2I$, vanishes as the angular momentum l goes to zero.

Now we raise the temperature of the system to $T > 0$. When $T^2 \ll \xi_0^2 = c^2/2\lambda$, not much of interest happens. There is an ideal gas of quasiparticles

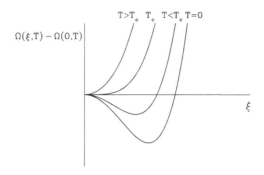

Fig. 7.3

with masses $\bar{m}_1 = \sqrt{2}c$ and $\bar{m}_2 = 0$. The thermal pressure is

$$P_{\text{thermal}} = P_0(T, \bar{m}_1^2) + P_0(T, \bar{m}_2^2) \tag{7.11}$$

where

$$P_0(T, \bar{m}^2) = -T \int \frac{d^3p}{(2\pi)^3} \ln\left(1 - e^{-\beta\omega}\right)$$

$$= \int \frac{d^3p}{(2\pi)^3} \frac{\mathbf{p}^2}{3\omega} \frac{1}{e^{\beta\omega} - 1}$$

$$\omega = \sqrt{\mathbf{p}^2 + \bar{m}^2} \tag{7.12}$$

When T is not small we must allow for the possibility that thermal fluctuations may change the equilibrium value of the condensate field ξ. If the interesting physics occurs when $T^2 \simeq c^2/\lambda \gg c^2$, then we make a high-temperature expansion of $P(T, m^2)$ as (see appendix Section A1.3)

$$P_0(T, m^2) = \frac{\pi^2}{90} T^4 - \frac{1}{24} m^2 T^2 + \frac{1}{12\pi} m^3 T$$

$$+ \frac{m^4}{64\pi^2} \left[\ln\left(\frac{m^2}{16\pi^2 T^2}\right) + 2\gamma_{\mathrm{E}} - \frac{3}{2} \right] + \cdots \tag{7.13}$$

Then, with $P = -\Omega$,

$$\Omega(\xi, T) = \lambda\xi^4 + \left(\frac{1}{3}\lambda T^2 - c^2\right)\xi^2 - \frac{\pi^2}{45}T^4 - \frac{1}{12}c^2 T^2 \tag{7.14}$$

Keeping only the first two terms in (7.13) yields (7.14). This is actually a very clever termination of the series, often used in the literature, since (i) it is correct when $T = 0$, (ii) it is a good approximation when $T > c$, and (iii) it is a remarkably transparent function of ξ and T. The isotherms of the thermodynamic potential are shown in Figure 7.3. The minimum shifts to smaller values of ξ and becomes less deep, as T increases. At

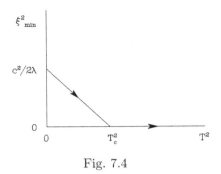

Fig. 7.4

$T_c^2 \equiv 3c^2/\lambda$, the coefficient of ξ^2 changes sign and the potential has a minimum at $\xi = 0$. The location of the minimum is

$$\xi_{min}^2(T) = \begin{cases} c^2/2\lambda - T^2/6 & T \le T_c \\ 0 & T \ge T_c \end{cases} \qquad (7.15)$$

This is shown in Figure 7.4. It can be seen that there is a phase transition at T_c. The spontaneously broken U(1) symmetry is restored!

Using (7.15) in (7.14), the pressures in the low- and high-temperature phases are, after normalizing the vacuum pressure and energy density to zero,

$$P_<(T) = \left(\frac{\pi^2}{45} + \frac{\lambda}{36}\right)T^4 - \frac{1}{12}c^2T^2$$

$$P_>(T) = \frac{\pi^2}{45}T^4 + \frac{1}{12}c^2T^2 - \frac{c^4}{4\lambda} \qquad (7.16)$$

The pressure and entropy are continuous at T_c,

$$P_<(T_c) = P_>(T_c)$$

$$\frac{dP_<(T_c)}{dT} = \frac{dP_>(T_c)}{dT} \qquad (7.17)$$

but the heat capacity is discontinuous,

$$\frac{d^2P_<(T_c)}{dT^2} - \frac{d^2P_>(T_c)}{dT^2} = \frac{2}{3}c^2 \qquad (7.18)$$

Hence this is a second-order phase transition. The physical origin of this symmetry-restoring phase transition is that the ordering inherent in the vacuum, and represented by the accumulation of an infinite number of particles into the zero-momentum state or condensate field ξ, is destroyed by thermal fluctuations at high temperatures. The second-order nature of the phase transition is expected from the general Landau theory of phase transitions (Landau and Lifshitz [1]). A first-order transition would arise if a term cubic in ξ were present in Ω, but this is not allowed by the U(1) symmetry.

There are potential flaws in the beautiful scenario just painted. First, the masses in the mean field approximation are

$$\bar{m}_1^2 = 6\lambda\xi_{\min}^2(T) - c^2 = \begin{cases} 2c^2 - \lambda T^2 & T \leq T_c \\ -c^2 & T \geq T_c \end{cases}$$
$$\bar{m}_2^2 = 2\lambda\xi_{\min}^2(T) - c^2 = \begin{cases} -\frac{1}{3}\lambda T^2 & T \leq T_c \\ -c^2 & T \geq T_c \end{cases}$$

$$(7.19)$$

We are burdened again by a negative mass-squared at $T > 0$. Also, where is the Goldstone boson when $0 < T \leq T_c$? Finally, what about the change in the zero-point energy in (7.8) as ξ varies with T? We shall return to these questions after a more general discussion of Goldstone's theorem.

7.2 Goldstone's theorem

Goldstone's theorem may be stated as follows:

If a continuous symmetry of the Lagrangian is spontaneously broken, and if there are no long-range forces, then there exists a zero-frequency excitation at zero momentum.

Here are some examples from nonrelativistic many-body systems [3].

- Ferromagnets. The absence of long-range forces, which may tend to couple spins at large distances, is necessary for the existence of a mode with $\omega \to 0$ as $k \to 0$.
- Superconductors. In the Bardeen–Cooper-Schrieffer (BCS) theory there is a spontaneous breaking of phase invariance associated with the conservation of electron number. However, there is an energy gap (equal to the mass of the Cooper pairs), so there is no Goldstone boson. The reason is that there are long-range electromagnetic forces.
- Superfluids. A low-temperature Bose system is a superfluid. The condensate field, at $T = 0$, is $\langle \Phi \rangle = \xi$, which is related to the particle number density by $n = |\xi|^2$. The phonon spectrum is

$$\omega^2 = \frac{k^2}{2m}\left(\frac{k^2}{2m} + 2nV(\mathbf{k})\right)$$

where $V(\mathbf{k})$ is the Fourier transform of the two-body potential. By definition, a short-range potential has the property that $V(\mathbf{k} = 0)$ is finite and positive. In that case, $\omega \to \sqrt{nV(\mathbf{k} = 0)/m}\,k$ as $k \to 0$. This is not so for a long-range potential. For the Coulomb force, $V(k) = e^2/k^2$ and, as $k \to 0$, $\omega \to e\sqrt{n/m} = \omega_P$, the plasma frequency.

We would like a nonperturbative proof of Goldstone's theorem. However, to be concrete, we will construct such a proof in the context of the U(1) scalar field theory discussed in the previous section.

The U(1) symmetry is $\Phi \to \Phi e^{-i\alpha}$, or $\delta\Phi = -i\alpha\Phi$ if $|\alpha| \ll 1$. The conserved current density may be recalled from (2.47). In terms of the shifted field, it is

$$j_\mu = \chi_2 \partial_\mu \chi_1 - \chi_1 \partial_\mu \chi_2 - \sqrt{2}\xi \partial_\mu \chi_2 \tag{7.20}$$

The total charge, $Q = \int d^3x\, j_0(\mathbf{x})$, is conserved; $\dot{Q} = 0$. The change in Φ due to an infinitesimal change in phase can also be expressed in operator form as

$$\delta\Phi = i\alpha[Q, \Phi] \tag{7.21}$$

That is, the total charge is the generator of the phase transformation. Taking the thermal, or ensemble, average of $\delta\Phi$, we find $\langle\delta\Phi\rangle = -i\alpha\langle\Phi\rangle = -i\alpha\xi$. Taking the thermal average of (7.21), we find an expression for the condensate field,

$$\xi = -\int d^3x\, \langle[j_0(\mathbf{x}, t),\ \Phi(\mathbf{0}, 0)]\rangle \tag{7.22}$$

Now we define the function

$$F^\mu(k_0, \mathbf{k}) = \int d^4x\, e^{ik\cdot x} \langle T\left[j^\mu(x), \Phi(0)\right]\rangle \tag{7.23}$$

Since $\partial_\mu j^\mu = 0$ and

$$T\left[j^\mu(x)\Phi(0)\right] = j^\mu(x)\Phi(0)\theta(x^0) + \Phi(0)j^\mu(x)\theta(-x^0) \tag{7.24}$$

it follows that

$$\begin{aligned} k_\mu F^\mu &= -i\int d^4x\, \partial_\mu\left\{e^{ik\cdot x}\langle T\left[j^\mu(x), \Phi(0)\right]\rangle\right\} \\ &+ i\int d^3x\, e^{-i\mathbf{k}\cdot\mathbf{x}}\langle[j_0(x), \Phi(0)]\rangle \end{aligned} \tag{7.25}$$

If the surface term in (7.25) vanishes then comparison with (7.22) shows that

$$\lim_{\mathbf{k}\to 0} k_\mu F^\mu = -i\xi \tag{7.26}$$

If $\xi \neq 0$, which means that the U(1) symmetry is spontaneously broken, then F has a pole at $k = 0$. This pole corresponds to a zero-frequency excitation at zero momentum.

It is not difficult to determine F^μ. Substituting (7.20) into (7.23) leads to

$$F^\mu(k) = -\xi k^\mu \int d^4x \, e^{ik\cdot x}\langle T\left[\chi_2(x)\chi_2(0)\right]\rangle$$
$$= -i\xi k^\mu D_2(k) \tag{7.27}$$

where D_2 is the real time Green's function. Combining (7.26) and (7.27) tells us that the imaginary part of the shifted field has a dispersion relation with the property that $\omega(\mathbf{k} = 0) = 0$. This is the Goldstone boson.

If the surface term in (7.25) is not zero then no conclusion may be drawn. This is often the case when there are massless spin-1 bosons in the theory. This is a gauge theory. We will discuss what happens in this case later on, focusing especially on the Higgs model and the Glashow–Weinberg–Salam model of the electroweak interaction.

7.3 Loop corrections

Now let us turn our attention to Ω and also to the self-energies of the fields.

In Section 7.1 we neglected the shift in the zero-point energy of the vacuum. Up to an (infinite) additive constant we can write

$$\int \frac{d^3p}{(2\pi)^3}\,\omega = \int \frac{d^4p}{(2\pi)^4}\,\ln(p^2 + m^2) \tag{7.28}$$

where $p = (\mathbf{p}, p_4)$ is a Euclidean four-vector. Our regularization procedure is simply to place an upper cutoff, Λ_c, on the integration over $|p|$. This is what we did in Section 3.4 (see also Chapter 4). Then

$$\int \frac{d^3p}{(2\pi)^3}\,\omega = \frac{1}{64\pi^2}\left[4m^2\Lambda_c^2 - 2m^4\ln\left(\frac{\Lambda_c^2}{m^2}\right) - m^4\right] + \text{constant} \tag{7.29}$$

plus terms that vanish as $\Lambda_c \to \infty$. We may add to the Lagrangian the counterterms

$$\delta c^2\Phi^*\Phi - \delta\lambda(\Phi^*\Phi)^2$$

In general, δc^2 and $\delta\lambda$ will depend on the other constants in the Lagrangian, and on c^2 and λ as well as Λ_c. The vacuum energy density is

$$\Omega(T = 0, \xi) = -(c^2 + \delta c^2)\xi^2 + (\lambda + \delta\lambda)\xi^4$$
$$+ \frac{1}{64\pi^2}\left[2(\bar{m}_1^2 + \bar{m}_2^2)\Lambda_c^2 - \bar{m}_1^4\ln\left(\frac{\Lambda_c^2}{\bar{m}_1^2}\right)\right.$$
$$\left. - \bar{m}_2^4\ln\left(\frac{\Lambda_c^2}{\bar{m}_2^2}\right) - \frac{1}{2}(\bar{m}_1^4 + \bar{m}_2^4)\right] \tag{7.30}$$

There is some freedom in choosing δc^2 and $\delta\lambda$. However, we should insist that $\Omega(T = 0, \xi)$ be finite (independent of Λ_c) and that Goldstone's theorem be satisfied ($\bar{m}_2 = 0$). The latter will occur only if $\Omega(T = 0, \xi)$ has its minimum at $\xi^2 = c^2/2\lambda$. A straightforward calculation yields

$$\delta c^2 = \frac{\lambda\Lambda_c^2}{4\pi^2} + \frac{\lambda c^2}{4\pi^2}\ln\left(\frac{\Lambda_c^2}{2c^2}\right) + c^2\frac{\delta'\lambda}{\lambda}$$

$$\delta\lambda = \frac{5\lambda^2}{8\pi^2}\ln\left(\frac{\Lambda_c^2}{2c^2}\right) + \delta'\lambda \tag{7.31}$$

Here $\delta'\lambda = \text{constant} \times \lambda^2$ is not determined by the above conditions. The renormalized vacuum energy density is

$$\Omega(T = 0, \xi) = -c^2\left(1 - \frac{\lambda}{8\pi^2} + \frac{\delta'\lambda}{\lambda}\right)\xi^2 + \lambda\left(1 - \frac{5\lambda}{16\pi^2} + \frac{\delta'\lambda}{\lambda}\right)\xi^4$$

$$+ \frac{\bar{m}_1^4}{64\pi^2}\ln\left(\frac{\bar{m}_1^2}{2c^2}\right) + \frac{\bar{m}_2^4}{64\pi^2}\ln\left(\frac{\bar{m}_2^2}{2c^2}\right) \tag{7.32}$$

There are several noteworthy points concerning (7.32). By construction it has its minimum at the same location as the classical energy density. Thus, in the true vacuum $\bar{m}_1^2 = 2c^2$ and $\bar{m}_2^2 = 0$, the same as in the classical approximation. Goldstone's theorem is obeyed. To (7.32) we may add any constant. Thus, not only the location of the minimum but also its depth can be made the same as in the classical approximation. Notice, however, that when $\xi^2 < c^2/2\lambda$ then $\bar{m}_2^2 < 0$ and Ω has an imaginary part. This is not unreasonable since in that region the system is unstable.

In the high-temperature expansion (7.13) there is also a term of order $m^4 \ln m^2$, with a coefficient of equal magnitude but opposite sign. Thus the order-$m^4 \ln m^2$ terms in the vacuum and high-temperature contributions cancel. Adding together (7.13) and (7.32) gives an improved high-temperature expression for the thermodynamic potential (for now we will neglect the term $-(m_1^3 + m_2^3)T/12$):

$$\Omega(T, \xi) = -\frac{\pi^2}{45}T^4 - \frac{c^2T^2}{12} + \frac{c^4}{32\pi^2}\ln\left(\frac{8\pi^2T^2}{c^2}e^{-2\gamma_E+3/2}\right)$$

$$- c^2\xi^2\left[1 + \frac{\delta'\lambda}{\lambda} + \frac{\lambda}{4\pi^2}\ln\left(\frac{8\pi^2T^2}{c^2}e^{-2\gamma_E+1}\right) - \frac{\lambda T^2}{3c^2}\right]$$

$$+ \lambda\xi^4\left[1 + \frac{\delta'\lambda}{\lambda} + \frac{5\lambda}{8\pi^2}\ln\left(\frac{8\pi^2T^2}{c^2}e^{-2\gamma_E+1}\right)\right] \tag{7.33}$$

The appearance of the logarithms is all that really distinguishes this improved potential from its predecessor. (The $\delta'\lambda/\lambda$ terms can be absorbed into the arguments of the logarithms if desired.) Now $\ln(T/c)$

is a slowly varying function compared with T^2 or T^4. So the shape of the potential is hardly affected. The critical temperature is determined as usual by the vanishing of the coefficient of ξ^2. To lowest order, $T_c^2 = 3c^2/\lambda$, as before. An improved formula is obtained by substituting the lowest-order result in the logarithm:

$$T_c^2 = \frac{3c^2}{\lambda} \left[1 + \frac{\delta'\lambda}{\lambda} + \frac{\lambda}{4\pi^2} \ln\left(\frac{24\pi^2}{\lambda} e^{-2\gamma_E + 1} \right) \right] \tag{7.34}$$

The correction is of relative order $\lambda \ln \lambda$. For instance, if we take $\delta'\lambda = 0$ and $\lambda = 0.1$ then the correction is only about 2%. It may seem as if the critical temperature depends on the rather arbitrary value of $\delta'\lambda$ but this is not so; the numerical values of c and λ depend on the renormalization prescription used to define them, which involves $\delta'\lambda$ through (7.31). In the end, T_c must be independent of the renormalization prescription.

The next problem we face in the mean field approximation is that $\bar{m}_2^2 < 0$ for $T > 0$ and $\bar{m}_1^2 < 0$ for $T > 2T_c^2/3$. Note that the finite-temperature corrections (7.19) to these masses are negative and proportional to λT^2 in the high-temperature limit ($T > c$). The one-loop contributions to the self-energies are of the same order. Therefore, they must be computed.

From the Lagrangian (7.5) we find the two-loop contributions to $\ln Z$ to be

$$3 \, \bigcirc\!\!\bigcirc + 3 \, (\!\bigcirc\!\!\bigcirc\!) + 2 \, \bigcirc\!\!(\bigcirc\!)$$

$$+ 3 \, \ominus + \ominus \tag{7.35}$$

A solid line represents the χ_1 propagator and a broken line represents the χ_2 propagator. There is a factor $-\lambda/4$ at each four-point vertex and a factor $-\sqrt{2}\lambda\xi$ at each three-point vertex. (Note that the 1PR diagrams do not appear on account of the stipulation that $\chi_0(0) = 0$. This can be shown by returning to the diagrammatic rules following from the functional integral in Section 3.2.) The self-energies are

$$\Pi_1 = -12 \, \bigcirc \,- 4 \, (\bigcirc\!) \,- 18 \,-\!\!\bigcirc\!\!- \,- 2 \,-\!(\bigcirc\!)\!-$$

$$\Pi_2 = -12 \, (\bigcirc\!) \,- 4 \, \bigcirc \,- 4 \,\cdots\!(\bigcirc\!)\!\cdots \tag{7.36}$$

The diagrams involving a three-point vertex, the so-called exchange diagrams, are momentum and frequency dependent. To renormalize, we must add the counterterms $-\delta c^2 + 6\xi^2\delta\lambda$ and $-\delta c^2 + 2\xi^2\delta\lambda$ to Π_1 and to Π_2, respectively. In the high-temperature approximation, and at low frequency and momentum, the exchange diagrams may be neglected. This follows simply from power counting. Both types of diagram involve one integration over the loop momentum, but the exchange diagrams involve

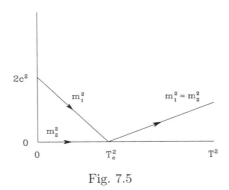

Fig. 7.5

two propagators instead of one. Then

$$\Pi_1^{\text{mat}} = \Pi_2^{\text{mat}} = \frac{1}{3}\lambda T^2 \qquad (7.37)$$

Adding these to (7.19), we obtain the masses

$$
\begin{aligned}
m_1^2 &= \bar{m}_1^2 + \Pi_1^{\text{mat}} =
\begin{cases}
2c^2\left(1 - \dfrac{T^2}{T_c^2}\right) & T \le T_c \\
\dfrac{1}{3}\lambda(T^2 - T_c^2) & T \ge T_c
\end{cases} \\[2mm]
m_2^2 &= \bar{m}_2^2 + \Pi_2^{\text{mat}} =
\begin{cases}
0 & T \le T_c \\
\dfrac{1}{3}\lambda(T^2 - T_c^2) & T \ge T_c
\end{cases}
\end{aligned}
\qquad (7.38)
$$

The behaviour of the masses as a function of temperature is shown in Figure 7.5. Thus the pathological behavior of the boson propagators has been cured. The vanishing of the masses at the critical point is characteristic of a second-order phase transition. Typically, one finds that the correlation lengths diverge at T_c. (The last diagram for Π_1 in (7.36) actually diverges if we let the external frequency and momentum go to zero and if $T < T_c$, because $m_2 = 0$. That is, the zero-mode contribution is proportional to $\lambda^2\xi^2 T \int dp/p^2$. This is of no physical importance since the mass is defined to be the location of the pole of the real time propagator at zero momentum. The relevant limit in (7.36) is $\Pi_1(\omega = m_1, \mathbf{k} = 0)$.)

The lesson learned is that the mean field approximation is not reliable in all respects. It turns out that it correctly predicts a second-order symmetry-restoring phase transition at $T_c^2 = 3c^2/\lambda$. However, it is incorrect in the finer details, such as the finite-temperature behavior of the correlation lengths (boson masses). This is a serious matter, since Goldstone's theorem is violated. At the very least, one should include all loop corrections to the same order in the coupling constants as is retained in the mean field approximation. The reason is that a loop expansion is essentially an expansion in powers of the Lagrangian. In order to respect the symmetries of the Lagrangian, one must retain all diagrams through a fixed number of loops.

A better approximation scheme would be to consider the thermodynamic potential Ω as function of the mean field $\langle \Phi \rangle = \xi$ and as a functional of the boson propagators $\mathcal{D}_1$ and $\mathcal{D}_2$. The mean field would be determined by the minimization condition $\partial\Omega/\partial\xi = 0$, and the propagators would be determined by the Schwinger–Dyson equations. To implement this idea, we would add to the quadratic part of the action S_0 the term

$$-\tfrac{1}{2}\beta^2 \sum_n \sum_{\mathbf{p}} \left[\chi_{1;-n}(-\mathbf{p})\Pi_1(\omega_n,\mathbf{p})\chi_{1;n}(\mathbf{p}) + \chi_{2;-n}(-\mathbf{p})\Pi_2(\omega_n,\mathbf{p})\chi_{2;n}(\mathbf{p}) \right]$$

(7.39)

and subtract the same quantity from S_{I}. In the S_0 case, (7.39) is to be treated as a counter-term. Recalling (2.36) and the steps leading up to it, we can write the thermodynamic potential as [4]

$$\Omega(T,\xi,\mathcal{D}_1,\mathcal{D}_2)$$

$$= \mathrm{U}(\xi) - \tfrac{1}{2}T \sum_n \int \frac{d^3p}{(2\pi)^3} \left[\ln(T^2\mathcal{D}_1) + \ln(T^2\mathcal{D}_2) - \frac{\mathcal{D}_1}{\bar{\mathcal{D}}_1^0} - \frac{\mathcal{D}_2}{\bar{\mathcal{D}}_2^0} + 2 \right]$$

$$+ \sum_{l=2}^{\infty} \Omega_l(\xi,\mathcal{D}_1,\mathcal{D}_2) + \text{subtractions}$$

(7.40)

Here

$$\bar{\mathcal{D}}_1^0 = \left(\omega_n^2 + \mathbf{p}^2 + \bar{m}_1^2 \right)^{-1}$$
$$\bar{\mathcal{D}}_2^0 = \left(\omega_n^2 + \mathbf{p}^2 + \bar{m}_2^2 \right)^{-1}$$

(7.41)

These are the mean field propagators, and Ω_l is the sum of all l-loop diagrams; in these loop diagrams, the bare propagators are to be replaced with the full propagators. Here, the potential Ω is an extremum with respect to independent functional variations of $\mathcal{D}_1$ and $\mathcal{D}_2$, on account of the Schwinger–Dyson equations

$$\mathcal{D}_1^{-1} - \bar{\mathcal{D}}_1^{0\,-1} = 2\sum_{l=2}^{\infty} \frac{\delta\Omega_l}{\delta\mathcal{D}_1}$$

$$\mathcal{D}_2^{-1} - \bar{\mathcal{D}}_2^{0\,-1} = 2\sum_{l=2}^{\infty} \frac{\delta\Omega_l}{\delta\mathcal{D}_2}$$

(7.42)

These equations determine Π_1 and Π_2 self-consistently, just as ξ is determined self-consistently from $\partial\Omega/\partial\xi = 0$.

As a practical matter, the loop sum must be terminated at a finite order. Then the momentum- and frequency-dependent self-energies must be determined self-consistently and substituted into (7.40) to compute Ω. The mean field is then determined by minimizing Ω. If only the two-loop diagrams (7.35) are retained and the high-temperature approximation is made, (7.37) and (7.38) follow. Then Ω may be computed from (7.40) straightforwardly since the propagators are both non-negative for all frequency and momentum. Minimization with respect to ξ will yield ξ as a function of T. One finds again that at $T_c^2 = 3c^2/\lambda$ there is a second-order symmetry-restoring phase transition, as predicted by the mean field approximation. This is left as an exercise.

7.4 Higgs model

The model discussed so far can be made more interesting by coupling the charged scalar field to the electromagnetic field. The Lagrangian density is

$$\mathcal{L} = (\partial^\mu - ieA^\mu)\Phi^*(\partial_\mu + ieA_\mu)\Phi + c^2\Phi^*\Phi - \lambda(\Phi^*\Phi)^2 - \tfrac{1}{4}F^{\mu\nu}F_{\mu\nu} \quad (7.43)$$

Anticipating the spontaneous breaking of the U(1) symmetry, which is now a local symmetry, we shift the field by setting

$$\Phi = \xi + \chi \quad (7.44)$$

and stipulate that $\langle\chi\rangle = 0$. Apart from terms linear in χ, we obtain

$$\mathcal{L} = -\mathrm{U}(\xi) + \mathcal{L}_0 + \mathcal{L}_\mathrm{I} \quad (7.45)$$

where

$$\mathcal{L}_0 = \tfrac{1}{2}(\partial_\mu\chi_1)(\partial^\mu\chi_1) - \tfrac{1}{2}\bar{m}_1^2\chi_1^2 + \tfrac{1}{2}(\partial_\mu\chi_2)(\partial^\mu\chi_2) - \tfrac{1}{2}\bar{m}_2^2\chi_1^2$$
$$\quad - \tfrac{1}{4}F^{\mu\nu}F_{\mu\nu} + e^2\xi^2 A^\mu A_\mu - \sqrt{2}e\xi\chi_2\partial_\mu A^\mu$$

$$\mathcal{L}_\mathrm{I} = -\sqrt{2}\lambda\xi(\chi_1^2 + \chi_2^2)\chi_1 - \tfrac{1}{4}\lambda(\chi_1^2 + \chi_2^2)^2$$
$$\quad + eA^\mu(\chi_1\partial_\mu\chi_2 - \chi_2\partial_\mu\chi_1)$$
$$\quad + e^2 A^\mu A_\mu\left[\sqrt{2}\xi\chi_1 + \tfrac{1}{2}(\chi_1^2 + \chi_2^2)\right]$$

Here $\bar{m}_1^2$, $\bar{m}_2^2$, and U(ξ) are as defined in Section 7.1. It would appear from $\mathcal{L}_0$ that the electromagnetic field has developed a mass $\sqrt{2}e\xi$. However, this must be carefully considered because of the mixing between χ_2 and A_μ.

To find the spectrum of excitations at $T = 0$ it is useful to make the change of variables

$$
\Phi = \left(\xi + 2^{-1/2}\phi\right) \exp\left(\frac{i\eta}{\sqrt{2}\xi}\right)
$$
$$
A'_\mu = A_\mu + \frac{\partial_\mu \eta}{\sqrt{2}e\xi}
\tag{7.46}
$$

where ϕ and η are two independent real fields. Substitution into (7.43) yields

$$
\mathcal{L} = -\mathrm{U}(\xi) + \mathcal{L}'_0 + \mathcal{L}'_\mathrm{I}
\tag{7.47}
$$

where

$$
\mathcal{L}'_0 = \tfrac{1}{2}\partial_\mu \phi\, \partial^\mu \phi - \tfrac{1}{2}\bar{m}_1^2 \phi^2 - \tfrac{1}{4}F'^{\mu\nu}F'_{\mu\nu} + e^2\xi^2 A'^\mu A'_\mu
$$
$$
\mathcal{L}'_\mathrm{I} = -\sqrt{2}\lambda\xi\phi^3 - \tfrac{1}{4}\lambda\phi^4 + e^2\left(\sqrt{2}\xi + \tfrac{1}{2}\phi\right)\phi A'^\mu A'_\mu
$$

Notice that all reference to the field η has gone! Minimizing the classical energy density $\mathrm{U}(\xi)$ gives an equilibrium condensate $\xi^2 = c^2/2\lambda$, the same as before. Thus, at $T = 0$, we have a real scalar field with mass $\sqrt{2}c$ and a vector field with mass $ec/\sqrt{\lambda}$. Counting the number of degrees of freedom, we have one for the former and three for the latter. This is the same as without spontaneous symmetry breaking, namely two for the Φ field and two for the massless A_μ field. There is no Goldstone boson; the Goldstone theorem does not apply, because A_μ is a vector field. The generation of mass for the vector field via spontaneous symmetry breaking is known as the Higgs mechanism. It is a central concept in modern gauge theories.

The choice of variables in (7.46) is not very appropriate for a mean field approximation at high temperature, because we expect ξ to decrease with increasing T and eventually to vanish above a critical temperature. Therefore we return to (7.45) to study the thermodynamics.

At $T = 0$ it can be shown that the χ_2 field in (7.45) does not represent an observable particle in scattering experiments [5]. In more picturesque language, it is said that the vector field increases its number of polarization degrees of freedom from two to three and becomes massive by eating the would-be Goldstone boson.

The partition function is

$$
Z = \int [dA_\mu]\,[d\Phi]\,[d\Phi^*]\,\delta(F)\, \det\left(\frac{\partial F}{\partial \alpha}\right)\exp\left(\int_0^\beta d\tau \int d^3x\, \mathcal{L}\right)
\tag{7.48}
$$

One convenient choice of gauge is the so-called R_ρ-gauge,

$$
F = \partial^\mu A_\mu - \sqrt{2}e\xi\rho\chi_2 - f(\mathbf{x}, \tau)
\tag{7.49}
$$

in the limit $\rho \to 0$. Under an infinitesimal gauge transformation

$$\Phi \to \Phi e^{i e \alpha} \approx \left(\xi + \frac{\chi_1 + i \chi_2}{\sqrt{2}} \right) (1 + i e \alpha)$$

$$A^\mu \to A^\mu - \partial^\mu \alpha$$

(7.50)

we have

$$\frac{\partial F}{\partial \alpha} = -\partial^2 - e^2 \xi \left(2 \xi + \sqrt{2} \chi_1 \right) \rho \to -\partial^2 \qquad (7.51)$$

Furthermore, multiplying the right-hand side of (7.48) by

$$\exp \left(-\frac{1}{2\rho} \int_0^\beta d\tau \int d^3 x \, f^2 \right)$$

and functionally integrating over f gives a β-independent correction. Hence

$$Z = \lim_{\rho \to 0} \det \left(-\partial^2 \right) \int [dA_\mu] \, [d\Phi] \, [d\Phi^*] \exp \left(\int_0^\beta d\tau \int d^3 x \, \mathcal{L}_{\text{eff}} \right) \qquad (7.52)$$

where

$$\mathcal{L}_{\text{eff}} = -\mathrm{U}(\xi) + \mathcal{L}_0 + \mathcal{L}_{\text{I}} - \frac{1}{2\rho} \left(\partial^\mu A_\mu - \sqrt{2} e \xi \rho \chi_2 \right)^2$$

Close scrutiny of (7.52) brings out the following points. The factor $\det \left(-\partial^2 \right)$ cancels two specious degrees of freedom. The gauge-fixing term has a part that is independent of ρ and that, in fact, cancels the mixing term between χ_2 and A_μ in (7.45). The limit $\rho \to 0$ ensures that only those gauge-field configurations with $\partial^\mu A_\mu = 0$ contribute to the partition function.

A high-temperature mean field approximation similar to (7.13) and (7.14) can be carried out, with the result

$$\Omega(\xi, T) = \lambda \xi^4 + \left[\left(\frac{\lambda}{3} + \frac{e^2}{4} \right) T^2 - c^2 \right] \xi^2 - \frac{2\pi^2}{45} T^4 - \frac{1}{12} c^2 T^2 \qquad (7.53)$$

This predicts a second-order symmetry-restoring phase transition at $T_{\text{c}}^2 = 12 c^2 / (4\lambda + 3 e^2)$. Of course, the particle masses exhibit the pathological behavior typical of the mean field approximation and it is necessary to calculate the one-loop self-energies to obtain a more respectable behavior.

Since the Higgs model contains two independent dimensionless coupling constants, new phenomena may occur. If $\lambda \gtrsim e^4$ then the qualitative behavior of the phase transition sketched above is not altered by higher-order loop corrections. If $\lambda \lesssim e^4$ then the mass of the vector meson is comparable with or greater than T_c, and the second-order phase transition may even become a first-order one. In fact, when $\lambda \to 3e^4/32\pi^2$, quantum corrections cause T_c to decrease to zero, and for $\lambda < 3e^4/32\pi^2$ there is no spontaneous symmetry breaking even at $T = 0$. The interested reader is referred to the review of Lindé [6].

If $c = 0$ then we are dealing with massless scalar electrodynamics, not the Higgs model. Surprisingly, spontaneous symmetry breaking occurs here also. It is driven by the one-loop quantum correction to the vacuum energy density, the shift in the zero-point energy of the vacuum. This phenomenon was discovered by Coleman and Weinberg [7]. The finite-temperature behavior of the Coleman–Weinberg model is left as an exercise.

7.5 Exercises

7.1 Choose $\delta'\lambda$ in (7.32) so that the depth of the minimum is the same as in the classical theory. Then plot the classical and one-loop quantum vacuum energy densities versus ξ for $\lambda = 0.1, 0.01, 0.001$.

7.2 Retaining the two-loop diagrams (7.35) and using the high-temperature approximation, as discussed at the end of Section 7.3, calculate T_c.

7.3 An alternative to the mean field expansion is an ordinary perturbative expansion based on the $T = 0$ value of the condensate field ξ_0. This scheme has the disadvantage that it is not self-consistent, but the advantage that one need not do an expansion in terms of full propagators since no tachyons appear in the perturbative expansion. In this case $\langle \chi \rangle$ will not vanish at $T > 0$. Using only the one-loop diagrams, show that $\langle \Phi \rangle = \xi_0 + \langle \chi \rangle$ vanishes at $T_c^2 = 3c^2/\lambda$.

7.4 Read the paper Coleman and Weinberg [7]. Verify their result that there is spontaneous symmetry breaking at $T = 0$ in massless scalar electrodynamics. Show that the symmetry is restored at high temperature, and calculate T_c.

References

1. Landau, L. D., and Lifshitz, E. M. (1959). *Statistical Physics* (Pergamon Press, Oxford).
2. Reif, F. (1965). *Fundamentals of Statistical and Thermal Physics* (McGraw-Hill, New York).
3. Fetter, A. L., and Walecka, J. D. (1971). *Quantum Theory of Many-Particle Systems* (McGraw-Hill, New York).
4. Lee, T. D., and Margulies, M., *Phys. Rev. D* **11**, 1591 (1975).
5. Abers, E. S., and Lee, B. W., *Phys. Rep.* **9**, 1 (1973).

6. Lindé, A. D., *Rep. Prog. Phys.* **42**, 389 (1979).
7. Coleman, S., and Weinberg, E., *Phys. Rev. D* **7**, 1888 (1973).

Bibliography

Spontaneous symmetry breaking, Goldstone theorem, quantization and renormalization of gauge theories

Abers, E. S., and Lee, B. W., *Phys. Rep.* **9**, 1 (1973).

Restoration of a spontaneously broken symmetry at high temperature in relativistic field theory

Kirzhnits, D. A., *Sov. Phys. JETP Lett.*, **15**, 529 (1972).
Kirzhnits, D. A., and Lindé, A. D., *Phys. Lett.* **B42**, 471 (1972).
Dolan, L., and Jackiw, R., *Phys. Rev. D* **9**, 3320 (1974).
Weinberg, S., *Phys. Rev. D* **9**, 3357 (1974).

Thermodynamics

Landau, L. D., and Lifshitz, E. M. (1959). *Statistical Physics* (Pergamon Press, Oxford).
Reif, F. (1965). *Fundamentals of Statistical and Thermal Physics* (McGraw-Hill, New York).

8

Quantum chromodynamics

The quark model of hadrons, developed by Gell-Mann and Zweig, began to be taken seriously in the mid to late 1960s. The discovery of scaling in deep inelastic electron–nucleon reactions in the late 1960s seemed to imply that at very short distances, or very high momentum transfers, the nucleon constituents (valence quarks) behaved like weakly interacting point particles. However, the interactions between quarks had to be very strong at long distances, or small momentum transfers, to confine them in hadrons and thus explain the non-observation of isolated quarks. Politzer [1] and Gross and Wilczek [2], who received the Nobel prize in 2004, showed that the only renormalizable field theory of quarks that had the property of an increasing force at long distance and a decreasing force at short distance was of the type discovered by Yang and Mills [3]. Quarks must be spin-1/2 fermions, with fractional electric charge, and must come in three colors (a new quantum charge akin to electric charge) in order to explain the systematics of hadron spectroscopy. Interactions between quarks are mediated by gluons (the glue which holds them together). Gluons are massless spin-1 bosons, as are photons, but unlike photons they interact among themselves directly (via point interactions) because they also carry a color charge. Such theories are called nonabelian gauge theories. This theory of quarks and gluons, quantum chromodynamics (QCD), is the accepted theory of the strong interactions. Unfortunately, it has been very difficult to make quantitative predictions with QCD, owing to its complexity and peculiar properties. For a more thorough discussion of the history of QCD and its experimental support see Close [4].

During the mid to late 1970s it was realized that there should be a qualitative change in the properties of hadronic matter as the temperature or density is increased. A dilute system could be described in terms of pions, nucleons, and other hadrons. In a very dense system such extended composite particles would overlap, and quarks and gluons would be free to roam. There might even be a color-deconfinement phase transition at a temperature of several hundred MeV or a baryon density of around ten times the normal nuclear density. A phase transition from hadron gas to quark–gluon plasma requires a very large energy density. Such a transition could have occurred in the very early universe during the first microsecond of the big bang, or it could occur in the interior of a neutron star or during the collisions of large nuclei at very high energy in terrestrial accelerators.

The outline of this chapter is as follows. In Section 8.1 the Lagrangian of QCD is discussed as well as the functional integral representation of the partition function, including ghosts. Section 8.2 contains a brief discussion of asymptotic freedom, whereby the effective coupling decreases to zero logarithmically at short distance. In Sections 8.3 and 8.4 the perturbative evaluation of the thermodynamic potential at high temperature and density is surveyed and all known results are summarized. Section 8.5 discusses various limits of the gluon propagator, in various gauges that are useful in linear response analyses. Instantons are nonperturbative, topological, excitations which contribute to the thermodynamic potential, and a short introduction to them is given in Section 8.6. Unresolved infrared problems which appear at high order in perturbation theory are discussed in Section 8.7. Strange cold quark matter is analyzed in Section 8.8. Finally, the very interesting problem of color superconductivity is studied in Section 8.9. Applications of QCD to neutron stars, the big bang, and high-energy heavy ion collisions will be made in later chapters.

8.1 Quarks and gluons

Quarks must come in three colors (color being a new, strong-interaction, quantum number) in order that we may construct the observed hadrons without violating the Pauli exclusion principle. The color gauge group of QCD is SU(3). However, we may base our analysis more generally on the group SU(N), $N = 2, 3, \ldots$. The generators of the group are written as G^a, where the index a runs in integral steps from 1 to $N^2 - 1$. The generators satisfy the commutation relations

$$[G^a, G^b] = i f^{abc} G^c \tag{8.1}$$

where the f^{abc} are the group structure constants. For example, for SU(2) the group generators may be represented by the 2×2 Pauli matrices and for SU(3) by the 3×3 Gell-Mann matrices.

The gauge field A_a^μ carries color with a color index $a = 1, \ldots, N^2 - 1$. The field strength is

$$F_a^{\mu\nu} = \partial^\mu A_a^\nu - \partial^\nu A_a^\mu - g f_{abc} A_b^\mu A_c^\nu \tag{8.2}$$

Here, the dimensionless coupling g enters. Under an infinitesimal gauge transformation $\alpha_a(\mathbf{x}, t)$, the gluon field transforms as

$$A_a^\mu \to A_a^\mu + g f_{abc} A_b^\mu \alpha_c - \partial^\mu \alpha_a \qquad (8.3)$$

The field strength is not invariant, unlike in QED, since

$$F_a^{\mu\nu} \to F_a^{\mu\nu} + g f_{abc} F_b^{\mu\nu} \alpha_c \qquad (8.4)$$

However, its square is invariant since $F_a^{\mu\nu} F_{\mu\nu}^a \to F_a^{\mu\nu} F_{\mu\nu}^a$.

The quarks come in N different colors, so the quark field ψ has a color index i which runs from 1 to N (where $N = 3$ for SU(3)). The QCD Lagrangian is

$$\mathcal{L} = \bar{\psi}(i\partial\!\!\!/ - M - g A\!\!\!/_a G^a)\psi - \tfrac{1}{4} F_a^{\mu\nu} F_{\mu\nu}^a \qquad (8.5)$$

The first term is the kinetic energy of the quarks. The second term is the quark mass matrix, which is diagonal in flavor space (that is, referring to the u, d, s, c, ... quarks). The third term is the minimal coupling of the quarks to the gluons. (Notice the suppression of the quark color indices in (8.5). If G^a is represented by an $N \times N$ matrix then ψ is represented by an N-dimensional column vector in color space.) In order for this coupling to be gauge invariant the quark field must transform as

$$\psi \to \exp(i g G^a \alpha_a)\psi \qquad (8.6)$$

The last term in (8.5) is gauge invariant by the construction of $F_a^{\mu\nu}$. When $g = 0$, (8.5) describes massive noninteracting quarks and $N^2 - 1$ massless noninteracting "photons".

The strong interactions conserve baryon number and electric charge. They also conserve quark flavor (such as strangeness), but the weak interactions allow for flavor change. Color charge is conserved by all known interactions. The color current density is

$$j_{(c)\mu}^a = g \left(\bar{\psi} \gamma_\mu G^a \psi + f^{abc} F_{\mu\nu}^b A_c^\nu \right) = \partial^\nu F_{\nu\mu}^a \qquad (8.7)$$

The second equality follows from the Lagrange equations of motion for A_a^μ. The conservation law $\partial^\mu j_{(c)\mu}^a = 0$ follows from the antisymmetry of the field strength in its two Lorentz indices. The color charge generators are

$$Q_{(c)}^a = \int d^3x \, j_{(c)0}^a \qquad (8.8)$$

The non-observation of isolated quarks or gluons leads us to postulate that only aggregates of quarks and gluons with zero net color charge, or color singlets, have finite energy. Aggregates with net color should have infinite energy. This would explain their absence. This color confinement is a generally accepted consequence of QCD but apparently has never been rigorously established.

Quantization proceeds in a way parallel to that of QED, discussed in Section 5.1. Equation (5.24) corresponds to the QCD formula

$$Z = \int [dA_a^\mu][d\bar\psi][d\psi]\delta(F^b) \det\left(\frac{\partial F^c}{\partial\alpha_d}\right) \exp\left(\int_0^\beta d\tau \int d^3x (\mathcal{L} + \bar\psi\mu\gamma^0\psi)\right)$$

(8.9)

The number of polarization degrees of freedom of the gluons is $2N_{\mathrm g} = 2(N^2 - 1)$ and F^b is the gauge fixing function; there is one for each $b = 1, \ldots, N^2 - 1$. Summation over quark color and flavor indices is implied.

One set of gauges that is often used is the set of covariant gauges

$$F^a = \partial^\mu A_\mu^a - f^a(\mathbf{x}, \tau) = 0$$

(8.10)

Under the infinitesimal gauge transformation (8.3),

$$F^a \to \partial^\mu \left(A_\mu^a + g f^{abc} A_\mu^b \alpha^c - \partial_\mu\alpha^a\right) - f^a$$

(8.11)

Then the argument of the determinant is

$$\frac{\partial F^c}{\partial\alpha^d} = -\partial^2\delta^{cd} + g f^{cbd}\partial^\mu A_\mu^b$$

(8.12)

As usual, we multiply Z by

$$\exp\left(-\frac{1}{2\rho}\int d\tau \int d^3x\, f_a^2\right)$$

and integrate over f^a to obtain

$$Z = \int [dA_a^\mu][d\bar\psi][d\psi] \det\left(-\partial^2\delta^{ac} + g f^{abc}\partial^\mu A_\mu^b\right)$$
$$\times \exp\left[\int d\tau \int d^3x \left(\mathcal{L} + \bar\psi\mu\gamma^0\psi - \frac{1}{2\rho}(\partial^\mu A_\mu^a)^2\right)\right] \quad (8.13)$$

As in (5.26) and (5.34), we introduce ghost fields $\bar C_a$ and C_a to represent the determinant in functional integral form:

$$Z = \int [dA_a^\mu][d\bar\psi][d\psi][d\bar C_a][dC_a]\exp\left(\int d\tau \int d^3x\,\mathcal{L}_{\mathrm{eff}}\right)$$

(8.14)

where

$$\mathcal{L}_{\mathrm{eff}} = \mathcal{L} - \frac{1}{2\rho}(\partial^\mu A_\mu^a)^2 + g f^{abc}\bar C_a \partial_\mu A_b^\mu C_c + \bar\psi\mu\gamma^0\psi + \partial_\mu\bar C_a\partial^\mu C_a$$

In the covariant gauges the ghost field does not decouple from the gluon field. The ghost field integration cannot be factored out.

In Table 8.1 the diagrammatic rules for QCD in the covariant gauges are listed. Table 8.2 contains a listing of the properties of the six quarks. The numerical values of the quark masses depend on the precise way in which they are defined, since freely propagating quarks do not exist. The three light quark masses are evaluated at an $\overline{\text{MS}}$ scale of 2 GeV, while the three heavy quark masses are evaluated at their own mass.

8.2 Asymptotic freedom

The renormalization-group running coupling for massless $\lambda\phi^4$ theory, λ_R, was discussed in Section 4.2. From (4.25) we saw that the effective coupling grows at high energy, or equivalently at short distance. The physical interpretation is that a point charge is shielded, or screened, by virtual pair production in the vacuum. As we approach the source of the charge, we penetrate the screening cloud surrounding it. The effective charge we see becomes larger due to the loss of screening. In a sense this is like penetrating the electron cloud surrounding an atomic nucleus. The difference is that the atomic electrons are real particles nearly on their mass shell. Electronic screening is essentially a classical effect. The increase in the renormalization-group charge is effected in the lowest approximation by virtual particles and so is a purely quantum effect.

To lowest order in the coupling constant, a β-function is either positive (the charge grows at short distance) or negative (the charge decreases at short distance). Until 1973, examples of only the former were known. The discovery that only nonabelian gauge theories allow for a negative β-function is credited to Politzer [1] and to Gross and Wilczek [2]. They showed that QCD yields a charge that decreases at short distance, an effect called asymptotic freedom that is required by experiment. This discovery was not anticipated by any simple intuitive reasoning. Let us examine the renormalization-group as it applies to QCD with massless quarks and in the set of covariant gauges.

The renormalization-group equation for the irreducible vertex function for n gluon and n' massless quark fields is (see the discussion leading to (4.13))

$$\left(M\frac{\partial}{\partial M} + \beta(g,\rho)\frac{\partial}{\partial g} + \delta(g,\rho)\frac{\partial}{\partial \rho} + n\gamma_A(g,\rho) + n'\gamma_\psi(g,\rho) \right)$$
$$\times \Gamma^{n,n'}(p_1,\ldots,p_{n+n'};g,\rho,M) = 0 \qquad (8.15)$$

Here δ is the "β-function" corresponding to the gauge parameter ρ.

Table 8.1. Bare propagators and vertices in QCD in covariant gauges

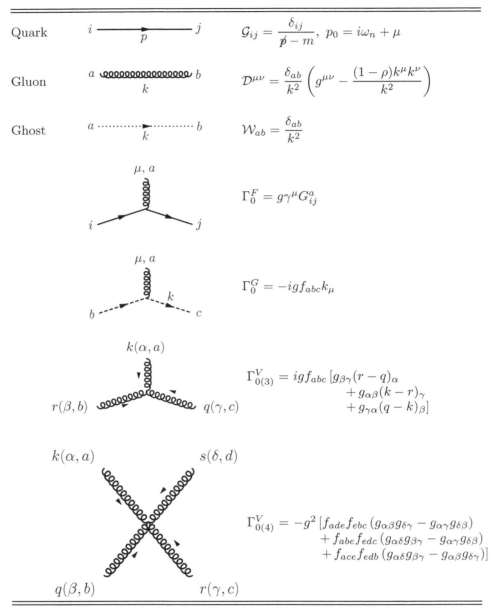

Quark $\qquad \mathcal{G}_{ij} = \dfrac{\delta_{ij}}{\not{p} - m}, \; p_0 = i\omega_n + \mu$

Gluon $\qquad \mathcal{D}^{\mu\nu} = \dfrac{\delta_{ab}}{k^2}\left(g^{\mu\nu} - \dfrac{(1-\rho)k^\mu k^\nu}{k^2}\right)$

Ghost $\qquad \mathcal{W}_{ab} = \dfrac{\delta_{ab}}{k^2}$

$\Gamma_0^F = g\gamma^\mu G_{ij}^a$

$\Gamma_0^G = -ig f_{abc} k_\mu$

$\Gamma_{0(3)}^V = ig f_{abc}\,[g_{\beta\gamma}(r-q)_\alpha$
$\qquad\qquad + g_{\alpha\beta}(k-r)_\gamma$
$\qquad\qquad + g_{\gamma\alpha}(q-k)_\beta]$

$\Gamma_{0(4)}^V = -g^2\,[f_{ade}f_{ebc}\,(g_{\alpha\beta}g_{\delta\gamma} - g_{\alpha\gamma}g_{\delta\beta})$
$\qquad\qquad + f_{abe}f_{edc}\,(g_{\alpha\delta}g_{\beta\gamma} - g_{\alpha\gamma}g_{\delta\beta})$
$\qquad\qquad + f_{ace}f_{edb}\,(g_{\alpha\delta}g_{\beta\gamma} - g_{\alpha\beta}g_{\delta\gamma})]$

Table 8.2.   Quark properties

Flavor	Electric charge	Baryon number	Mass
u (up)	2/3	1/3	3 MeV
d (down)	−1/3	1/3	7 MeV
s (strange)	−1/3	1/3	120 MeV
c (charm)	2/3	1/3	1.2 GeV
b (bottom)	−1/3	1/3	4.25 GeV
t (top)	2/3	1/3	175 GeV

There is a Ward identity for QCD which states that the longitudinal part of the inverse gluon propagator is not altered by interactions. That is,

$$\Gamma_{\mathrm{L}}^{2,0} = \frac{p^{\mu}p^{\nu}}{\rho} \tag{8.16}$$

This is the same as in QED (see (5.46)). Application of (8.15) to (8.16) then yields the relation

$$\delta(g,\rho) = 2\rho\gamma_A(g,\rho) \tag{8.17}$$

This points to the advantage of the Landau gauge $\rho = 0$; in this gauge $\delta(g,0) = 0$. Hence, starting with $\rho = 0$ we are guaranteed that after renormalization $\rho = 0$ will remain true, on account of the renormalization group equation

$$M\frac{\partial\bar{\rho}}{\partial M} = \delta(\bar{g},\bar{\rho}) \tag{8.18}$$

Otherwise, we must keep ρ arbitrary in our equations. For example, $\rho = 1$ will not remain as such under application of the renormalization group.

The γ's may be obtained most directly from $\Gamma_{\mathrm{T}}^{2,0}$ and $\Gamma^{0,2}$. To lowest order in g we must evaluate the following diagrams:

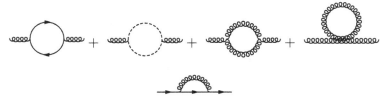

If these are normalized to have their free-field values at $p^2 = -M^2$ (according to Euclidean momentum subtraction) then

$$\Gamma_T^{2,0} = (p^2 g^{\mu\nu} - p^\mu p^\nu) \left\{ 1 + \left[\left(\frac{13}{6} - \frac{1}{2}\rho \right) c_1 - \frac{4}{3} c_2 \right] \frac{g^2}{16\pi^2} \ln\left(-\frac{p^2}{M^2} \right) \right\}$$
(8.19)

$$\Gamma^{0,2} = \not{p} \left[1 - \rho\, c_1 \frac{g^2}{16\pi^2} \ln\left(-\frac{p^2}{M^2} \right) \right]$$
(8.20)

The c's are given by

$$f_{acd} f_{bcd} = c_1 \delta_{ab} = N \delta_{ab}$$

$$N_f\, \mathrm{Tr}\, G_a G_b = c_2\, \delta_{ab} = \frac{1}{2} N_f\, \delta_{ab}$$
(8.21)

and N_f is the number of quark flavors. If we apply (8.15) to (8.19) and (8.20), and after differentiation set $p^2 = -M^2$, we can solve for the γ's:

$$\gamma_A = \frac{g^2}{16\pi^2} \left[\left(\frac{13}{6} - \frac{1}{2}\rho \right) c_1 - \frac{4}{3} c_2 \right]$$
(8.22)

$$\gamma_\psi = -\frac{g^2}{16\pi^2} \rho c_1$$
(8.23)

It is not possible to determine the β-function in these covariant gauges with knowledge of the two-point functions (propagators) alone.

Knowledge of a three-point function would suffice to determine β. From the following diagrams,

we compute

$$\Gamma^{3,0} = -ig f_{abc}(g_{\beta\gamma} p_\alpha + g_{\alpha\beta} p_\gamma - 2g_{\gamma\alpha} p_\beta)$$

$$\times \left\{ 1 + \left[\left(\frac{17}{12} - \frac{3}{4}\rho \right) c_1 - \frac{4}{3} c_2 \right] \frac{g^2}{16\pi^2} \ln\left(-\frac{p^2}{M^2} \right) \right\} \quad (8.24)$$

and from

we compute

$$\Gamma^{1,2} = -g\gamma^\mu G_a \left[1 - \left(\frac{3}{4} + \frac{5}{4}\rho \right) c_1 \frac{g^2}{16\pi^2} \ln\left(-\frac{p^2}{M^2} \right) \right] \qquad (8.25)$$

These are computed with external momenta $(p_1, p_2, p_3) = (0, -p, p)$ and normalized at $p^2 = -M^2$. Application of (8.15) to either (8.24) or (8.25) yields the lowest-order renormalization-group β-function,

$$\beta = -\frac{g^2}{48\pi^2}(11N - 2N_f) \qquad (8.26)$$

This will be negative, and the running coupling g will decrease with increasing energy, as long as $N_f < 5.5N$. This condition is fulfilled for SU(3), with six quark flavors.

It is worthwhile remarking that knowledge of the gluon two-point function in the Coulomb gauge ($\nabla \cdot \mathbf{A}_a = 0$) and the axial gauge ($n \cdot A = 0$, where n is a fixed four-vector) is sufficient to determine β. The reason is that the noncovariance of these gauges provides a tensorial structure for the gluon propagator and self-energy that requires two independent scalar functions, even in the vacuum (see Sections 5.4 and 6.3). These two independent scalar functions then allow the determination of both γ_A and β. The result is identical to (8.26).

The renormalization-group running coupling is determined by (see Section 4.2)

$$M\frac{\partial \bar{g}}{\partial M} = \beta(\bar{g}) \qquad (8.27)$$

with solution

$$\bar{\alpha} = \frac{\bar{g}^2}{4\pi} = \frac{12\pi}{(11N - 2N_f)\ln(M^2/\Lambda^2)} \qquad (8.28)$$

This explicitly displays asymptotic freedom: $\bar{\alpha} \to 0$ as $M \to \infty$. Notice the absence of any intrinsic coupling "constant" on the right-hand side of (8.28). In its place as the free parameter of the theory is the QCD energy scale Λ. The numerical value of Λ is, however, dependent on the gauge and on the renormalization scheme chosen (for example, this might be the choice used in (8.24) and (8.25)). This is seen in higher order.

Finite quark masses can be incorporated into the renormalization-group analysis by adding to the differential operator in (8.15) a term

$$\gamma_m \left(g, \rho, \frac{m_f}{M} \right) m_f \frac{\partial}{\partial m_f}$$

for each quark flavor f. That is, m_f/M is treated as a dimensionless coupling constant. The quark mass may be defined by

$$\mathcal{G}^{-1}|_{p^2=-M^2} = \not{p} - m \tag{8.29}$$

This is one possible renormalization prescription, but there exist others. A direct computation of β and γ_{m} in the Landau gauge yields [5]

$$M\frac{\partial g}{\partial M} = \beta = -\frac{g^3}{48\pi^2}\left[11N - \frac{2}{3}\sum_f B_0\left(\frac{m_f^2}{M^2}\right)\right] \tag{8.30}$$

where

$$B_0(x) = 1 - 6x + 12\left(\frac{x^2}{y}\right)\ln\left(\frac{y+1}{y-1}\right)$$
$$y = \sqrt{1+4x} \tag{8.31}$$

and

$$\frac{M}{m}\frac{\partial m}{\partial M} = \gamma_{\mathrm{m}} = -\frac{g^2}{2\pi^2}C_0\left(\frac{m^2}{M^2}\right) \tag{8.32}$$

$$C_0(x) = 1 - x\ln(1+x^{-1}) \tag{8.33}$$

Good approximations for B_0 and C_0 are

$$B_0(x) \simeq (1+5x)^{-1}$$
$$C_0(x) \simeq (1+2x)^{-1} \tag{8.34}$$

(We have now removed the overbar from g and m and will denote the running coupling and mass by g and m for notational simplicity.)

In general, (8.30) and (8.32) form a set of $N_{\mathrm{f}}+1$ coupled first-order nonlinear differential equations that must be solved numerically. The basic features of these equations are readily understood in the following way. The running coupling can be written as

$$\frac{g^2}{4\pi} = \frac{12\pi}{\left[11N - 2N_{\mathrm{f}}^{\mathrm{eff}}(M)\right]\ln(M^2/\Lambda^2)} \tag{8.35}$$

where

$$N_{\mathrm{f}}^{\mathrm{eff}}(M) \simeq \frac{1}{\ln(M^2/\Lambda^2)}\sum_f \frac{M^2 + m_f^2(M)}{\Lambda^2 + m_f^2(M)} \tag{8.36}$$

is the effective number of quark flavors at the energy scale M. Equations (8.35) and (8.36) form a solution to (8.30) valid to the lowest order in g. If $m_{\mathrm{f}}(M)$ is small then it contributes to β, but if it is large then it decouples. That is, if the quark mass is large compared with the energy scale of interest then there is insufficient energy for pair production, so that flavor does not add to the charge screening.

As an example, consider the first three quark flavors with $m_u = m_d = 0$ but $m_s \neq 0$. We look at high energy where $M \gg m_s$. Then $g^2/4\pi \simeq 2\pi/[9\ln(M/\Lambda)]$ can be inserted into (8.32):

$$\frac{dm_s}{dM} = -\frac{4}{9\ln(M/\Lambda)}\frac{m_s}{M} \tag{8.37}$$

This has the solution

$$m_s(M) = m_{s0}\left[\frac{\ln(M_0/\Lambda)}{\ln(M/\Lambda)}\right]^{4/9} \tag{8.38}$$

where m_{s0} is the mass at the scale M_0. The monotonic decrease in quark mass with increasing energy is in fact a general feature of (8.32), since $\gamma_m < 0$.

The β-function has been computed to two loops, that is, to order g^5 (the reader is referred to [6] for a more detailed discussion of massive quarks). For massless quarks [7],

$$\beta = -(11N - 2N_f)\frac{g^3}{48\pi^2} - \left(34N^2 - 13NN_f + 3\frac{N_f}{N}\right)\frac{g^5}{768\pi^4}$$
$$\equiv -\beta_0 g^3 - \beta_1 g^5 \tag{8.39}$$

which is still gauge and prescription independent. An approximate solution of the renormalization-group equation is

$$\alpha_2(M) = \alpha_1(M) - 4\pi\left(\frac{\beta_1}{\beta_0}\right)\alpha_1^2(M)\ln\left(\frac{1}{\alpha_1(M)}\right) \tag{8.40}$$

where $\alpha_1(M) = 1/[4\pi\beta_0 \ln(M^2/\Lambda^2)]$ is the lowest-order solution. Corrections to (8.40) are of order $\alpha_1^3(M) \sim [1/\ln(M^2/\Lambda^2)]^3$. For QCD with $N_f = 3$, $\alpha_2(M) = \alpha_1(M) + 0.0354\,\alpha_1^2(M)\ln\alpha_1(M)$. Thus, when $\alpha_1(M) \ll 1$ we have $\alpha_2(M) \simeq \alpha_1(M)$ to rather good accuracy.

The thermodynamic potential Ω must be independent of gauge and of renormalization prescription since it is a measurable quantity. However, the way this works in practice can be rather subtle. For example, if we work in a covariant gauge then

$$\frac{d}{d\rho}\Omega(g(\rho),\rho) = \left(\frac{\partial}{\partial\rho} + \frac{\partial g}{\partial\rho}\frac{\partial}{\partial g}\right)\Omega(g(\rho),\rho) = 0 \tag{8.41}$$

must hold, not $\partial\Omega(g,\rho)/\partial\rho = 0$. The reason is that g depends on the gauge and on the renormalization prescription used to render it finite from its bare value.

8.3 Perturbative evaluation of partition function

Since the effective QCD coupling goes to zero logarithmically at short distances, it is reasonable to attempt a perturbative expansion of the thermodynamic potential at high energy density [8, 9, 10]. In this and the next section we summarize the results so far obtained. Possible limits to the usefulness of perturbation theory will be discussed in later sections, as will some applications of the formulae obtained here. In the following discussion we quote perturbative results for the pressure $P(T, \mu)$. The entropy density $s = \partial P / \partial T$, flavor densities $n_f = \partial P / \partial \mu_f$, and energy density $\epsilon = -P + Ts + \sum_f \mu_f n_f$ are computed straightforwardly.

To zero order in the coupling, the QCD plasma is an ideal gas of gluons and quarks. The pressure can be written down immediately from (1.31) and (1.32):

$$P_0 = \frac{\pi^2}{45} N_g T^4 + \frac{N}{3\pi^2} \sum_f \int_0^\infty \frac{dp \, p^4}{E_p} N_F(p) \tag{8.42}$$

where $N_g = N^2 - 1$ is the number of gluons, which is eight for SU(3). When $m_f = 0$ the integral in (8.42) can be evaluated in closed form. The contribution to the pressure is

$$P_{0f}(m_f = 0) = N \left(\frac{7\pi^2 T^4}{180} + \frac{\mu_f^2 T^2}{6} + \frac{\mu_f^4}{12\pi^2} \right) \tag{8.43}$$

The exchange corrections to the ideal gas pressure are of order g^2. The relevant diagrams are shown below:

$$-\frac{1}{2} \, \bigcirc \quad - \quad \frac{1}{2} \, \bigcirc \quad + \quad \frac{1}{12} \, \bigcirc \quad + \quad \frac{1}{8} \, \bigcirc\bigcirc \tag{8.44}$$

The diagram with the quark loop is analyzed exactly as the QED diagram in (5.39) but with the replacement $e^2 \to g^2 \, \mathrm{Tr} \, G_a G_a = \frac{1}{2} g^2 N_g$. Then (5.58) to (5.61) can be taken over straightforwardly. The ghost diagram and the two pure gluon diagrams can be evaluated by means that should now be familiar. Since these are two-loop diagrams, the unrenormalized contributions will have parts that are quadratic and linear in the massless boson occupation probability $(e^{\beta \omega} - 1)^{-1}$. (Parts that are T-independent only renormalize the vacuum energy and these are discarded.) The subtraction procedure eliminates the linear parts. The three diagrams contribute, in respective order,

$$P_2^{\text{gluon}} = g^2 N N_{\text{g}} \left(\int \frac{d^3 p}{(2\pi)^3} \frac{1}{\omega} \frac{1}{e^{\beta\omega} - 1} \right)^2 \left(-\frac{1}{4} + \frac{9}{4} - 3 \right)$$

$$= -\frac{g^2}{144} N N_{\text{g}} T^4 \qquad (8.45)$$

This result is gauge invariant, although the individual diagrams are not.

As in QED and in massless $\lambda\phi^4$ theory, the next contributions are not of order g^4; they are of order $g^4 \ln g^2$ and g^3. These come from the set of ring diagrams

$$\frac{1}{2} \left[\frac{1}{2} \, \bigcirc \, - \, \frac{1}{3} \, \bigcirc \, + \, \cdots \right]$$

where

$$\bigcirc\!\!\Pi\!\!\bigcirc = \bigcirc + \bigcirc - \frac{1}{2} \, \bigcirc - \frac{1}{2} \, \bigcirc$$

$$(8.46)$$

The analysis proceeds exactly in parallel with that in subsection 5.5.2. What is needed is the static infrared limit of $\Pi^{\mu\nu}$. This will be discussed more thoroughly in a later section, and for now we simply quote the result at $T > 0$,

$$P_{\text{ring}}^{(1)} = \frac{N_{\text{g}}}{12\pi} T m_{\text{el}}^3 \qquad (8.47)$$

where

$$m_{\text{el}}^2 = F(n = 0, \mathbf{k} \to \mathbf{0}) = -\Pi^{00}(n = 0, \mathbf{k} \to \mathbf{0})$$

$$= g^2 \left(\frac{1}{3} N T^2 + \frac{1}{2\pi^2} \sum_f \int_0^\infty \frac{dp}{E_p} (p^2 + E_p^2) N_{\text{F}}(p) \right) \qquad (8.48)$$

is the square of the inverse screening length for color charge. When all quark masses can be neglected, we have

$$m_{\text{el}}^2 = g^2 \left[\left(\frac{1}{3} N + \frac{1}{6} N_{\text{f}} \right) T^2 + \frac{1}{2\pi^2} \sum_f \mu_f^2 \right] \qquad (8.49)$$

It should be noted that (8.47) and (8.48) have been obtained in the covariant gauges, in the Coulomb gauge, and in the temporal axial gauge ($A_0^a = 0$). If, in addition, the lowest-order momentum dependence of Π^{00}

is retained,

$$-\Pi^{00} = m_{\text{el}}^2 - \tfrac{1}{4}Ng^2|\mathbf{k}|T + \cdots \qquad (8.50)$$

then one obtains a $g^4 \ln g^2$ term not present in QED [11],

$$P_{\text{ring}}^{(2)} = \frac{NN_{\text{g}}}{65\pi^2}T^2 m_{\text{el}}^2 \, g^2 \ln g^2 \qquad (8.51)$$

However, this term is not precisely defined until the full order-g^4 contribution at finite temperature is determined. This will be discussed in the next section.

At $T = 0$, the three loop diagrams that are not already included in the ring sum are

(8.52)

The first two diagrams are analogous to those of QED, (5.77), but the last is peculiar to QCD on account of the three-gluon coupling. These diagrams are technically quite involved, owing to overlapping ultraviolet divergences. The interested reader is referred to Freedman and McLerran [12] and Baluni [13] for their evaluations. The result of summing the ring diagrams (8.46) together with (8.52) is

$$
\begin{aligned}
P_{\text{ring}} + P_4 = {} & \frac{1}{4\pi^2}\Bigg\{ \sum_f \mu_f^4 \left[N_{\text{g}} \frac{11N - 2N_{\text{f}}}{3} \left(\frac{\alpha(M)}{4\pi}\right)^2 \ln\left(\frac{\mu_f^2}{M^2}\right) \right. \\
& + N_{\text{g}}\left(-2.250N + 0.409N_{\text{f}} - 3.697 - \frac{(4.236)}{N}\right) \\
& \left. \times \left(\frac{\alpha(M)}{4\pi}\right)^2 \right] - (\boldsymbol{\mu}^2)^2 N_{\text{g}} \left[2\ln\left(\frac{\alpha(M)}{4\pi}\right) - 0.476 \right] \\
& \times \left(\frac{\alpha(M)}{4\pi}\right)^2 - N_{\text{g}}\bar{F}(\boldsymbol{\mu})\left(\frac{\alpha(M)}{4\pi}\right)^2 \Bigg\}
\end{aligned}
\qquad (8.53)
$$

where $\boldsymbol{\mu} = (\mu_u, \mu_d, \mu_s, \ldots)$ and

$$
\begin{aligned}
\bar{F}(\boldsymbol{\mu}) = {} & -2\boldsymbol{\mu}^2 \sum_f \mu_f^2 \ln\left(\frac{\mu_f^2}{\boldsymbol{\mu}^2}\right) + \frac{2}{3}\sum_{i>j}\Bigg[(\mu_i - \mu_j)^4 \ln\left(\frac{|\mu_i^2 - \mu_j^2|}{\mu_i\mu_j}\right) \\
& + 4\mu_i\mu_j(\mu_i^2 + \mu_j^2)\ln\left(\frac{(\mu_i + \mu_j)^2}{\mu_i\mu_j}\right) - (\mu_i^4 - \mu_j^4)\ln\left(\frac{\mu_i}{\mu_j}\right) \Bigg]
\end{aligned}
\qquad (8.54)
$$

These formulae were obtained in the Landau gauge using the momentum subtraction scheme; that is, the gluon self-energy was renormalized in such a way that $F(\bar{k}^2 = M^2, \boldsymbol{\mu} = \mathbf{0}) = G(\bar{k}^2 = M^2, \boldsymbol{\mu} = \mathbf{0}) = 0$. The corresponding formulae for nonzero quark masses have not been computed.

It should be noticed that the pressure in (8.53) depends explicitly on the renormalization energy scale M. To avoid the large logarithms, $\ln(\mu_f^2/M^2)$, that would appear if $\mu_f \to \infty$ while M is fixed, we should choose M in an optimum way. There is an arbitrariness in this, but a natural choice would be $M^2 = \boldsymbol{\mu}^2$ and another would be $M^2 = \boldsymbol{\mu}^2/N_{\mathrm{f}}$. Of course, if we could sum all orders of perturbation theory it would not matter. Truncating at a finite order means that we should choose an optimum M to reduce the importance of the terms neglected.

The QCD coupling g is not a fixed quantity. It depends on the gauge and on the renormalization prescription. This dependence is not apparent at order g^2 but first arises at order g^4. Thus, consider two gauges and/or prescriptions labeled i and j. One can show that (see for example [14])

$$g_i^2 = g_j^2(1 + A_{ij}g_j^2 + \cdots)$$
(8.55)

where A_{ij} is a computable number. The QCD scale Λ thus also depends on the gauge and/or prescription. Putting together (8.28), (8.39), and (8.55) we find that

$$\frac{\Lambda_i}{\Lambda_j} = \exp\left(\frac{A_{ij}}{2\beta_0}\right)$$
(8.56)

These features of QCD must be kept in mind when using high-order perturbation theory. For example, the numerical coefficient of α^2 in (8.53) is gauge and prescription dependent, in just such a way that when (8.55) is used the pressure is independent of gauge and prescription to this order (see also Section 8.2).

8.4 Higher orders at finite temperature

As we have seen previously, the simplest possible interaction yields a contribution to the pressure that is of order g^2. Owing to the summation implied by the ring diagrams, there are then contributions of order g^3 and $g^4 \ln g^2$. By now it should be clear that one cannot determine the order of a diagram by simply counting the number of interaction vertices, if the diagram requires resummed gluons. This resummation procedure has the great advantage of curing potential infrared divergences, as already seen in Chapters 3 and 5, because in effect the resummation induces a mass which is the static infrared limit of the self-energy. Calculations of the pressure to order g^4 and order g^5 have used the following strategy in order to improve the convergence of the perturbation expansion. One redefines the Lagrangian according to

$$\mathcal{L} \to (\mathcal{L} + \tfrac{1}{2} m_{\text{el}}^2 A_0^a A_0^a \delta_{p_0,0}) - \tfrac{1}{2} m_{\text{el}}^2 A_0^a A_0^a \delta_{p_0,0} \tag{8.57}$$

where $\mathcal{L}$ is the original Lagrangian in frequency–momentum space, A_0 is the zeroth component of the color gauge field, and $p_0 = 2\pi n T i$ is the zeroth component of its momentum. With this redefinition, the term in parentheses becomes the unperturbed Lagrangian and the other term creates a thermal counterterm, necessary to avoid double-counting. Following this scheme, the g^4 term receives contributions from the sub-leading part of the two-loop diagrams as well as from the leading part of the three-loop diagrams. The complete finite-temperature g^4 result for gauge fields with fermions was obtained by Arnold and Zhai [14] and it is (for zero chemical potential)

$$
\begin{aligned}
P = {} & d_A T^4 \frac{\pi^2}{9} \left\{ \frac{1}{5} \left(1 + \frac{7 d_F}{4 d_A} \right) \right. \\
& - \left(\frac{g}{4\pi} \right)^2 \left(C_A + \frac{5}{2} S_F \right) + \frac{16}{\sqrt{3}} \left(\frac{g}{4\pi} \right)^3 (C_A + S_F)^{3/2} \\
& + 48 \left(\frac{g}{4\pi} \right)^4 C_A (C_A + S_F) \ln \left(\frac{g}{2\pi} \sqrt{\frac{C_A + S_F}{3}} \right) \\
& - \left(\frac{g}{4\pi} \right)^4 C_A^2 \left[\frac{22}{3} \ln \left(\frac{M}{4\pi T} \right) + \frac{38}{3} \frac{\zeta'(-3)}{\zeta(-3)} - \frac{148}{3} \frac{\zeta'(-1)}{\zeta(-1)} - 4\gamma_E + \frac{64}{5} \right] \\
& - \left(\frac{g}{4\pi} \right)^4 C_A S_F \left[\frac{47}{3} \ln \left(\frac{M}{4\pi T} \right) + \frac{1}{3} \frac{\zeta'(-3)}{\zeta(-3)} - \frac{74}{3} \frac{\zeta'(-1)}{\zeta(-1)} \right. \\
& \qquad\qquad\qquad\qquad\qquad \left. - 8\gamma_E + \frac{1759}{60} + \frac{37}{5} \ln 2 \right] \\
& - \left(\frac{g}{4\pi} \right)^4 S_F^2 \left[-\frac{20}{3} \ln \left(\frac{M}{4\pi T} \right) + \frac{8}{3} \frac{\zeta'(-3)}{\zeta(-3)} - \frac{16}{3} \frac{\zeta'(-1)}{\zeta(-1)} \right. \\
& \qquad\qquad\qquad \left. \left. - 4\gamma_E - \frac{1}{3} + \frac{88}{5} \ln 2 \right] - \left(\frac{g}{4\pi} \right)^4 S_{2F} \left(-\frac{105}{4} + 24 \ln 2 \right) \right\}
\end{aligned}
\tag{8.58}
$$

In the equation above, ζ is Riemann's zeta function, γ_E is Euler's constant, and M is the renormalization scale in the modified minimal subtraction scheme, $\overline{\text{MS}}$. For SU(N) with N_f fermions one may write $d_A = N^2 - 1$, $C_A = N$, $d_F = N N_f$, $S_F = N_f/2$, $S_{2F} = (N^2 - 1) N_f / 4N$.

The extension of those techniques to one order higher by Zhai and Kastening [16] yields the g^5 term:

$$
\begin{aligned}
P_5 = \left(\frac{g}{4\pi}\right)^5 & \left(\frac{C_A + S_F}{3}\right)^{1/2} \\
\times & \left\{ C_A^2 \left[176\ln\left(\frac{M}{4\pi T}\right) + 176\gamma_E - 24\pi^2 - 494 + 264\ln 2 \right] \right. \\
& + C_A S_F \left[112\ln\left(\frac{M}{4\pi T}\right) + 112\gamma_E + 72 - 128\ln 2 \right] \\
& \left. + S_F^2 \left[-64\ln\left(\frac{M}{4\pi T}\right) - 64\gamma_E + 32 - 128\ln 2 \right] - 144 S_{2F} \right\}
\end{aligned}
$$

(8.59)

As will be discussed in Section 8.7, the hopes of pursuing an order-by-order expansion in finite-temperature QCD are too optimistic. The analytic expansion has serious infrared problems. Postponing a discussion of these aspects, it suffices here to say that for the pressure, this problem is met at order g^6. Kajantie *et al.* [17] evaluated perturbatively the last calculable contribution, that of order $g^6 \ln(1/g^2)$. This result is partly a conjecture, as this order receives a contribution from the complete $\mathcal{O}(g^6)$ term. However, without going into the details, general arguments based on the pattern of singularity cancellation order by order can be given in order to make progress. The interested reader may consult the quoted reference for a discussion of these technical aspects. The pressure at order $g^6 \ln(1/g^2)$ with N_f flavors is given by these authors as

$$
\begin{aligned}
P_6 = \frac{8\pi^2}{45} T^4 \left(\frac{\alpha_s(M)}{\pi}\right)^3 & \left\{ \left[-659.2 - 65.89 N_f - 7.653 N_f^2 + 742.5 \left(1 + \frac{1}{6} N_f\right) \right. \right. \\
& \left. \times \left(1 - \frac{2}{33} N_f\right) \ln\left(\frac{M}{2\pi T}\right) \right] \ln\left[\frac{\alpha_s}{\pi}\left(1 + \frac{1}{6} N_f\right)\right] \\
- 475.6 \ln\left(\frac{\alpha_s}{\pi}\right) & + q_a(N_f) \ln^2\left(\frac{M}{2\pi T}\right) + q_b(N_f) \ln\left(\frac{M}{2\pi T}\right) + q_c(N_f) \right\}
\end{aligned}
$$

(8.60)

where $q_a(N_f), q_b(N_f), q_c(N_f)$ are polynomials in N_f. The polynomials $q_{a,b}$ may be written down using the cancellation pattern alluded to earlier:

$$
q_a(N_f) = -\frac{1815}{16}\left(1 + \frac{5}{12} N_f\right)\left(1 - \frac{2}{33} N_f\right)^2
$$
$$
q_b(N_f) = 2932.9 + 42.83 N_f - 16.48 N_f^2 + 0.2767 N_f^3 \qquad (8.61)
$$

The last polynomial, $q_c(N_f)$, is the one that receives a nonperturbative contribution and is as yet uncalculated.

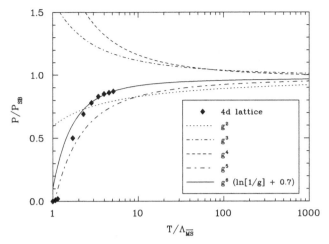

Fig. 8.1. Perturbative results for the pressure at various orders, including g^6 with an optimal constant, normalized to the noninteracting Stefan–Boltzmann value P_{SB} (Kajantie *et al.* [17]), against the scaled temperature.

It is instructive to examine the convergence of the perturbative expansion term by term. This is shown in Figure 8.1 for the pure gluon case with $N = 3$ and $N_{\text{f}} = 0$. The ratio of the pressure and its value in the Stefan–Boltzmann limit is plotted against the reduced temperature. At high temperatures, the pressure tends to the Stefan–Boltzmann limit.

8.5 Gluon propagator and linear response

In applying linear response theory to nonabelian gauge theories one must be careful to distinguish between gauge-invariant, physically observable, quantities and gauge-noninvariant quantities. The latter may still be relevant, though, provided that we can construct some observable out of them. This can be demonstrated with color electric screening.

The components of the color electric field,

$$E_i^a = F_{i0}^a = \partial_i A_0^a - \partial_0 A_i^a - g f^{abc} A_i^b A_0^c \qquad (8.62)$$

are not gauge invariant, unlike the electric field of QED, which is gauge invariant. Thus color electric screening as a physical phenomenon cannot be demonstrated on the basis of the color electric field alone. However, screening can be examined by computing the free energy V of a static, color-singlet (total color charge zero), quark–antiquark pair as a function of separation R. This can be done most directly in the temporal axial gauge (TAG) $A_0^a = 0$, for in this gauge the electric field is

$$\boldsymbol{E}_a = -\frac{\partial \boldsymbol{A}_a}{\partial t} \qquad \text{TAG} \qquad (8.63)$$

Then the analysis of Section 6.3 can be applied, with the result that the static dielectric function is (cf. (6.64))

$$\epsilon(\mathbf{q}) = 1 + \frac{F(0, \mathbf{q})}{\mathbf{q}^2}, \quad \text{TAG} \tag{8.64}$$

The scalar function F must be computed in TAG. Referring to (6.58) and (8.46) (except that in the latter there is no ghost diagram) one finds that, for the pure TAG gluon contribution[†] [19],

$$
\Pi_{00}^{\text{mat}}(q_0, q) = -\frac{g^2 N}{4\pi^2} \int_0^\infty \mathrm{d}k \; k N_{\mathrm{B}}(k)
$$
$$
\times \mathrm{Re} \left[4 - \frac{(q^2 - 2kq_0 - q_0^2)(2k + q_0)^2}{2k^2(k + q_0)^2} + \frac{(2k + q_0)^2}{2kq} \right.
$$
$$
\left. + \left(1 + \frac{(k^2 + (k + q_0)^2 - q^2)^2}{4k^2(k + q_0)^2} \right) \ln\left(\frac{R_+}{R_-}\right) \right] \tag{8.65}
$$

$$
\Pi_{ii}^{\text{mat}}(q_0, q) = -\frac{g^2 N}{4\pi^2} \int_0^\infty \mathrm{d}k \; k N_{\mathrm{B}}(k)
$$
$$
\times \mathrm{Re} \left[12 - \frac{2}{(k + q_0)^2} \left(8k^2 - q^2 - \frac{q^4}{4k^2} + 9q_0(q_0 + 2k) \right. \right.
$$
$$
\left. - \frac{5q^2}{4k^2}(q_0 + 2k)q_0 + \frac{3q_0^2}{2k^2}(q_0 + 2k)^2 \right)
$$
$$
+ \frac{1}{2kq} \left\{ -q^2 + 10k^2 + 10(k + q_0)^2 \right.
$$
$$
- \frac{1}{2k^2(k + q_0)^2}[k^2 + (k + q_0)^2 - q^2]^2
$$
$$
\left. \left. \times [3k^2 + 3(k + q_0)^2 + \tfrac{1}{2}q^2] \right\} \ln\left(\frac{R_+}{R_-}\right) \right] \tag{8.66}
$$

where $q = |\mathbf{q}|$, $q_0 = 2\pi n T i$, $R_\pm = q^2 - 2kq_0 - q_0^2 \pm 2kq$, and Re means that the even part of the following function of q_0 shoud be taken. The quark matter contributions are given by (5.51) with the substitution

---

[†] The axial-gauge pole $1/(n \cdot p)$ can be handled in one-loop diagrams with the principal value (PV) prescription. For example,

$$
\text{P.V.} \frac{1}{n \cdot p} = \lim_{\epsilon \to 0} \frac{1}{2} \left(\frac{1}{n \cdot p + i\epsilon} + \frac{1}{n \cdot p - i\epsilon} \right)
$$

see [18]. This makes sense in TAG at $T > 0$ only after analytic continuation of p_0 and replacement of frequency sums by contour integrals.

$e^2 \to \frac{1}{2}g^2$. In the following discussion, we will need only the static limit of the vacuum contribution,

$$F^{\text{vac}}(0, \text{q}) = \frac{g^2}{48\pi^2}(11N - 2N_{\text{f}})\text{q}^2 \ln\left(\frac{\text{q}^2}{M^2}\right) \qquad \text{TAG} \qquad (8.67)$$

Recall that $\Pi^{\mu\nu}$ is related to F and G in TAG just as in (5.46).

Consider the vacuum dielectric function. Inserting (8.67) into (8.64), we obtain the vacuum-polarization-corrected effective charge

$$\begin{aligned} \bar{g}^2(q) &= \frac{g^2}{\epsilon(q)} = \frac{g^2}{1 + (g^2/48\pi^2)(11N - 2N_{\text{f}})\ln(q^2/M^2)} \\ &= \frac{48\pi^2}{(11N - 2N_{\text{f}})\ln(q^2/\Lambda^2)} \end{aligned} \qquad (8.68)$$

which is the same as the renormalization-group charge.

The situation in other gauges is not so simple. Consider the set of covariant gauges (COVG). Then from (8.62) the color electric field has terms that are linear or quadratic in the vector potential. To find the linear response to an applied color electric field (such as that due to stationary quarks) we need to compute the correlation function between two electric field operators, and that entails knowledge of not only the propagator but also the three- and four-point gluon functions

$$\langle A_\mu A_\nu A_\alpha \rangle \qquad \langle A_\mu A_\nu A_\sigma A_\gamma \rangle$$

There is also the complication of the ghost field. What happens if we neglect the nonlinearity in (8.62) and naively apply (8.64)? From (8.19) we have

$$F^{\text{vac}}(0, q) = \frac{g^2}{48\pi^2}\left[\left(\frac{13}{12} - \frac{3}{2}\rho\right)N - 2N_{\text{f}}\right]q^2 \ln\left(\frac{q^2}{M^2}\right) \qquad \text{COVG} \quad (8.69)$$

This does not yield the correct renormalization-group-improved charge, nor does it yield the correct vacuum-polarization-corrected potential between stationary quarks. This should be expected. In Section 8.2 we found that knowledge of the gluon propagator alone was not sufficient to determine the β-function in the covariant gauges, although it was sufficient in the axial gauges.

Computation of the free energy, as a function of separation, of the static quark–antiquark pair at $T > 0$ (in TAG) proceeds just as in QED. At large separation,

$$V(R) = \frac{Q_1 \cdot Q_2}{4\pi} \frac{e^{-m_{\text{el}}R}}{R} \qquad (8.70)$$

In the color-singlet state, the product of charges is $Q_1 \cdot Q_2 = -g^2 N_{\mathrm{g}}/2N$. Since (8.70) is physically measurable (at least in principle!), m_{el} must be gauge invariant. In TAG, it is given by

$$m_{\mathrm{el}}^2 = F(0, \mathbf{q} \to 0) \quad \text{TAG} \tag{8.71}$$

which is an exact relation.

To one-loop order, all gauges receive the same contribution to $\Pi^{\mu\nu}$ from dynamical quarks. To focus on the essentials, we shall consider a quark-free world in the remainder of this section.

In the Feynman gauge (FG, $\rho = 1$), the $T > 0$ contribution to the gluon self-energy is

$$\Pi_{00}^{\mathrm{mat}}(q_0, q) = -\frac{g^2 N}{4\pi^2} \int_0^\infty \mathrm{d}k \; k N_{\mathrm{B}}(k)$$
$$\times \operatorname{Re} \left\{ 4 + \frac{1}{qk} [(q_0 + 2k)^2 - 2q^2] \ln\left(\frac{R_+}{R_-}\right) \right\} \tag{8.72}$$

$$\Pi_{ii}^{\mathrm{mat}}(q_0, q) = \frac{g^2 N}{4\pi^2} \int_0^\infty \mathrm{d}k \; k N_{\mathrm{B}}(k)$$
$$\times \operatorname{Re} \left[4 + \frac{1}{qk} \left(4q_0^2 - 4kq_0 - 4k^2 - 3q^2 \right) \ln\left(\frac{R_+}{R_-}\right) \right] \tag{8.73}$$

These are not the same as (8.65) and (8.66). Thus $\Pi^{\mu\nu}$, and the functions F and G, are not gauge invariant in nonabelian gauge theories.

From the perspective of screening, the interesting limit is $q_0 = 0$, $|\mathbf{q}| = q \to 0$. One finds that in TAG

$$F(0, q \to 0) = -\Pi_{00}(0, q \to 0)$$
$$= \tfrac{1}{3} g^2 N T^2 - \tfrac{1}{4} g^2 N T q - \tfrac{11}{48} \frac{g^2}{\pi^2} N q^2 \ln\left(\frac{q^2}{T^2}\right) + \cdots \tag{8.74}$$

$$G(0, q \to 0) = \tfrac{1}{2} \Pi_{ii}(0, q \to 0) = -\tfrac{5}{16} g^2 N T q + \cdots \tag{8.75}$$

and in FG

$$F(0, q \to 0) = -\Pi_{00}(0, q \to 0) = \tfrac{1}{3} g^2 N T^2 - \tfrac{1}{4} g^2 N T q + \cdots \tag{8.76}$$

$$G(0, q \to 0) = \tfrac{1}{2} \Pi_{ii}(0, q \to 0) = -\tfrac{3}{16} g^2 N T q + \cdots \tag{8.77}$$

There are a number of interesting aspects to these results. The first two terms of (8.74) and (8.76) are identical. This would not have been expected on the basis of our earlier discussion of electric screening. The reason that they are, and must be, the same is that these first two terms of F give rise to the order g^3 and order $g^4 \ln g^2$ terms in the pressure via summation of the ring diagrams (recall Section 6.5). The coefficients of all terms in the pressure up to (but not including) g^4 must be gauge independent, on account of (8.55). The third term of (8.74) combines with the vacuum contribution to yield

$$\tfrac{11}{48}\frac{g^2}{\pi^2}Nq^2\ln\left(\frac{T^2}{M^2}\right)$$

So again we see that we should choose M proportional to T to eliminate potentially large logarithms at high temperature. The first nonzero term of G is gauge dependent. This will be discussed further in Section 8.7.

Plasma oscillations may be discussed in a manner parallel to the discussion in Section 6.6. In TAG, a physical gauge with the proper number of gluon-polarization degrees of freedom and no ghosts, one finds that the long-wavelength dispersion relations for transverse and longitudinal oscillations are

$$\omega_{\mathrm{T}}^2 = \omega_{\mathrm{P}}^2 + \tfrac{6}{5}k^2 + \cdots$$
$$\omega_{\mathrm{L}}^2 = \omega_{\mathrm{P}}^2 + \tfrac{3}{5}k^2 + \cdots \tag{8.78}$$

where $\omega_{\mathrm{P}}^2 = g^2NT^2/9$. These waves are damped with damping constant $\gamma_{\mathrm{T}} = \gamma_{\mathrm{L}} = g^2NT/24\pi$. The short-wavelength longitudinal oscillations are overdamped and do not propagate. The transverse oscillations have the spectrum

$$\omega_{\mathrm{T}}^2 = k^2 + \tfrac{3}{2}\omega_{\mathrm{P}}^2 + \cdots \tag{8.79}$$

and to order g^2 are not damped by thermal effects.

A proper linear response analysis has also been done in another gauge, the Coulomb gauge [20]. The results are identical to (8.78) and (8.79). If one tries to do a cheap analysis in an unphysical gauge by simply searching for the poles of the gluon propagator, one obtains certain erroneous results. For example, in the Feynman gauge one recovers (8.78) and (8.79), but the damping constant is a factor 5 too large and of the opposite (wrong) sign. In addition, short-wavelength longitudinal waves propagate with $\omega_{\mathrm{L}}^2 = k^2 + \cdots$, which is unphysical.

It must be acknowledged that, at this time, color plasma waves are an enigma. Whether they represent physically observable phenomena has not been rigorously established.

8.6 Instantons

Instantons are nonperturbative solutions of the classical field equations which carry topological charge. After their discovery by Belavin *et al.*

[21] it was hoped that they would provide a means of understanding confinement. That has turned out not to be the case. In QCD their effects are reliably computed only at short distance (or high temperature). In this domain, it has been found that they are always dominated by perturbative corrections. For this reason, and because the mathematics of instantons can become quite involved, this brief section will present only an overview.

Instantons contribute to the partition function *in addition* to all perturbative contributions. Although not quantitatively important in their own right, these nonperturbative solutions are of course interesting in principle. They have also been used in a more phenomenological way to understand various aspects of chiral symmetry breaking and restoration and hadronic structure.

Consider an SU(2) gauge field theory without quarks. It is advantageous in this context to work in Euclidean space, with Greek indices running from 1 to 4. Define the matrix functions

$$\begin{aligned}
A_\mu &= -ig\left(\tfrac{1}{2}\sigma^a\right)A_\mu^a \\
F_{\mu\nu} &= -ig\left(\tfrac{1}{2}\sigma^a\right)F_{\mu\nu}^a
\end{aligned} \tag{8.80}$$

The action is

$$S = \frac{1}{2g^2}\int d^4x\,\mathrm{Tr}(F_{\mu\nu}F^{\mu\nu}) \tag{8.81}$$

and the classical equations of motion are

$$\partial^\mu F_{\mu\nu} + [A^\mu, F_{\mu\nu}] = 0 \tag{8.82}$$

We make the ansatz that

$$A_\mu = i\bar{\sigma}_{\mu\nu}a^\nu \tag{8.83}$$

where a^ν is spacetime dependent, and we define the following objects:

$$\begin{aligned}
\sigma_{ij} &= -i\left[\tfrac{1}{2}\sigma_i, \tfrac{1}{2}\sigma_j\right] \\
\sigma_{i4} &= \tfrac{1}{2}\sigma_i \\
\sigma_{\mu\nu} &= -\sigma_{\nu\mu} \\
\bar{\sigma}_{ij} &= \sigma_{ij} \\
\bar{\sigma}_{i4} &= -\sigma_{i4}
\end{aligned} \tag{8.84}$$

With a dual defined by

$$^*\sigma_{\mu\nu} = \tfrac{1}{2}\epsilon_{\mu\nu\alpha\beta}\sigma^{\alpha\beta}$$

we find that

$$
\begin{aligned}
{}^{*}\sigma_{\mu\nu} &= \sigma_{\mu\nu} & \text{self-dual} \\
{}^{*}\bar{\sigma}_{\mu\nu} &= -\bar{\sigma}_{\mu\nu} & \text{antiself-dual}
\end{aligned}
\tag{8.85}
$$

The equations of motion are satisfied when

$$
\begin{aligned}
a_\mu &= \partial_\mu \ln \phi \\
\partial^2 \phi &= 0
\end{aligned}
\tag{8.86}
$$

This solution is said to be self-dual because ${}^{*}F_{\mu\nu} = F_{\mu\nu}$. An antiself-dual solution (with ${}^{*}F_{\mu\nu} = -F_{\mu\nu}$) is obtained with

$$
A_\mu = i\sigma_{\mu\nu}a^\nu
\tag{8.87}
$$

In both cases the classical action can be expressed as

$$
S = \frac{1}{2g^2} \int d^4x \, \partial^2 \partial^2 \ln \phi
\tag{8.88}
$$

For the solution of Laplace's equation we have

$$
\phi(x) = 1 + \sum_{i=1}^{n} \frac{\lambda_i^2}{(x - y_i)^2}
\tag{8.89}
$$

where each λ_i is a real number and each y_i is a fixed vector. Clearly y_i represents the position of some object and λ_i its size. When this solution is used in the self-dual ansatz, it is said to represent n instantons; when it is used in the antiself-dual ansatz it is said to represent n anti-instantons. The instantons and anti-instantons represent tunnelings between different states.

These field configurations can be characterized by a topological charge q, a gauge invariant, called a Pontryagin index:

$$
q = \frac{1}{16\pi^2} \int d^4x \, \mathrm{Tr} \left({}^{*}F_{\mu\nu}F^{\mu\nu}\right)
\tag{8.90}
$$

A direct calculation shows that $q = n$ for the n-instanton solution and $q = -n$ for the n-anti-instanton solution. There is no known exact solution for n instantons and n' anti-instantons. It is not possible to change the Pontryagin index by a smooth deformation of the gauge field. Since perturbative calculations always start with $A_\mu = 0$ and $q = 0$, the instantons and anti-instantons make topologically distinct contributions to the functional integral.

When computing the partition function at $T = 0$ (useful for calculating vacuum correlation functions), the most straightforward approach is to treat the instantons and anti-instantons as individual, noninteracting objects (the dilute gas approximation, DGA). Only instantons with $q = 1$

and anti-instantons with $q = -1$ are included. One-loop quantum corrections can be included by writing

$$A_\mu = A_\mu^{\text{cl}} + A_\mu'$$

where A_μ^{cl} is the classical solution, expanding the Lagrangian in powers of A_μ' and dropping terms cubic and quartic in A_μ'. That is, the quantum fluctuations must be calculated in the presence of a background instanton (or anti-instanton) field. For SU(N), the SU(2) instantons must be embedded in the appropriate fashion. The calculations of t'Hooft [22], in particular, are a *tour de force* of mathematical physics. The result is simple and elegant. It is (assuming no state mixing)

$$\ln Z_{\text{DGA}} = 2C_{\text{N}}V\beta \int_0^\infty \frac{d\lambda}{\lambda^5} \left(\frac{4\pi^2}{g^2}\right)^{2N} \exp\left(-\frac{8\pi^2}{\bar{g}^2}\right) \tag{8.91}$$

We make the following remarks.

1 The exponential of the classical action is evident.
2 The factor $V\beta$ is the total spacetime volume.
3 The factor 2 arises because both instantons and anti-instantons are included.
4 The factor C_N is group-theoretic in origin. In the Pauli–Villars regularization scheme,

$$C_N = \frac{4}{\pi^2} \frac{\exp\left[-0.433 - 0.292(N - 2 - N_{\text{f}})\right]}{(N - 1)!(N - 2)!} \tag{8.92}$$

5 Integration over scale size λ must be done. The power -5 of λ arises from the scale size and from the four components of the position coordinate. It also is required so that $\ln Z$ is dimensionless.
6 Quantum fluctuations amount to replacing the coupling constant g^2 with the renormalization-group running coupling

$$\bar{g}^2 = \frac{24\pi^2}{(11N - 2N_{\text{f}})\ln(1/\lambda\Lambda_{\text{R}})} \tag{8.93}$$

in the exponential factor, although this replacement is presumed to happen (at the next order) in the pre-exponential factor as well. Here Λ_{R} is the QCD scale parameter in the Pauli–Villars scheme.
7 There should be an additional factor in $\ln Z_{\text{DGA}}$, which is

$$\prod_f (m_f\lambda)$$

for each light quark ($m_f < \lambda^{-1}$) flavor. Light quarks greatly suppress instantons.

8 The integral over λ does not exist: it diverges for large λ. Thus one must go beyond the dilute gas approximation in the QCD vacuum and confront the infrared confinement problem.

The way to avoid point 8 is to focus on a physical circumstance which provides a natural cutoff on instanton size λ. For example, we could consider computing instanton-induced corrections to the process $e^+e^- \to$ hadron jets at high energy. A cutoff would then be supplied by the center-of-mass energy $\sqrt{s}$; the dominant contribution should come from instanton scale sizes $\lambda \approx 1/\sqrt{s}$. Another circumstance, of interest to us, is the contribution of instantons to the thermodynamic potential of a high-temperature quark–gluon plasma. At high temperatures, color electric fields should be screened just as in QED plasma. The temperature should provide an infrared cutoff on instanton sizes.

It is possible to generalize these solutions to finite temperature. We still work in Euclidean space but x_4 is replaced by the variable τ. The instanton solutions must now be periodic in τ with period β. This is accomplished by making the field ϕ periodic [23]. The $n = 1$ solution goes over into

$$
\phi = 1 + \lambda^2 \sum_{k=-\infty}^{\infty} \left[(\mathbf{x} - \mathbf{y}_1)^2 + (\tau - \tau_1 - k\beta)^2 \right]^{-1}
$$

$$
= 1 + \frac{\pi T \lambda^2}{|\mathbf{x} - \mathbf{y}_1|} \frac{\sinh(2\pi T |\mathbf{x} - \mathbf{y}_1|)}{\cosh(2\pi T |\mathbf{x} - \mathbf{y}_1|) - \cos[2\pi T (\tau - \tau_1)]} \tag{8.94}
$$

Here, $\mathbf{y}_1$ and $0 \le \tau_1 \le \beta$ represent the position of the instanton, while the summation over k replicates it periodically along the imaginary time axis. Surprisingly, the finite-temperature instanton and anti-instanton have exactly the same classical action, $8\pi^2/g^2$, as the $T = 0$ instanton and anti-instanton.

It is necessary to compute the one-loop quantum correction in the background field of an instanton or anti-instanton at finite temperature. This is a formidable task but has been done by Pisarski and Yaffe [24]. The result is that the integrand in (8.91) is multiplied by a cutoff factor

$$
\exp\left[-\tfrac{1}{3}(2N + N_{\mathrm{f}})\pi^2 T^2 \lambda^2\right] \tag{8.95}
$$

at large λ. This means that the λ integration is now both infrared and ultraviolet convergent. Finite temperature suppresses large instantons as expected. The lack of appearance of the coupling constant in the cutoff is simply understood as follows. At $T = 0$, quantum corrections replace the coupling constant in the classical action with the renormalization-group running coupling. Therefore we may postulate that at finite temperature the running coupling would be replaced by the static screened charge

$$\frac{8\pi^2}{\bar{g}^2} = \frac{8\pi^2}{g^2} + \left(\frac{11N}{6} - \frac{N_f}{3}\right) \ln\left(\frac{\mathbf{q}^2}{M^2}\right) + \frac{8\pi^2 m_{el}^2}{g^2 \mathbf{q}^2} \qquad (8.96)$$

where m_{el}^2 is given by (8.49). The screening factor (8.95) is reproduced by the m_{el}^2/g^2 term in (8.96) if we make the replacement $\mathbf{q}^2 \to 4/\lambda^2$.

Instanton effects are greatest in a world without light quarks, on account of point 7 above. Then the contribution to the pressure is

$$P_{DGA} = 2C_N \int_0^\infty \frac{d\lambda}{\lambda^5} \left(\frac{4\pi^2}{\bar{g}^2}\right)^{2N} \exp\left(-\frac{8\pi^2}{\bar{g}^2} - \frac{2}{3}N(\pi T\lambda)^2\right) \qquad (8.97)$$

which can be integrated to give

$$P_{DGA} = T^4 \left(\frac{\Lambda_R}{T}\right)^{11N/3} \sum_{l=0}^{2N} a_l(N) \left[\ln\left(\frac{T}{\Lambda_R}\right)\right]^l \qquad (8.98)$$

The coefficients $a_l(N)$ depend on N and must be computed numerically. The most noteworthy feature of P_{DGA} is that it decreases dramatically with increasing temperature. For instance, for SU(3) it falls as Λ_R^{11}/T^7, modulo logarithms. Comparison of these results with the perturbation theory results is left as an exercise.

Extensive numerical studies have been performed of an instanton-liquid description of QCD at zero and finite temperature. The reader is referred to the review of Schäfer and Shuryak [25].

8.7 Infrared problems

It would seem that if only we had the strength and willpower, we could continue to calculate corrections to P and $\Pi^{\mu\nu}$ to arbitrary order in g. However, a barrier that arises at order g^6 for P and at order g^4 for $\Pi^{\mu\nu}$ was identified by Lindé [26].

Let us investigate the infrared convergence of the $(l + 1)$-loop diagram ($l > 0$)

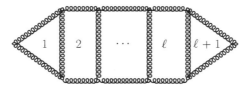

There are $2l$ three-gluon vertices and $3l$ propagators. The dominant infrared behavior arises from the $n = 0$ mode sums. To estimate, we dispense with the complicated tensorial structure of the propagator and the vertex and write

$$g^{2l} \left(T \int d^3 p \right)^{l+1} p^{2l} (p^2 + m^2)^{-3l} \tag{8.99}$$

The first and third factors arise from the vertices, the second factor from the loop integration, and the last factor from the propagators. We have introduced a possible static infrared cutoff m. We may wish to identify m with the "electric mass" $m_{\text{el}}^2 = F(0, \mathbf{0})$ or with the "magnetic mass" $m_{\text{mag}}^2 = G(0, \mathbf{0})$. In any case, (8.99) is of order

$$
\begin{array}{ll}
g^{2l} T^4 & \text{for } l = 1, 2 \\
g^6 T^4 \ln(T/m) & \text{for } l = 3 \\
g^6 T^4 (g^2 T/m)^{l-3} & \text{for } l > 3
\end{array}
\tag{8.100}
$$

We have placed an ultraviolet cutoff T on the momentum integration. This cutoff should arise automatically when summing over all modes n.

The interesting aspect of (8.100) is that if $m = 0$ and $l > 2$, then the diagram is infrared divergent. Now it may happen that when all diagrams of the same order are added together the coefficients of the infrared divergent parts are zero, although there is no symmetry to suggest that this is the case. The possibility is difficult to verify or deny, owing to the complexity of the diagrams. If we take $m = m_{\text{el}} \sim gT$ then no problem arises. At one-loop order, m_{mag} vanishes in all gauges; the next possibility is that $m_{\text{mag}} \sim g^2 T$. Substitution in (8.100) then suggests that all loops with $l > 3$ contribute to order g^6! It is not known how to sum all such diagrams, thus making it impractical even in principle to calculate analytically the coefficient of the order-g^6 term in P.

The same difficulty arises if we attempt to compute the static infrared limit of the gluon self-energy. For example, the diagram

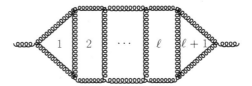

at $q_0 = 0$, $\mathbf{q} \to \mathbf{0}$ is of order

$$
\begin{array}{ll}
g^4 T^2 \ln(T/m) & \text{for } l = 1 \\
g^4 T^2 (g^2 T/m)^{l-1} & \text{for } l > 1
\end{array}
\tag{8.101}
$$

So, the infrared problem arises for $\Pi^{\mu\nu}$ at order g^4. Suppose, for the purpose of illustration, that $m_{\text{mag}}^2 = cg^4 T^2$. Then (8.101) suggests that to compute c we must sum an infinite set of diagrams. The constant c would then arise self-consistently. The magnetic contribution to the sum of ring diagrams would be proportional to $m^3_{\text{mag}}T \sim g^6 T^4$. This is another way of viewing the qualitatively different infrared effects that may arise at order g^6 in the pressure.

The static infrared problem in the above diagrams occurs when the momentum $p \leq g^2 T$. Another way to see this is to examine the factor $[p^2 + G(0, p)]^{-1}$ in the propagator. This changes sign as $p \to 0$; in TAG, $G(0, p) \to -(5/16)g^2 NTp$ and for the case of an arbitrary COVG, $G(0, p) \to -\{[8 + (\rho + 1)^2]/64\}g^2 NTp$.

The resolution of this problem, as suggested by Braaten [27], involves effective-field-theory methods coupled with lattice gauge calculations. However, it probably does not have much quantitative impact on thermodynamic functions like P at extremely high temperatures since it first occurs at order g^6 and $g(T) \to 0$ as $T \to \infty$.

8.8 Strange quark matter

Strange particles, like kaons and hyperons, do not play any role in daily life; that is to say, they are not stable particles and they are not found in atomic nuclei. Generally, they are only produced in high-energy reactions, and subsequently decay into nonstrange particles via the weak interactions. Could the situation be different in cold and dense quark matter? For cold neutron matter the baryon density is approximately

$$n = 2 \int \frac{d^3 p}{(2\pi)^3} \theta(p_{\text{F}} - p) = \frac{p^3_{\text{F}}}{3\pi^2} \tag{8.102}$$

where the Fermi momentum is $p^2_{\text{F}} = \mu^2 - m^2_N$. One may estimate the density at which neutrons overlap in coordinate space by multiplying this density by the volume of a nucleon, taking the nucleon radius to be 0.8 fm. Although very crude, this estimate determines the critical chemical potential as 1050 MeV, where a qualitative change in the nature of hadronic matter ought to occur. Since each quark carries one-third of a baryon charge, the quark chemical potential would be 350 MeV. This is larger than the generally accepted strange quark mass (see Table 8.2) and so allows for the possibility of the existence of strange quarks even when $T = 0$. Strange quarks might be produced and eventually come to equilibrium via the weak interactions $d \leftrightarrow u + e + \bar{\nu}_e$, $s \leftrightarrow u + e + \bar{\nu}_e$, and $s + u \leftrightarrow u + d$, provided that the circumstances are right. Indeed, it may very well be energetically favorable for some u and d quarks to be converted into s quarks at high density. The situation would be analogous to the presence of neutrons in nuclei. In free space, a neutron decays weakly into a proton, an electron, and an antineutrino. That does not happen in radioactively stable nuclei or in nuclear matter, because the

Pauli exclusion principle forbids the addition of a proton with an energy below the Fermi energy.

Let us assume chemical equilibrium under the weak interactions among u, d, s quarks and electrons. Then the aforementioned reactions imply that

$$\mu_u + \mu_e = \mu_d = \mu_s \tag{8.103}$$

For the present discussion the neutrinos may be neglected. We require that bulk matter be electrically neutral. Then we have the constraint

$$\tfrac{2}{3}n_u - \tfrac{1}{3}n_d - \tfrac{1}{3}n_s - n_e = 0 \tag{8.104}$$

The densities are functions of the chemical potentials. Together, (8.103) and (8.104) allow only one independent chemical potential.

For simplicity, first analyze the thermodynamics neglecting perturbative interactions among the quarks. For the large chemical potentials of interest it is reasonable to set $m_e = m_u = m_d = 0$. However, m_s is not so small and must be kept nonzero. The thermodynamic potential is a sum of contributions from each species:

$$\Omega_e = -\frac{\mu_e^4}{12\pi^2} \quad \Omega_u = -\frac{\mu_u^4}{4\pi^2} \quad \Omega_d = -\frac{\mu_d^4}{4\pi^2}$$

$$\Omega_s = -\frac{1}{4\pi^2}\left[\mu_s\sqrt{\mu_s^2 - m_s^2}(\mu_s^2 - 2.5m_s^2)\right. \tag{8.105}$$

$$\left. + 1.5m_s^4\ln\left(\frac{\mu_s + \sqrt{\mu_s^2 - m_s^2}}{m_s}\right)\right]$$

The energy density carried by the fermions is added to that associated with the vacuum, sometimes referred to as the MIT bag model constant, B, yielding a total energy density

$$\epsilon = \sum_i (\Omega_i + \mu_i n_i) + B \tag{8.106}$$

The baryon number density is

$$n_{\mathrm{B}} = \tfrac{1}{3}(n_u + n_d + n_s) \tag{8.107}$$

The quark matter is in stable mechanical equilibrium when $P = 0$. Including the bag pressure, this means

$$P = \sum_i P_i - B = -\sum_i \Omega_i - B = 0 \tag{8.108}$$

With the set of equations above, all parameters can be calculated for a given choice of m_s and B. The result of such calculations is shown in Figure 8.2.

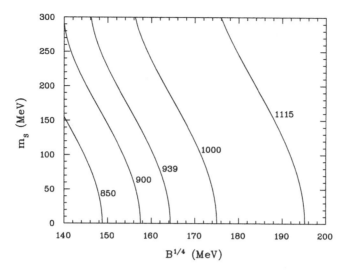

Fig. 8.2. Contours of fixed energy per baryon (in MeV) for strange quark matter in the $B^{1/4}-m_s$ plane; B is the bag model constant.

It is known that nonstrange quark matter is unbound. It must have an energy per baryon of at least 930 MeV $+ \Delta$ in order that ordinary atomic nuclei do not decay into nonstrange quark matter, which has never been observed. A detailed calculation suggests $\Delta = 4$ MeV [28]. A straightforward calculation then leads to a minimum value $B_{\min} = (145 \text{ MeV})^4$: this is the minimum value of the bag constant needed for atomic nuclei to be stable (in the $T = 0$ case, and neglecting interaction between the quarks). Considering Figure 8.2, normal atomic nuclei do not exist for values of $B < B_{\min}$. To the left of the 939 MeV contour, strange quark matter would be stable against decay into nucleons. The analysis outlined above is for degenerate, noninteracting, quark matter in bulk. Calculations including exchange corrections to order α_s have also been done. Farhi and Jaffe [28] found that the inclusion of exchange interactions up to order α_s effectively lowers $B_{\min}$ to smaller values.

The question arises of why ordinary nuclei have not decayed into strange quark matter, if it is more stable? The answer is that the conversion of many u and d quarks into s quarks requires a very high order of the weak interaction; thus the probability for this to happen is essentially zero. It is for this reason that nuclei may have been mistakenly taken to be the ground state of hadronic matter. Searches for strange quark matter in terrestrial experiments and in astrophysical observations have been ongoing. The effects of finite temperature and of finite size on the stability have been evaluated [29]. The fact that strange matter might be self-bound is in itself a fascinating proposition. The theoretical uncertainties surrounding it will surely decrease as our ability to perform numerical lattice gauge calculations at finite density grows.

8.9 Color superconductivity

The experimental discovery of superconductivity by Kamerlingh Onnes in 1911 was totally unexpected. It defied fundamental theoretical understanding until the Nobel-prize-winning work of Bardeen, Cooper, and Schrieffer (BCS) in 1957. The discovery of high-T_c materials in 1986 was also totally unexpected; its theoretical understanding is still a topic of research. Superconductivity has many applications nowadays, primarily in magnets used in research and in medicine. Conventional superconductivity arises from the pairing of electrons with equal but nearly opposite momentum near the Fermi surface. This pairing occurs because of a very weak attraction originating in phonon exchange, despite the fact that electrons experience a repulsive Coulomb interaction. This is one of the reasons why it took so long to work out a fundamental theoretical description of superconductivity. In QCD the situation is different. In a cold quark gas, single-gluon exchange is attractive for two quarks in a state that is antisymmetric in color, the $\bar{3}$ channel. The possibility of color superconductivity therefore exists, and indeed it happens.

Color superconductivity was first studied by Barrois [30] and Frautschi [31]. Further studies were reported by Bailin and Love [32], but it was not until 1998 that the field exploded in a flood of research papers led by Alford, Rajagopal, and Wilczek [33], and Rapp *et al.* [34]. These studies can be categorized into one of two classes: weak coupling methods using the fundamental QCD Lagrangian, valid at asymptotically high densities; and phenomenological methods using four-quark interactions, some motivated by instantons, which are intended for application at densities not much greater than those in ordinary atomic nuclei.

In order to allow for the pairing of quarks, we follow the path pioneered by Nambu and by Gorkov [35]. An eight-component Dirac field is introduced as

$$\Psi = \left(\psi, \bar{\psi}_{\mathrm{T}} \right)$$

where T denotes the transpose. The inverse propagator is an 8×8 matrix in Dirac space:

$$\mathcal{G}^{-1}(p) = \mathcal{G}_0^{-1}(p) + \Sigma(p) = \begin{pmatrix} \not{p} - m + \mu\gamma_0 & \bar{\Delta} \\ \Delta & (\not{p} + m - \mu\gamma_0)_{\mathrm{T}} \end{pmatrix} \quad (8.109)$$

Here $\bar{\Delta} = \gamma_0 \Delta^\dagger \gamma_0$ and Δ is an object with color, flavor, and Dirac indices, which have been suppressed. Setting $\Delta = 0$ yields the free propagator for this eight-component field. The self-energy contribution to the block diagonal components is neglected in order to focus on the coupling term, which gives rise to a gap and to superconductivity. In this section the chemical potential is separated out explicitly and is not subsumed into p_0.

In order to demonstrate the existence of superconductivity at high density we first focus on two flavors of massless u and d quarks with common chemical potential μ. This is referred to as the 2SC phase. The assumptions that are usually made are: (i) the gap matrix is antisymmetric in both flavor and color, which is the channel in which single-gluon exchange is attractive; (ii) condensation occurs in the channel with zero angular momentum, $J = 0$; (iii) the gap has positive parity, which is favored by the relatively weak instanton-induced interactions; (iv) chiral symmetry-breaking condensates coupling left- and right-handed quarks are neglected. Given these assumptions, the gap matrix takes the form

$$\Delta_{ij}^{ab}(p) = (\lambda_2)^{ab} \, (\tau_2)_{ij} \, C \, \gamma_5 \left[\Delta_+(p) P_+(p) + \Delta_-(p) P_-(p) \right] \quad (8.110)$$

where $C = i\gamma^0\gamma^2$ is essentially the charge conjugation operator and makes the operand into a scalar rather than a pseudoscalar; a, b are color indices, i, j are flavor indices, and Dirac indices are suppressed. The operators P_+ and P_- project onto particles and antiparticles, respectively:

$$P_\pm(p) = \tfrac{1}{2}(1 \pm \gamma_0 \boldsymbol{\gamma} \cdot \hat{\mathbf{p}}) \quad (8.111)$$

Thus Δ_+ describes the modification of the propagator due to particle–particle pairing, whereas Δ_- describes that due to antiparticle–antiparticle pairing. Particles and antiparticles are in this situation distinguished by the sign of the chemical potential.

The self-energy satisfies the Schwinger–Dyson equation,

$$\Sigma(k) = -g^2 T \sum_n \int \frac{d^3p}{(2\pi)^3} \, \Gamma_\mu^a(k,p) \mathcal{G}(p) \Gamma_\nu^b(k,p) \mathcal{D}_{ab}^{\mu\nu}(k-p) \quad (8.112)$$

which is written in Minkowski space; the factor $\Gamma_\mu^a(k,p)$ comes from the fully dressed quark–gluon vertex. At very high densities, where the running coupling becomes arbitrarily small, $\Gamma_\mu^a(k,p)$ can be replaced by the bare vertex:

$$\Gamma_\mu^a = - \begin{pmatrix} \tfrac{1}{2}\lambda^a\gamma_\mu & 0 \\ 0 & -(\tfrac{1}{2}\lambda^a\gamma_\mu)_{\mathrm{T}} \end{pmatrix} \quad (8.113)$$

Then the Schwinger–Dyson equation determines the gap function:

$$\Delta(k) = g^2 T \sum_n \int \frac{d^3p}{(2\pi)^3} \left(\gamma_\mu \frac{\lambda^a}{2} \right)_{\mathrm{T}} \mathcal{G}_{21}(p) \left(\gamma_\nu \frac{\lambda^a}{2} \right) \mathcal{D}^{\mu\nu}(k-p) \quad (8.114)$$

The 2, 1 component of the quark propagator has entered here. With the given ansatz for the gap matrix, (8.110), we get

$$\mathcal{G}_{21}(p) = -\lambda_2 \, \tau_2 \, C \, \gamma_5 \left[\frac{\Delta_+(p) P_-(p)}{p_0^2 - (|\mathbf{p}| - \mu)^2 - \Delta_+^2(p)} \right.$$
$$\left. + \frac{\Delta_-(p) P_+(p)}{p_0^2 - (|\mathbf{p}| + \mu)^2 - \Delta_-^2(p)} \right] \qquad (8.115)$$

The flavor factor τ_2 cancels on both sides of the gap equation; so does the color factor λ_2, because

$$\left(\tfrac{1}{2}\lambda^a \right)_{\mathrm{T}} \lambda_2 \left(\tfrac{1}{2}\lambda^a \right) = -\frac{N+1}{2N} \lambda_2$$

After substitution one finds a pair of coupled gap equations,

$$\Delta_\pm(k) = -\frac{g^2}{3} T \sum_n \int \frac{d^3p}{(2\pi)^3} \, \mathcal{D}^{\mu\nu}(k-p)$$
$$\times \left\{ \mathrm{Tr} \left[\gamma_\mu P_-(p) \gamma_\nu P_\pm(k) \right] \frac{\Delta_+(p)}{p_0^2 - (|\mathbf{p}| - \mu)^2 - \Delta_+^2(p)} \right.$$
$$\left. + \mathrm{Tr} \left[\gamma_\mu P_+(p) \gamma_\nu P_\mp(k) \right] \frac{\Delta_-(p)}{p_0^2 - (|\mathbf{p}| + \mu)^2 - \Delta_-^2(p)} \right\}$$
$$(8.116)$$

to be solved for the gaps $\Delta_\pm$. In order to take into account static or dynamic screening of the color fields, the one-loop dressed gluon propagator in a covariant gauge is used, as given in Section 8.5.

For the scattering of quarks near the Fermi surface, which is relevant for determining the gaps, the energy transfer is negligible compared with the momentum transfer. With $q \equiv k - p$, this means that $|q_0| \ll |\mathbf{q}|$. Then q_0 may be taken to zero wherever possible in the numerators of the gap equations, but not in the denominators since there may be a near singularity in the infrared. The Landau- and Coulomb-gauge gluon propagators give the same answer in this limit:

$$\Delta_\pm(k) = \frac{g^2}{3} T \sum_n \int \frac{d^3p}{(2\pi)^3}$$
$$\times \left\{ \frac{\Delta_\pm(p)}{p_0^2 - (|\mathbf{p}| \mp \mu)^2 - \Delta_\pm^2(p)} \left(\frac{3 - \hat{\mathbf{k}} \cdot \hat{\mathbf{p}}}{q^2 - G(q)} + \frac{1 + \hat{\mathbf{k}} \cdot \hat{\mathbf{p}}}{q^2 - F(q)} \right) \right.$$
$$\left. + \frac{\Delta_\mp(p)}{p_0^2 - (|\mathbf{p}| \pm \mu)^2 - \Delta_\mp^2(p)} \left(\frac{1 + \hat{\mathbf{k}} \cdot \hat{\mathbf{p}}}{q^2 - G(q)} + \frac{1 - \hat{\mathbf{k}} \cdot \hat{\mathbf{p}}}{q^2 - F(q)} \right) \right\}$$
$$(8.117)$$

These gaps are gauge independent only in this kinematic limit. This is a consequence of the fact that the gaps are determined by the scattering of quarks that are almost on-shell. Of course, any physical observable must be gauge independent. If one is working to higher order in the interactions, the approximations made above would not be acceptable.

Only the first term in the gap equation has a singularity on the Fermi surface, and so we keep it but drop the second term. This gives rise to a single integral equation to be solved for $\Delta \equiv \Delta_+$, the gap for quasi-particles and their holes near the Fermi surface (we are not interested in the gap for the antiparticles). In order to solve the integral equation for the gap we make a further physically motivated approximation. Since the participating quarks are those on or very near the Fermi surface, they all have essentially the Fermi momentum. Therefore we neglect the very weak three-momentum dependence of the gap and write it as $\Delta(k_0)$. We also write $\mathbf{p} = \mathbf{p}_F + \mathbf{l}$, where $\mathbf{p}_F$ is on the Fermi surface and $\mathbf{l}$ is perpendicular to it. For very large chemical potential and vanishingly small temperature it is adequate to use $l \ll \mu$ and write the integration measure as $\mu^2 dl\, d(\cos\theta) d\phi$; furthermore, $|\mathbf{q}| = |\mathbf{k} - \mathbf{p}| \approx \sqrt{2}\mu(1 - \cos\theta)$. The Matsubara sum can be replaced by an integral over Euclidean momentum p_4 (see (3.71)). The integral over ϕ is trivial, and the integral over l can be done by contour integration, picking up the pole of the diquark propagator:

$$
\Delta(k_4) = \frac{g^2\mu^2}{12\pi^2} \int_{-\infty}^{\infty} dp_4 \int_{-1}^{1} d\cos\theta \ \frac{\Delta(p_4)}{\sqrt{p_4^2 + \Delta^2(p_4)}}
$$
$$
\times \left[\frac{3 - \cos\theta}{q_4^2 + 2\mu^2(1 - \cos\theta) + G(q)} \right.
$$
$$
\left. + \frac{1 + \cos\theta}{q_4^2 + 2\mu^2(1 - \cos\theta) + F(q)} \right] \tag{8.118}
$$

Here F and G are evaluated with $q_4 = k_4 - p_4$ and $|\mathbf{q}| = \sqrt{2}\mu(1 - \cos\theta)$. In principle this gap equation should now be solved with no further approximations.

To get an idea of how the solution depends on the parameters we use the approximate forms for F and G,

$$
F(q) = m_{\text{el}}^2 \qquad G(q) = i\frac{\pi q_4}{4|\mathbf{q}|} m_{\text{el}}^2 \tag{8.119}
$$

in the limit $0 \le q_4 \ll |\mathbf{q}|$. This means that the electric part of the interaction is screened on the momentum scale $q_{\text{el}} = m_{\text{el}}$ while the magnetic part is screened on the scale $q_{\text{mag}} = (\pi m_{\text{el}}^2 \Delta/4)^{3/2}$. Integration over the angle θ gives the simplified gap equation

$$\Delta(k_4) = \frac{g^2}{18\pi^2} \int dp_4 \ \frac{\Delta(p_4)}{\sqrt{p_4^2 + \Delta^2(p_4)}}$$

$$\times \left[\ln\left(1 + \frac{32\mu^3}{\pi m_{\rm el}^2 |k_4 - p_4|} \right) + \frac{3}{2}\ln\left(1 + \frac{4\mu^2}{m_{\rm el}^2} \right) \right]$$

$$(8.120)$$

This integral equation can be converted to a differential equation and solved in the small-g approximation. The asymptotic solution is

$$\Delta(k_4) \approx \Delta_0 \sin\left[\frac{g}{3\sqrt{2}\pi} \ln\left(\frac{c\mu}{k_4} \right) \right] \qquad k_4 > \Delta_0 \qquad (8.121)$$

where

$$\Delta_0 = 2c\mu \exp\left(-\frac{3\pi^2}{\sqrt{2}g} \right)$$

$$2c = \frac{512}{\pi} \left(\frac{\mu}{m_{\rm el}} \right)^5$$

$$= 512\pi^4 g^{-5} \quad (N_{\rm f} = 2)$$

The amazing feature about this result is that the gap depends exponentially on $1/g$, not on $1/g^2$ as it does in ordinary superconductivity. This feature emerges from the longer-range nature of the color magnetic field compared with the color electric field. This result was first obtained by Son [36]. It has important implications for the numerical value of the gap, and the critical temperature, since g should be small compared to unity for the whole analysis to make sense.

Equation (8.120) is an approximation of (8.118). Numerical solution of the latter equation for small g yields a gap that is well described by (8.121) but with an overall coefficient that is smaller by a factor 0.28. However, there are several approximations that would need to be relaxed in order to obtain an accurate value of the coefficient of $g^{-5}\exp(-3\pi^2/\sqrt{2}g)$ for the gap. These include the diagonal contribution to the diquark self-energy (which modifies the quasiparticle dispersion relation), a renormalization-group improvement to obtain the proper choice of scale at which to evaluate the running coupling $g \to \bar{g}(\mu)$, and the use of dressed vertices in the Schwinger–Dyson equation. The first two of these have been done individually, and the third not at all. If these effects are ignored, Δ_0 at first decreases with increasing μ, reaches a minimum of about 10 MeV at $\mu = 1$ TeV, and then increases logarithmically with μ. A plot of Δ_0 versus μ (Figure 8.3) gives the scale of the gap, although its absolute magnitude is uncertain by a factor 2–4, increasingly so as μ decreases.

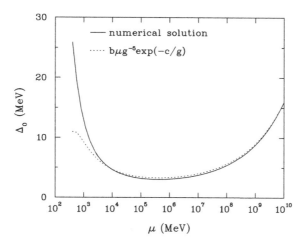

Fig. 8.3. The gap for two-flavor color superconductivity.

When color superconductivity occurs, the thermodynamic potential is lowered when compared with the case where pairing is absent. One may readily write an approximate expression for the thermodynamic potential that reproduces the Schwinger–Dyson equation used to obtain the gap equations. It is [37]

$$\Omega = \Omega_0 + \frac{1}{2}T\sum_n \int \frac{d^3p}{(2\pi)^3} \operatorname{Tr}\left[\ln\left(\frac{\mathcal{G}}{\mathcal{G}_0}\right) - \mathcal{G}\mathcal{G}_0^{-1} + 1\right]$$
$$+ \frac{1}{4}T^2\sum_{n,n'} \int \frac{d^3p\,d^3p'}{(2\pi)^6} \Gamma_\mu^a(p,p')\mathcal{G}(p)\Gamma_\nu^b(p',p)\mathcal{G}(p')\mathcal{D}_{ab}^{\mu\nu}(p'-p)$$

$$(8.122)$$

where Ω_0 is the potential in the absence of pairing. Treating Ω as a functional of $\mathcal{G}$ and requiring that it be an extremum results in $\mathcal{G}^{-1} - \mathcal{G}_0^{-1} = \Sigma$. When evaluated at the extremum, the shift in the potential (relative to no pairing) is

$$\delta\Omega = \frac{1}{2}T\sum_n \int \frac{d^3p}{(2\pi)^3} \operatorname{Tr}\left[\ln\left(\frac{\mathcal{G}}{\mathcal{G}_0}\right) - \frac{1}{2}\mathcal{G}\mathcal{G}_0^{-1} + \frac{1}{2}\right] \qquad (8.123)$$

Using the explicit form of the propagator and integrating over spatial momentum, the zero-temperature shift in the energy density is

$$\delta\Omega = 4\frac{\mu^2}{\pi^2}\int dp_4\left[\sqrt{p_4^2 + \Delta^2(p_4)} - p_4 - \frac{\Delta^2(p_4)}{2\sqrt{p_4^2 + \Delta^2(p_4)}}\right] \qquad (8.124)$$

This expression requires numerical solution, but a very good approximation is

$$\delta\Omega(\text{2SC}) = -4\left(\frac{\mu^2\Delta_0^2}{4\pi^2}\right) \tag{8.125}$$

This was first obtained by Miransky, Shovkovy, and Wijewardhana [38]. The overall factor 4 comes from the pairing of four quarks in the 2SC state.

Now we turn to the case of three flavors of massless quarks. Numerous studies have shown that the energetically most favorable state is one in which rotations of SU(3) color and SU(3) flavor are locked together. This is referred to as color–flavor locking (CFL) [39]. The simplest ansatz for the gap matrix is

$$\Delta_{ij}^{ab}(p) = (\lambda_{\text{I}})^{ab}\,(\lambda_{\text{I}})_{ij}\,C\,\gamma_5\left[\Delta_+(p)P_+(p) + \Delta_-(p)P_-(p)\right] \tag{8.126}$$

where λ_{I} is one of the three antisymmetric SU(3) matrices. The analysis then closely parallels the case of 2SC. It turns out that there are eight color–flavor combinations of quarks with gap $\Delta/2^{1/3}$ and one with gap $2\Delta/2^{1/3}$, where Δ is the same function as in 2SC. To logarithmic accuracy the CFL phase is favored over the 2SC phase at asymptotically high density, as long as $m_s^2 \ll 2\mu\Delta_0$:

$$\delta\Omega(\text{CFL}) = -\frac{12}{2^{2/3}}\left(\frac{\mu^2\Delta_0^2}{4\pi^2}\right) \tag{8.127}$$

For these weak coupling estimates to be valid, one may require for example that $g < 0.8$. This translates to $\mu > 100$ GeV, a density not relevant for any known terrestrial, astrophysical, or cosmological environment. What is of most interest for neutron stars and high-energy heavy ion collisions is the region of μ in the range from several hundred MeV to several GeV (remember that μ is one-third of the baryon chemical potential) and temperatures up to several hundred MeV. In this region of phase space, weak-coupling calculations may be a guide but cannot provide quantitative predictions. Therefore model studies have been done using the Nambu–Jona–Lasinio (NJL) model and instanton models for quark interactions. Figure 8.4 shows four likely phase diagrams depending on the number of quark flavors and the values of their masses. Panel (a) shows $N_{\text{f}} = 2$ flavors of massless quarks. At low density and temperature there is a nuclear liquid–gas phase transition, to be discussed in Chapter 11: a curve showing a line of first-order phase transition terminates in a second-order transition at the dot. A second curve separates hadronic and nuclear matter from quark–gluon plasma (QGP) and cold quark matter; the phase transition is second order above a critical point indicated by

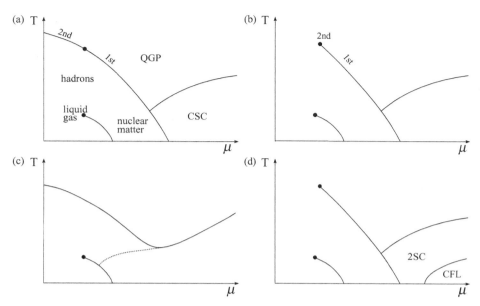

Fig. 8.4.   A model study of the phase diagram (*T* as a function of μ) for strongly interacting matter. The two upper panels are for two flavors of quarks; in (a) both quarks are massless whereas in (b) their masses have a nonzero common value. The two lower panels are for three flavors of quarks. In (c) they all are massless, in (d) the up and down quark masses are equal and the strange quark is given a heavier mass. (From Schäfer [40].)

the dot and first order below it. Color superconductivity (CSC) exists at high density and small temperatures and is separated from an unpaired quark–gluon plasma by a third curve, a line of second-order phase transition.

Panel (b) is like panel (a) except that the up and down quark masses are given a nonzero common value. In this case the line of first-order phase transition terminates at the critical point; there are paths along which one can go from nuclear or hadronic matter to quark–gluon plasma without undergoing a phase transition.

Panel (c) shows a scenario for three massless quark flavors. There is a line of first-order phase transition starting at $\mu = 0$ and extending to infinite density. At a given density, there is a high-temperature phase of quark–gluon plasma and a low-temperature phase which is either nuclear or hadronic or a CFL superconductor. Whether there is a sharp transition between dense nuclear matter and the CFL phase is unclear, as indicated by the dotted line.

Panel (d) shows a scenario for equal nonzero up and down quark masses and a heavier strange quark mass. The main difference between this panel and panel (b) is that, for a given temperature, the 2SC phase is favored at first and then the CFL phase is favored as the density increases. The non-superconducting phases and their phase transitions will be addressed more extensively in later chapters. The structure of strongly interacting matter is very rich and interesting!

8.10 Exercises

8.1 Prove that the QCD field strength tensor is not invariant under an infinitesimal gauge transformation, but that its square is.

8.2 Verify (8.26) by either of the two methods suggested.

8.3 Solve the renormalization-group equation for the covariant gauge parameter $\bar{\rho}$. Use (8.17), (8.18) with the one-loop results (8.22), (8.28).

8.4 Make graphs of (8.31) and (8.33), and then plot equations (8.34) to see how good an approximation they are.

8.5 Evaluate at least one of the non-quark two-loop diagrams for the pressure and show that it contributes to (8.45) as stated.

8.6 In the calculation of the pressure at order g^4 with fermions, enumerate all the diagrams that contribute and determine their individual combinatoric factors.

8.7 When evaluating the static ($k_0 = 0$) limit of the gluon polarization tensor $\Pi_{\mu\nu}$, one encounters integrals of the type

$$\int_0^\infty dx \left(\frac{1}{e^{ax} - 1} \right) \ln \left(\frac{1 + x}{|1 - x|} \right) \tag{E8.1}$$

Derive the asymptotic $a \to 0$ expansion for this integral, which is $\frac{\pi^2}{2a} + \ln\left(\frac{a}{2\pi}\right) + \gamma_E + \mathcal{O}(a)$ *Hint*: Divide the range of integration into $(0, 1)$ and $(1, \infty)$. Expand the logarithm in the integrand for each range appropriately and integrate term by term. Take the leading terms for $a \to 0$ and sum the resulting series.

8.8 (8.100) was determined on the basis of loop diagrams containing only three-point vertices. Can you find diagrams containing only four-point vertices which give the same behavior?

8.9 Plot the pressure of pure gluons to order g^3, with g^2 running with T according to the one-loop β-function. How much do the results change when the two-loop β-function is used instead?

8.10 Compare numerically the contribution to the pressure from instantons to the perturbative contribution at order g^2.

8.11 Calculate the dispersion relations for quarks in the 2SC phase.

References

1. Politizer, H. D., *Phys. Rev. Lett.* **30**, 1346 (1973).
2. Gross, D. J., and Wilczek, F., *Phys. Rev. Lett.* **30**, 1343 (1973).
3. Yang, C. N., and Mills, R., *Phys. Rev.* **96**, 191 (1954).
4. Close, F. (1979). *Introduction to Quarks and Partons* (Academic Press, New York).
5. Georgi, H., and Politizer, H. D., *Phys. Rev. D* **14**, 1829 (1976).
6. Hagiwara, K., and Yoshino, T., *Z. Phys.* **C24**, 185 (1984).
7. Caswell, W. E., *Phys. Rev. Lett.* **33**, 244 (1974).
8. Collins, J. C., and Perry, J. M., *Phys. Rev. Lett.* **34**, 1353 (1975).
9. Shuryak, E. V., *Sov. Phys. JETP* **47**, 212 (1978).
10. Kapusta, J. I., *Nucl. Phys.* **B148**, 461 (1979).
11. Toimela, T., *Phys. Lett.* **B124**, 407 (1983).
12. Freedman, B. A., and McLerran, L. D., *Phys. Rev. D* **16**, 1130, 1147, 1169 (1977).
13. Baluni, V., *Phys. Rev. D* **17**, 2092 (1978).
14. Celmaster, W., and Sivers, D., *Phys. Rev. D* **23**, 227 (1981).
15. Arnold, P., and Zhai, C., *Phys. Rev. D* **51**, 1906 (1995); **50**, 7603 (1994).
16. Zhai, C., and Kastening, B., *Phys. Rev. D* **52**, 7322 (1995).
17. Kajantie, K., Laine, M., Rummukainen, K., and Schröder, Y., *Phys. Rev D* **67**, 105008 (2003).
18. Leibbrandt, G., *Rev. Mod. Phys.* **59**, 1067 (1987).
19. Kajantie, K., and Kapusta, J. I., *Ann. Phys. (NY)* **160**, 477 (1985).
20. Heinz, U., Kajantie, K., and Toimela, T., *Ann. Phys. (NY)* **176**, 218 (1987).
21. Belavin, A. A., Polyakov, A. M., Schwartz, A. S., and Tyupkin, Yu. S., *Phys. Lett.* **B59**, 85 (1975).
22. 't Hooft, G., *Phys. Rev. D* **14**, 3432 (1976); Err. **18**, 2199 (1978).
23. Harrington, B. J., and Sheppard, H. K., *Phys. Rev. D* **17**, 2122 (1978).
24. Pisarski, R. D., and Yaffe, L. G., *Phys. Lett.* **B97**, 110 (1980).
25. Schäfer, T., and Shuryak, E. V., *Rev. Mod. Phys.* **70**, 323 (1998).
26. Lindé, A. D., *Phys. Lett.* **B96**, 289 (1980).
27. Braaten, E., *Phys. Rev. Lett.* **74**, 2164 (1995).
28. Farhi, E., and Jaffe, R. L., *Phys. Rev. D* **30**, 2379 (1984).
29. Madsen, J., *Lect. Notes Phys.* **516**, 162 (1999).
30. Barrois, B. C., *Nucl. Phys.* **B129**, 390 (1977).
31. Frautschi, S. (1980). *Proceedings of the Workshop on Hadronic Matter at Extreme Energy Density, Erice 1978* (Plenum Press, New York).
32. Bailin, D., and Love, A., *Nucl. Phys.* **B205**, 119 (1982).
33. Alford, M. G., Rajagopal, K., and Wilczek, F., *Phys. Lett.* **B422**, 247 (1998).
34. Rapp, R., Schäfer, T., Shuryak, E. V., and Velkovsky, M., *Phys. Rev. Lett.* **81**, 53 (1998).
35. Nambu, Y., *Phys. Rev.* **117**, 648 (1960); Gorkov, L. P., *Sov. Phys. JETP* **9**, 1364 (1959).
36. Son, D. T., *Phys. Rev. D* **59**, 094019 (1999).

37. Schäfer, T., *Nucl. Phys.* **B575**, 269 (2000).
38. Miransky, V. A., Shovkovy, I. A., and Wijewardhana, L. C. R., *Phys. Lett.* **B468**, 270 (1999).
39. Alford, M. G., Rajagopal, K., and Wilczek, F., *Nucl. Phys.* **B537**, 443 (1999).
40. Schäfer, T., *Int. J. Mod. Phys.* **B15**, 1474 (2001).

Bibliography

Reviews of QCD (at zero and finite temperature)

Politzer, H. D., *Phys. Rep.* **141**, 129 (1974).
Close, F. (1979). *Introduction to Quarks and Partons* (Academic Press, New York).
Shuryak, E. V., *Phys. Rep.* **61**, 71 (1980).
Gross, D. Pisarski, R., and Yaffe, L., *Rev. Mod. Phys.* **53**, 43 (1981).
Smilga, A. V. (2001). Hot and Dense QCD, in *At the Frontiers of Particle Physics: Handbook of QCD* (World Scientific, Singapore).

9

Resummation and hard thermal loops

We saw in the last chapter that QCD perturbation theory has problems at finite temperature, when taken into the infrared domain. These problems are the cause of the breakdown of the perturbative expansion of the pressure beyond $\mathcal{O}(g^6)$. We shall see here that these divergences are also responsible for the need to resum an infinite set of Feynman diagrams in order to compute a physical quantity at a given order in the coupling constant. These concepts were discussed before, in Chapter 3, when the set of ring diagrams was evaluated. In order to make the discussion as simple as possible, let us revisit the case of scalar $\lambda\phi^4$ theory.

Recalling our evaluation of the one-loop self-energy diagram, we had

$$\Pi_1 = 12\lambda T \sum_n \int \frac{d^3k}{(2\pi)^3} \frac{1}{\omega_n^2 + \omega^2} \tag{9.1}$$

As previously the vacuum contribution is renormalized by a mass counterterm and the complete self-energy, after analytic continuation to real energies, at finite temperature and at first order in the coupling is

$$\Pi_1^{\mathrm{ren}} = 12\lambda \int \frac{d^3k}{(2\pi)^3} \frac{1}{\omega} \frac{1}{\mathrm{e}^{\beta\omega} - 1} \tag{9.2}$$

In the high-temperature limit all masses are negligible, and then one can write $\Pi_1 \to \lambda T^2$. Therefore, at one-loop order thermal fluctuations generate a mass for the scalar field, $m_{\mathrm{eff}} = \sqrt{\lambda}T$. Notice that in the massless limit the integral in (9.2) is dominated by momenta of the order of the temperature, $k \sim T$. In our high-temperature limit these momenta would be referred to as "hard". The effects of the thermal mass can be incorporated by defining an effective propagator which, in frequency–momentum space, would be given by

$$\mathcal{D}^*(\omega_n, \mathbf{k}) = \frac{1}{\omega_n^2 + \mathbf{k}^2 + \lambda T^2} \tag{9.3}$$

This simple example tells us that if momenta are of the order of the temperature, or hard, the self-energy correction to the propagator is a perturbative correction and can be neglected. However, if the momentum is "soft", so that $k \sim \sqrt{\lambda}T$, then the thermal mass term is as large as the inverse bare propagator and certainly must be included. In this limit, the correction is as big as the leading term. The previous discussion also suggests that it is useful to define hard, $k \simeq T$, and soft, $k \simeq \sqrt{\lambda}T$, scales of momenta. Here k means indiscriminately energy or momentum. An instructive exercise consists of recalculating the self-energy, only this time using the effective propagator defined above. The only change from the previous evaluation of the self-energy is that now one has the energy appropriate for a massive field, $\omega = \sqrt{k^2 + m_{\text{eff}}^2}$. We examine the behavior of the integrand, which is largely dictated by the distribution function, and recall that $m_{\text{eff}}/T = \sqrt{\lambda}$. The contribution to the integral from hard momenta is small and generates corrections of order λ to the self-energy. The contribution to the integral from soft momenta allows the distribution function to be approximated by $N_{\text{B}}(\omega) \approx T/\omega$. Keeping in mind that the upper limit in this case is of order $\sqrt{\lambda}T$, we get a contribution to Π of order

$$\Pi \sim \frac{\lambda T}{m_{\text{eff}}^2} \int_0^{\sqrt{\lambda}T} dk \, k^2$$

The quantitative result is

$$m_*^2 = m_{\text{eff}}^2 \left(1 - \frac{3m_{\text{eff}}}{\pi T} + \cdots\right) \tag{9.4}$$

The improved effective mass is the same as m_{eff} to leading order but also contains a correction of order $\sqrt{\lambda}$ which is given entirely by the soft momenta in the loop integral. Importantly, this correction is obtained if one uses the effective propagator (9.3), which represents a resummation of an infinite set of higher-order diagrams. It is instructive to note that even though each of these diagrams is infrared divergent, their sum is finite. We have encountered this situation previously in the form of the ring diagrams in Chapter 3.

The scalar field application is considerably simpler than that of gauge theories, but it conveys the essential part of the message: in order to calculate systematically amplitudes with soft lines, it is necessary to resum perturbation theory by including all possible hard thermal loops. We shall see that, in general, this procedure involves effective propagators and vertices. In $\lambda\phi^4$ theory, it is sufficient to consider only the effective propagator as defined above. Since the coupling depends on the temperature only logarithmically, the use of bare vertices is adequate. The case of gauge theories is more involved technically because the self-energy is generally energy and momentum dependent, and because there are vertices that are energy and momentum dependent. Also, there is a rich set of important physical scales in weakly coupled gauge theories. The next section outlines how to pick out the contributions of hard thermal loops from a diagram with soft external four-momenta.

9.1 Isolating the hard thermal loop contribution

We will concentrate here on one-loop diagrams and generalize later. We follow the original treatment of Pisarski and Braaten [1, 2]. As discussed previously, the evaluation of one-loop self-energies involves a sum over discrete frequencies as well as an integral over three-momenta. This sum may be evaluated using the following technique. Let us first define a Fourier-transformed propagator with respect to $\omega_n = 2n\pi T$, for bosons, as

$$\Delta_{\mathrm{B}}(\tau, \mathbf{k}) = T \sum_{n=-\infty}^{\infty} \mathrm{e}^{-i\omega_n \tau} \mathcal{D}_0(\omega_n, \mathbf{k}) \qquad (9.5)$$

The sum over discrete frequencies is easily done by using the contour integration technique of Chapter 3. The result is

$$\Delta_{\mathrm{B}}(\tau, \mathbf{k}) = \frac{1}{2|\mathbf{k}|} \left\{ [1 + N_{\mathrm{B}}(\mathbf{k})] \mathrm{e}^{-|\mathbf{k}|\tau} + N_{\mathrm{B}}(\mathbf{k}) \mathrm{e}^{|\mathbf{k}|\tau} \right\} \qquad (9.6)$$

where $N_{\mathrm{B}}(\mathbf{k}) = 1/\left[\exp(|\mathbf{k}|/T) - 1\right]$. The inverse of (9.5) is

$$\mathcal{D}_0(\omega_n, \mathbf{k}) = \int_0^\beta d\tau \mathrm{e}^{i\omega_n \tau} \Delta_{\mathrm{B}}(\tau, \mathbf{k}) \qquad (9.7)$$

It is easy to verify that the boson propagator in imaginary time has the following properties:

$$\Delta_{\mathrm{B}}(\tau - \beta, \mathbf{k}) = \Delta_{\mathrm{B}}(-\tau, \mathbf{k}) = \Delta_{\mathrm{B}}(\tau, \mathbf{k}) \qquad (9.8)$$

A similar analysis for fermions, with $\omega_n = (2n + 1)\pi T$, yields the intuitive result

$$\Delta_{\mathrm{F}}(\tau, \mathbf{k}) = T \sum_{n=-\infty}^{\infty} \frac{\mathrm{e}^{-i\omega_n \tau}}{\omega_n^2 + \mathbf{k}^2} = \frac{1}{2|\mathbf{k}|} \left\{ [1 - N_{\mathrm{F}}(\mathbf{k})] \mathrm{e}^{-|\mathbf{k}|\tau} - N_{\mathrm{F}}(\mathbf{k}) \mathrm{e}^{|\mathbf{k}|\tau} \right\}$$

$$(9.9)$$

where $N_{\mathrm{F}}(\mathbf{k}) = 1/\left[\exp(|\mathbf{k}|/T) + 1\right]$.

The usefulness of this approach is easily illustrated for the case of one-loop integrals. There, each internal propagator (we concentrate on bosons for simplicity) is written as an integral over τ as in (9.7). Evaluating the sum over discrete frequencies will create a delta function in τ. The overall integral over imaginary time can be done directly. Then the contribution of order T^2 is easy to pick out. A few examples will help to illustrate the procedure.

In a tadpole self-energy one has to evaluate

$$
T \sum_{n=-\infty}^{\infty} \int \frac{d^3k}{(2\pi)^3} \mathcal{D}_0(\omega_n, \mathbf{k}) = T \sum_{n=-\infty}^{\infty} \int \frac{d^3k}{(2\pi)^3} \int_0^\beta d\tau e^{i\omega_n \tau} \Delta_{\mathrm{B}}(\tau, \mathbf{k})
$$

$$
= \int \frac{d^3k}{(2\pi)^3} \Delta_{\mathrm{B}}(\tau = 0, \mathbf{k})
$$

$$
= \int \frac{d^3k}{(2\pi)^3} \frac{1}{2|\mathbf{k}|} [1 + 2N_{\mathrm{B}}(\mathbf{k})] \qquad (9.10)
$$

The first term corresponds to the usual $T = 0$ ultraviolet divergence and is removed by renormalization. Rewriting the result for the remaining term we get

$$
T \sum_{n=-\infty}^{\infty} \int \frac{d^3k}{(2\pi)^3} \mathcal{D}_0(\omega_n, \mathbf{k}) \approx \frac{1}{12} T^2 \qquad (9.11)
$$

The approximation sign means that equality holds "in the hard thermal loop (HTL) limit". Here this is actually an exact result.

Another example is that of the photon self-energy in scalar QED. Let us write the Lagrangian that governs the behavior of the scalar field ϕ and of the photon field A_μ as

$$
\mathcal{L} = (D_\mu \phi)^* D^\mu \phi - \frac{1}{4} F_{\mu\nu} F^{\mu\nu} - \frac{1}{2\rho} (\partial^\mu A_\mu)^2 \qquad (9.12)
$$

where ρ is the gauge-fixing parameter, discussed in Chapter 5. Recall that $D_\mu = \partial_\mu + ieA_\mu$. The Feynman diagrams that contribute to the first-order self-energy are

In Euclidean space the photon self-energy is

$$
\Pi^{\mu\nu}(\omega_m, \mathbf{p}) = -e^2 T \sum_n \int \frac{d^3k}{(2\pi)^3} \frac{(2k + p)^\mu (2k + p)^\nu}{(\omega_n^2 + \mathbf{k}^2)[(\omega_m + \omega_n)^2 + |\mathbf{p} + \mathbf{k}|^2]}
$$

$$
+ 2\,\delta^{\mu\nu} e^2 T \sum_n \int \frac{d^3k}{(2\pi)^3} \frac{1}{\omega_n^2 + \mathbf{k}^2} \qquad (9.13)
$$

To facilitate the usage of noncovariant propagators, this may be written as

$$\Pi^{\mu\nu}(\omega_m, \mathbf{p}) = -e^2 T \sum_n \int \frac{d^3k}{(2\pi)^3} \frac{(k-q)^\mu (k-q)^\nu}{(\omega_n^2 + \mathbf{k}^2)(\omega_q^2 + \mathbf{q}^2)}$$
$$+ 2\delta^{\mu\nu} e^2 T \sum_n \int \frac{d^3k}{(2\pi)^3} \frac{1}{\omega_n^2 + \mathbf{k}^2} \tag{9.14}$$

where $\mathbf{q} = \mathbf{p} - \mathbf{k}$ and $\omega_q = \omega_m - \omega_n$.

Concentrating on the finite-temperature contributions, and recalling that the temperature-independent divergent parts are regulated using the same techniques that operate at zero temperature, we may write

$$\Pi^{\mu\nu} = F P_L^{\mu\nu} + G P_T^{\mu\nu} \tag{9.15}$$

which is now in Minkowski space. The analysis and the results are very similar to those of electronic QED that were derived in Chapters 5 and 6. The quantities P_L and P_T are the familiar longitudinal and transverse projection tensors. The scalar functions F and G are inferred as follows:

$$F = \Pi^{\mu\nu} P_{L\,\mu\nu} , \tag{9.16}$$

$$G = \frac{1}{2} \Pi^{\mu\nu} P_{T\,\mu\nu} \tag{9.17}$$

Note that the transversality of $\Pi^{\mu\nu}$ ($k_\mu \Pi_L^{\mu\nu} = k_\mu \Pi_T^{\mu\nu} = 0$) is manifest, as required by current conservation. Writing $F = (p^2/\mathbf{p}^2) \Pi^{00}$ and using (9.15) and the fact that the integral defining the scalar function will be dominated by the hard momentum scale $k \sim T$, we get

$$F \approx \frac{e^2 T^2}{3} \left(1 - \frac{p_0^2}{\mathbf{p}^2}\right) \left[1 - \frac{p_0}{2|\mathbf{p}|} \ln\left(\frac{p_0 + |\mathbf{p}|}{p_0 - |\mathbf{p}|}\right)\right] \tag{9.18}$$

where for this discussion $p_0 = i\omega_m$, and $p^2 = p_0^2 - \mathbf{p}^2 = -(\omega_m^2 + \mathbf{p}^2)$. Here again the approximation sign is to be interpreted as meaning "in the HTL limit". Similarly, one may show that

$$G \approx \frac{1}{2} \left(\frac{e^2 T^2}{3} - F\right) \tag{9.19}$$

From this analysis of the HTL contribution in scalar QED, some new aspects are immediately apparent. Unlike $\lambda\phi^4$ theory, the self-energy is not only temperature dependent but now also momentum dependent. Also, the self-energy can develop an imaginary part when the kinematics are such that $|\mathbf{p}| > p_0 > -|\mathbf{p}|$. This situation corresponds to that of Landau damping, where one particle is emitted from the thermal medium and another is absorbed.

The reader will undoubtedly have noticed that writing propagators in frequency–momentum space and then performing a Fourier transform to imaginary time is but another method of doing the frequency sums. In some cases, it has advantages over the direct contour integral technique. However, the trick turns out to be useful only for loop diagrams; for tree-level diagrams little is gained by going over to the imaginary time–momentum domain.

The examples considered here can be used to extract some rules for the evaluation of the HTLs. Following Braaten and Pisarski, one might generalize the procedure to the evaluation of N-point functions in one-loop amplitudes for QCD at finite temperature. Before one proceeds to the more general case, it is instructive to evaluate explicitly the HTL contribution to the gluon self-energy. This calculation was first performed by Kalashnikov and Klimov [3] and Weldon [4].

We shall perform the QCD calculation in the Coulomb gauge $(\nabla \cdot \mathbf{A}^a = 0)$, with a as color index. Even though this gauge is a little awkward for many applications, owing partly to the fact that it is noncovariant, it has certain advantages at finite temperature. We will see some of those shortly. Of course, the result of a calculation of any physical quantity should be gauge invariant. Recall that the bare gluon propagator in the Coulomb gauge is (omitting the color indices)

$$\mathcal{D}^{\mu\nu} = -\frac{1}{p^2}P_{\mathrm{T}}^{\mu\nu} - \frac{1}{\mathbf{p}^2}u^\mu u^\nu \tag{9.20}$$

A gauge-fixing term $(\nabla \cdot \mathbf{A}^a)^2/2\rho$ could be added but would not change the analysis that follows. At one-loop order, the gluon self-energy is obtained by computing the Feynman diagrams in the following figure:

Note that the ghost propagator in the Coulomb gauge is $1/\mathbf{p}^2$, omitting color indices. Thus, ghost fields are static in the Coulomb gauge: they do not propagate. The same is true of the longitudinal gluons, another convenient feature. Hence there are advantages in using the Coulomb gauge in an application like the one considered here. In gauges with propagating unphysical degrees of freedom, the contributions from ghosts and longitudinal gluons cancel each other only in the final stages of a calculation. Therefore, the choice of the Coulomb gauge makes the computation of the second diagram in the figure above unnecessary.

The first diagram in the figure generates a contribution to the self-energy that is

$$\Pi = -\frac{g^2 N}{2}T\sum_n \int \frac{d^3k}{(2\pi)^3}\Gamma\mathcal{D}\Gamma\mathcal{D} \tag{9.21}$$

The prefactor is easily understood. The factor $1/2$ comes from a combination of the coefficient in the perturbative expansion of the thermodynamic potential, the numerical factors associated with the triple-gluon vertex, and combinatorics. The factor N comes from a color trace. In the HTL limit, the loop momentum constitutes a hard scale. In that limit, the vertex functions can be rewritten as

$$p(\alpha, a)$$

$$\approx i g f_{abc} \left(g_{\beta\gamma}\, 2k_\alpha - g_{\alpha\beta}\, k_\gamma - g_{\gamma\alpha}\, k_\beta \right)$$

$$(k-p)(\beta, b) \qquad\qquad k(\gamma, c)$$

where k is the hard loop momentum. Inserting this vertex and using the high-temperature limit, one obtains the HTL limit of the contribution of the first self-energy diagram in frequency–momentum space:

$$\Pi^{\mu\nu}(\omega_m, \mathbf{p}) \approx 4g^2 N T \sum_n \int \frac{d^3 k}{(2\pi)^3} k^\mu k^\nu \mathcal{D}_0(k) \mathcal{D}_0(p-k)$$

$$+ g^2 N T \sum_n \int \frac{d^3 k}{(2\pi)^3} \delta^{\mu i} \delta^{\nu j} \mathcal{D}_{ij}(k) \tag{9.22}$$

In a similar fashion one may compute the self-energy corresponding to the four-gluon vertex and to the quark–antiquark loop. The sum of these different contributions is written in Euclidean space as

$$\Pi^{\mu\nu}(\omega_m, \mathbf{p}) \approx 4g^2 \left(N + \frac{1}{2} N_{\mathrm{f}} \right) \left(T \sum_n \int \frac{d^3 k}{(2\pi)^3} k^\mu k^\nu \mathcal{D}_0(k) \mathcal{D}_0(p-k) \right.$$

$$\left. - \frac{1}{2} \delta^{\mu\nu} T \sum_n \int \frac{d^3 k}{(2\pi)^3} \mathcal{D}_0(k) \right) \tag{9.23}$$

The high-temperature limit of the gluon self-energy takes exactly the same form as (9.18), (9.19) with the replacement of the overall factor $e^2 T^2/3$ by the square of the color electric mass m_{el}^2. The latter is

$$m_{\mathrm{el}}^2 = \frac{1}{3} g^2 \left[N T^2 + \frac{1}{2} \sum_f \left(T^2 + \frac{3}{\pi^2} \mu_f^2 \right) \right] \tag{9.24}$$

where the sum refers to the quark flavors f, which may have differing chemical potentials. The energy and momentum dependence of the gluon self-energy, in this limit, is also identical to that of electronic QED, as analyzed in Section 6.7. It follows that the functional form of the dispersion relation is the same as for photons, at least to lowest order in the coupling constants.

Upon generalizing our result from self-energies to an arbitrary N-point function, considerable progress is made by observing that the momentum-independent term in (9.22) is specific to the self-energy topology. For example, it will be absent in the HTL limit of the three-point functions.

Next consider the N-gluon amplitude in the Coulomb gauge. One of the Feynman diagrams is shown in the following figure. (The complete HTL calculation also needs a diagram with an internal quark loop.)

In the usual notation this N-point amplitude is proportional to

$$T \sum_n \int \frac{d^3k}{(2\pi)^3} k^{\mu_1} \cdots k^{\mu_N} \mathcal{D}_0(k) \mathcal{D}_0(p_1 - k) \cdots \mathcal{D}_0(p_{N-1} - k) \quad (9.25)$$

Insert the noncovariant propagators and do the frequency sum. One of the resulting terms is

$$\int d^3k \, \frac{k^{\mu_1} \cdots k^{\mu_N}}{|\mathbf{k}||\mathbf{p}_1 - \mathbf{k}| \cdots |\mathbf{p}_{N-1} - \mathbf{k}|} \left[N_B(\mathbf{k}) - N_B(\mathbf{p}_1 - \mathbf{k}) \right]$$
$$\times \left[(p_1^0 - |\mathbf{k}| + |\mathbf{p}_1 - \mathbf{k}|) \cdots (p_{N-1}^0 - |\mathbf{k} - \mathbf{p}_{N-1}| + |\mathbf{k}|) \right]^{-1} \quad (9.26)$$

The structure of the integrand can be understood as follows. The N momenta in the first denominator come from the denominators of the Fourier transform of the Euclidean propagators, (9.6); the N gluon momenta in the numerator come from the triple-gluon coupling, which is linear in momentum. The energy denominators come from integrating over the different imaginary time variables associated with the use of (9.7). There are only $N - 1$ of them, as the first integral was used in conjunction with the delta function generated by the frequency sum. As argued previously, hard thermal loops occur when the integrating region is hard, of order T. We may get an estimate of the magnitude of (9.26) when the external momentum is soft, of order gT. In that limit

$$k \sim T \qquad |\mathbf{p}_i - \mathbf{k}| \sim T \qquad |\mathbf{k}| - |\mathbf{p} - \mathbf{k}| \sim |\mathbf{p}|$$

Now we use $N_B(\mathbf{k}) - N_B(\mathbf{p} - \mathbf{k}) \approx |\mathbf{p}| z dN_B(\mathbf{k})/dk$, with $z = \hat{\mathbf{p}} \cdot \hat{\mathbf{k}}$. Putting all this together, and recalling that there is one power of the coupling constant at each vertex, the amplitude for N external gluons is $g^N T^2/|\mathbf{p}|^{N-2}$. The tree-level diagram for the N-point gluon amplitude is easier to estimate: it contains $N - 2$ vertices and $N - 3$ propagators. Its magnitude is thus $g^{N-2}/|\mathbf{p}|^{N-4}$. Clearly, when $|\mathbf{p}| \sim gT$ in the one-loop N-gluon amplitude, its magnitude is that of the tree-level contribution and therefore has to be included in a consistent calculation. Therefore a resummation is required.

A set of rules for the power-counting of one-loop diagrams, first established by Braaten and Pisarski, may be inferred from the above analysis. They are summarized here for the case where all external momenta are soft.

1 The measure of the integral over the loop momentum is of magnitude T^3.
2 One propagator does not contribute an energy denominator since it is used in the integral over the delta function in imaginary time. Thus, there is a factor $1/T$ for the first propagator from the denominator of (9.6), and a factor $1/gT$ for each additional propagator owing to Landau damping contributions.
3 Each k^μ in the numerator, from vertices or fermion propagators, is replaced by T.
4 For loops with at least two propagators, if the latter represent fields of the same statistics, an extra factor of p/T appears owing to the cancellation of distribution functions.

Note that the $N = 2$ case does require separate consideration. This can be seen upon examination of the tadpole diagram in the scalar QED example, and also in the computation of the gluon self-energy in the Coulomb gauge. It is now also clear why loops with ghost fields will not contribute to the HTL term in the Coulomb gauge: because they are nonpropagating they cannot generate the term $1/gT$ associated with Landau damping. More specifically, the transverse gluon propagator will have a contribution $1/T \times 1/gT$, whereas a field with propagator $\sim 1/\mathbf{k}^2$ will have a contribution $1/T^2$, suppressed by one power of g. It is also useful to note here that these rules assume that the N-point functions are linear in the thermal distribution functions, whereas from (9.6) it would appear that powers of the distribution function would arise. This power would be the same as the number of propagators. However, in the final result a cancellation always yields a single power of N_{B} or N_{F}. This fact is most easily seen when the frequency sum is performed by the technique of contour integration. Indeed, considering (3.40) one sees that each pole residue gets multiplied by a single distribution function. Finally, we note that there is no HTL amplitude with external ghost fields.

9.2 Hard thermal loops and Ward identities

In the case of a gauge theory we know that N-point functions are related to $(N - 1)$-point functions by Ward identities. At high temperature, where HTLs give the leading contribution, we shall verify that Ward identities are indeed satisfied. This is a useful check on the method.

As an exercise, we can first check whether the HTL limit of the photon self-energy in scalar QED satisfies a Ward identity; verification is immediate, $p^\mu \Pi_{\mu\nu}(\omega_m, \mathbf{p}) = 0$. The same observation can be made for the gluon self-energy we obtained previously. It suffices to calculate explicitly three-point and four-point functions in order to generalize to a given topology at some order of the coupling. For example, the three-gluon amplitude will receive contributions from a pure gluon loop and from a quark loop. With the rules, the HTL limit of their sum can be obtained, and it is

$$\Gamma^{\mu\nu\sigma} \approx - 8g^2 \left(N + \frac{1}{2} N_{\mathrm{f}} \right) T$$
$$\times \sum_n \int \frac{d^3 k}{(2\pi)^3} k^\mu k^\nu k^\sigma \mathcal{D}_0(k) \mathcal{D}_0(p_1 - k) \mathcal{D}_0(p_2 + k) \quad (9.27)$$

Note that the momentum-labeling convention in the vertex is such that $p_1 + p_2 + p_3 = 0$. Similarly, the HTL limit of the two-quark one-gluon vertex is obtained through the evaluation of

and is

$$\Gamma^\mu_{2q-1g} \approx -4g^2 C_{\mathrm{f}} \gamma_\nu T \sum_n \int \frac{d^3 k}{(2\pi)^3} k^\mu k^\nu \mathcal{D}_0(k) \mathcal{D}_0(p_1 - k) \mathcal{D}_0(p_2 + k) \quad (9.28)$$

where $C_{\mathrm{f}} = (N^2 - 1)/2N$ and N is the number of colors.

The Ward identities can be derived in a straightforward fashion from the properties of the three-point and four-point functions in the HTL limit. Besides the transversality condition already mentioned, they are

$$p_{3\gamma} \Gamma^{\alpha\beta\gamma}(p_1, p_2, p_3) = \Pi^{\alpha\beta}(p_1) - \Pi^{\alpha\beta}(p_2)$$
$$p_{3\mu} \Gamma^\mu_{2q-1g}(p_1, p_2, p_3) = \Sigma(p_1) - \Sigma(p_2) \quad (9.29)$$

where $\Sigma(p)$ is the high-temperature quark self-energy. Similarly, the four- and three-point functions, in the HTL limit, are related by

$$p_{4\delta} \Gamma^{\alpha\beta\gamma\delta}(p_1, p_2, p_3, p_4) = \Gamma^{\alpha\beta\gamma}(p_1 + p_4, p_2, p_3)$$
$$- \Gamma^{\alpha\beta\gamma}(p_1, p_2 + p_4, p_3)$$
$$p_{4\beta} \Gamma^{\alpha\beta}_{2q-2g}(p_1, p_2, p_3, p_4) = \Gamma^\alpha_{2q-1g}(p_1 + p_4, p_2, p_3)$$
$$- \Gamma^\alpha_{2q-1g}(p_1, p_2 + p_4, p_3) \quad (9.30)$$

where a trace over the color indices of the two gluons has been taken. It is now apparent that effective vertices can and will exist in cases where no bare vertex is defined. For example, the 1PI vertex between a pair of quarks and a pair of gluons, $\Gamma^{\mu\nu}_{2q-2g}$, will consist solely of one-loop HTL contributions, as is obvious from the figure:

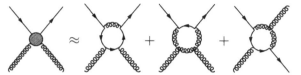

A consequence of the fact that hard thermal loops obey Ward identities similar to those of tree amplitudes is that their generating functional is a gauge-invariant functional of the quark and gauge fields.

9.3 Hard thermal loops and effective perturbation theory

We have seen that, owing to hard thermal loops, some Feynman diagrams that are superficially higher order in the coupling constant will have the same magnitude as tree-level diagrams in finite-temperature field theories. We have also seen how to evaluate the HTL contributions. We can employ this knowledge to resum HTLs into an effective theory. In this formalism bare vertices and propagators will be replaced by effective vertices and propagators, which are obtained via a HTL resummation.

Using the Schwinger–Dyson equation one may define an effective gluon inverse propagator in terms of the bare one as

$$(\mathcal{D}^*)^{-1}_{\mu\nu} = p^2 g_{\mu\nu} - p_\mu p_\nu + \Pi^*_{\mu\nu} \tag{9.31}$$

where the self-energy is evaluated in the HTL limit. An equivalent expression exists for quarks, the inverse propagator being related to the self-energy. The effective three-gluon vertex can be constructed similarly:

$$\Gamma^{*\mu\nu\sigma}(p_1, p_2, p_3) = \Gamma^{\mu\nu\sigma}_0(p_1, p_2, p_3) + \delta\Gamma^{\mu\nu\sigma}(p_1, p_2, p_3) \tag{9.32}$$

where the finite-temperature contribution (the second term on the right-hand side) is evaluated in the HTL limit. The contributions to the three-point function, in a ghost-free gauge, are represented by

This procedure is generalized to more complicated topologies. As noted in the previous section, HTL effective vertices can exist in the absence of their bare counterparts. If all the external momenta are of order gT then the HTL self-energies are of the same order as the bare inverse propagators; the same statement holds true for vertices. Therefore, in the evaluation of a loop contribution, propagators and vertices need to be of the effective kind for the kinematical region where the loop momentum is soft. An example is that of the one-loop quark self-energy

where the blobs denote effective quantities.

One might formalize the effective perturbation theory one step further by starting with effective Lagrangians. One can show [5] that the effective Lagrangian for gluonic hard thermal loops is

$$\mathcal{L} = -\frac{1}{2}m_{el}^2 \,\mathrm{Tr}\left[F_{\mu\nu}(x) \int \frac{d\Omega}{4\pi} \frac{\hat{k}^\nu \hat{k}^\lambda}{(\hat{k}\cdot D)^2} F_\lambda{}^\mu(x) \right] \tag{9.33}$$

Here the trace runs over color indices, $F_{\mu\nu} = F_{\mu\nu}^a G_a$ where the G_a are the generators of the group, and $\hat{k} = (-i, \hat{\mathbf{k}})$ (in Minkowski space). The integration over solid angle refers to the direction $\hat{\mathbf{k}}$. Also, $D^\mu = \partial^\mu + igA^\mu$ is the covariant derivative ($A^\mu = A_a^\mu G^a$). Similarly the effective Lagrangian for fermionic hard thermal loops is

$$\mathcal{L} = m_q^2 \bar{\psi}(x) \gamma_\mu \int \frac{d\Omega}{4\pi} \frac{\hat{k}^\mu}{\hat{k}\cdot D} \psi(x) \tag{9.34}$$

with

$$m_q^2 = \frac{N^2 - 1}{16N} g^2 \left(T^2 + \frac{\mu^2}{\pi^2} \right) \tag{9.35}$$

9.4 Spectral densities

It is interesting to know where the spectral weights are concentrated for various operators within the hard thermal loop approximation. Here we shall focus on the quark spectral densities since they will be used in Chapter 14 to compute the rate of photon emission from the quark–gluon plasma formed in high-energy heavy ion collisions.

The quark self-energy in the HTL limit may be immediately inferred from the electron self-energy given in Section 6.8. The only difference is the change in the numerical factor in the fermion–vector meson vertex. The quark propagator is

$$\mathcal{G}^*(p) = \mathcal{G}_+^*(p)\frac{\gamma_0 - \hat{\mathbf{p}}\cdot\boldsymbol{\gamma}}{2} + \mathcal{G}_-^*(p)\frac{\gamma_0 + \hat{\mathbf{p}}\cdot\boldsymbol{\gamma}}{2} \tag{9.36}$$

where

$$\mathcal{G}_\pm^*(p) = \left\{ -p_0 \pm |\mathbf{p}| + \frac{m_q^2}{|\mathbf{p}|} \left[Q_0 \left(\frac{p_0}{|\mathbf{p}|} \right) \mp Q_1 \left(\frac{p_0}{|\mathbf{p}|} \right) \right] \right\}^{-1} \tag{9.37}$$

The functions Q_0 and Q_1 are the Legendre functions of the second kind, namely

$$Q_0(z) = \frac{1}{2} \ln \left(\frac{1+z}{1-z} \right) \qquad Q_1(z) = zQ_0(z) - 1 \tag{9.38}$$

The effective quark mass was given in (9.35). In the limit $g \to 0$ we recover the bare quark propagator.

It is a straightforward exercise to compute the spectral densities for the functions $\mathcal{G}_\pm^*$. They are

$$\rho_\pm^*(\omega, \mathbf{p}) = \frac{\omega^2 - \mathbf{p}^2}{2m_q^2} \left[\delta(\omega - \omega_\pm(\mathbf{p})) + \delta(\omega + \omega_\mp(\mathbf{p})) \right] + \beta_\pm(\omega, \mathbf{p}) \theta(\mathbf{p}^2 - \omega^2) \tag{9.39}$$

with

$$\beta_\pm(\omega, \mathbf{p}) = \frac{1}{2} m_q^2 (|\mathbf{p}| \mp \omega) \left(\left\{ |\mathbf{p}|(\omega \mp |\mathbf{p}|) - m_q^2 \left[Q_0 \left(\frac{\omega}{|\mathbf{p}|} \right) \mp Q_1 \left(\frac{\omega}{|\mathbf{p}|} \right) \right] \right\}^2 \right.$$
$$\left. + \left[\frac{1}{2} \pi m_q^2 \left(1 \mp \frac{\omega}{|\mathbf{p}|} \right) \right]^2 \right)^{-1} \tag{9.40}$$

The $\omega_\pm(\mathbf{p})$ represent the two branches of the dispersion relation for quarks, essentially as discussed in Section 6.8. They are, of course, determined by the poles of $\mathcal{G}_\pm^*(\omega, \mathbf{p})$. The functions $\beta_\pm$ represent branch cuts that give rise to Landau damping, which is possible when $|\omega| < |\mathbf{p}|$.

9.5 Kinetic theory

The connection between kinetic theory and the HTL formalism is at first sight surprising and mysterious. However, once we realize that small deviations from local thermal equilibrium may be described by either kinetic theory or by linear response theory, the connection may be viewed as different manifestations of the same physics.

The connection can be initiated by considering the elementary example of an ensemble of charged classical particles. Assuming that hard collisions can be neglected, and that the particles thus interact only through average electric and magnetic fields, one can write an equation for the time evolution of the single-particle phase-space distribution $f(\mathbf{x}, \mathbf{p}, t)$:

$$\frac{\partial f}{\partial t} + \mathbf{v} \cdot \frac{\partial f}{\partial \mathbf{x}} + \mathbf{F} \cdot \frac{\partial f}{\partial \mathbf{p}} = 0 \tag{9.41}$$

Here $\mathbf{v}$ is a velocity and $\mathbf{F}$ is the Lorentz force. Note that in general the single-particle distribution function depends on time, on the position and on the momentum. This transport-type equation can be derived by requiring the total time derivative to vanish in the absence of hard collisions: a statement of Liouville's theorem. This equation is the collisionless Boltzmann or Vlasov equation. The derivation is completed by using the appropriate Lagrangian for electromagnetic interactions as well as Hamilton's equations. For an electromagnetic plasma in equilibrium the distribution functions will not depend on the spacetime coordinates and will be isotropic in momentum space. Keeping those facts in mind, let us slightly perturb the distribution function $f^{(0)}$:

$$f(\mathbf{x}, \mathbf{p}, t) = f^{(0)}(|\mathbf{p}|) + \delta f(\mathbf{x}, \mathbf{p}, t) \tag{9.42}$$

Then, to first order in the modification of the distribution function,

$$\left(\frac{\partial}{\partial t} + \mathbf{v} \cdot \frac{\partial}{\partial \mathbf{x}} \right) \delta f(\mathbf{x}, \mathbf{p}, t) = -e \mathbf{E} \cdot \mathbf{v} \frac{df^{(0)}(|\mathbf{p}|)}{d\epsilon} \tag{9.43}$$

where ϵ is the energy of the particle of charge e. Assuming that the perturbation is switched on adiabatically, one may solve for the out-of-equilibrium part of the distribution function:

$$\delta f(\mathbf{x}, \mathbf{p}, t) = -e \frac{df^{(0)}(|\mathbf{p}|)}{d\epsilon} \int_{-\infty}^{t} dt' e^{-\eta(t-t')} \mathbf{v} \cdot \mathbf{E}(\mathbf{x} - \mathbf{v}(t - t'), t') \tag{9.44}$$

This leads to an induced current

$$j_{\text{ind}}^{\mu}(\mathbf{x}, t) = e \int \frac{d^3 p}{(2\pi)^3} v^{\mu} \delta f(\mathbf{x}, \mathbf{p}, t) \tag{9.45}$$

where $v^{\mu} = (1, \mathbf{v})$. Finally, relating the polarization tensor to the induced current via

$$j_{\text{ind}}^{\mu}(x) = \int d^4 y \, \Pi^{\mu\nu}(x - y) A_{\nu}(y) \tag{9.46}$$

one obtains the following results in frequency–momentum space:

$$\Pi_{00}(\omega_m, \mathbf{p}) = -\frac{e^2 T^2}{3} \left(1 - \int \frac{d\Omega}{4\pi} \frac{i\omega_m}{i\omega_m - \mathbf{q} \cdot \hat{\mathbf{v}}} \right)$$
$$\Pi_{ij}(\omega_m, \mathbf{p}) = \frac{e^2 T^2}{3} \int \frac{d\Omega}{4\pi} \frac{i\omega_m \hat{v}_i \hat{v}_j}{i\omega_m - \mathbf{q} \cdot \hat{\mathbf{v}}} \tag{9.47}$$

We have made the assumption that the particles are massless, in which case their velocity vector is a unit vector $\hat{\mathbf{v}}$. The integrals in (9.47) are then over the orientation of the unit velocity vector. Remarkably, the result above is in fact the HTL contribution to the one-loop photon polarization tensor in QED. A direct calculation to show this is straightforward.

An important feature emerges here that generalizes to both the quantum domain and to nonabelian quantum field theories. This feature is that the Vlasov equation is an effective equation of motion for the soft modes of the plasma and corresponds to the fact that the hard thermal loops are obtained by isolating the leading-order contributions to one-loop diagrams with soft external lines. Put another way, the induced current calculated from the solutions of the Vlasov equation generates directly the HTL contribution.

Since in QED the HTL in the vacuum polarization tensor could be obtained by using classical transport theory, one could attempt a similar treatment for QCD. This approach appears promising, as HTLs represent the interaction of energetic quanta with weak mean fields. The hard thermal effects should then be driven by thermal fluctuations that can be cast in a classical framework. The starting point consists of considering a set of particles carrying nonabelian SU(N) color charge Q^a. One may write down the time evolution equations for the space-momentum coordinates of those particles. An important difference arises immediately in QCD: the particles may exchange color with the fields with which they interact. There needs to be an equation of motion for the color quantum number. The set (x, p, Q^a) can be thought of as an augmented phase space. Note that, except for Q^a, the elements of this set are now four-vectors.

The dynamical evolution of these phase-space variables is dictated by S. K. Wong's equations [6]. They can be derived by starting with the Dirac equation, suitably generalized to include QCD, finding the equations of motion for the operators, and then letting $\hbar \to 0$. One obtains

$$m\frac{dx^\mu}{d\tau} = p^\mu \qquad m\frac{dp^\mu}{d\tau} = gQ^a F_a^{\mu\nu} p_\nu \tag{9.48}$$

$$m\frac{dQ^a}{d\tau} = -gf^{abc}p^\mu A_\mu^b Q^c$$

As usual $F_{\mu\nu}^a$ is the field strength tensor, g is the strong coupling constant, and the f^{abc} are the structure constants of the group. These equations can be generalized to include spin, but this is not important for the present discussion. In the collisionless case, the proper-time total derivative of the phase-space density should vanish: $df(x, p, Q)/d\tau = 0$. Using the equations of motion presented above, one obtains the Boltzmann equation in the absence of collisions,

$$p^\mu \left(\frac{\partial}{\partial x^\mu} - gQ_a F_{\mu\nu}^a \frac{\partial}{\partial p_\nu} - gf_{abc}A_\mu^b Q^c \frac{\partial}{\partial Q_a} \right) f(x, p, Q) = 0 \tag{9.49}$$

Together with the Yang–Mills equation, $(D_\nu F^{\mu\nu})^a = J^{\mu a}$ (where the covariant derivative is $D_\mu^{ac} = \partial_\mu \delta^{ac} + gf^{abc}A_\mu^b$), one obtains a self-consistent set of nonabelian Vlasov equations. The net current is $J^{\mu a} = \sum j^{\mu a}$, where the sum runs over all species and spins. The space-time coordinates are implicit. More explicitly,

$$j^{\mu a}(x, p) = g \int dQ\, p^\mu Q^a f(x, p, Q) \qquad (9.50)$$

Physical states are guaranteed if the appropriate constraints are incorporated in the measure of the augmented phase space. In the limit of vanishing masses,

$$dQ = d^8 Q\, \delta(Q^a Q_a - q_2)\, \delta(d_{abc} Q^a Q^b Q^c - q_3) \qquad (9.51)$$

$$dP = \frac{d^4 p}{(2\pi)^3} 2\theta(p_0)\delta(p^2) \qquad (9.52)$$

The first equation is specific to SU(3) and ensures the invariance of the Casimir constants. The d_{abc} are the totally symmetric group constants. The second equation makes positivity manifest along with the on-shell requirement.

Specializing in small departures from equilibrium, one may write

$$f = f^{(0)} + g f^{(1)} + g^2 f^{(2)} + \cdots \qquad (9.53)$$

To first order in the coupling, the transport equation reduces to

$$p^\mu \left(\frac{\partial}{\partial x^\mu} - g f^{abc} A_\mu^b Q_c \frac{\partial}{\partial Q^a} \right) f^{(1)} = p^\mu Q_a F_{\mu\nu}^a \frac{\partial}{\partial p_\nu} f^{(0)} \qquad (9.54)$$

Arguments that are phase-space variables are once again left implicit.

Integrating by parts, using the definition of the current at a given order in terms of the distribution function, (9.50), and summing over the $N_{\rm f}$ quarks, $N_{\rm f}$ antiquarks, and the $N^2 - 1$ gluons and their physical spin states, one gets

$$[p \cdot DJ^\mu(x, p)]^a = 2g^2 p^\mu p^\nu F_{\nu 0}^a \frac{d}{dp_0} [N N_{\rm B}(p_0) + N_{\rm f} N_{\rm F}(p_0)] \qquad (9.55)$$

Kelly *et al.* [7], as well as Taylor and S. M. H. Wong [8] have shown that a solution of the above can be obtained in a functional form: $J^\mu(x) = -\delta\Gamma(A)/\delta A_\mu = -G^a \delta\Gamma(A)/\delta A_\mu^a$. The generating functional is

$$\Gamma = \frac{m_{\rm el}^2}{2} \int d^4 x A_0^a(x) A_0^a(x) - \int \frac{d\Omega}{(2\pi)^3} W(A) \qquad (9.56)$$

where an explicit expression for W is given in [8]. This generating functional is consistent with the effective Lagrangians we wrote down earlier.

A field-theoretic procedure can also be invoked to derive the results above. Following Blaizot and Iancu [9], the field equations of motion may be obtained by functional differentiation of the nonabelian generating functional. However, by definition this procedure does not produce gauge-invariant equations of motion. Indeed, the original Lagrangian in the action has to be gauge-fixed. This is not a problem, as physical results will not depend on the choice of gauge. However, intermediate steps that are gauge independent do permit a clearer physical interpretation. A method that circumvents this annoyance is that of the background gauge field [10], where the gauge field is split into a classical background field, identifiable as a mean field, and a fluctuating quantum field. Then a mean field approximation, where hard degrees of freedom interact with softer mean fields, together with an extraction of terms of leading order in g, is performed. Care has to be taken to preserve the gauge symmetry in those procedures. As in the classical limit, one allows first-order fluctuations in the density matrices, the Wigner transforms of which have many of the properties of classical phase-space distributions. The transport equations thus obtained yield (9.55). Note that practical applications typically involve the evaluation of quantities like the polarization tensor. This may be obtained from the current in the case of weak fields or, equivalently, in the linear response limit by a functional derivative of (9.46). Finally, the formal manipulations in [9] have greatly clarified the physical nature of hard thermal loops. As already mentioned, the high-temperature limit does permit an ordering of scales. One starts with an identification of plasma particles that have typical momenta of order T. Soft collective degrees of freedom then appear, which carry the same quantum numbers as the primordial constituents but which have typical momenta of the order of gT. This scale separation allows for the derivation of a kinetic equation for the plasma particles, the solution of which provides a generating functional for the hard thermal loops. Therefore, hard thermal loops describe long-wavelength collective excitations of the thermal particles. A natural consequence of this fact is that HTL perturbation theory is useful for the evaluation of physical quantities that are only sensitive to scales of the order of gT. Many other observables will be sensitive to scattering processes whose treatment will go beyond hard thermal loops.

9.6 Transport coefficients

In Section 6.9 we discussed the general Kubo formulae for transport coefficients. For completeness we quote here the values for the shear viscosity and flavor diffusion constant for QCD at high temperature. They were computed to lowest order in the gauge coupling but to all orders in the logarithm of the coupling by Arnold, Moore and Yaffe [11]. For the pure gauge theory without dynamical quarks the results are

$$D = \frac{0.203}{\alpha_s^2 \ln(0.580/\alpha_s)} \frac{1}{T} \qquad \eta = \frac{0.344}{\alpha_s^2 \ln(0.608/\alpha_s)} T^3 \qquad (9.57)$$

while for two flavors they are

$$D = \frac{0.165}{\alpha_s^2 \ln(0.497/\alpha_s)} \frac{1}{T} \qquad \eta = \frac{1.095}{\alpha_s^2 \ln(0.521/\alpha_s)} T^3 \qquad (9.58)$$

and for three flavors of massless quarks they are

$$D = \frac{0.150}{\alpha_s^2 \ln(0.461/\alpha_s)} \frac{1}{T} \qquad \eta = \frac{1.351}{\alpha_s^2 \ln(0.464/\alpha_s)} T^3 \qquad (9.59)$$

Here D refers to quark flavor diffusion. These QCD expressions have an extra logarithmic factor arising from the Debye screening of the long-range color Coulomb force.

9.7 Exercises

9.1 Derive (9.6) and (9.10).

9.2 Obtain the polarization tensor for QED in the HTL limit, starting with the effective Lagrangian of (9.33)–(9.35).

9.3 Derive the formulae for the spectral densities in (9.39).

9.4 Derive the formulae for gluon spectral densities that are analogous to those for quarks.

9.5 Verify that (9.44) satisfies (9.43).

9.6 Obtain the polarization tensor for QED in the HTL limit starting with (9.47).

References

1. Pisarski, R. D., *Nucl. Phys.* **B309**, 476 (1988); *Phys. Rev. Lett.* **63**, 1129 (1989).
2. Braaten, E., and Pisarski, R. D., *Nucl. Phys.* **B337**, 569 (1990).
3. Kalashnikov, O. K., and Klimov, V. V., *Yad. Phys.* **31**, 1357 (1980) (*Sov. J. Nucl. Phys.* **31**, 699 (1980)); *Phys. Lett.* **B88**, 328 (1979).
4. Weldon, H. A., *Phys. Rev. D* **26**, 1394 (1982).
5. Braaten, E., and Pisarski, R. D., *Phys. Rev. D* **45**, 1827 (1992).
6. Wong, S. K., *Nuovo Cimento* **65A**, 689 (1970).
7. Kelly, P. F., Liu, Q., Lucchesi, C., and Manuel, C., *Phys. Rev. D*, **50** 4209 (1994).
8. Taylor, J. C., and Wong, S. M. H., *Nucl. Phys.* **B346**, 115 (1990).
9. Blaizot, J.-P., and Iancu, E., *Phys. Rep.* **359**, 355 (2002).
10. De Witt, B. S., *Phys. Rev.* **162**, 1195, 1239 (1967).
11. Arnold, P., Moore, G. D., and Yaffe, L. G., *JHEP* **05**, 051 (2003).

10
Lattice gauge theory

Perturbation theory applied to QCD predicts that the normally strong interactions among quarks and gluons become weak at high temperatures and densities on account of asymptotic freedom. This leads to a state known as quark–gluon plasma. The perturbative analysis of QCD was the subject of the last two chapters. At low temperatures and densities quarks and gluons are not observed individually but only as color-neutral objects, hadrons, on account of confinement. Then hadrons are the relevant degrees of freedom, just as atoms and molecules are the relevant degrees of freedom in biological physics. Nuclear matter and hot hadronic matter are the subjects of the next two chapters. The standard computational method for studying QCD in the transitional region is lattice gauge theory.

Lattice gauge theory is a field of intellectual study in itself. It is not possible in one chapter to cover it in all detail, not the least reason being that it is numerically quite involved. We will introduce the basic theoretical ideas and the main numerical results. As the field is evolving owing to rapid increases in computational power, these results will no doubt be superseded in the near future. Nevertheless, the main conclusions should stand the test of time.

The formulation of nonabelian gauge theories on a spacetime lattice in Euclidean space was introduced by Wilson [1] with the purpose of studying quark confinement. The infinite-dimensional functional integral that defines a quantum field theory becomes a finite-dimensional integral when the lattice has a finite extent in space and time and is therefore unambiguously defined. It is natural to expect that there is a unique continuum limit when the lattice spacing a goes to zero, at least for asymptotically free theories. The argument is that the bare coupling $g(a)$ becomes small in this limit, and the long-distance properties of the theory should be insensitive to the details of the ultraviolet cutoff introduced by the lattice.

After Creutz [2] demonstrated that the functional integral of lattice gauge theory could be evaluated with the help of Monte Carlo numerical techniques, lattice gauge theory became the method of choice for the calculation of observables that are beyond the reach of perturbation theory. This includes most properties of individual hadrons as well as their interactions at low energies.

It is natural to apply the lattice technique to the study of deconfinement and chiral symmetry restoration at finite temperature. The first such studies were made by Polyakov [3] and by Susskind [4] using the Hamiltonian formulation of lattice gauge theory. They could show that deconfinement disappears at high temperature but were unable to prove that this phenomenon persists in the continuum limit, nor could they compute a critical temperature. This was first achieved in SU(2) gauge theory by means of numerical calculations by McLerran and Svetitsky [5] and by Kuti, Polónyi, and Szlachányi [6].

10.1 Abelian gauge theory

As a warm-up to full QCD let us consider how to define a Hamiltonian on a discrete spatial lattice for a pure gauge theory that has Abelian gauge theory as its continuum limit. The procedure is not unique since one can always add terms to the discretized theory which vanish in the limit that the lattice spacing goes to zero. In fact, this arbitrariness can be both a boon and a bane, as we shall see.

Consider a cubic lattice with spacing a. Label each *site* of a lattice with a vector $\mathbf{x} = (x_1, x_2, x_3)$. There are six unit lattice vectors: $\mathbf{n}_1$, $\mathbf{n}_2$, $\mathbf{n}_3$, $\mathbf{n}_{-1}$, $\mathbf{n}_{-2}$, $\mathbf{n}_{-3}$, where $\mathbf{n}_1$ points in the positive x_1 direction, $\mathbf{n}_{-1}$ points in the negative x_1 direction, and so on. A directed *link* is defined by the pair of vectors $(\mathbf{x}, \mathbf{n})$; it starts from the site $\mathbf{x}$ and goes to the neighboring site $\mathbf{x} + a\mathbf{n}$ (see Figure 10.1). The lattice may be finite or infinite in extent, meaning that there are either a finite or a countably infinite number of degrees of freedom. A continuum field theory, by contrast, has an uncountably infinite number of degrees of freedom no matter whether the box is finite or infinite in extent.

It is natural to associate links with dynamical degrees of freedom. Let us define this degree of freedom to be

$$U(\mathbf{x}, \mathbf{n}) = \exp[i\phi(\mathbf{x}, \mathbf{n})] \tag{10.1}$$

Each link has a mate: $(\mathbf{x}, \mathbf{n}) \Leftrightarrow (\mathbf{x} + a\mathbf{n}, -\mathbf{n})$. The link and its mate should not have independent degrees of freedom associated with them, and so it is natural to require that

$$U(\mathbf{x} + a\mathbf{n}, -\mathbf{n}) = U^\dagger(\mathbf{x}, \mathbf{n}) \tag{10.2}$$

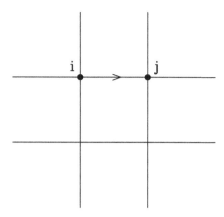

Fig. 10.1. Neighboring lattice sites i and j connected by a directed link.

and so

$$\phi(\mathbf{x} + a\mathbf{n}, -\mathbf{n}) = -\phi(\mathbf{x}, \mathbf{n}) \qquad (10.3)$$

It is also natural to associate each link with an electric flux $E(\mathbf{x}, \mathbf{n})$. Clearly we should require that

$$E(\mathbf{x} + a\mathbf{n}, -\mathbf{n}) = -E(\mathbf{x}, \mathbf{n}) \qquad (10.4)$$

The theory is quantized by demanding that E be the momentum canonically conjugate to the variable ϕ:

$$[\phi(\mathbf{x}, \mathbf{n}), \ E(\mathbf{x}, \mathbf{n})] = i \qquad (10.5)$$

Since ϕ is an angle, E has integers for its spectrum of eigenvalues. We are dealing with a compact U(1) gauge theory.

The electric contribution to the Hamiltonian H_{electric} must be proportional to

$$\sum_{\text{links}} \frac{E^2}{2a}$$

where the factor $1/a$ gives the term the proper dimension, and the factor one-half is inserted for the conventional reasons of normalization.

The magnetic field energy is less obvious. The independent degrees of freedom have already been defined and so it must be expressed in terms of them. The coordinate ϕ is the natural starting point. One defines a *plaquette* Γ to be a square made of four links connected head to tail (see Figure 10.2). The variable U associated with this plaquette is

$$U(\Gamma) = U(i)U(j)U(k)U(l) = \exp\left\{i[\phi(i) + \phi(j) + \phi(k) + (l)]\right\} \qquad (10.6)$$

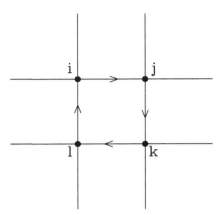

Fig. 10.2.   A plaquette.

where the four sides of the plaquette are labeled $ijkl$, with the head of l connected to the tail of i. By going around in a closed loop like this, one will obtain a curl in the limit $a \to 0$.

The Hamiltonian is defined to be

$$H = \sum_{\text{links}} \frac{g^2 E^2}{2a} - \sum_{\text{plaquettes}} \frac{1}{2ag^2} \left[U(\Gamma) + U^\dagger(\Gamma) - 2 \right] \qquad (10.7)$$

Notice that a dimensionless coupling constant g has been used in this definition of H. The fact that the coefficient of the electric and magnetic field energies can depend on g^2 should not be surprising. It does represent the strength of interactions because the Hamiltonian is not quadratic in the independent dynamical variables ϕ but has terms to all orders in ϕ. The particular normalization is chosen, with hindsight, to reproduce the Abelian theory in the continuum limit.

To calculate the continuum limit of the magnetic energy, consider a single plaquette with corners labeled $abcd$. The quantity

$$U_{ab} = \mathrm{e}^{i\phi_{ab}} \qquad (10.8)$$

is associated with the link ab. Define a vector potential A_i via

$$\phi_{ab} = g(\mathbf{x}_b - \mathbf{x}_a)_i \, A_i \left(\frac{\mathbf{x}_a + \mathbf{x}_b}{2} \right) \qquad (10.9)$$

We do the same for the four links comprising the plaquette, namely, ab, bc, cd, da. In the limit that the lattice spacing becomes very small we can expand the exponentials in a power series:

$$H_{\text{magnetic}}(\Gamma) = -\frac{1}{2ag^2}\left[U(\Gamma) + U^\dagger(\Gamma) - 2\right]$$

$$= -\frac{1}{2ag^2}\left[i(\phi_{ab} + \phi_{bc} + \phi_{cd} + \phi_{da})\right.$$

$$\left. -\frac{1}{2}(\phi_{ab} + \phi_{bc} + \phi_{cd} + \phi_{da})^2 + \cdots + \text{ c.c.}\right]$$

$$\approx \frac{1}{2a}\left[(\mathbf{x}_b - \mathbf{x}_a)_i A_i\left(\frac{\mathbf{x}_a + \mathbf{x}_b}{2}\right) + (\mathbf{x}_c - \mathbf{x}_b)_i A_i\left(\frac{\mathbf{x}_b + \mathbf{x}_c}{2}\right)\right.$$

$$\left. + (\mathbf{x}_d - \mathbf{x}_c)_i A_i\left(\frac{\mathbf{x}_c + \mathbf{x}_d}{2}\right) + (\mathbf{x}_a - \mathbf{x}_d)_i A_i\left(\frac{\mathbf{x}_a + \mathbf{x}_d}{2}\right)\right]^2$$

$$(10.10)$$

Now Taylor-expand the vector potentials about the center of the plaquette. For example, if the link ab points in the direction $\mathbf{n}_2$ and the link bc points in the direction $\mathbf{n}_1$ then

$$A_i\left(\frac{\mathbf{x}_a + \mathbf{x}_b}{2}\right) \approx A_i\left(\frac{\mathbf{x}_a + \mathbf{x}_b + \mathbf{x}_c + \mathbf{x}_d}{4}\right) - \frac{1}{2}a\frac{\partial A_i}{\partial x_1}\left(\frac{\mathbf{x}_a + \mathbf{x}_b + \mathbf{x}_c + \mathbf{x}_d}{4}\right)$$

$$(10.11)$$

and similarly for the other terms. The result is that the magnetic Hamiltonian for this plaquette is

$$H_{\text{magnetic}}(\Gamma) = \frac{1}{2}a^3\left(\frac{\partial A_1}{\partial x_2} - \frac{\partial A_2}{\partial x_1}\right)^2 \qquad (10.12)$$

which is proportional to the square of the third component of the curl of the vector potential, otherwise known as the magnetic field.

Summing over all plaquettes results in the continuum limit for the magnetic part of the Hamiltonian,

$$H_{\text{magnetic}} = \frac{1}{2}\int d^3x\,\mathbf{B}^2(\mathbf{x}) \qquad (10.13)$$

since $a^3\sum_{\mathbf{x}} \to \int d^3x$ in the continuum limit. The physical electric field $\mathbf{E}$ associated with the link ab is defined to be $g\mathbf{n}_2 E(\mathbf{x}_a, \mathbf{n}_2)/a^2$, which has both the correct dimensions and direction. The continuum limit of the electric part of the Hamiltonian is therefore

$$H_{\text{electric}} = \frac{1}{2}\int d^3x\,\mathbf{E}^2(\mathbf{x}) \qquad (10.14)$$

The resulting continuum theory is a free-field theory in the absence of electric charges. Note, however, that the original lattice theory is a fully interacting theory with interactions to all orders in the vector potential. The range of these interactions is of the order of the lattice spacing a.

Since this is a Hamiltonian formulation, only physical states obeying Gauss's law should be included when calculating the partition function:

$$Z(\beta) = \sum_{\substack{\text{physical} \\ \text{states } \psi}} \langle \psi | e^{-\beta H} | \psi \rangle \tag{10.15}$$

These states $|\psi\rangle$ should satisfy

$$\sum_{\mathbf{n}} E(\mathbf{x}, \mathbf{n}) |\psi\rangle = 0 \tag{10.16}$$

for each site $\mathbf{x}$, assuming that there are no electric charges in the system. To impose Gauss's law, we insert a factor

$$\delta\left(\sum_{\mathbf{n}} E(\mathbf{x}, \mathbf{n}) \right) = \int_{-\pi}^{\pi} \frac{d\alpha(\mathbf{x})}{2\pi} \exp\left(i\alpha(\mathbf{x}) \sum_{\mathbf{n}} E(\mathbf{x}, \mathbf{n}) \right) \tag{10.17}$$

at each site. This will take care of the restriction to physical states automatically.

Let us study the theory in the strong-coupling limit, $g^2 \gg 1$. This is the extreme opposite of the weak-coupling limit, where perturbation theory can be applied. In strong coupling we can drop the magnetic energy and keep only the electric. Imposing Gauss's law by use of the Dirac delta function leads to the expression

$$Z = \prod_{\mathbf{x}} \int_{-\pi}^{\pi} \frac{d\alpha(\mathbf{x})}{2\pi} \prod_{\text{links at } \mathbf{x}} \left(\sum_{E} \exp\left\{ -\frac{\beta g^2}{2a} E^2(\mathbf{x}, \mathbf{n}) \right. \right.$$
$$\left. \left. + i[\alpha(\mathbf{x}) - \alpha(\mathbf{x} + \mathbf{n})]E(\mathbf{x}, \mathbf{n}) \right\} \right) \tag{10.18}$$

Here and from now on E represents the eigenvalues (integers) of the operator. To understand the nature of the strong-coupling phase, insert a pair of static immobile charges, one of charge g located at $\mathbf{x} = \mathbf{0}$ and the other of charge $-g$ located at $\mathbf{x} = \mathbf{R}$. Then Gauss's law becomes

$$\sum_{\mathbf{n}} E(\mathbf{0}, \mathbf{n}) = 1$$
$$\sum_{\mathbf{n}} E(\mathbf{R}, \mathbf{n}) = -1 \tag{10.19}$$
$$\sum_{\mathbf{n}} E(\mathbf{x}, \mathbf{n}) = 0 \qquad \text{for } \mathbf{x} \neq \mathbf{0}, \mathbf{R}$$

This leads to an extra factor in Z of $e^{i\alpha(\mathbf{0})}e^{-i\alpha(\mathbf{R})}$:

$$Z(\beta, \mathbf{R}) = Z(\beta) \left\langle e^{i\alpha(\mathbf{0})}e^{-i\alpha(\mathbf{R})} \right\rangle \tag{10.20}$$

The free energy of this configuration is

$$\Delta F(\beta, \mathbf{R}) = -[T \ln Z(\beta, \mathbf{R}) - T \ln Z(\beta)] = -T \ln \left\langle e^{i\alpha(\mathbf{0})} e^{-i\alpha(\mathbf{R})} \right\rangle$$

$$(10.21)$$

First consider the low-temperature limit, $\beta g^2/2a \gg 1$. Then only the eigenvalues $E = 0, \pm 1$ matter and

$$\left\langle e^{i\alpha(\mathbf{0})} e^{-i\alpha(\mathbf{R})} \right\rangle$$

$$= \prod_{\mathbf{x}} \int_{-\pi}^{\pi} \frac{d\alpha(\mathbf{x})}{2\pi} \prod_{\text{links at } \mathbf{x}} \left\{ e^{i[\alpha(\mathbf{0})-\alpha(\mathbf{R})]} + e^{-\beta g^2/2a} e^{i[\alpha(\mathbf{x})-\alpha(\mathbf{x}+\mathbf{n})+\alpha(\mathbf{0})-\alpha(\mathbf{R})]} \right.$$
$$\left. + e^{-\beta g^2/2a} e^{i[-\alpha(\mathbf{x})+\alpha(\mathbf{x}+\mathbf{n})+\alpha(\mathbf{0})-\alpha(\mathbf{R})]} \right\} \quad (10.22)$$

The first of the three exponentials integrates to zero. So do the second and third, except for those paths that connect the two charges. For simplicity, choose $\mathbf{R}$ to lie on an axis running through the origin. Then

$$\left\langle e^{i\alpha(\mathbf{0})} e^{-i\alpha(\mathbf{R})} \right\rangle = 2 \left(e^{-\beta g^2/2a} \right)^{N_{\text{links}}(0,R)} = 2 e^{-\beta g^2 R/2a^2} \quad (10.23)$$

where $N_{\text{links}}(0, R) = R/a$ is the number of links connecting the charges. Therefore

$$\Delta F = \frac{g^2}{2a^2} R \quad (10.24)$$

The potential energy is linear and thus confining.

The high-temperature limit, $\beta g^2/2a \ll 1$, is left as an exercise. It should be no surprise that the answer is a Coulombic potential,

$$\Delta F = -\frac{g^2}{R} \quad (10.25)$$

Since the low- and high-temperature limits have completely opposite behavior, one should expect a phase transition separating them. The critical temperature is estimated as $\beta_c g^2/2a \approx 1$ or

$$T_c \approx \frac{g^2}{2a} \quad (10.26)$$

This depends very strongly on the lattice spacing. In fact, since this is a quantum field theory, quantum corrections will cause the effective coupling constant to depend on a, namely, $g^2(a)$ will replace g^2 in the estimate for the critical temperature. This being an Abelian theory it does not have the property of asymptotic freedom. Therefore $g^2(a)$ will grow with decreasing a. Hence T_c will grow without bound as $a \to 0$. The low-temperature confining phase does not smoothly extrapolate to the continuum limit but is separated from it by a phase transition; confinement exists only in the discretized lattice version of the theory. This is well and good since we know that QED is not a confining theory.

10.2 Nonabelian gauge theory

Both QCD and electroweak theory involve nonabelian gauge groups, SU(3) in the former case and SU(2) in the latter. Essentially all modern numerical calculations in these theories use the Lagrangian formulation, not the Hamiltonian one. Calculations are done on a finite discrete lattice of volume $V = L^3$, with

$$L = N_\mathrm{s} a \tag{10.27}$$

where a is the lattice spacing and N_s is the number of sites in each of the three spatial directions. The imaginary time variable is also discrete: $0 \leq \tau \leq \beta$ as usual with

$$\beta = \frac{1}{T} = N_\tau a \tag{10.28}$$

where N_τ is the number of sites in the imaginary time direction. The unit directional vectors $\mathbf{n}$ in three spatial dimensions must be extended to unit directional vectors n_α in four Euclidean dimensions. It is then convenient to define $x_4 = \tau$. The lattice spacings in the space and time directions need not be the same, and sometimes they are chosen differently, but equal spacing is the norm.

The notions of site, link, and plaquette all carry over from the lattice version of the Abelian theory. The generalization of the link variable from U(1), as given in (10.1), to SU(N) is straightforward, but for definiteness we specialize to SU(2) for the rest of this section:

$$U(x; n_\alpha) = \exp\left[i a \sigma_j A_\alpha^j(x)\right] = u_4 I + \boldsymbol{\sigma} \cdot \mathbf{u} \tag{10.29}$$

Here the link begins at the site $x = (\mathbf{x}, x_4)$ and goes in the direction n_α. The σ_j with $j = 1, 2, 3$ are the Pauli matrices while I is the identity matrix. These link variables are elements of the group SU(2). In the continuum limit the A_α^j will be identified as $1/g$ times the four-vector potential. (It is conventional to factor out the coupling constant.) Compare with (10.9). By the definition of the link variables there is a constraint

$$u_4^2 + \mathbf{u}^2 = 1 \tag{10.30}$$

This is a compact gauge group.

The action should be defined so that (i) it reduces to the continuum expression, (ii) it is gauge invariant even on the lattice, and (iii) it is as simple as possible. Requirement (iii) means that an infinite number of extra terms that all vanish in the continuum limit $a \to 0$ could be added. Actually it could be advantageous to add such extra terms if it means that the continuum limit is approached more rapidly and hence more efficiently in terms of computer time and memory. This goes under the title of improved actions, and will be discussed in Section 10.4. Motivated by the lattice action for the Abelian theory, the simplest possible action for SU(2) is

$$S(U) = \frac{4}{g^2} \sum_{\text{plaquettes } abcd} \left(1 - \tfrac{1}{2}\text{Tr}\, U_{ab}U_{bc}U_{cd}U_{da}\right) \tag{10.31}$$

That this reduces to the proper continuum action is left as an exercise. This action is invariant under the gauge transformation

$$U(x; n_\alpha) \to V(x)U(x; n_\alpha)V^{-1}(x + an_\alpha) \tag{10.32}$$

where

$$V(x) = \exp[ia\sigma_j \Lambda_j(x)] \tag{10.33}$$

This invariance is obvious at a glance.

The functional integral expression for the partition function involves integration over all possible field configurations:

$$Z = \int \prod_{\text{links } ab} dU_{ab} \exp[-S(U)] \tag{10.34}$$

When integrating over the link variables U it must be remembered that they are unitary matrices in the group SU(2) and therefore one must use the appropriate Haar measure. One could integrate over the u_0 and $\mathbf{u}$ subject to the constraint (10.30), or one could integrate over three angles in four-dimensional Euclidean space.

At this point perturbation theory could be used to compute physical observables on the finite lattice at finite temperature. However, it is much more interesting to attempt to evaluate the large but finite-dimensional integral for Z using Monte Carlo techniques. The results of such numerical work are the subject of Sections 10.4 and 10.5.

10.3 Fermions

Introducing fermionic fields on a lattice has been a challenge. The most used techniques result in a multiplication of the number of fermion species in the continuum limit and/or the breaking of chiral symmetry on the lattice when the fermions are massless. Much technical work has been

done to overcome these problems. In this section we introduce the reader
to the most commonly used techniques, as originally formulated.

The staggered-fermion approach was invented by Kogut and Susskind
[7]. In order to define Dirac fields ψ with finite derivatives in the continuum limit, they introduced two separate two-component spinors residing
on alternate lattice sites. In the Hamiltonian formalism on a cubic lattice, a lattice site $\mathbf{x}$ is defined to be even or odd according to whether
$s \equiv (x_1 + x_2 + x_3)/a$ is an even or odd integer. The upper two components of a four-component Dirac field, ψ_{upper}, reside on even lattice sites
while the lower two components, ψ_{lower}, reside on odd lattice sites. The
Hamiltonian for free fermions is taken to be

$$H = \frac{1}{ia} \sum_{\mathbf{x},\mathbf{n}} \psi^\dagger(\mathbf{x}) \, \boldsymbol{\sigma} \cdot \mathbf{n} \, \psi(\mathbf{x} + a\mathbf{n}) + m \sum_{\mathbf{x}} (-1)^s \psi^\dagger(\mathbf{x}) \psi(\mathbf{x}) \quad (10.35)$$

Imposition of the canonical commutation relations

$$\left\{ \psi_\alpha(\mathbf{x}), \, \psi_\beta^\dagger(\mathbf{x}') \right\} = \delta_{\alpha,\beta} \delta_{\mathbf{x},\mathbf{x}'} \quad (10.36)$$

leads to the equation of motion

$$i\frac{\partial \psi(\mathbf{x})}{\partial t} = [\psi(\mathbf{x}), H]$$

$$= \frac{1}{ia} \sum_{\mathbf{n}} \boldsymbol{\sigma} \cdot \mathbf{n} \, \psi(\mathbf{x} + a\mathbf{n}) + m(-1)^s \psi(\mathbf{x})$$

$$= \frac{1}{2ia} \sum_{\mathbf{n}} \boldsymbol{\sigma} \cdot \mathbf{n} \, [\psi(\mathbf{x} + a\mathbf{n}) - \psi(\mathbf{x} - a\mathbf{n})]$$

$$+ m(-1)^s \psi(\mathbf{x}) \quad (10.37)$$

In the continuum limit the finite differences become derivatives. Remembering that the upper and lower components of ψ reside on even and odd
lattice sites, the equation of motion becomes

$$i\frac{\partial \psi_{\text{upper}}}{\partial t} = -i\boldsymbol{\sigma} \cdot \nabla \psi_{\text{lower}} + m\psi_{\text{upper}}$$

$$i\frac{\partial \psi_{\text{lower}}}{\partial t} = -i\boldsymbol{\sigma} \cdot \nabla \psi_{\text{upper}} - m\psi_{\text{lower}} \quad (10.38)$$

This is the Dirac equation for free fermions.

Coupling to the gauge field can be done in such a way as to render
the Hamiltonian gauge invariant. Inspection of (10.35) suggests that we
rotate the Dirac field according to

$$\psi(\mathbf{x}) \rightarrow V(\mathbf{x})\psi(\mathbf{x}) \quad (10.39)$$

The kinetic energy term in the free Hamiltonian that involves neighboring
sites x and $x + an_\alpha$ requires us to use the parallel-transporter or link

variable U to connect them in a gauge-invariant way:

$$H = \frac{1}{ia} \sum_{\mathbf{x},\mathbf{n}} \psi^\dagger(\mathbf{x}) \boldsymbol{\sigma} \cdot \mathbf{n}\, U(\mathbf{x};\mathbf{n})\psi(\mathbf{x}+a\mathbf{n}) + m \sum_{\mathbf{x}} (-1)^s \psi^\dagger(\mathbf{x})\psi(\mathbf{x})$$

(10.40)

Recalling (10.32) we see immediately that this Hamiltonian is minimally coupled and gauge invariant. Taking the zero-lattice-spacing limit in the usual way reproduces the correct continuum equations.

The staggered-fermion approach of Kogut and Susskind can also be expressed in Lagrangian form and on a lattice. After some work one finds the action

$$S_{\text{fermion}}^{\text{KS}} = \frac{1}{a} \sum_{xx'} \bar\psi(x) \left[D^{\text{KS}}(x,x') + am\delta(x,x') \right] \psi(x')$$

(10.41)

where in this expression ψ has only one Dirac component. The matrix is

$$D^{\text{KS}}(x,x')$$
$$= \frac{1}{2} \sum_{j=1}^{3} \text{sign}(x,j) \left[\delta(x+an_j,\, x')U(x;n_j) - \delta(x,\, x'+an_j)U^\dagger(x';n_j) \right]$$
$$+ \frac{1}{2} \left[\delta(x+an_4,\, x')U(x;n_4)\mathrm{e}^{a\mu} - \delta(x,\, x'+an_4)U^\dagger(x';n_4)\mathrm{e}^{-a\mu} \right]$$

(10.42)

and the sign factor is given by

$$\text{sign}(x,j) = (-1)^{x_4/a} \begin{cases} 1 & \text{if } j=1 \\ (-1)^{x_1/a} & \text{if } j=2 \\ (-1)^{(x_1+x_2)/a} & \text{if } j=3 \end{cases}$$

(10.43)

The delta functions appearing in (10.42) are Kronecker not Dirac. A chemical potential μ has also been added. Integrating over the fermion field gives the usual determinant of the operator:

$$Z = \int \prod_{\text{links } ab} dU_{ab} \exp[-S(U)] \det\left[D^{\text{KS}}(U) + am \right]$$

(10.44)

where the action $S(U)$ is due to the gauge fields alone. If one takes the continuum limit with zero mass one finds not one but four species of fermion. This is an illustration of the fermion doubling (better to say multiplication) problem on the lattice. Sometimes the $(N_f/4)$th root of the fermion determinant is taken to represent N_f species of fermion; for example, $N_f = 1$ for one species.

Wilson [1] introduced fermions on the lattice in a different way. Every lattice site is associated with a four-component Dirac field. The action is taken to be

$$S_{\text{fermion}}^{\text{W}} = -\frac{1}{2a} \sum_{x,n_\alpha} \bar{\psi}(x)\gamma_\alpha \left[U(x;n_\alpha)\psi(x + an_\alpha) \right.$$

$$\left. - U^\dagger(x - an_\alpha;n_\alpha)\psi(x - an_\alpha) \right] + m \sum_x \bar{\psi}(x)\psi(x) \tag{10.45}$$

This is gauge invariant and reduces to the correct action in the continuum limit. For example, for free fermions $U = 1$ and the term

$$\frac{1}{2a} \left[\psi(x + an_\alpha) - \psi(x - an_\alpha) \right] \to \partial_\alpha \psi(x) \tag{10.46}$$

in the continuum limit. There is a corresponding matrix $D^{\text{W}}(U)$ that replaces $D^{\text{KS}}(U)$ in (10.44). In this case one finds that the number of species increases by 16 in the continuum limit and when the fermion is massless. The reason is easy to see for free fermions. Instead of the expression (2.94) for $\ln Z$ one finds the replacement ($m = 0$ and $\mu = 0$)

$$\omega_n^2 + \mathbf{p}^2 \to \frac{1}{a^2}\sin^2(ap_4) + \frac{1}{a^2}\sum_{i=1}^3 \sin^2(ap_i) \tag{10.47}$$

The lattice propagator has poles not only at zero momentum but also at all the corners of the Brillouin zone, namely, $p_j = \pm\pi/a$, $p_4 = \pm\pi/a$. The way out of this is to introduce another term in the action proportional to $a\bar{\psi}\partial_\alpha^2\psi$ that vanishes in the $a \to 0$ limit. However, with any finite lattice spacing chiral symmetry is broken, and so the chiral condensate cannot serve as an order parameter on the lattice.

Specific calculations with quarks will be reviewed in Section 10.5.

10.4 Phase transitions in pure gauge theory

The best-understood lattice gauge theories are the pure gauge theories without quarks. Extensive numerical calculations have been done for SU(2) and SU(3). Results for the equation of state in the vicinity of T_c for SU(3) are shown in Figure 10.3. The SU(3) theory undergoes a first-order phase transition. It has been found that any SU(N) theory, with N equal to or greater than 3, undergoes a first-order transition [8], while SU(2) undergoes a second-order transition. This was predicted on the basis of universality arguments [9]. The essential degrees of freedom below T_c may be thought of as glueballs, while above T_c they may be thought of as gluons. In either region the degrees of freedom certainly do interact amongst themselves to a greater or lesser extent.

One must ask just how big a lattice ought to be in order to obtain results that are truly representative of the continuum limit for temperatures of the order of one to several hundred MeV. The necessary size

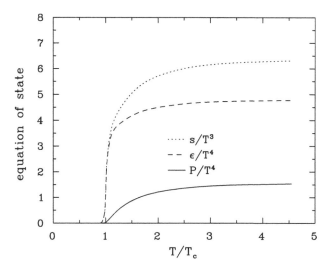

Fig. 10.3. The equation of state of pure SU(3) gauge theory with no quarks. The results shown are extrapolations to the continuum limit, with an estimated uncertainty of order ± 0.1. The latent heat is about $1.5T_c^4$. The data were taken from [10].

may be estimated as follows. Hadrons, including glueballs, have a spatial extent of the order of 1 fm. Thus the size of the system should be at least 5 to 10 fm on a side in order to contain enough particles that the thermodynamic limit is approximately attained. Because the boundary conditions are usually chosen to be periodic in space, the effective size of the box is somewhat reduced due to surface effects. Therefore we should be conservative and require a box with sides of length 10 fm. At the other end of the scale, a hadron has internal structure characterized by a length of 0.1 fm. If changes in the hadronic structure, such as deconfinement, due to finite temperature are to be seen then the lattice spacing should be no larger than about 0.05 fm. Taken together, this implies that the lattice should be at least 100 to 200 sites per spatial dimension. The temporal dimension has length $\beta = 1/T$. Taking a lattice spacing of 0.05 fm and a temperature of 200 MeV requires about 20 sites in the temporal direction. Numerical calculations with lattices of size up to $N_s = 64$ or 128 and $N_\tau = 16$ or 32 have been done. Extensive work on scaling with system size shows that this is probably large enough to obtain reasonable results.

Rather than going to larger lattices, it can be advantageous to add additional terms to the simplest actions described in the previous sections. These terms vanish in the continuum limit, being higher order in a, but can noticeably improve the approach to the continuum. For example, suppose that the thermal average of some observable has the Taylor series expansion in the lattice spacing

$$\frac{\langle \mathcal{O} \rangle_{\text{lattice}}}{\langle \mathcal{O} \rangle_{\text{continuum}}} = 1 + \sum_{n=1}^{\infty} c_{2n} a^{2n} \tag{10.48}$$

for the simplest lattice action. By the addition of judicious terms to the action it is possible to cancel the term $c_2 a^2$, bringing about a faster convergence to the continuum limit. Of course, the remaining coefficients are likely to be modified as a result, $c_{2n} \to c'_{2n}$. The modified coefficients may be larger or smaller or even of opposite sign. One approach is to add six planar link terms to the action, which is then called a tree-level improved 1×2 action. Another approach is to change certain coefficients in the action to correspond to the renormalization group; the action is then termed RG-improved. Yet another approach is to recognize that the coefficients c_{2n} can be expanded in powers of g^2. The coefficients in the action can be adjusted to make c_{2n} equal to zero to some order in g^2; this is the Symanzik improvement program [11].

If there is a deconfinement phase transition then the free energy of a heavy quark–antiquark pair should grow linearly with separation below T_c and be Debye screened above. This free energy can be calculated using the Wilson line

$$W(\mathbf{x}, \beta) = T_\tau \, \exp \left(i \int_0^\beta d\tau \, \lambda_a A_4^a(\mathbf{x}, \tau) \right) \tag{10.49}$$

and Polyakov loop

$$L(\mathbf{x}) = \frac{1}{N} \operatorname{Tr} W(\mathbf{x}, \beta) \tag{10.50}$$

Here the λ_a are the Gell-Mann matrices for SU(3) (for SU(2) they would be the Pauli matrices), Tr is the trace with respect to the indices of those matrices, and T_τ denotes time ordering. This is useful because a static, immovable, quark field evolves in imaginary time according to

$$\psi(\mathbf{x}, \tau) = W(\mathbf{x}, \tau)\psi(\mathbf{x}, 0) \tag{10.51}$$

which solves the Dirac equation. The free energy of a system that contains one quark with color index c located at $\mathbf{x}$ and one antiquark with color index c' located at $\mathbf{x}'$ is then determined by

$$
\begin{aligned}
\exp(-\beta F_{q\bar{q}}) = {} & \frac{1}{N^2} \sum_{a,a'} \sum_s \langle s | \psi_a(\mathbf{x}, 0) \psi_{a'}^c(\mathbf{x}', 0) \\
& \times \exp(-\beta H) \, \psi_a^\dagger(\mathbf{x}, 0) \psi_{a'}^{c\dagger}(\mathbf{x}', 0) | s \rangle \\
= {} & \frac{1}{N^2} \sum_{a,a'} \sum_s \langle s | \exp(-\beta H) \, \psi_a(\mathbf{x}, \beta) \psi_a^\dagger(\mathbf{x}, 0) \\
& \times \psi_{a'}^{c\dagger}(\mathbf{x}', 0) \psi_{a'}^c(\mathbf{x}', 0) | s \rangle
\end{aligned} \tag{10.52}
$$

where the superscript c indicates the operation of charge conjugation. The states $|s\rangle$ do not include any quarks; these must be created by the field operators acting on $|s\rangle$. This can be expressed in terms of the Polyakov loop

$$\exp(-\beta F_{q\bar{q}}) = \text{Tr} \left[\exp(-\beta H)\, L(\mathbf{x}) L^\dagger(\mathbf{x}') \right] \qquad (10.53)$$

This is the free energy of the entire system of gluons plus quark and anti-quark. To obtain the free energy $\Delta F_{q\bar{q}}$ associated with the quark and antiquark only we must divide by the partition function $Z = \text{Tr}\,\exp(-\beta H)$ for a system of gluons only, obtaining

$$\exp(-\beta \Delta F_{q\bar{q}}) = \langle L(\mathbf{x}) L^\dagger(\mathbf{x}') \rangle \qquad (10.54)$$

The generalization to an assembly of N_q quarks and $N_{\bar{q}}$ antiquarks is straightforward:

$$\exp\left(-\beta \Delta F_{N_q N_{\bar{q}}}\right) = \langle L(\mathbf{x}_1) \cdots L(\mathbf{x}_{N_q}) L^\dagger(\mathbf{x}'_1) \cdots L^\dagger(\mathbf{x}'_{N_{\bar{q}}}) \rangle \qquad (10.55)$$

The transcription of the Polyakov loop to the lattice is

$$L(\mathbf{x}) = \frac{1}{N} \text{Tr} \prod_{j=0}^{N_\tau - 1} U(\mathbf{x}, ja; n_4) \qquad (10.56)$$

which is the trace of the product of the U matrices along the time axis.

The link variables U are required to be periodic in time, but the class of allowable gauge transformations is not restricted to those that are periodic. Among them is a special set of gauge transformations that obey

$$V(\mathbf{x}, \beta) = V(\mathbf{x}, 0) e^{i2\pi n/N} \qquad (10.57)$$

where n is an integer. The action of the pure gauge theory is invariant too. This is a global $Z(N)$ symmetry. However, the Polyakov loop is changed:

$$L(\mathbf{x}) \to L(\mathbf{x})\, e^{i2\pi n/N} \qquad (10.58)$$

and so is the free energy of a system of static quarks and antiquarks:

$$\exp\left(-\beta \Delta F_{N_q N_{\bar{q}}}\right) \to \exp\left(-\beta \Delta F_{N_q N_{\bar{q}}}\right) e^{i2\pi n(N_q - N_{\bar{q}})/N} \qquad (10.59)$$

Unless $N_q - N_{\bar{q}}$ is an integral multiple of N, the free energy of this assembly of quarks and antiquarks is infinite. This is one manifestation of quark confinement. For SU(3) this means that the number of quarks minus antiquarks must be an integer multiple of 3, whereas for SU(2) it must be an integer multiple of 2.

If the $Z(N)$ symmetry of the pure gauge theory is spontaneously broken then there ought to be N distinct possible values of $\langle L \rangle$, with

$$\langle L \rangle = e^{i2\pi n/N} L_0 \qquad n = 0, 1, \ldots, N-1 \qquad (10.60)$$

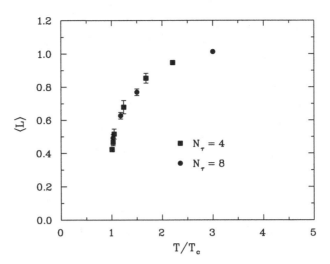

Fig. 10.4. The average value of the renormalized Polyakov loop as a function of temperature on lattices of spatial size $N_s = 32$. It is zero below a critical temperature. Systematic errors are not included. It can go above unity because it is normalized to the short-distance perturbative result on the lattice. The data are from [12].

Therefore $\langle L \rangle$ is an order parameter analogous to the magnetization in a $Z(N)$ spin system. Numerical calculations with the latter systems show a second-order phase transition for $N = 2$ and a first-order transition for $N \geq 3$, in agreement with explicit calculations for SU(N) gauge theories. Calculation of the mean value of L as a function of temperature for SU(3) does indeed show the expected behavior of an order parameter, as may be seen in Figure 10.4.

The static quark–antiquark free energy has contributions from the color singlet and octet potentials:

$$\exp(-\beta \Delta F_{1,1}(r,T)) = \tfrac{1}{9} \exp[-\beta \Delta F_1(r,T)] + \tfrac{8}{9} \exp[-\beta \Delta F_8(r,T)]$$

(10.61)

The octet is repulsive and the singlet is attractive. The latter is usually of most interest. It can be separated out via

$$\exp[-\beta \Delta F_1(r,T)] = \tfrac{1}{3} \mathrm{Tr} \langle W(\mathbf{x}, \beta) W^\dagger(\mathbf{0}, \beta) \rangle$$

(10.62)

This requires us to fix the gauge in order to obtain a physically relevant observable. Monte Carlo calculations for the color-singlet potential at various temperatures near T_c are shown in Figure 10.5. At large values of the separation r the free energy is independent of separation, indicating that the linear confining potential characteristic of the low-temperature phase is screened.

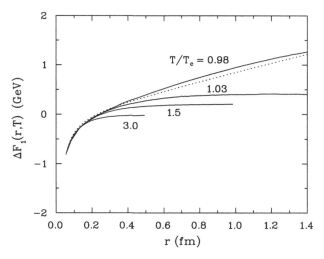

Fig. 10.5. The free energy of a heavy quark–antiquark pair in the color-singlet state as a function of separation for various temperatures. The calculations were done for $N_s = 32$ and $N_\tau = 4, 8,$ and 16. The data are from [13]. The dotted line represents a zero-temperature potential $V(r) = -4\alpha_s/3r + \sigma r$ with $\alpha_s = 0.18$.

In the pure gauge theories all physical observables are expressed in terms of the lattice spacing. The renormalization group relates the coupling, the lattice spacing, and the scale parameter, Λ_L. To two-loop order the relationship is

$$a\Lambda_L = \frac{1}{[\beta_0 g^2(a)]^{\beta_1/2\beta_0^2}} \exp\left(\frac{-1}{2\beta_0 g^2(a)}\right) \qquad (10.63)$$

where the coefficients β_0 and β_1 were given in (8.39). The scale parameter Λ_L can be related to the scale parameters in other schemes, such as $\bar\Lambda_{\rm MS}$.

The best approach to obtaining physically relevant numbers from lattice calculations is to express the results as dimensionless ratios. Then the explicit dependence on the lattice spacing drops out, and hopefully the sensitivity to nonzero a is reduced. For example, pure SU(3) gauge theory at $T = 0$ does not have the usual assortment of hadrons, such as pions and nucleons. It only has glueballs. A relevant ratio then would be $T_c/\sqrt{\sigma}$, where σ is the string tension, which may be obtained from the asymptotic part of the color singlet potential at $T = 0$. An average of several existing calculations yields $T_c/\sqrt{\sigma} = 0.632 \pm 0.002$. Now, there is no absolute argument that says that the string tension in quarkless SU(3) must be the same as in the real world with its six flavors of quarks with various masses. Still, one needs some scale to compare with and it might as well be the string tension. From the phenomenology of heavy quark systems and from the observed linear Regge trajectories it is known that

$\sqrt{\sigma} \approx 420$ MeV. In this case $T_c \approx 265$ MeV. If one chose a different comparison scheme then one would undoubtedly get a different number for the critical temperature in MeV because the relationship among observables depends on the number of quarks and their masses.

10.5 Lattice QCD

Inclusion of quarks on the lattice intensifies the numerical difficulty tremendously. A straightforward evaluation of the fermion determinant involves $(N_s^3 N_\tau)!$ terms. Since most entries in the fermion matrix are zero, clever techniques can be used to reduce this number significantly, but the task is still formidable. As in the pure gauge theory, additional terms can be added to the fermion part of the action that quicken the approach to the continuum, albeit at the expense of computational time.

The quark masses used in lattice calculations to date are not constants but scale with either the lattice spacing, $am = $ constant, or with the temperature, $m/T = $ constant. In principle the values of the up, down, and strange quark masses should be adjusted to yield the experimentally observed values of the pion and kaon masses. Then all other hadron masses, and all calculations done at finite temperature, would be absolute predictions of the theory. However, it turns out that the computational time grows quickly with decreasing quark mass, so that this goal has not been achieved yet.

Figure 10.6 shows the energy density versus temperature for two flavors of light quarks, two flavors of light quarks and one heavier quark, and three flavors of light quarks. Light and heavy mean $m_{\text{light}} = 0.4T$ and $m_{\text{heavy}} = T$, roughly corresponding to up and down quarks and strange quarks. There is a big jump in the energy density centered at a temperature defined to be T_c. Finite size and lattice spacing ($N_\tau = 4$) prevent one from concluding whether there is a first- or second-order phase transition or only a very rapid crossover. Extrapolation to the physical value of the ratios of pion to rho and omega vector meson masses suggests that $T_c = 172 \pm 9$ MeV for two light quarks, where the quoted uncertainty is statistical only. The systematic uncertainty is comparable in magnitude. The value of T_c is reduced by 15 to 20 MeV for a world of three light quarks.

The Polyakov loop cannot be used as an order parameter when dynamical quarks are included because the quark part of the action is not invariant under Z(3) transformations. This can be understood intuitively by remembering that the Polyakov loop is used to measure the free energy of any configuration of static quarks and antiquarks. When a quark and antiquark are pulled apart the potential ceases to be linear in the separation,

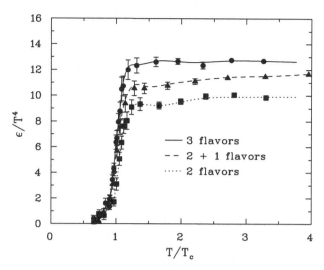

Fig. 10.6. Energy density in units of T^4 as a function of T/T_c for two flavors of light quarks, two flavors of light quarks and one flavor of heavier quark, and three flavors of light quarks. The lattice size is $16^3 \times 4$. The data are from [14].

because the string can break owing to the creation of a light dynamical quark and antiquark. So what then could the order parameter be? If some of the quarks have zero mass then the theory has chiral symmetry, and the quark condensate $\langle \bar{\psi}\psi \rangle$ serves as an order parameter. Figure 10.7 shows the temperature dependence of this condensate for a sequence of ever decreasing light quark masses, for two light flavors and one heavier flavor. The quark condensate goes to zero if the light quark mass is light enough, but only decreases monotonically without ever reaching zero for more massive light quarks. This is also reflected in the hysteresis behavior when the system is numerically cooled and heated and cooled again by Monte Carlo.

To study the effects of varying the number of quark flavors and their masses, Pisarski and Wilczek [16] constructed an effective Lagrangian for an order-parameter field taken to be $\Phi_{ij} = f\bar{q}_i(1 + \gamma_5)q_j$, where f is a constant. The Lagrangian should reflect the symmetries of the QCD Lagrangian. For N_f flavors of massless quarks the symmetry group is

$$G_f = U(1)_A \times SU(N_f) \times SU(N_f) \rightarrow G'_f = Z(N_f)_A \times SU(N_f) \times SU(N_f)$$

Here the classical axial $U(1)_A$ symmetry is broken to $Z(N_f)_A$ symmetry, owing to the quantum axial anomaly. The form of the effective

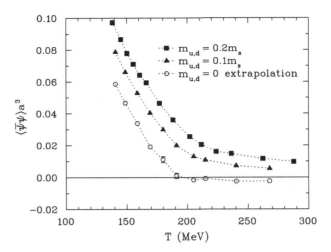

Fig. 10.7. The quark condensate, measured in lattice units, versus temperature. The strange quark mass is fixed in such a way that the mass of the ϕ meson takes its physical value as calculated on the lattice. The lattice size is $16^3 \times 8$. The data are from [15].

(renormalizable) Lagrangian is

$$
\begin{aligned}
\mathcal{L}_{\text{eff}} = {} & \tfrac{1}{2}\operatorname{Tr}\left(\partial_\mu \Phi^\dagger \partial^\mu \Phi\right) - \tfrac{1}{2}m_\Phi^2 \operatorname{Tr}\left(\Phi^\dagger \Phi\right) \\
& - \tfrac{1}{3}\pi^2 g_1 \left[\operatorname{Tr}\left(\Phi^\dagger \Phi\right)\right]^2 - \tfrac{1}{3}\pi^2 g_2 \operatorname{Tr}\left[\left(\Phi^\dagger \Phi\right)^2\right] \\
& + c\left(\det \Phi^\dagger + \det \Phi\right) + \operatorname{Tr}\left[M\left(\Phi^\dagger + \Phi\right)\right]
\end{aligned}
\tag{10.64}
$$

The determinants originate in the anomaly and are sometimes associated with instanton effects. In hadronic phenomenology they are necessary to give the η' its large observed mass. The last term involving the matrix M represents the effect of nonzero quark masses. Pisarski and Wilczek then used universality to infer the behavior of the QCD system from studies of simpler systems with the same symmetry. Those systems were studied using an ϵ expansion in $4 - \epsilon$ dimensions. Assuming that all quarks are massless, they found that for $N_{\text{f}} = 1$ there is no true phase transition, for $N_{\text{f}} = 2$ the phase transition may be first or second order depending on the strength of the anomaly at T_{c} as reflected in the coefficient c, and for $N_{\text{f}} \geq 3$ the phase transition is first order.

The likely phase diagram in the m_s versus $m_u = m_d$ plane is shown in Figure 10.8. When all quarks are infinitely heavy it is as if they do not exist at the temperatures of interest, and there is a first-order deconfining phase transition as in the pure SU(3) gauge theory. When all three

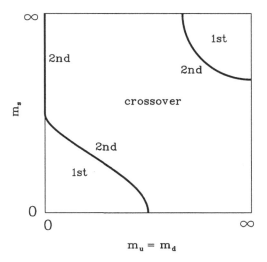

Fig. 10.8. A possible phase diagram for QCD in the strange quark mass versus light quark mass plane. The lower left-hand corner exhibits a first-order chiral-symmetry phase transition, the upper right-hand corner a deconfinement phase transition. These are separated from an intermediate region of rapid crossover between phases.

flavors are massless there is a first-order chiral-symmetry-restoring phase transition. When the strange quark is heavy and the up and down quarks have zero mass there is a second-order chiral-symmetry-restoring phase transition. When all three flavors have masses of the order of several hundred MeV, there is no true thermodynamic phase transition but only a sharp crossover with a jump in the energy density over a small range of temperatures. A variety of lattice calculations seem to support this general picture, but it should still be considered merely as a reasonable conjecture.

The application of lattice QCD to the study of finite-density matter is in its infancy. The difficulty lies in the fact that the chemical potential acts as a constant imaginary time component of the vector potential; see (5.18), (5.19), and (10.42). As such the fermion determinant is complex. This should not lead to a complex partition function; the imaginary part must average to zero. However, it does mean that straightforward Monte Carlo sampling techniques cannot be applied. Two interesting approaches are (i) a Taylor series expansion in powers of $(\mu_B/T)^2$ and (ii) calculation with an imaginary chemical potential followed by analytic continuation to a real baryon chemical potential μ_B. The latter approach was followed by de Forcrand and Philipsen. For two flavors of light quarks they found

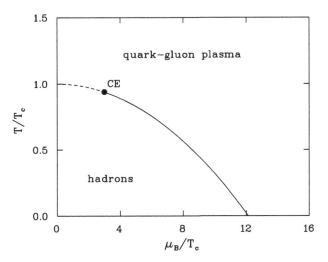

Fig. 10.9. A possible critical line for QCD. The solid curve represents a first-order phase transition terminating at a second-order phase transition at the critical endpoint labeled CE. The broken line represents a rapid crossover.

the critical line [17]

$$\frac{T}{T_c} = 1 - 0.0056 \pm 0.0004 \left(\frac{\mu_B}{T_c}\right)^2 \qquad (10.65)$$

and for three flavors of light quarks [18]

$$\frac{T}{T_c} = 1 - 0.0068 \pm 0.0001 \left(\frac{\mu_B}{T_c}\right)^2 \qquad (10.66)$$

The stated range of validity is $|\mu_B| < 500$ MeV; the systematic uncertainties are at least as large as the quoted statistical uncertainties. If these are optimistically extrapolated to zero temperature, then taking $T_c \approx 160$ MeV one gets $\mu_B \approx 2$ GeV. This is quite reasonable. The above relations were shown not to be sensitive to the precise numerical value of the quark mass. The critical line for $N_f = 3$ is shown in Figure 10.9. The lattice calculations did not establish conclusively whether the critical line represents a true phase transition or just a rapid crossover between hadronic matter and quark–gluon plasma. If the quark masses are not small enough to yield a first-order phase transition at zero baryon density but finite temperature, there may exist a critical point along the critical line. At lower baryon density there is a rapid crossover and at higher baryon density there is a first-order phase transition. This would connect nicely with the color superconductivity analysis; see Figure 8.4. At the present time the existence of a critical point is not well established.

10.6 Exercises

10.1 Show that (10.5) leads to the correct commutation relation between the vector potential and the electric field in the continuum limit.

10.2 Derive the Coulomb potential (10.25) in the high-temperature limit of the lattice gauge theory.

10.3 Show that (10.31) is gauge invariant.

10.4 Calculate the continuum limit of (10.31).

10.5 Consider a free neutral boson on a $N_s^3 N_\tau$ lattice of spacing a in all four directions in the limit $N_s \gg N_\tau \gg 1$. Since the action is quadratic, the functional integral expression for the partition function can be evaluated exactly. Following the analysis in Chapter 2 show that the propagator can be written as

$$\mathcal{D}^{-1}(\mathbf{p}, p_4) = \frac{4}{a^2} \sin^2 \left(\frac{ap_4}{2} \right) + \frac{4}{a^2} \sum_{i=1}^{3} \sin^2 \left(\frac{ap_i}{2} \right) + m^2$$

where $-\pi/a \leq p_4 \leq \pi/a$ and $-\pi/a \leq p_i \leq \pi/a$, which defines the Brillouin zone. This reduces to the usual scalar propagator in the $a \to 0$ limit. This propagator has only one minimum, which is located at $\mathbf{p} = \mathbf{0}$, $p_4 = 0$.

10.6 Using periodic boundary conditions in the spatial directions, and antiperiodic boundary conditions in the temporal direction, compute the partition function for massless staggered fermions in the limit of large N_s and N_τ.

10.7 Repeat Exercise 10.5 for Wilson fermions.

10.8 Show formally that the thermodynamic identities are obeyed with the action (10.42). This implies that the chemical potential in (10.42) has been implemented correctly.

10.9 Work out the details leading from (10.51) to (10.52).

10.10 Derive the result (10.60), which holds when $Z(N)$ symmetry is spontaneously broken.

10.11 Construct a function $P_{\text{fit}}(T)$ that parametrizes the results of the pure SU(3) lattice results. Be sure to compare with entropy and energy densities too.

10.12 Are there other terms that could be added to (10.64) for two flavors of massless quarks? If so, what are they?

References

1. Wilson, K. G., *Phys. Rev. D* **10**, 2445 (1974).
2. Creutz, M., *Phys. Rev. D* **21**, 2308 (1980).
3. Polyakov, A. M., *Phys. Lett.* **B72**, 477 (1978).
4. Susskind, L., *Phys. Rev. D* **20**, 2610 (1979).

5. McLerran, L. D., and Svetitsky, B. *Phys. Lett.* **B98**, 195 (1981); *Phys. Rev. D* **24**, 450 (1981).

6. Kuti, J., Polónyi, J., and Szlachányi, K., *Phys. Lett.* **B98**, 199 (1981).

7. Kogut, J., and Susskind, L., *Phys. Rev. D* **11**, 395 (1975).

8. Lucini, B., Teper, M., and Wenger, U., *Phys. Lett.* **B545**, 197 (2002); *JHEP* **0401**, 061 (2004).

9. Svetitsky, B., and Yaffe, L. G., *Nucl. Phys.* **B210**, 423 (1982); Yaffe, L. G., and Svetitsky, B., *Phys. Rev. D* **26**, 963 (1982).

10. Boyd, G., Engels, J., Karsch, F., Laermann, E., Legeland, C., Luetgemeier, M., and Petersson, B., *Nucl. Phys.* **B469**, 419 (1996).

11. Symanzik, K., *Nucl. Phys.* **B226**, 187 (1983); *ibid.* 205 (1983).

12. Kaczmarek, O., Karsch, F., Petreczky, P., and Zantow, F., *Phys. Lett.* **B543**, 41 (2002).

13. Kaczmarek, O., Karsch, F., Zantow, F., and Petreczky, P., *Phys. Rev. D* **70**, 074505 (2004).

14. Karsch, F., Laermann, E., and Peikert, A., *Phys. Lett.* **B478**, 447 (2000).

15. Bernard, C., *et al.* (MILC collaboration), *Nucl. Phys. Proc. Suppl.* **119**, 523 (2003), and Gottlieb, S. (private communication).

16. Pisarski, R. D., and Wilczek, F., *Phys. Rev. D* **29**, 338 (1984).

17. de Forcrand, P., and Philipsen, O., *Nucl. Phys.* **B642**, 290 (2002).

18. de Forcrand, P., and Philipsen, O., *Nucl. Phys.* **B673**, 170 (2003).

Bibliography

Creutz, M. (1983). *Quarks, Gluons, and Lattices* (Cambridge University Press, Cambridge).

Montvay, I., and Münster, G. (1994). *Quantum Fields on a Lattice* (Cambridge University Press, Cambridge).

Rothe, H. J. (1997). *Lattice Gauge Theories* (World Scientific, Singapore).

Kogut, J. B. (2004). *Milestones in Lattice Gauge Theory* (Kluwer, Amsterdam).

11

Dense nuclear matter

One aspect of nuclear physics is the study of nuclear matter. Up until the mid 1970s, nearly all studies used a nonrelativistic potential to describe the nucleon–nucleon interaction. The results were not entirely satisfactory. It was difficult to obtain simultaneously the saturation density (about 0.153 nucleons per fm^3) and the binding energy (about 16.3 MeV per nucleon with the Coulomb force turned off) in a microscopic nonrelativistic approach. Part of the discrepancy was ascribed to three-body interactions. However, relativity can also play a small but significant role at normal nuclear matter density. The importance of relativity may be judged by comparing the Fermi momentum p_F with the nucleon mass. The baryon density is

$$n = \frac{2p_F^3}{3\pi^2} \tag{11.1}$$

At normal nuclear density $p_F = 259$ MeV, and at four times normal nuclear density $p_F = 411$ MeV. These should be compared with the vacuum nucleon mass $m_N = 939$ MeV and the Fermi kinetic energy

$$K_F = m_N \left[(1 - v_F^2)^{-1/2} - 1\right] = \tfrac{1}{2} m_N v_F^2 + \tfrac{3}{8} m_N v_F^4 + \cdots \tag{11.2}$$

Although one might think that the relativistic correction at normal nuclear density, which is of order v_F^4 and numerically about 2 MeV, is rather small, still it is not insignificant compared with the binding energy of 16.3 MeV. Of course, at higher densities, relativity certainly cannot be ignored. As we shall learn, relativity plays an even greater role in the interactions among nucleons. The relativistic approach to nuclear matter was pioneered by Johnson and Teller [1], Duerr [2], and Walecka [3].

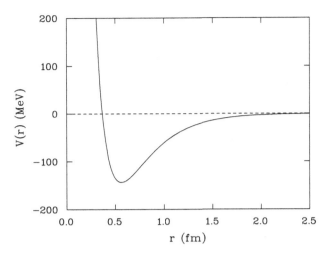

Fig. 11.1.   Model of the nucleon–nucleon potential illustrating long-range attraction and short-range repulsion. The parameters are given in the text.

11.1 Walecka model

The force between nucleons is conventionally thought of as mediated by the exchange of mesons. The long-range part of the nuclear force comes from one-pion exchange. Its range is $1/m_\pi = 1.4$ fm. However, this averages to zero unless parity is broken. The force mediated by exchange of the ρ meson vanishes in isospin-symmetric matter (equal numbers of protons and neutrons); the dominant one-meson exchanges in isospin-symmetric nuclear matter come from the omega meson (ω) and a scalar meson (σ). The ω is a vector meson, is electrically neutral, and has a mass of about 783 MeV. The σ meson represents a very broad resonance in $\pi\pi$ scattering at 500–600 MeV. The exchange of the electrically neutral σ is usually thought of as simulating some part of two-pion exchange. With single-ω and single-σ exchange, the static nonrelativistic potential between two nucleons is the sum of two Yukawa interactions:

$$V(r) = \frac{g_\omega^2}{4\pi}\frac{\mathrm{e}^{-m_\omega r}}{r} - \frac{g_\sigma^2}{4\pi}\frac{\mathrm{e}^{-m_\sigma r}}{r} \tag{11.3}$$

If $g_\omega > g_\sigma$ and $m_\omega > m_\sigma$ then the potential looks like that shown in Figure 11.1. It is attractive at long distances and repulsive at short distances and so has the structure necessary to bind nuclear matter.

The Lagrangian that contains the Yukawa couplings of the nucleon to the ω and to the σ is

$$\begin{aligned}
\mathcal{L}_{\mathrm{W}} = {} & \bar\psi(i\,\slashed\partial - m_N + g_\sigma\sigma - g_\omega\,\slashed\omega)\psi \\
& + \tfrac{1}{2}\left(\partial_\mu\sigma\partial^\mu\sigma - m_\sigma^2\sigma^2\right) - \tfrac{1}{4}F^{\mu\nu}F_{\mu\nu} + \tfrac{1}{2}m_\omega^2\omega_\mu\omega^\mu
\end{aligned} \tag{11.4}$$

where

$$F_{\mu\nu} = \partial_\mu \omega_\nu - \partial_\nu \omega_\mu$$

Now of course we know that baryons and mesons are not elementary point particles. They are composite structures of quarks and gluons. The idea of an effective-nuclear-field theory is to write down a Lagrangian that contains the low-lying baryons and mesons as relativistic fields. This allows us to use the standard machinery of relativity, quantum mechanics, and statistical mechanics. This approach will break down if we attempt to probe the theory at very short distances, say a few tenths of a fermi, since then the quarks and gluons must manifest themselves.

Let us investigate the properties of dense nuclear matter using the Lagrangian $\mathcal{L}_W$. The partition function is

$$Z = \int \left[d\bar{\psi}_p \right] \left[d\psi_p \right] \left[d\bar{\psi}_n \right] \left[d\psi_n \right] \left[d\sigma \right] \left[d\omega_\mu \right]$$

$$\times \exp \left(\int_0^\beta d\tau \int d^{3x} \left(\mathcal{L}_W + \mu_p \psi_p^\dagger \psi_p + \mu_n \psi_n^\dagger \psi_n \right) \right) \quad (11.5)$$

where μ_p and μ_n are the proton and neutron chemical potentials. For isospin-symmetric matter $\mu_p = \mu_n$ and for pure neutron matter $\mu_p = 0$, $\mu_n \neq 0$ (although the presence of electrons in a neutron star allows $\mu_p \neq 0$). For the remainder of this section, we concentrate on symmetric matter and write $\mu = \mu_n = \mu_p$.

The nucleons act as sources in the meson field equations. This suggests that a net baryon density will generate scalar and vector meson condensates. This can be checked by allowing σ and ω_μ to have nonzero expectation values. Thus we write

$$\sigma = \bar{\sigma} + \sigma'$$
$$\omega_\mu = \delta_{\mu 0} \bar{\omega}_0 + \omega'_\mu \quad (11.6)$$

where the bar indicates the ensemble average value of the field and the prime indicates the fluctuation about the average. (Note that $\bar{\omega}_i = 0$ on account of rotational symmetry.) In the *mean field approximation*, one neglects fluctuations in the meson fields. This means that the nucleons are taken to move independently in the mean fields $\bar{\sigma}$ and $\bar{\omega}_0$, which themselves are generated self-consistently by the nucleons. Based on the success of the nuclear shell model, we anticipate that this will provide a reasonable first-order estimate of the properties of dense nuclear matter. The Lagrangian $\mathcal{L}_W$ is commonly referred to as the Walecka Lagrangian, and when used in conjunction with the mean field approximation is referred to as the Walecka model.

The partition function may be evaluated exactly in the mean field approximation because the functional integral is just a product of Gaussian integrals. The argument of the exponential in (11.5) is thus approximated by

$$\bar{\psi}\left[i\,\partial\!\!\!/ - (m_N - g_\sigma\bar{\sigma}) + (\mu - g_\omega\bar{\omega}_0)\gamma_0\right]\psi - \tfrac{1}{2}m_\sigma^2\bar{\sigma}^2 + \tfrac{1}{2}m_\omega^2\bar{\omega}_0^2 \qquad (11.7)$$

This means that the nucleon develops an effective mass

$$m_N^* = m_N - g_\sigma\bar{\sigma} \qquad (11.8)$$

and an effective chemical potential

$$\mu^* = \mu - g_\omega\bar{\omega}_0 \qquad (11.9)$$

Using (11.5) together with (11.7), we obtain the pressure as

$$P(\mu, T) = P_{\mathrm{FG}}(\mu^*, T) - \tfrac{1}{2}m_\sigma^2\bar{\sigma}^2 + \tfrac{1}{2}m_\omega^2\bar{\omega}_0^2 \qquad (11.10)$$

where P_{FG} is the Fermi-gas expression for nucleons with the quoted effective mass and chemical potential.

We now must determine the mean fields $\bar{\sigma}$ and $\bar{\omega}_0$. If we allow $\bar{\sigma}$ and $\bar{\omega}_0$ to vary, the equilibrium configuration will be attained when P is an extremum. Thus

$$\bar{\sigma} = -\left(\frac{g_\sigma}{m_\sigma^2}\right)\frac{\partial P_{\mathrm{FG}}}{\partial m_N^*} \qquad (11.11)$$

$$\bar{\omega}_0 = \left(\frac{g_\omega}{m_\omega^2}\right)\frac{\partial P_{\mathrm{FG}}}{\partial \mu^*} \qquad (11.12)$$

Of these, the vector condensate can be determined directly in terms of the baryon density n:

$$\bar{\omega}_0 = \frac{g_\omega}{m_\omega^2}n \qquad (11.13)$$

A quick calculation utilizing (2.99) shows that the scalar condensate is proportional to the scalar density n_{s}:

$$\bar{\sigma} = \frac{g_\sigma}{m_\sigma^2}n_{\mathrm{s}} \qquad (11.14)$$

where

$$n_{\mathrm{s}} \equiv 4\int\frac{d^3p}{(2\pi)^3}\frac{m_N^*}{E^*}\left(\frac{1}{e^{\beta(E^*-\mu^*)}+1} + \frac{1}{e^{\beta(E^*+\mu^*)}+1}\right)$$
$$E^* = \sqrt{p^2 + m_N^{*2}}$$

$$(11.15)$$

Equation (11.14) is a self-consistent equation to be solved for m_N^*, as may be seen from the alternate form

$$m_N^* = m_N - \left(\frac{g_\sigma^2}{m_\sigma^2}\right) n_s \tag{11.16}$$

Now let us focus on cold nuclear matter and delay the discussion of finite temperature to a later section. Using the above equations and the standard thermodynamic identities, we have

$$P = \frac{1}{4\pi^2}\left[\frac{2}{3}E_F^* p_F^3 - m_N^{*2}E_F^* p_F + m_N^{*4}\ln\left(\frac{E_F^* + p_F}{m_N^*}\right)\right]$$
$$+ \frac{1}{2}\left(\frac{g_\omega^2}{m_\omega^2}\right)n^2 - \frac{1}{2}\left(\frac{g_\sigma^2}{m_\sigma^2}\right)n_s^2 \tag{11.17}$$

$$\epsilon = \frac{1}{4\pi^2}\left[2E_F^{*3}p_F - m_N^{*2}E_F^* p_F - m_N^{*4}\ln\left(\frac{E_F^* + p_F}{m_N^*}\right)\right]$$
$$+ \frac{1}{2}\left(\frac{g_\omega^2}{m_\omega^2}\right)n^2 + \frac{1}{2}\left(\frac{g_\sigma^2}{m_\sigma^2}\right)n_s^2$$

where

$$n = \frac{2}{3\pi^2}p_F^3$$

$$E_F^* = \mu^* = \sqrt{p_F^2 + m_N^{*2}}$$

$$m_N^* = m_N - \left(\frac{g_\sigma^2}{m_\sigma^2}\right)n_s$$

$$n_s = \frac{m_N^*}{\pi^2}\left[E_F p_F - m_N^{*2}\ln\left(\frac{E_F^* + p_F}{m_N^*}\right)\right]$$

In these equations it is natural to take the Fermi momentum p_F as the one independent variable. Notice that this equation of state is given essentially in analytic form with only the equation for m_N^* to be solved self-consistently.

Before investigating the equation of state in detail, it is worthwhile to consider the extremes of low and high density. At low density $p_F \to 0$, and we recover the equation of state of a nonrelativistic ideal Fermi gas.

The quantities in (11.17) have the following limits:

$$P \to \frac{2}{15\pi^2} \frac{p_F^5}{m_N}$$

$$\epsilon \to \left(m_N + \frac{3}{10} \frac{p_F^2}{m_N} \right) n \qquad (11.18)$$

$$m_N^* \to m_N$$

$$n_s \to n$$

At high density, $p_F \to \infty$, the effective nucleon mass goes to zero:

$$m_N^* \to \frac{m_N}{1 + (g_\sigma^2/\pi^2)(p_F^2/m_\sigma^2)} \qquad (11.19)$$

The pressure and energy density are dominated by the vector mean field:

$$P \to \epsilon \to \frac{1}{2} \left(\frac{g_\omega^2}{m_\omega^2} \right) n^2 \qquad (11.20)$$

Thus the speed of sound, $c_s^2 = \partial P / \partial \epsilon$, approaches the speed of light at very high density. This is to be compared with the speed in sound in a massless Fermi gas, which is $1/\sqrt{3}$.

In these equations there are only two parameters at our disposal, g_ω^2/m_ω^2 and g_σ^2/m_σ^2. The nucleon mass is $m_N = 939$ MeV and the vector meson mass is $m_\omega = 783$ MeV. For definiteness we take $m_\sigma = 550$ MeV, corresponding to the scalar–isoscalar resonance in π–π scattering. Then the choice of couplings $g_\omega^2/4\pi = 14.717$ and $g_\sigma^2/4\pi = 9.537$ leads to a binding energy of 16.3 MeV per nucleon and a saturation density of 0.153 nucleons per fm^3. The curve of energy per nucleon versus density is shown in Figure 11.2. The energy per nucleon rises rather dramatically with density. In fact the compressibility of nuclear matter at saturation density turns out to be

$$K \equiv p_F^2 \frac{d^2(\epsilon/n)}{dp_F^2} = 563 \text{ MeV} \qquad (11.21)$$

The generally accepted value, based on measurements of the isoscalar giant monopole resonance in heavy nuclei, is 250 ± 30 MeV [4–7]. The Walecka model predicts $m_N^* = 0.57 m_N$, which is somewhat smaller than some estimated values of the effective nucleon mass at nuclear saturation density [8] but quite consistent with others [9, 10]. Nevertheless, we should not expect to fit a large body of nuclear-matter properties to high accuracy for several reasons: (i) the Lagrangian $\mathcal{L}_W$ is too simple to represent accurately the complicated nuclear forces, and (ii) the mean field approximation neglects nucleon–nucleon correlations.

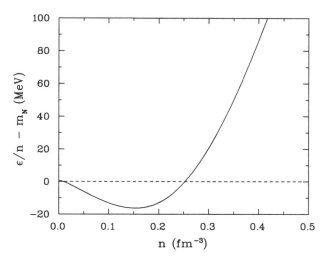

Fig. 11.2. The average energy per nucleon minus the nucleon mass as a function of the baryon density in the mean field approximation to the Walecka model.

One may ask whether the numerical values obtained for the two Yukawa couplings are reasonable. Fitting nucleon–nucleon phase shifts up to 300 MeV typically yields similar values. For example, Machleidt, Holinde, and Elstor [11] used a boson-exchange model employing the π, ρ, ω, and σ mesons, and found that $g_\omega^2/4\pi = 20$ and $g_\sigma^2/4\pi = 9.2$. This agreement is satisfactory considering that the Yukawa couplings used in the mean field approach are really effective couplings that are fine-tuned to mimic all the many-body effects not included. Fine-tuning is actually required in the Walecka model, where the delicate cancelation between short-range vector repulsion and medium-range scalar attraction is really a relativistic effect. This may be seen in the following way. In the nonrelativistic mean field approximation the average potential energy felt by a nucleon is

$$\langle V \rangle = n \int d^3r \, V(r) \tag{11.22}$$

The average kinetic energy is $(3/5)(p_F^2/2m_N)$. Since $n \propto p_F^3$ this means that $\int d^3r \, V(r) > 0$ for the energy to be bounded from below. Hence both the average kinetic and potential energies must be positive at all densities, and an equilibrium bound state cannot arise. Relativity plays an important role in the sense that the baryon (vector) density and the scalar density are not the same; they differ by a velocity factor m_N^*/E^* in the relevant integrands at zero temperature. In this vein it is illustrative to expand the energy per nucleon as a power series in the Fermi velocity at $T = 0$. The difference between the scalar and baryon densities shows

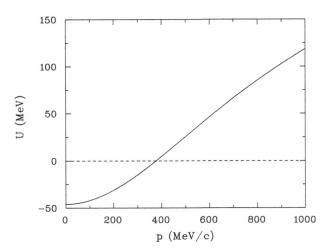

Fig. 11.3. The nuclear optical potential as a function of momentum at nuclear saturation density in the Walecka model.

up at order v_F^5, which is just one power more than the first relativistic correction to the kinetic energy. See, for example, Serot and Walecka [12].

The nuclear optical potential U is defined by

$$E(p, p_F) = \sqrt{p^2 + m_N^2} + U(p, p_F) \qquad (11.23)$$

Here E is the single-particle energy of a nucleon. The optical potential is both density and momentum dependent. From (11.10), (11.14), and (11.16),

$$E(p, p_F) = \sqrt{p^2 + m_N^{*2}} + \left(\frac{g_\omega^2}{m_\omega^2}\right) n \qquad (11.24)$$

The optical potential at saturation density is plotted in Figure 11.3. Various phenomenological optical potentials of this form are widely used in interpreting proton–nucleus scattering [13] and nucleus–nucleus scattering [14, 15] at energies of several hundred MeV. The optical potential in the Walecka model rises too rapidly at high momentum compared with the data. Such disagreement should not be a surprise, for the reasons given above.

11.2 Loop corrections

The mean field used in the previous section has two great advantages: it is relativistically and thermodynamically self-consistent, neither of which is a trivial achievement. By fine-tuning only two input parameters, the binding energy and the density of cold isospin-symmetric nuclear matter, one may extrapolate to both lower and higher densities, to isospin-asymmetric matter (after the ρ meson is included, see Chapter 16), and to moderate temperatures. However, the coupling constants are large, of the order of 10, making a convergent loop expansion highly unlikely. Both one-loop and two-loop corrections have been computed in the Walecka model, and we shall present the results in this section. The two-loop corrections, in particular, are very large, as expected. However, nucleons are composite objects, constructed from quarks and gluons. Therefore any effective theory using baryons and mesons as the degrees of freedom will necessarily bring form factors into play. It will turn out that with reasonable choices of the form factors, the sum of the two-loop contributions to the energy per nucleon is surprisingly small. This does not, of course, imply that other physical observables are also small.

11.2.1 Relativistic Hartree

The Walecka model is renormalizable even though it has a massive vector boson because it couples to the conserved baryon current. Regarding the scalar boson, the Walecka model truncates the Lagrangian at order σ^2. However, all terms that keep the theory renormalizable and respect the symmetries can and should be kept. This means powers of σ up to and including 4. In fact, these are required in order to cancel divergences coming from the shift in the zero-point energy of the nucleons. Relative to the vacuum, the shift in the zero-point energy is [16]

$$\epsilon_{\text{ZP}}(m_N^*) = -2 \int \frac{d^3p}{(2\pi)^3} \left(\sqrt{p^2 + m_N^{*2}} - \sqrt{p^2 + m_N^2} \right) - \sum_{n=1}^{4} \frac{c_n}{n!} \sigma^n$$

(11.25)

The coefficients c_n of the counterterms are dependent upon the regularization scheme. In momentum-cutoff schemes they diverge as the cutoff goes to infinity, and in dimensional regularization schemes they diverge as four dimensions are approached. The minimal procedure is to choose the c_n so as to cancel the first four powers of σ arising from the integration over momentum. (Recall that $m_N^* = m_N - g_\sigma \bar{\sigma}$.) Although this procedure is not unique, it has the feature of minimizing the many-body forces arising from this vacuum correction. The result is

$$\epsilon_{\text{ZP}}(m_N^*) = -\frac{1}{4\pi^2} \left[m_N^{*4} \ln \left(\frac{m_N^*}{m_N} \right) + m_N^3(m_N - m_N^*) - \frac{7}{2} m_N^2 (m_N - m_N^*)^2 \right.$$
$$\left. + \frac{13}{3} m_N(m_N - m_N^*)^3 - \frac{25}{12}(m_N - m_N^*)^4 \right]$$

(11.26)

This formula assumes an isospin degeneracy factor 2.

The pressure and energy density in the one-loop relativistic Hartree approximation are related to those in the relativistic mean field approximation by

$$P_{\mathrm{RH}} = P_{\mathrm{MF}} - \epsilon_{\mathrm{ZP}}$$
$$\epsilon_{\mathrm{RH}} = \epsilon_{\mathrm{MF}} + \epsilon_{\mathrm{ZP}}$$

(11.27)

This equation of state is also thermodynamically consistent. Minimizing the energy at fixed density leads to a modification of the mean field self-consistency condition for the scalar condensate, namely

$$m_N^* = m_N - \left(\frac{g_\sigma^2}{m_\sigma^2}\right) n_{\mathrm{s}} + \frac{g_\sigma^2}{m_\sigma^2} \frac{1}{\pi^2} \left[m_N^{*3} \ln\left(\frac{m_N^*}{m_N}\right) - m_N^2 (m_N - m_N^*) \right.$$
$$\left. - \frac{5}{2} m_N (m_N - m_N^*)^2 - \frac{11}{6}(m_N - m_N^*)^3 \right]$$

(11.28)

A small change to the parameters in (11.4) again reproduces the saturation density and binding energy of nuclear matter. Numerical results will be shown in the next subsection, where we consider two-loop contributions.

11.2.2 Two loops

Two-loop contributions to the partition function can be performed in the usual fashion. The general form is

$$\ln Z_2 = -\frac{1}{2} \sum_{n_1 n_2} \int \frac{d^3 p_1}{(2\pi)^3} \frac{d^3 p_2}{(2\pi)^3} \, \mathrm{Tr}\left[\mathcal{G}(p_1) \Gamma(p_1, p_2, k) \, \mathcal{G}(p_2) \Gamma(p_2, p_1, k) \mathcal{D}(k) \right]$$

(11.29)

Here $\mathcal{G}$ is the nucleon propagator, $\mathcal{D}$ is the boson propagator (for either the scalar or vector meson), Γ is the relevant vertex, and $k = p_1 - p_2$. Lorentz and Dirac indices are suppressed.

These contributions have been evaluated at zero temperature by Furnstahl, Perry, and Serot [17]. The two-loop diagram has several physical contributions. One contribution originates from the exchange of momentum between two nucleons in the Fermi sea. A second contribution comes from the Lamb shift, the change in the properties of a nucleon as it propagates in the medium. The third contribution is a shift in the zero-point vacuum fluctuations owing to the presence of nuclear matter. Unfortunately the results cannot be expressed in terms of elementary functions because the nucleon and meson masses are all nonzero. The diagrams are as follows:

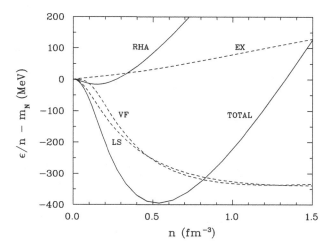

Fig. 11.4. The energy per nucleon as a function of density. The solid curve labeled RHA is the relativistic Hartree approximation. The solid curve labeled TOTAL includes the exchange, Lamb shift, and two-loop vacuum fluctuations as well. Point vertices are used.

Figure 11.4 shows the equation of state in the relativistic Hartree approximation (RHA). In this figure the values of the parameters are $g_\sigma^2/4\pi = 4.32$, $m_\sigma = 458$ MeV, $g_\omega^2/4\pi = 8.18$, and $m_\omega = 783$ MeV. Also shown are the contributions from the exchange term, the Lamb shift, and the vacuum fluctuations. They are computed as if they were perturbations, with no change in the numerical values of the coupling constants. The exchange term is relatively modest but the Lamb shift and the vacuum fluctuations are enormous. When all are added, the binding energy changes to nearly 400 MeV from 16 MeV at a density of 3.7 times the empirical value. The two-loop contributions are not perturbatively small, quite the contrary. This is not unexpected, owing to the large values of the coupling constants. Undoubtedly higher-loop contributions are important too, and the whole calculational scheme breaks down.

11.2.3 Form factors

The integrals in the two-loop terms receive contributions from internal momenta as high as 5 GeV. But nucleons and mesons are not point particles. Their finite spatial size should soften these contributions significantly. Prakash, Ellis, and Kapusta [18] introduced form factors at both vertices with the philosophy that they do not arise from interactions within the confines of the Walecka model. The origin of these form factors runs deeper, back to the quark and gluon substructure of hadrons. Of course, the full meson–nucleon vertex function will include dressings from the hadronic degrees of freedom too. A consistent treatment of form factors is not yet available. In general, they will involve several scalar functions, the number of which depends on the Lorentz structure of the vertex. Each one may depend on the invariants p_1^2, p_2^2, and k^2. The minimal assumption is that the scalar and vector form factors are functions of k^2 only and have the simple monopole form

$$f(k^2) = \frac{1}{1 - k^2/\Lambda^2} \tag{11.30}$$

when $k^2 < 0$ (spacelike) and where Λ is a cutoff of order 1 GeV. Then the vertices to be used in the two loop diagrams are just the point vertices multiplied by $f(k^2)$.

Three comments are in order. First, the selection of relativistic monopole form factors with a cutoff of this order is consistent with the relativistic dipole structure of the on-shell nucleon electromagnetic form factor. In that case, one monopole factor arises from the finite size of the nucleon while the other arises from the ρ meson propagator in the context of the vector-meson-dominance model [19]. Second, if one were to include the off-mass-shell p_1^2 and p_2^2 dependences as well then it might be possible to get an even greater suppression of the two-loop contributions than that displayed in Figure 11.5. Third, since the vector meson couples to the baryon current there is a generalized type of Ward identity. This identity is different from that in QED because in the strong-interaction case the vector–current coupling is nonlocal. Since the form factor $f(k^2)$ is taken to be intrinsic to the nucleon rather than generated by the Yukawa interactions of the nucleons with the meson fields, there is no inconsistency in using the mean field propagators in the loop expansion. To lowest order in g_σ and g_ω the identity must be such that it is satisfied by the free-field form of the nucleon and meson propagators.

Multiplying the bare point-particle vertices by $f(k^2)$, it can easily be shown that the energy density is obtained from the identity

$$\epsilon = \frac{\Lambda^2}{\Lambda^2 - m^2} \left\{ \frac{\Lambda^2}{\Lambda^2 - m^2} \left[\epsilon_{\text{pt}}(m^2) - \epsilon_{\text{pt}}(\Lambda^2) \right] + \frac{d\epsilon_{\text{pt}}(\Lambda^2)}{d \ln \Lambda^2} \right\} \tag{11.31}$$

Here m is the mass of the exchanged meson and ϵ_{pt} is the two-loop energy density with point vertices.

Figure 11.5 shows what happens when form factors are inserted at each vertex with the cutoff chosen as 1 GeV for both the scalar and vector meson vertices. There is a tremendous reduction compared with the case of point vertices. Near the equilibrium density, both the scalar and vector meson exchange terms are reduced by 10%–15%. For the Lamb shift the reduction is by a factor 5 for scalar mesons and by a factor 10 for the vector mesons. The vacuum fluctuation contributions are reduced

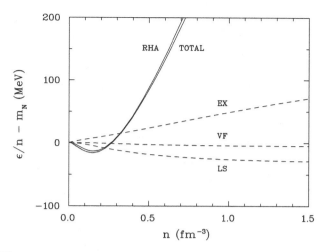

Fig. 11.5. The same as Figure 11.4 except that form factors with $\Lambda = 1$ GeV have been used at each vertex.

by similar amounts. Because the exchange terms are generally positive while the other two are negative in the density range shown in the figure, the final result is a reduction in the two-loop contribution by a factor of more than 100. Now the largest contribution is the exchange, followed by the Lamb shift and vacuum fluctuations. This is satisfying, in the sense that the exchange term survives in the quantum nonrelativistic limit whereas the Lamb shift and vacuum fluctuations are truly field theoretic in origin. With the form factors included, the two-loop contributions really are perturbative additions to the relativistic Hartree equation of state. A minimization of the Hartree equation plus two-loop contributions with respect to the effective nucleon mass at each density gives results nearly identical to those neglecting the two-loop contributions.

As the cutoff Λ increases, the two-loop contributions increase in magnitude, of course. When Λ is increased to 1.5 GeV the binding energy is increased by 11 MeV and the equilibrium density increases by about 30%. A small change in the coupling constants will restore the location of the empirical minimum in the equation of state.

It still remains a great challenge in strong-interaction physics to understand in detail the nature and structure of these form factors. The point of view presented here is that form factors represent the quark and gluon substructure of hadrons and cannot be calculated within the boundaries of hadronic degrees of freedom alone; one unfortunate consequence is that this renders the theory unrenormalizable. This is not the only point of view to which one may subscribe. For example, Serot and Tang [20] computed the effects of vertex corrections within the Walecka model itself instead of using imposed form factors. Whatever the point of view, however, it seems that the relativistic Hartree approximation, or even the mean field approximation, may not be an unreasonable approach to parametrizing the nuclear equation of state. It is consistent with our empirical knowledge near the equilibrium point and with relativity and the thermodynamic identities.

11.3 Three- and four-body interactions

The Lagrangian given in (11.5) represents a renormalizable theory. Even though a low-energy effective theory need not be renormalizable (form factors should cut off unphysical short distance contributions) one may desire to keep this property. Then one may add the cubic and quartic terms $-\frac{1}{3}bm_N(g_\sigma\sigma)^3 - \frac{1}{4}c(g_\sigma\sigma)^4$ to the Lagrangian. Just as σ^2 represents a two-body interaction, σ^3 represents a three-body interaction (a vertex with three σ lines emanating from it and attached to external nucleon lines) and σ^4 represents a four-body interaction (a vertex with four σ lines emanating from it and attached to external nucleon lines). It has been found that a phenomenological three-body interaction is necessary to describe bound nuclear matter in a nonrelativistic-potential approach. There is less information available on a microscopic four-body interaction.

In the relativistic mean field approach, the cubic and quartic terms can be used to fit more of the empirically known properties of nuclear matter. These include the following:

the saturation density [4]

$$n_0 = 0.153 \text{ fm}^{-3} \tag{11.32}$$

the binding energy [4]

$$\frac{\epsilon}{n} - m_N = -16.3 \text{ MeV} \tag{11.33}$$

the Landau mass [8]

$$m_\mathrm{L} = \sqrt{m_N^{*2} + p_\mathrm{F}^2} = 0.83 m_N \tag{11.34}$$

the compressibility [4–7]

$$K = p_\mathrm{F}^2 \frac{d^2}{dp_\mathrm{F}^2}\left(\frac{\epsilon}{n}\right) = 250 \text{ MeV} \tag{11.35}$$

Of these, the compressibility has the greatest uncertainty, approximately ± 30 MeV. With the freedom of two additional parameters, b and c, in the Lagrangian it is possible to fit all four of the above numbers.

The consequences of adding the cubic and quartic terms are to add to the energy density the quantity $\frac{1}{3}bm_N(g_\sigma\bar\sigma)^3 + \frac{1}{4}c(g_\sigma\bar\sigma)^4$ and to subtract

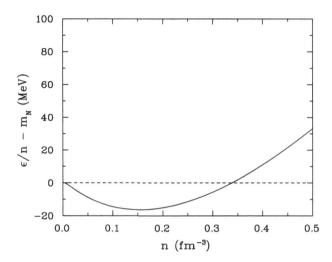

Fig. 11.6.   The average energy per nucleon minus the nucleon mass as a function of baryon density in the mean field approximation when three- and four-body forces are included.

the same amount from the pressure. The self-consistency condition for the mean scalar field that replaces (11.14) is

$$m_\sigma^2 \bar\sigma + b m_N g_\sigma^3 \bar\sigma^2 + c g_\sigma^4 \bar\sigma^3 = g_\sigma n_s \qquad (11.36)$$

where once again n_s is the scalar density. One deduces from a numerical calculation that $g_\sigma^2/4\pi = 6.003$, $g_\omega^2/4\pi = 5.948$, $b = 7.950 \times 10^{-3}$, and $c = 6.952 \times 10^{-4}$. The resulting equation of state is plotted in Figure 11.6. This may be viewed as a means of quantifying the nuclear equation of state in such a way that known nuclear properties are fitted at saturation density and also that the extrapolation to both lower and higher densities is consistent with the principles of relativity and is thermodynamically consistent.

11.4 Liquid–gas phase transition

It is generally true that any system of fermions that is self-bound, in three space dimensions, will undergo a liquid–gas phase transition. This phase transition is essentially of the Van der Waals type. The reason for a phase transition is easy to understand intuitively. First, consider nuclear matter at $T = 0$ and density $n < n_0$. The binding energy curves of Figures 11.2 and 11.6 suggest that it is energetically favorable for the nucleons to form isolated clumps or droplets with a local density n_0 rather than to be distributed homogeneously throughout space. The space surrounding the isolated droplets is simply a vacuum. As the temperature is turned up from zero, two things happen. Nucleons within a droplet have an increased kinetic energy owing to the finite temperature, and so the droplet swells in size and is reduced in local density. Finite temperature also means that the droplets will evaporate nucleons into what was formerly the vacuum. Thus we have a phase mixture: isolated droplets of nuclear liquid with local density $n_L < n_0$ are surrounded by a nuclear gas with density $n_G < n_L$. As T increases, n_G increases and n_L decreases. Eventually, at some temperature T_c we reach a critical point where $n_G = n_L$ and the distinction between liquid and gas disappears.

The liquid–gas phase transition is readily studied in the relativistic mean field model of nuclear matter with cubic and quartic interactions. The equation of state is

$$P(\mu, T) = P_{FG} + \frac{1}{2}\left(\frac{g_\omega^2}{m_\omega^2}\right)n^2 - \frac{1}{2}m_\sigma^2\bar{\sigma}^2 - \frac{1}{3}bm_N(g_\sigma\sigma)^3 - \frac{1}{4}cm_N(g_\sigma\sigma)^4$$

(11.37)

where

$$P_{FG} = 4T\int\frac{d^3p}{(2\pi)^3}\left[\ln\left(1 + e^{-\beta(E^*-\mu^*)}\right) + \ln\left(1 + e^{-\beta(E^*+\mu^*)}\right)\right]$$

$$n = 4\int\frac{d^3p}{(2\pi)^3}\left(\frac{1}{e^{\beta(E^*-\mu^*)}+1} - \frac{1}{e^{\beta(E^*+\mu^*)}+1}\right)$$

$$n_s = 4\int\frac{d^3p}{(2\pi)^3}\frac{m_N^*}{E^*}\left(\frac{1}{e^{\beta(E^*-\mu^*)}+1} + \frac{1}{e^{\beta(E^*+\mu^*)}+1}\right)$$

$$E^* = \sqrt{p^2 + m_N^{*2}}$$

$$g_\sigma n_s = m_\sigma^2\bar{\sigma} + bm_N g_\sigma^3\bar{\sigma}^2 + cg_\sigma^4\bar{\sigma}^3$$

$$m_N^* = m_N - g_\sigma\bar{\sigma}$$

$$\mu^* = \mu - \left(\frac{g_\omega^2}{m_\omega^2}\right)n$$

(11.38)

From these it is possible to verify the thermodynamic identity $n = \partial P(\mu, T)/\partial\mu$, and to compute the entropy density and energy density according to $s = \partial P(\mu, T)/\partial T$, $\epsilon = -P + Ts + \mu n$. Strictly speaking, the contribution of thermal mesons should be added. However, for the temperatures of interest here, $T < 30$ MeV, the σ and ω mesons contribute very little since $T \ll m_\sigma, m_\omega$.

Some isotherms of pressure versus density are plotted in Figure 11.7. Consider moving along the $T = 10$ MeV isotherm. For very small n, $0 < n < n_A$, only the gas phase is present. When $n > n_D$, only the liquid

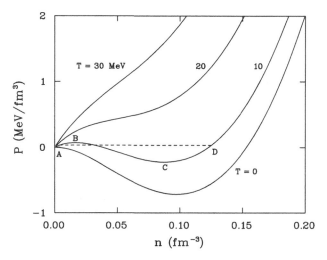

Fig. 11.7. Isotherms of pressure versus baryon density in the mean field approximation with the inclusion of three- and four-body forces. The horizontal line is the Maxwell construction for phase equilibrium. The critical temperature is 16.4 MeV.

phase is present. The points A and D are defined by the condition that they have the same value of the chemical potential μ. The straight line connecting A and D is the Maxwell construction. For densities $n_A < n < n_D$ the equilibrium configuration is a mixture of the liquid phase (with local density n_D) and the gas phase (with local density n_A). The reason is that the Gibbs criterion of equal P, T, and μ is satisfied. From A to B it is possible for the system to remain in the gas phase, but it is metastable and will not survive indefinitely. Similarly the liquid phase is metastable from C to D. The portion of the curve between B and C is unstable. Recall the stability condition [21] $\partial P(n, T)/\partial n > 0$. If the inequality does not hold then the isothermal speed of sound is imaginary and isothermal perturbations will grow exponentially.

For $T < T_c$, the phase transition is first order. At T_c, the points A, B, C, D merge into one point, an inflection point, also called the critical point. At the critical point, the line of first-order phase transitions terminates in a second-order one. For $T > T_c$, there is no distinction between gas and liquid and no phase transition.

The critical temperature in this model is 16.4 MeV. Other models of nuclear matter typically yield T_c in the range 14–19 MeV [22–25]. Generally, T_c is a monotonically increasing function of the compressibility K.

There have been attempts, which have met with some success, to find experimental evidence of a nuclear liquid–gas phase transition in heavy ion collisions. Unlike in the theory, in the experiments the Coulomb force cannot be turned off and this complicates the analysis. Interested readers are referred to the reviews by Csernai and Kapusta [26] and by Das Gupta, Mekjian, and Tsang [27].

11.5 Summary

The relativistic field theories used to describe dense nuclear matter can only be effective because nucleons and mesons are composite objects, constructed in a complicated way from quark and gluon fields. Nevertheless, nucleons and mesons are the relevant degrees of freedom for densities up to perhaps four to eight times nuclear saturation density and temperatures less than about 150 MeV. One should not think of the effective Lagrangians as providing a fundamental theory that must be solved to all orders in the coupling constants. On the contrary, this is bound to fail because of the large numerical values of the coupling constants. Explicit two-loop calculations with point vertices show just how large these contributions can be. Inserting physically plausible form factors at the vertices softens these contributions considerably, resulting in only minor corrections to the relativistic mean field or relativistic Hartree approximations. The practical view, which we espouse, is that the relativistic mean field approximation is the simplest way to parametrize the nuclear equation of state. It does so in a way that embodies as much of our empirical knowledge as possible (binding energy and density, compressibility, etc.) while being consistent with special relativity and the thermodynamic identities. The approach is flexible enough to allow such additional degrees of freedom and additional interaction terms as are necessary to bring about agreement with new data on nuclear matter properties.

In this brief introduction to the subject of dense nuclear matter, we have focused on relatively simple, renormalizable, Lagrangians with only a vector meson and a scalar meson. Much work has been and continues to be done with theories involving more mesonic degrees of freedom, such as ρ mesons, pions, and kaons, and more baryonic degrees of freedom, such as hyperons and delta resonances. An extension of this sort is presented in Chapter 16 for the purpose of obtaining the equation of state to be used in computing the structure of neutron stars. Since these theories are effective-field theories there is no reason why they should be restricted to normalizable interactions. In principle, all low-lying degrees of freedom and all interactions consistent with the symmetries of QCD ought to be allowed. Such low-energy expansions have been worked out by Furnstahl,

Serot, and Tang [28]. Not only must the effective Lagrangian be consistent with the symmetries of QCD, so also must the approximations one employs to calculate the properties of nuclear matter. The problem of the pion self-energy in nuclear matter [29, 30] is a good illustration of how chiral symmetry can be violated by the mean field approximation when used in conjunction with certain representations of the pion–nucleon interaction.

11.6 Exercises

11.1 Evaluate the integrals that result in (11.17).

11.2 Calculate $\int d^3r\, V(r)$ using (11.3) and evaluate it numerically with the parameters of the Walecka model. Show that the nonrelativistic limit of the Walecka equation of state is equivalent to the sum of the average kinetic and potential energies calculated directly from the nonrelativistic Hamiltonian in the mean field approximation.

11.3 Verify the Hugenholtz–Van Hove theorem [31], which states that the single-particle energy at the Fermi surface equals the binding energy per nucleon at saturation density, for the Walecka model.

11.4 Derive the formula (11.31).

11.5 Calculate the nuclear optical potential with the inclusion of three- and four-body interactions. Compare the result with the cited literature on the optical potential used in proton–nucleus and nucleus–nucleus scattering experiments.

11.6 Estimate the dependence of the critical temperature and density of the nuclear liquid–gas phase transition on the binding energy, compressibility and other nuclear matter properties as follows. Near the saturation point of cold nuclear matter the energy per nucleon may be parametrized as

$$E_0(n) = \frac{K}{18}\left(\frac{n}{n_0} - 1\right)^2 - B$$

where $B = 16.3$ MeV is the binding energy. If the thermal excitation energy is taken to be that of a degenerate Fermi gas then the pressure may be written as

$$P(n,T) = \frac{K}{9}\frac{n^2}{n_0}\left(\frac{n}{n_0} - 1\right) + \frac{1}{3}\left(\frac{2\pi}{3}\right)^{2/3} mn^{1/3}T^2$$

See [32] for more details, especially regarding the entropy.

References

1. Johnson, M. H., and Teller, E., *Phys. Rev.* **98**, 783 (1955).
2. Duerr, H.-P., *Phys. Rev.* **103**, 469 (1956).
3. Walecka, J. D., *Ann. Phys. (NY)* **83**, 491 (1974).
4. Möller, P., Myers, W. D., Swiatecki, W. J., and Treiner, J., *At. Data Nucl. Data Tables* **39**, 225 (1988); Myers, W. D. and Swiatecki, W. J., *Phys. Rev. C* **57**, 3020 (1998).
5. Blaizot, J.-P., *Phys. Rep.* **64**, 171 (1980); Blaizot, J.-P., Berger, J. F., Decharge, J., and Girod, M., *Nucl. Phys.* **A591**, 435 (1991).
6. Khoa, Dao T., Satchler, G. R., and Von Oertzen, W., *Phys. Rev. C* **56**, 954 (1997).
7. Youngblood, D. H., Clark, H. L., and Lui, Y. W., *Phys. Rev. Lett.* **82**, 691 (1999).
8. Johnson, C. H., Horen, D. J., and Mahaux, C., *Phys. Rev. C* **36**, 2252 (1987).
9. Jaminon, M., and Mahaux, C., *Phys. Rev. C* **40**, 354 (1989).
10. Furnstahl, R. J., Rusnak, J. J., and Serot, B. D., *Nucl. Phys.* **A632**, 607 (1998).
11. Machleidt, R., Holinde, K., and Elstor, Ch., *Phys. Rep.* **149**, 1 (1987).
12. Serot, B. D., and Walecka, J. D., *Int. J. Mod. Phys. E* **6**, 515 (1997).
13. Arnold, L. G., *et al.*, *Phys. Rev. C* **25**, 936 (1982).
14. Gale, C., Bertsch, G., and Das Gupta, S., *Phys. Rev. C* **35**, 1666 (1987).
15. Aichelin, J., Rosenhauer, A., Peilert, G., Stöcker, H., and Greiner, W., *Phys. Rev. Lett.* **58**, 1926 (1987).
16. Chin, S. A., *Ann. Phys. (NY)* **108**, 301 (1977).
17. Furnstahl, R. J., Perry, R. J., and Serot, B. D., *Phys. Rev. C* **40**, 321 (1989).
18. Prakash, M., Ellis, P. J., and Kapusta, J. I., *Phys. Rev. C* **45**, 2518 (1992).
19. Sakurai, J. J. (1969). *Currents and Mesons* (University of Chicago Press).
20. Serot, B. D., and Tang, H.-B., *Phys. Rev. C* **51**, 969 (1995).
21. Landau L. D., and Lifshitz, E. M. (1959). *Statistical Physics* (Addison-Wesley).
22. Walecka, J. D., *Phys. Lett.* **B59**, 109 (1975); Freedman, R. A., *ibid.* **B71**, 369 (1977).
23. Sauer, G., Chandra, H., and Mosel, U., *Nucl. Phys.* **A264**, 221 (1976).
24. Lattimer, M., and Ravenhall, D. G., *Astrophys. J.* **223**, 314 (1978).
25. Friedman, B., and Pandharipande, V. R., *Nucl. Phys.* **A361**, 502 (1981).
26. Csernai, L. P., and Kapusta, J. I., *Phys. Rep.* **131**, 223 (1986).
27. Das Gupta, S., Mekjian, A. Z., and Tsang, M. B., *Adv. Nucl. Phys.* **26**, 89 (2001).
28. Furnstahl, R. J., Serot, B. D., and Tang, H.-B., *Nucl. Phys.* **A615**, 441 (1997); **A640**, 505 (E) (1998).
29. Kapusta, J. I., *Phys. Rev. C* **23**, 1648 (1981).
30. Matsui, T., and Serot, B. D., *Ann. Phys. (NY)* **144**, 107 (1982).
31. Van Hove, L., and Hugenholtz, N. M., *Physica* **24**, 363 (1958).
32. Kapusta, J. I., *Phys. Rev. C* **29**, 1735 (1984).

Bibliography

Relativistic nuclear field theories

Serot, B. D., and Walecka, J. D., *Adv. Nucl. Phys.* **16**, 1 (1986).

Serot, B. D., *Rep. Prog. Phys.* **55**, 1855 (1992).

Serot, B. D., and Walecka, J. D., *Int. J. Mod. Phys. E* **6**, 515 (1997).

12

Hot hadronic matter

We know that QCD is the formal theory of the strong interaction. In principle, its solution should yield the complete particle spectrum as well as produce the interaction terms that regulate how different particle species interact. However, this complete solution is at present impossible, partly owing to the fact that at the scale of the lighter degrees of freedom QCD is strongly coupled. To describe the interaction and the properties of hot and dense hadronic ensembles, one must turn to effective approaches. They vary in character and in philosophy. In this chapter, we shall discuss some of these techniques. They comprise effective Lagrangian theories, which aim to represent in a simple way the dynamical content of a theory in the low-energy limit. The heavier fields are integrated out, leaving a set of constants to be determined by experiment. In the specific case of QCD, the choice of low-energy effective Lagrangian is dictated by general symmetry principles, and chiral symmetry will be seen to play a special role.

A remarkably successful effective Lagrangian approach to low-energy QCD is that of chiral perturbation theory. We consider this first and study its finite-temperature behavior. Next, we will use the fact that the spectrum of strongly interacting particles is quite well known experimentally to outline a technique that enables an evaluation of in-medium self-energies directly from experimental data input. The rest of the chapter will be devoted to a discussion of the Weinberg sum rules at nonzero temperatures [1] and to investigations of the characteristics of the linear and nonlinear σ models [2].

12.1 Chiral perturbation theory

Chiral perturbation theory draws its power from the observation that the light pseudoscalar degrees of freedom in the spectrum of the confined sector of QCD can be explained in terms of a spontaneously broken symmetry. Let us first elaborate how and why this statement is true. Consider for example the QCD Lagrangian for massless quarks, assuming a generic quark field, ψ, for simplicity:

$$\mathcal{L} = i\bar{\psi}\not{D}\psi \tag{12.1}$$

Based on a comparison with the QCD scale, the massless approximation is a good one for u and d quarks, is less so for s quarks, and is simply bad for c, t, and b (see Table 8.2). The free-particle Dirac equation for massless fermions is

$$\not{k}\psi = 0 \tag{12.2}$$

Using the fact that $\{\gamma_\mu, \gamma_5\} = 0$, a solution is also $\gamma_5\psi$. Consequently, two solutions are $\psi_{L/R} = \frac{1}{2}\left(1 \mp \gamma_5\right)\psi$, and this establishes γ_5 as a chirality operator: $\gamma_5\psi_{L/R} = \mp\,\psi_{L/R}$. The subscripts L and R refer to the left- and right-handed solutions, respectively. Finally, this labeling is made more explicit by manipulating the free massless-particle Dirac equation (12.2) into the form

$$\boldsymbol{\sigma} \cdot \hat{\mathbf{k}}\psi = \pm\gamma_5\psi \tag{12.3}$$

where $\gamma_5\gamma_0\boldsymbol{\gamma} = \boldsymbol{\sigma}$. Therefore, for right-handed solutions, the helicity and the sign of the energy (identified by the $\pm$ symbols) are correlated, whereas they are anticorrelated for left-handed solutions.

One can then rewrite (12.1) as

$$\mathcal{L} = i\bar{\psi}_L\not{D}\psi_L + i\bar{\psi}_R\not{D}\psi_R \tag{12.4}$$

and it is seen that the L and R sectors decouple. Consequently, symmetry transformations of the type

$$\psi_{L/R} \to \exp\left(-i\sum_j \alpha^j_{L/R}\lambda^j\right)\psi_{L/R} \tag{12.5}$$

will leave the Lagrangian invariant. Note that it is also invariant with respect to $U(1)_A$, but there is an anomaly which we will not discuss here. In the case of SU(2), the matrices λ^j are Pauli matrices and $\psi_{L/R}$ are the chiral projections of the light $\begin{pmatrix} u \\ d \end{pmatrix}$ doublet. Similarly, for SU(3) the matrices λ^j are then the Gell-Mann matrices and the chiral projections involved are those obtained from the u, d, and s fields. The elements $\alpha^j_{L/R}$ are the components of arbitrary constant vectors. Using the case of SU(2) as an example, the invariance of the Lagrangian under the symmetry transformations (12.5) is usually labeled chiral SU(2), $SU(2)_L \times SU(2)_R$, or $SU(2)_V \times SU(2)_A$; in the latter case, we have defined $\alpha^j_{V/A} =$

$(\alpha_R^j \pm \alpha_L^j)/2$. Thus, in the case of SU(2), if chiral symmetry were realized in the conventional fashion in Nature then one would expect to have three time-independent vector charges and three time-independent axial charges. Since those charges are proportional to number operators, this leads to the prediction of parity doublets (owing to the transformation properties of γ_5), which are not observed. What has gone wrong? It turns out that this is another case where a symmetry of the Lagrangian is broken by the ground state of the theory. As we have seen in Chapter 7, this leads to the appearance of Goldstone bosons.

In the case at hand the breaking is dynamical, meaning that the Noether current associated with the axial sector is not divergenceless but receives contributions from quantum corrections. This was discussed early on by Adler and by Bell and Jackiw [3]. Another puzzling fact is that there is indeed a triplet of light particles, the pions, but these are not massless. This is to be understood in terms of the fact that our original assumption that the quarks have no bare mass is in fact incorrect. If u and d quarks were strictly massless, the pion would be a genuine Goldstone boson, with $m_\pi = 0$. To first order in the explicit symmetry breaking, the finite pion mass can be traced back to the u and d quark condensates [4].

The aim of chiral perturbation theory is to provide an effective theory that possesses the symmetries of the complete theory, QCD, and is applicable at low energies where the exact theory is strongly coupled. Then the effective theory of QCD is formulated in terms of the lightest hadron fields, the pions. Bearing in mind that the chiral symmetry is not manifest in the ground state of QCD, there is a procedure to implement a spontaneously broken symmetry in a quantum field theory [5]. In the special case of chiral symmetry, a convenient way to collect the Goldstone fields is the exponential parametrization. For SU(3) it is $U(\phi) = \exp\left(i \sum_1^8 \lambda_a \phi^a / F\right)$, λ_a being a Gell-Mann matrix and F a constant. Specifically,

$$\frac{1}{\sqrt{2}} \sum_{a=1}^{8} \lambda_a \phi^a = \begin{pmatrix} \frac{1}{\sqrt{2}}\pi^0 + \frac{1}{\sqrt{6}}\eta_8 & \pi^+ & K^+ \\ \pi^- & -\frac{1}{\sqrt{2}}\pi^0 + \frac{1}{\sqrt{6}}\eta_8 & K^0 \\ K^- & \bar{K}^0 & -\frac{2}{\sqrt{6}}\eta_8 \end{pmatrix} \quad (12.6)$$

Strictly speaking, the Lagrangian of the standard model is not chirally invariant. The chiral symmetry of the strong interactions is broken by the electroweak interaction owing to the quark Yukawa coupling, which generates nonzero quark masses. The basic assumption of chiral perturbation theory is that the chiral limit is a viable starting point for a perturbative expansion. This expansion is in fact a double expansion, in powers of both the momentum and the quark masses. The Goldstone bosons will decouple from each other in the low-energy limit.

An elegant technique that enables one to calculate Green's functions of quark currents is that associated with the introduction of external fields. Following Gasser and Leutwyler [6], the chirally invariant QCD Lagrangian is extended by coupling the quarks to external Hermitian matrix fields $s(x)$, $p(x)$, $v_\mu(x)$, and $a_\mu(x)$:

$$\mathcal{L} = \mathcal{L}_{\mathrm{QCD}} + \bar{\psi}\gamma^\mu(v_\mu + a_\mu\gamma_5)\psi - \bar{\psi}(s - ip\gamma_5)\psi \qquad (12.7)$$

The external fields transform under parity as a scalar, a pseudoscalar, a vector, and an axial vector, respectively. They are color-neutral 3×3 matrices, where the matrix character with respect to the flavor indices u, d, and s can be illustrated, for example, by the vector field

$$v^\mu = v_0^\mu + \sum_{j=1}^{8} \tfrac{1}{2}\lambda_j v_j^\mu \qquad (12.8)$$

As before, chiral fields can be defined: $r_\mu = v_\mu + a_\mu$, $l_\mu = v_\mu - a_\mu$.

The usual QCD Lagrangian is recovered in the limit $p = v_\mu = a_\mu = 0$ and $s = \mathrm{diag}(m_u, m_d, m_s)$. The physically relevant Green's functions are functional derivatives of the usual zero-temperature generating functional $Z(s, p, v, a)$. For example,

$$\langle 0|\bar{\psi}(x)\psi(x)|0\rangle = i\, \frac{\delta \ln Z}{\delta s_0(x)}\bigg|_{p=v=a=0,\, s=m} \qquad (12.9)$$

where, as in the vector example above, the subscript 0 identifies the singlet component. Similarly, various currents can be obtained directly from the Lagrangian, such as the left-handed current $j_\mu^{l,a}(x)$ derived from $\partial\mathcal{L}/\partial l_a^\mu$.

Inclusion of the external fields transforms the global chiral symmetry to a local one. The invariance requirements are now contained in the following set of transformation rules. For any $g_{R/L}$ in SU(3) such that

$$\begin{aligned} \psi_R &\to g_R\psi_R \\ \psi_L &\to g_L\psi_L \end{aligned} \qquad (12.10)$$

the invariance is preserved if the external fields transform as gauge fields,

$$\begin{aligned} r_\mu &\to g_R r_\mu g_R^\dagger + i g_R \partial_\mu g_R^\dagger \\ l_\mu &\to g_L l_\mu g_L^\dagger + i g_L \partial_\mu g_L^\dagger \\ s + ip &\to g_R(s + ip)g_L^\dagger \end{aligned} \qquad (12.11)$$

and if $U \to g_R U g_L^\dagger$. The covariant derivative which, by definition, has the same transformation properties as the object on which it is acting, is $D_\mu U = \partial_\mu U - i r_\mu U + i U l_\mu$.

We are now in a position to formulate the basic premises of chiral perturbation theory. At zero temperature, one can write a generating functional of QCD as

$$Z(s, p, v, a) = \int [dA_a^\mu][d\bar{\psi}][d\psi] \exp\left(i \int d^4x\, \mathcal{L}\right) \qquad (12.12)$$

For simplicity, the possible ghost fields have been omitted. At low energy, the propagating modes are the Goldstone modes. In the language of effective field theory, all the heavy degrees of freedom are integrated out and are absorbed into the parameters of the effective action. Specifically,

$$Z(s, p, v, a) = \int [dU] \exp\left(i \int d^4x\, \mathcal{L}_{\text{eff}}\right) \qquad (12.13)$$

One starts by constructing an effective Lagrangian in terms of derivatives and of the external fields. The limit where the external fields vanish is that of low-energy QCD. Whereas it is plausible that this procedure does reproduce the Green's functions of QCD at low energy, its formal validity has not been proven here. This was done by Leutwyler [7].

Searching for an interaction that would constitute the leading-order term in a momentum expansion, one realizes that that there are no candidates with the required invariance properties that have no derivatives and no external fields. In fact, the only candidate is $\text{Tr}(UU^\dagger) = 3$, which is simply a constant. The most general chirally invariant effective Lagrangian with the minimum number of derivatives is

$$\mathcal{L}_2 = \tfrac{1}{4}F^2 \text{Tr}(D_\mu U D^\mu U^\dagger + \chi U^\dagger + \chi^\dagger U) \qquad (12.14)$$

where $\chi = 2B(s + ip)$. Thus at this order there are two parameters, F and B. Observe that in order to reproduce the kinetic term in the free-pion Lagrangian, the constant F above needs to be the same as that in the definition of the field matrix, $U(\phi)$ (just above (12.6)). The two constants F and B are related to the pion decay constant and to the quark condensate, up to chiral corrections [6]:

$$\begin{aligned} f_\pi &= F + \mathcal{O}(m_q) \\ \langle 0|\bar{u}u|0\rangle &= -F^2 B + \mathcal{O}(m_q) \end{aligned} \qquad (12.15)$$

Using our definition for $U(\phi)$ and setting the external scalar field s equal to the quark mass matrix, one can read off from $\mathcal{L}_2$ the pseudoscalar meson masses, again up to leading order in chiral corrections. For example,

$$\begin{aligned} m_\pi^2 &= 2\bar{m}B \\ m_{K^+}^2 &= (m_u + m_s)B \\ m_{K^0}^2 &= (m_d + m_s)B \end{aligned} \qquad (12.16)$$

with $\bar{m} = \tfrac{1}{2}(m_u + m_d)$. Those relations are consistent with the chiral counting rules, which stipulate the dimensions of the operators and of the fields in the chiral expansion in terms of the momentum k:

$$U : \mathcal{O}(1)$$
$$D_\mu U, \, v_\mu, \, a_\mu : \mathcal{O}(k)$$
$$s, \, p : \mathcal{O}(k^2)$$

With these rules, the meaning of the subscript in $\mathcal{L}_2$ is clear. The effective Lagrangian is then expanded up to order k^4 [6].

We now formulate the problem at finite temperature and calculate the pressure of a pion gas using chiral perturbation theory. The symmetry requirements will translate into exact statements for the coefficients of the expansion in powers of the temperature. Furthermore, we shall make use of the fact that pions are considerably less massive than any of their other SU(3) partners. Therefore these lighter degrees of freedom will be excited first and should play a leading role: we then restrict our discussion to SU(2). We can rewrite $\mathcal{L}_2$ in terms of the nonlinear σ model:

$$\mathcal{L}_2 = \tfrac{1}{4}F^2 \mathrm{Tr}[\partial_\mu U \partial^\mu U^\dagger - M^2(U + U^\dagger)] \tag{12.17}$$

with $M^2 = (m_u + m_d)B$. The effective Lagrangian to order k^4 is written down by identifying all the independent terms to this order that have the required symmetry properties (Lorentz invariance, P, C, and chiral symmetry):

$$
\begin{aligned}
\mathcal{L}_4 = &-\tfrac{1}{4}l_1 \left[\mathrm{Tr}(\partial_\mu U \partial^\mu U^\dagger)\right]^2 - \tfrac{1}{4}l_2 \, \mathrm{Tr}(\partial_\mu U \partial^\mu U^\dagger) \, \mathrm{Tr}(\partial_\mu U \partial^\mu U^\dagger) \\
&+ \tfrac{1}{8}l_4 M^2 \, \mathrm{Tr}(\partial_\mu U \partial^\mu U^\dagger) \, \mathrm{Tr}(U + U^\dagger) \\
&- \tfrac{1}{16}(l_3 + l_4)M^4 \, \mathrm{Tr}(U + U^\dagger) - h_1 M^4
\end{aligned}
\tag{12.18}
$$

The contact term h_1 is a vacuum contribution, and isospin-breaking effects are ignored. The evaluation of the finite-temperature contribution to the thermodynamic potential proceeds as in preceding chapters, only now the chiral effective Lagrangian is used:

$$Z \approx \int_{\mathrm{periodic}} [dU] \exp\left(\int_0^\beta d\tau \int d^3x \, (\mathcal{L}_2 + \mathcal{L}_4) \right) \tag{12.19}$$

The complete expansion in loop topologies that yield terms up to T^8 was worked out by Gerber and Leutwyler [8]. A few of those diagrams are shown in Figure 12.1. The diagrams so obtained fall into three categories.

1 Those that generate temperature-independent contributions. These only renormalize the vacuum contribution.
2 The genuine temperature-dependent terms that will generate the thermal pressure. These are shown in Figure 12.1.

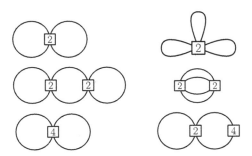

Fig. 12.1. Some of the diagrams that occur in the finite-temperature expansion of the thermodynamic potential in chiral perturbation theory, up to order T^8. The labeling of the vertices refers to the order of the chiral Lagrangian that provided the vertex.

3 Up to $\mathcal{O}(T^8)$, the thermodynamic potential will contain some diagrams with vertices coming from chiral Lagrangians of order higher than 4. These temperature-dependent contributions renormalize the bare mass M in the free-gas term, in such a way that

$$M \to M_1$$

with

$$M_1 = M^2 + 2l_3 \frac{M^4}{F^2} + c_0 \frac{M^6}{F^4} \qquad (12.20)$$

where c_0 is a constant.

The divergences present in the zero-temperature theory are isolated using dimensional regularization then subtracted away by appropriate counterterms. At low temperatures, the pressure will be of order $\exp(-m_\pi/T)$. Using this prescription in the expansion of the pressure enables identification of the physical pion mass with parameters of the theory:

$$m_\pi^2 = M^2 + (2l_3 + \lambda)\frac{M^4}{F^2} + c\frac{M^6}{F^4} + \mathcal{O}(M^8) \qquad (12.21)$$

The constant c is a linear combination of some regularization counterterms and l_3. When the physical pion mass is used in the theory, the parametrical dependence on counterterms disappears. Also, λ isolates the pole appearing when $d \to 4$ in the zero-temperature part of the bare-mass coordinate-space propagator, $D(x)$, when its argument vanishes:

$$\lim_{x \to 0} D(x) = 2M^2\lambda$$
$$\lambda = \frac{1}{2}(4\pi)^{-d/2}\Gamma(1 - d/2)M^{d-4} \qquad (12.22)$$

Putting all these ingredients together, the pressure can be written as [8]

$$P = 3P_0 + \frac{4}{\pi^3}aT^4h_3^2 + \frac{24}{\pi^3}T^6h_5\left(8T^2h_5 + m_\pi^2h_3\right)\left(b - \frac{I}{\pi^3F^4}\right) + \mathcal{O}(T^{10})$$
(12.23)

The first term is the pressure of a noninteracting Bose gas of pions (the factor 3 arising from the three charged states of the pion) with

$$P_0 = \frac{4}{\pi^2}T^4h_5\left(\frac{m_\pi}{T}\right)$$
(12.24)

The functions $h_n(m/T)$ are discussed in the Appendix. It is amusing to note that h_3 is proportional to the field fluctuations of noninteracting bosons:

$$\langle\phi^2\rangle = \frac{\partial P_0(T,m)}{\partial m^2} = \int \frac{d^3k}{(2\pi)^3}\frac{1}{\omega}\frac{1}{e^{\beta\omega}-1} = \frac{T^2}{\pi^2}h_3\left(\frac{m}{T}\right)$$
(12.25)

The dimensionless function $I(m_\pi/T)$ in (12.23) represents a three-dimensional integral that must be calculated numerically. Its low-temperature limit is

$$I(x) = 0.6x^{-1} + \mathcal{O}(x^{-2})$$

while the high-temperature limit is

$$I(x) = -\frac{5}{8}\ln x + 0.6360 + 0.1289x^2 + \mathcal{O}(x^3)$$

Some constants, such as c in (12.21), are absent in the final result as they are absorbed into the physical pion mass. The two constants that do appear explicitly in the pressure, a and b, are functions of the renormalized Lagrangian parameters:

$$a = -\frac{3M^2}{32\pi F^2} + \frac{5M^4}{128\pi^3F^4}\left(\bar{l}_1 + 2\bar{l}_2 - \frac{3}{10}\bar{l}_3 + \frac{9}{8}\right)$$
(12.26)

$$b = \frac{1}{16\pi^3F^4}\left(\bar{l}_1 + 4\bar{l}_2 - \frac{29}{24}\right)$$

where

$$l_i = \gamma_i\left(\lambda + \frac{1}{32\pi^2}\bar{l}_i\right) \quad \gamma_1 = \frac{1}{3} \quad \gamma_2 = \frac{2}{3} \quad \gamma_3 = -\frac{1}{2} \quad \gamma_4 = 2 \quad (12.27)$$

The quantity λ, defined in (12.21), contains the singularity.

It is extremely satisfying to verify that the expression for the pressure, derived in finite-temperature chiral perturbation theory, agrees with a treatment based on the virial expansion [8]. This represents an important consistency check.

12.2 Self-energy from experimental data

At this point, it is clear that any quantum field in interaction with other fields will see its vacuum properties modified. A rigorous formalism for calculating these changes was set up in the previous chapters and, given an interaction Lagrangian, it mainly consists of calculating an in-medium self-energy. This can in turn be related to in-medium masses and decay widths, through the real and imaginary parts, respectively. Up to now we have seen that a satisfactory way of organizing the perturbation expansion was to follow the topology of the multiloop diagrams. However, this procedure becomes questionable when large coupling constants are involved, as the very validity of the perturbation expansion is called into question. Owing to asymptotic freedom, QCD in its nonperturbative sector will involve just such large constants, and a calculation of hadronic properties from first principles becomes prohibitively difficult. Nevertheless, data does exist on the scattering of the different QCD bound states among themselves. As those measurements carry some information on the underlying interaction, it should be possible to infer from them how the fundamental characteristics of a specific field get changed in a strongly interacting medium. Relying on experimental measurements to the extent that they are available will help to develop a procedure that is as model independent as possible. A method that is applicable to dilute media is described in what follows.

For a particle of type a traversing a medium with a de Broglie wavelength less than the interparticle spacing of target particles of type b, there is a direct proportionality between the scattering amplitude and the energy. The dispersion relation of a boson is determined by

$$E^2 = m^2 + p^2 + \Pi \tag{12.28}$$

In the nonrelativistic limit we may wish to express the energy in terms of an optical potential U as

$$E = m + \frac{p^2}{2m} + U \tag{12.29}$$

The optical potential will in general have both real and imaginary parts. This leads to real and imaginary parts of the energy: $E = E_{\mathrm{R}} - i\Gamma/2$. The imaginary part is related to the mean free path $1/\rho\sigma$, where σ is the scattering cross section and ρ is the density of scatterers, and to the velocity, as $\Gamma = v\rho\sigma$. Using the forward scattering amplitude f and the optical theorem $p\sigma = 4\pi f$ gives

$$\operatorname{Im}\Pi = 2m\operatorname{Im}U = -4\pi\rho\operatorname{Im}f \tag{12.30}$$

In the low-energy limit the mean potential energy of the particle is

$$\mathrm{Re}\, U = \rho \int d^3x\, V(\mathbf{x}) \tag{12.31}$$

where V is the two-body potential. In this limit the Born approximation gives

$$\mathrm{Re}\, f = -\frac{m}{2\pi} \int d^3x\, V(\mathbf{x}) \tag{12.32}$$

Hence both the real and imaginary parts fit a simple formula,

$$\Pi = -4\pi \rho f \tag{12.33}$$

This formula has a wider range of applicability than this derivation might suggest; it is the leading term in a multiple-scattering expansion [9].

The generalization to target particles that are moving, and to relativistic kinematics, is straightforward. For meson a scattering from hadron b in the medium, the contribution to the self-energy is:

$$\begin{aligned}
\Pi_{ab}(E,p) &= -4\pi \int \frac{d^3k}{(2\pi)^3}\, n_b(\omega)\, \frac{\sqrt{s}}{\omega}\, f_{ab}^{(\mathrm{cm})}(s) \\
&= -\frac{1}{2\pi p} \int_{m_b}^{\infty} d\omega\, n_b(\omega) \int_{s_-}^{s_+} ds\, \sqrt{s} f_{ab}^{(\mathrm{cm})}(s)
\end{aligned} \tag{12.34}$$

where E and p are the energy and momentum of the particle, $\omega^2 = m_b^2 + k^2$,

$$s_{\pm} = E^2 - p^2 + m_b^2 + 2(E\omega \pm pk) \tag{12.35}$$

n_b is either a Bose–Einstein or Fermi–Dirac occupation number, and f_{ab} is the forward scattering amplitude. The normalization of the amplitude corresponds to the standard form of the optical theorem,

$$\sigma = \frac{4\pi}{q_{\mathrm{cm}}} \mathrm{Im}\, f^{(\mathrm{cm})}(s) \tag{12.36}$$

where q_{cm} is the momentum in the cm frame. The dispersion relation is determined by the poles of the propagator after summing over all target species and including the vacuum contribution to the self-energy:

$$E^2 - m_a^2 - p^2 - \Pi_a^{\mathrm{vac}}(E,p) - \sum_b \Pi_{ab}(E,p) = 0 \tag{12.37}$$

The applicability of (12.34) is limited to those cases where interference between sequential scatterings is negligible.

Taking various limits of (12.34) is instructive. First of all, we note that the cross section is invariant under longitudinal boosts. It is convenient to know how the scattering amplitude transforms. They are related to each other as follows.

$$m_a f_{ab}^{(a\text{'s rest frame})} = m_b f_{ab}^{(b\text{'s rest frame})} = \sqrt{s} f_{ab}^{(\text{cm})} \qquad (12.38)$$

In the limit that the target particles b move nonrelativistically we can approximate ω in the first line of (12.34) by m_b, in which case

$$\Pi_{ab} = -4\pi f_{ab}^{(b\text{'s rest frame})} \rho_b \qquad (12.39)$$

where ρ_b is the spatial density. Next consider the chiral limit when pions serve as the target particles, relevant for low-temperature baryon-free matter. From (12.38) $\sqrt{s} f_{a\pi}^{(\text{cm})} = m_a f_{a\pi}^{(a\text{'s rest frame})}$. Since $f_{a\pi}^{(a\text{'s rest frame})}$ involves two derivative couplings of the pion to the massive state a (Adler's theorem) one sees from (12.34) that $\Pi_{a\pi} \sim T^4$. Finally, if the self-energy is evaluated in the rest frame of a it is possible to do all the integrations but one:

$$\Pi_{ab}(E, p) = -\frac{m_a^2 T}{\pi p} \int_{m_b}^{\infty} d\omega \ln \left(\frac{1 - \exp(-\omega_+/T)}{1 - \exp(-\omega_-/T)} \right) f_{ab}^{(a\text{'s rest frame})}(\omega) \qquad (12.40)$$

Here $\omega_\pm = (E\omega \pm pk)/m_a$. This assumes that b is a boson; a similar formula ensues if it is a fermion.

As a specific application, we will estimate the ρ meson dispersion relation for finite temperature and baryon density and for momenta up to 1 GeV/c. This is of special interest, as vector mesons can couple directly to the photon [10] and therefore the in-medium modification of vector meson properties can in principle be inferred from the measurement of electromagnetic observables. This direct conversion of a vector meson to a photon (real or virtual) is often referred to as vector meson dominance (VMD). The low-energy part of the ρ meson scattering amplitude will be dominated by coupling to resonances. The physical context assumed here is that ρ mesons are formed during the last stage of the evolution of hadronic matter created in a heavy ion collision. The matter there is approximated as a weakly interacting gas of pions and nucleons. This stage is formed when the local temperature is of the order of 100 to 150 MeV and when the local baryon density is of the order of the normal nucleon density in a nucleus. The main ingredients of the calculation are $\rho\pi$ and ρN forward scattering amplitudes and total cross sections.

We will consider the momentum p to be real and evaluate the scattering amplitudes on-shell, that is, evaluate the self-energy at $E = \sqrt{p^2 + m_\rho^2}$. In this case (12.37) takes the form

$$E^2 = m_\rho^2 + p^2 + \Pi_\rho^{\text{vac}} + \Pi_{\rho\pi}(p) + \Pi_{\rho N}(p) \qquad (12.41)$$

Since the self-energy has real and imaginary parts, so does $E(p) = E_R(p) - i\Gamma(p)/2$. In the narrow-width approximation the dispersion relation is determined from

$$E_R^2(p) = p^2 + m_\rho^2 + \operatorname{Re}\Pi_{\rho\pi}(p) + \operatorname{Re}\Pi_{\rho N}(p)$$
$$\Gamma(p) = -\left[\operatorname{Im}\Pi_\rho^{\text{vac}} + \operatorname{Im}\Pi_{\rho\pi}(p) + \operatorname{Im}\Pi_{\rho N}(p)\right]/E_R(p) \qquad (12.42)$$

The width of the ρ meson in vacuum, $\Gamma_\rho^{\text{vac}} = -\operatorname{Im}\Pi_\rho^{\text{vac}}/m_\rho$, is 150 MeV.

We can also define a mass shift and an optical potential:

$$\Delta m_\rho(p) = \sqrt{m_\rho^2 + \operatorname{Re}\Pi_{\rho\pi}(p) + \operatorname{Re}\Pi_{\rho N}(p)} - m_\rho$$
$$U(p) = E_R(p) - \sqrt{m_\rho^2 + p^2} \qquad (12.43)$$

These will be evaluated for temperatures of 100 and 150 MeV and nucleon densities of 0, 1, and 2 times the normal nuclear matter density (0.155 nucleons per fm^3). Recall that one needs a Bose–Einstein distribution for pions and a Fermi–Dirac distribution for nucleons. The pion chemical potentials are zero and the nucleon chemical potentials are 745 and 820 MeV for densities of 1 and 2 times normal at $T = 100$ MeV, and 540 and 645 MeV at $T = 150$ MeV. Antinucleons are not considered here. For a ρ meson scattering from a particle a and going to a resonance R, the forward scattering amplitude can be written in its usual nonrelativistic form, in the center of mass:

$$f_{\rho a}^{\text{cm}}(s) = \frac{1}{2q_{\text{cm}}}\sum_R W_{\rho a}^R \frac{\Gamma_{R\to\rho a}}{M_R - \sqrt{s} - \frac{1}{2}i\Gamma_R} - \frac{q_{\text{cm}}r_P^{\rho a}}{4\pi s}\frac{(1 + \exp^{-i\pi\alpha_P})}{\sin\pi\alpha_P}s^{\alpha_P}$$

$$(12.44)$$

In familiar notation, the subscript P refers to the Pomeron, $\sqrt{s}$ is the total cm energy and the magnitude of the cm momentum is

$$q_{\text{cm}} = \frac{1}{2\sqrt{s}}\sqrt{[s - (m_\rho + m_a)^2][s - (m_\rho - m_a)^2]} \qquad (12.45)$$

The statistical averaging factor for spin and isospin is

$$W_{\rho a}^R = \frac{(2s_R + 1)}{(2s_\rho + 1)(2s_a + 1)}\frac{(2t_R + 1)}{(2t_\rho + 1)(2t_a + 1)} \qquad (12.46)$$

The second part of the forward scattering amplitude is a nonresonant background contribution, a description of which goes beyond this text. See, for example, Collins [11] for a detailed discussion. It suffices here to state that the parameters are determined by high-energy scattering phenomenology. Also, the real and imaginary parts of the scattering amplitude are related by a dispersion relation. This constraint turns out to be better satisfied in the presence of the background term [12].

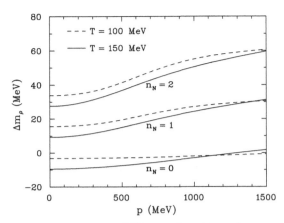

Fig. 12.2.   The vector meson mass shift as a function of momentum for various temperatures and nucleon densities n_N (expressed in units of equilibrium nuclear matter density).

For the case of ρN scattering, the intermediate resonance can be one of several species of N^* or Δ resonances. One then needs to know the width of that resonance in the channel where there is a ρ meson and a nucleon. Because of kinematical constraints, this width is often not measured, but the radiative decays often are. These can be related to the width one is after, using the VMD relationship of the scattering amplitudes:

$$f_{\gamma N} = 4\pi\alpha \left(\frac{1}{g_\rho^2} f_{\rho N} + \frac{1}{g_\omega^2} f_{\omega N} + \frac{1}{g_\phi^2} f_{\phi N} \right) \qquad (12.47)$$

where α is the fine structure constant. From measurements of ϕ photoproduction, the last term is small and can be neglected. In the spirit of the quark model, one further assumes that $f_{\omega N} \approx f_{\rho N}$. This assumption may in fact be examined more closely [13] The direct vector-meson–photon coupling can be deduced from $V \to l^+l^-$ measurements. With these ingredients, the widths in the ρN channel can be directly extracted from the radiative decay widths. The details of the procedure outlined here, along with specific parameter values and relevant references, can be found in Eletsky *et al.* [12] Note that the calculation of the real and imaginary parts of the in-medium self-energy of any species can proceed in the same way, provided that enough experimental data can map its interaction with other fields. The mass shift and width of the ρ meson, as defined in (12.43) and (12.42), are shown in Figures 12.2 and 12.3 for different temperatures and densities (n_N is in units of n_0, the equilibrium nuclear matter density). The width is systematically larger at larger temperatures and densities. The change in mass is numerically less important. Any interaction

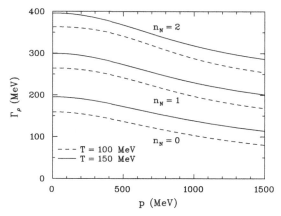

Fig. 12.3. The vector meson width as a function of momentum for various temperatures and densities.

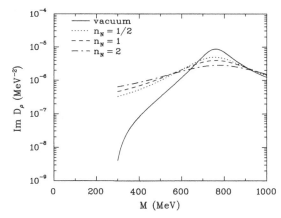

Fig. 12.4. The imaginary part of the vector meson propagator as a function of invariant mass at a momentum 300 MeV and temperature 150 MeV.

will contribute to a larger width, but the real part of the self-energy can be less affected owing to cancellations between different channels. The information in the mass shift and in the width is also contained in a plot of the imaginary part of the ρ propagator, shown in Figure 12.4, and is directly related to the in-medium spectral density. Note that since the thermal medium constitutes a preferred rest frame (that in which temperature is defined), the self-energy in general depends on the energy and the momentum separately. Alternatively, one may fix the momentum at a specific value (300 MeV here) and study the self-energy as a function of invariant mass, since $E = \sqrt{p^2 + M^2}$.

Alternatively, a method complementary to the one presented here consists of using effective hadronic Lagrangians (i.e., those whose basic symmetries are consistent with that of QCD), with parameters fitted to measured properties [14–16]. Because they are both constrained by experimental data, the two techniques should of course yield comparable results unless ones deviates significantly from the on-shell condition for the vector field.

12.3 Weinberg sum rules

Spectral sum rules were in use before the advent of QCD as the theory of the strong interaction. Weinberg had in fact proposed two sum rules based on current algebra, relating moments of the spectral density of vector and axial-vector currents [17]. These relied on the validity of chiral symmetry. It is instructive to revisit these sum rules in the language of QCD and then to pursue a finite-temperature extension, in order to explore the implications of the approach to chiral symmetry restoration at finite temperature that follow from sum rules of the Weinberg type [1]. Note that the up and down quark masses are then implicitly assumed to be zero, so that chiral symmetry is indeed exact.

12.3.1 Sum rules at zero temperature

One first defines vector and axial-vector currents (using an explicit notation for the current operators):

$$V_\mu^a = \bar{\psi}\gamma_\mu(\tau^a/2)\psi \tag{12.48}$$

$$A_\mu^a = \bar{\psi}\gamma_\mu\gamma_5(\tau^a/2)\psi \tag{12.49}$$

where $\tau^a/2$ is the isospin generator. With this normalization the current algebra of charges obeys the equal-time commutation relations

$$\left[Q_V^a, Q_V^b\right] = i\varepsilon^{abc}Q_V^c \tag{12.50}$$

$$\left[Q_V^a, Q_A^b\right] = i\varepsilon^{abc}Q_A^c \tag{12.51}$$

$$\left[Q_A^a, Q_A^b\right] = i\varepsilon^{abc}Q_V^c \tag{12.52}$$

Each charge is the volume integral of the zeroth component of the corresponding current operator. We now write the vector and axial-vector spectral densities. They are positive definite quantities defined for positive s:

$$\langle 0|V_a^\mu(x)V_b^\nu(0)|0\rangle = -\frac{\delta^{ab}}{(2\pi)^3}\int d^4p\,\theta(p^0)\,e^{ip\cdot x}\left(g^{\mu\nu}-\frac{p^\mu p^\nu}{p^2}\right)\rho_V(s)$$
$$(12.53)$$

$$\langle 0|A_a^\mu(x)A_b^\nu(0)|0\rangle = -\frac{\delta^{ab}}{(2\pi)^3}\int d^4p\,\theta(p^0)\,e^{ip\cdot x}\left[\left(g^{\mu\nu}-\frac{p^\mu p^\nu}{p^2}\right)\rho_A(s)\right.$$
$$\left.+f_\pi^2\delta(s)p^\mu p^\nu\right]$$
$$(12.54)$$

The dimension of the spectral densities is energy-squared. Note that the pion contribution to the axial-vector correlator has been written out explicitly in the second term in (12.54).

Imaginary time is used, so that all distances are space-like, or Euclidean: $x^2 = t^2 - r^2 = -\tau^2$. In this domain the spectral representation of the correlation functions is as follows:

$$\Delta D_\mu^{ab\mu}(\tau) \equiv \langle 0|T_\tau\left[V^{a\mu}(x)V_\mu^b(0) - A^{a\mu}(x)A_\mu^b(0)\right]|0\rangle$$
$$= -\frac{\delta^{ab}}{4\pi^2\tau}\int_0^\infty ds\,\sqrt{s}\left[3\rho_V(s) - 3\rho_A(s) - s\,f_\pi^2\delta(s)\right]K_1(\sqrt{s}\tau)$$
$$(12.55)$$

and

$$\Delta D_{ab}^{00}(\tau) \equiv \langle 0|T_\tau\left[V_a^0(x)V_b^0(0) - A_a^0(x)A_b^0(0)\right]|0\rangle$$
$$= -\frac{\delta_{ab}}{4\pi^2\tau}\int_0^\infty ds\,\sqrt{s}\left[\rho_V(s) - \rho_A(s) - s\,f_\pi^2\delta(s)\right]$$
$$\times\left[\frac{K_0(\sqrt{s}\tau)}{\sqrt{s}\tau} + \left(\frac{2}{s\tau^2}+1\right)K_1(\sqrt{s}\tau)\right]$$
$$(12.56)$$

Notice that the integrands essentially involve the standard Feynman propagator for a particle of mass m, which, in the Euclidean domain, is

$$D(m,\tau)_{\text{free scalar}} = \frac{m}{4\pi^2\tau}K_1(m\tau)$$
$$(12.57)$$

Exponential decay of the Bessel function K_1 at large values of the argument ensures the convergence of such integrals for any QCD correlation functions, except probably at $\tau = 0$.

Each sum rule will correspond to a particular term in the small-distance asymptotic expansion of the correlation function. In the limit $\tau \to 0$ the product of currents can be expanded according to the operator product expansion (OPE), a very successful means of connecting vacuum expectation values (VEVs) of quark and gluon operators with experimentally observable hadronic properties. We will refer the reader to the original literature for a discussion of this powerful theoretical method, but a general description can be given as follows. Consider for example the current–current correlator in real time and its expansion:

$$
i \int d^4x \; e^{iq \cdot x} T_t \{ \bar{\psi}(x) \gamma_\mu \psi(x), \; \bar{\psi}(0) \gamma_\nu \psi(0) \} = (q_\mu q_\nu - q^2 g_{\mu\nu}) \sum_d C_d(q^2) O_d
$$

(12.58)

The O_d are local operators and the $C_d(q^2)$ are c-numbers called Wilson coefficients. The operator expansion is organized according to dimension. When considering a vacuum matrix element of the current–current correlator, one might simply expect all operators except the unit operator to have a vanishing expectation value. However, long-distance nonperturbative effects will make this expectation unrealized. In principle, all vacuum expectation values, often called vacuum condensates, should be calculable in lattice gauge theory. The initial terms in this expansion were first computed perturbatively by Shifman, Vainshtein, and Zakharov [18]. For the contracted polarization tensor the result is

$$
D_\mu^{ab\mu}(\tau) \equiv \langle 0 | T_\tau \left[V^{a\mu}(x) V_\mu^b(0) \right] | 0 \rangle
$$

$$
= -\frac{3\delta^{ab}}{\pi^4 \tau^6} \left(1 + \frac{\alpha_s(\tau)}{\pi} - \frac{\langle 0 | \left(g F_{\mu\nu}^c \right)^2 | 0 \rangle \tau^4}{3 \times 2^7} \right.
$$

$$
\left. - \frac{\pi^2 \tau^6}{8} \ln(\mu\tau) \langle 0 | \mathcal{O}_\rho | 0 \rangle + \cdots \right)
$$

(12.59)

where, in the argument of the logarithm, $\mu \ll 1/\tau$ is the renormalization scale, and $\mathcal{O}_\rho$ is a complicated four-quark operator. There is a similar expression for the correlator of two axial-vector currents but it has a different four-quark operator $\mathcal{O}_{a_1}$. For our purposes we only need their difference, which is given below.

Since chiral symmetry breaking is a long-wavelength phenomenon, at very short distances or at very high energies the difference between vector and axial-vector correlators should go to zero. Indeed, taking this difference one finds that all terms except for the four-quark operators in (12.59) drop out. One can now look for consequences of this statement for the spectral density. Expanding the Bessel function in (12.55) for small values of τ we get

$$
\Delta D_\mu^{ab\mu}(\tau) = -\frac{3\delta^{ab}}{4\pi^2} \int_0^\infty ds \, [\rho_V(s) - \rho_A(s)]
$$

$$
\times \left[\frac{1}{\tau^2} + \frac{s}{2} \ln \left(\frac{\sqrt{s}\tau}{2} e^{\gamma_E - 1/2} \right) + \mathcal{O}(\tau^2, \tau^2 \ln \tau) \right]
$$

(12.60)

where γ_E is Euler's constant. The OPE has no power divergence in τ in the difference $\Delta D_\mu^{ab\mu}$. Therefore the coefficient of $1/\tau^2$ in (12.60) must vanish. This gives the second Weinberg sum rule (see below). In the OPE framework it simply follows from the observation that the first covariant operators which are not chirality blind are four-quark ones that have dimension 6 or more. Similarly expanding (12.56) for small τ and applying the observation of chirality blindness we get

$$\int_0^\infty \frac{ds}{s} \left[\rho_V(s) - \rho_A(s) - s\, f_\pi^2 \delta(s) \right] \left(\frac{1}{\tau^4} + \frac{s}{4\tau^2} \right) = 0 \qquad (12.61)$$

The first and second terms in the last parentheses give

$$\mathrm{I} \qquad \int_0^\infty \frac{ds}{s} \left[\rho_V(s) - \rho_A(s) \right] = f_\pi^2 \qquad (12.62)$$

and

$$\mathrm{II} \qquad \int_0^\infty ds \left[\rho_V(s) - \rho_A(s) \right] = 0 \qquad (12.63)$$

respectively. These are Weinberg's first and second sum rules.

The phenomenological implications of the zero-temperature sum rules have been discussed numerous times in the literature and we will therefore not do so here.

12.3.2 Sum rules at finite temperature

Weinberg's two sum rules can be extended to finite temperature using essentially the same methods as he used without any specific reference to QCD. As seen in other applications, earlier in this text, the introduction of a thermal medium will complicate some expressions as Lorentz invariance is no longer manifest. This preferred rest frame will cause functions that previously depended only on $\sqrt{s}$ to depend separately on energy and momentum, and the number of Lorentz tensors will increase because there is a new vector available, namely, the vector $u_\mu = (1, 0, 0, 0)$ that specifies the rest frame of the matter.

We now define the longitudinal and transverse spectral densities for the vector current as

$$\langle V_a^\mu(x) V_b^\nu(0) \rangle = \frac{\delta^{ab}}{(2\pi)^3} \int d^4p\, \mathrm{e}^{ip\cdot x} [1 + N_\mathrm{B}(p_0)] \big(\rho_V^\mathrm{L} P_\mathrm{L}^{\mu\nu} + \rho_V^\mathrm{T} P_\mathrm{T}^{\mu\nu} \big) \tag{12.64}$$

and for the axial-vector current as

$$\langle A_a^\mu(x) A_b^\nu(0) \rangle = \frac{\delta^{ab}}{(2\pi)^3} \int d^4p\, \mathrm{e}^{ip\cdot x} [1 + N_\mathrm{B}(p_0)] \big[\rho_A^\mathrm{L} P_\mathrm{L}^{\mu\nu} + \rho_A^\mathrm{T} P_\mathrm{T}^{\mu\nu} \big] \tag{12.65}$$

In these expressions the angle brackets refer to the thermal average. The longitudinal and transverse projection tensors were defined in Chapter 5. These spectral densities are the ρ^{n} discussed in Section 6.2. In general the spectral densities depend on p^0 and $\mathbf{p}$ separately as well as on the temperature (and chemical potential). In the vacuum we can always go to the rest frame of a massive particle and in this frame there can be no difference between longitudinal and transverse polarizations, so that $\rho_{\mathrm{L}} = \rho_{\mathrm{T}} = \rho$. Since $P_{\mathrm{L}}^{\mu\nu} + P_{\mathrm{T}}^{\mu\nu} = -(g^{\mu\nu} - p^{\mu}p^{\nu}/p^2)$, (12.64) and (12.65) collapse to (12.53) and (12.54). The pion, being a massless Goldstone boson, is special. It contributes to the longitudinal axial spectral density and not to the transverse one. In fact, we could write

$$f_{\pi}^2 \delta(p^2) p^{\mu} p^{\nu} \ = \ f_{\pi}^2 p^2 \delta(p^2) P_{\mathrm{L}}^{\mu\nu} \qquad (12.66)$$

This should not be done at finite temperature because the contribution of the pion to the longitudinal spectral density cannot be assumed to be a delta function in p^2. In general the pion's dispersion relation will be more complicated and will develop a width at nonzero momentum. Therefore, we do not try to separate out the pionic contribution but subsume it into the spectral density ρ_A^{L}, without any loss of generality.

Following Weinberg, we define a three-point function by

$$-i\epsilon_{abc} M^{\mu\nu\lambda}(q,p) \ = \ \int d^4x \, d^4y \, \mathrm{e}^{-i(q\cdot x + p\cdot y)} \left\langle T_t \left[A_a^{\mu}(x) A_b^{\nu}(y) V_c^{\lambda}(0) \right] \right\rangle \qquad (12.67)$$

We multiply both sides by q_{μ}. On the right-hand side we can use

$$q_{\mu} \mathrm{e}^{-i(q\cdot x + p\cdot y)} \ = \ i\frac{\partial}{\partial x^{\mu}} \mathrm{e}^{-i(q\cdot x + p\cdot y)} \qquad (12.68)$$

Both the vector and axial-vector currents are conserved. We assume that we can integrate by parts and that the surface term is zero. The nonzero contribution comes from

$$\frac{\partial}{\partial x^{\mu}} \left\{ T_t \left[A_a^{\mu}(x) A_b^{\nu}(y) V_c^{\lambda}(0) \right] \right\}$$

$$= \delta(x^0 - y^0) \left\{ \theta(x^0) \left[A_a^0(x), A_b^{\nu}(y) \right] V_c^{\lambda}(0) + \theta(-x^0) V_c^{\lambda}(0) \left[A_a^0(x), A_b^{\nu}(y) \right] \right\}$$

$$+ \, \delta(x^0) \left\{ \theta(y^0) A_b^{\nu}(y) \left[A_a^0(x), V_c^{\lambda}(0) \right] + \theta(-y^0) \left[A_a^0(x), V_c^{\lambda}(0) \right] A_b^{\nu}(y) \right\} \qquad (12.69)$$

From this expression we see the need for knowledge of the equal-time commutators. Consistently with the normalization of (12.50)–(12.52) we have

$$
\delta(z^0) \left[A_a^0(x), A_b^\nu(y) \right] = i\epsilon_{abd} V_d^\nu(x)\delta(\mathbf{z}) + S_{Vab}^{\nu j}(\mathbf{x})\frac{\partial}{\partial z^j}\delta(\mathbf{z})
$$

$$
\delta(z^0) \left[A_a^0(x), V_b^\nu(y) \right] = i\epsilon_{abd} A_d^\nu(x)\delta(\mathbf{z}) + S_{Aab}^{\nu j}(\mathbf{x})\frac{\partial}{\partial z^j}\delta(\mathbf{z})
$$

(12.70)

Here $z = x - y$ and the S's denote the Schwinger terms. These terms do not vanish, in general, and they need to appear to guarantee the self-consistency of the current algebra.

Consider now the contribution of the Schwinger terms to the thermal average. Generically they will be of the form

$$
\langle SJ \rangle = Z^{-1} \sum_{m,n} e^{-K_n/T} \langle n|S|m\rangle\langle m|J|n\rangle
$$

(12.71)

where $K = H - \mu N$ is the Hamiltonian minus the chemical potential times the conserved particle number, the states are chosen to be eigenstates of H, N, and isospin, and J is either the vector or the axial-vector current. J has isospin 1, so we get zero if either (i) S is a c-number, or (ii) S is an operator with no isospin-1 component. We shall assume that one of these holds. Then

$$
\frac{\partial}{\partial x^\mu} \left\langle T_t \left[A_a^\mu(x)A_b^\nu(y)V_c^\lambda(0) \right] \right\rangle = i\epsilon_{abd}\delta(x-y) \left\langle T_t \left[V_d^\nu(x)V_c^\lambda(0) \right] \right\rangle
$$

$$
+ i\epsilon_{acd}\delta(x) \left\langle T_t \left[A_b^\nu(y)A_d^\lambda(0) \right] \right\rangle \quad (12.72)
$$

It is now a simple matter to show that

$$
\tfrac{1}{2}q_\mu M^{\mu\nu\lambda}(q,p) = D_V^{\nu\lambda}(q+p) - D_A^{\nu\lambda}(p)
$$

(12.73)

where the D's are the propagators for the currents; for example,

$$
\delta_{ab}D_A^{\nu\lambda}(p) = \int d^4y \, e^{-ip\cdot y} \left\langle T_t \left[A_a^\nu(y)A_b^\lambda(0) \right] \right\rangle
$$

(12.74)

Similarly, one can show that

$$
\tfrac{1}{2}(q+p)_\lambda M^{\mu\nu\lambda}(q,p) = D_A^{\mu\nu}(q) - D_A^{\mu\nu}(p)
$$

(12.75)

These Ward identities have exactly the same form as at zero temperature [17].

With a similar consideration of the three-point function

$$
-i\epsilon_{abc}N^{\mu\nu\lambda}(q,p) = \int d^4x \, d^4y \, e^{-i(q\cdot x + p\cdot y)} \left\langle T_t \left[V_a^\mu(x)V_b^\nu(y)V_c^\lambda(0) \right] \right\rangle
$$

(12.76)

one can prove two more Ward identities,

$$
\tfrac{1}{2}q_\mu N^{\mu\nu\lambda}(q,p) = D_V^{\nu\lambda}(q+p) - D_V^{\nu\lambda}(p)
$$

(12.77)

and

$$\tfrac{1}{2}(q+p)_\lambda N^{\mu\nu\lambda}(q,p) = D_V^{\mu\nu}(q) - D_V^{\mu\nu}(p) \qquad (12.78)$$

Multiply (12.75) by $(q+p)_\lambda$ and (12.77) by q_μ. Doing the same for the other two Ward identities, one obtains the constraints

$$(q+p)_\lambda D_V^{\nu\lambda}(q+p) = q_\lambda D_V^{\nu\lambda}(q) + p_\lambda D_V^{\nu\lambda}(p) = q_\lambda D_A^{\nu\lambda}(q) + p_\lambda D_A^{\nu\lambda}(p) \qquad (12.79)$$

The equation above holds for all values of q and p. This implies

$$k_\lambda D_V^{\nu\lambda}(k) = k_\lambda D_A^{\nu\lambda}(k) = C^{\nu\lambda} k_\lambda \qquad (12.80)$$

where $C^{\nu\lambda}$ is momentum independent (but can depend on temperature) and is the same for the vector and axial-vector channels. By taking the Fourier transform of these relations we can find the thermal averages of the equal-time commutators,

$$\delta(x^0)\big\langle\, [V_a^\nu(x), V_b^0(0)]\,\big\rangle = \delta(x^0)\big\langle\, [A_a^\nu(x), A_b^0(0)]\,\big\rangle = \delta_{ab} C^{\nu\lambda} \frac{\partial}{\partial x^\lambda}\delta(x) \qquad (12.81)$$

The commutators above can be expressed in terms of the spectral densities from (12.64) and (12.65). Taking their difference one obtains the finite-temperature generalization of the first Weinberg sum rule,

$$\text{I} \qquad \int_0^\infty \frac{d\omega\,\omega}{\omega^2 - \mathbf{p}^2}\left[\rho_V^{\mathrm{L}}(\omega, \mathbf{p}) - \rho_A^{\mathrm{L}}(\omega, \mathbf{p})\right] = 0 \qquad (12.82)$$

Here (6.44) has been used to write the integral over positive ω only. Notice that this sum rule involves only the longitudinal spectral densities and not the transverse ones. At zero temperature the spectral densities depend only on $p^2 = s = \omega^2 - \mathbf{p}^2$. Then this equation reduces to (12.62) once we remember to separate out the pion part of ρ_A^{L}, namely, $s f_\pi^2 \delta(s)$. At finite temperature, the spectral densities in general will depend on ω and $\mathbf{p}$ separately and not just on the combination s. Then this sum rule must be satisfied at each value of the momentum.

For the second sum rule, we follow a method due to Das, Mathur, and Okubo [19]. Omitting the index V or A the explicit expressions for the time-ordered propagator are

$$D^{00}(p^0, \mathbf{p}) = \mathbf{p}^2 D_{\mathrm{L}}(p^0, \mathbf{p}) \qquad (12.83)$$
$$D^{0j}(p^0, \mathbf{p}) = p^0 p^j D_{\mathrm{L}}(p^0, \mathbf{p}) \qquad (12.84)$$
$$D^{ij}(p^0, \mathbf{p}) = \left(\delta^{ij} - \frac{p^i p^j}{\mathbf{p}^2}\right) D_{\mathrm{T}}(p^0, \mathbf{p}) + \frac{p^i p^j}{\mathbf{p}^2} D_{\mathrm{L}}'(p^0, \mathbf{p}) \qquad (12.85)$$

where

$$D_{\text{L}}(p^0, \mathbf{p}) = 2i \int_{-\infty}^{\infty} \frac{d\omega\, \omega}{\omega^2 - \mathbf{p}^2} \left[\frac{\rho^{\text{L}}(\omega, \mathbf{p})}{(\omega + i\epsilon)^2 - p_0^2} \right] [1 + N_{\text{B}}(\omega)] \quad (12.86)$$

$$D_{\text{L}}'(p^0, \mathbf{p}) = 2i \int_{-\infty}^{\infty} \frac{d\omega\, \omega^3}{\omega^2 - \mathbf{p}^2} \left| \frac{\rho^{\text{L}}(\omega, \mathbf{p})}{(\omega + i\epsilon)^2 - p_0^2} \right| [1 + N_{\text{B}}(\omega)] \quad (12.87)$$

$$D_{\text{T}}(p^0, \mathbf{p}) = 2i \int_{-\infty}^{\infty} d\omega\, \omega \left| \frac{\rho^{\text{T}}(\omega, \mathbf{p})}{(\omega + i\epsilon)^2 - p_0^2} \right| [1 + N_{\text{B}}(\omega)] \quad (12.88)$$

and for the Schwinger term they are

$$C^{00} = C^{0j} = C^{j0} = 0 \qquad C^{ij}(\mathbf{p}) = \delta^{ij} D_{\text{S}}(\mathbf{p}) \quad (12.89)$$

where

$$D_{\text{S}}(\mathbf{p}) = 2i \int_{-\infty}^{\infty} \frac{d\omega\, \omega}{\omega^2 - \mathbf{p}^2} \rho^{\text{L}}(\omega, \mathbf{p}) [1 + N_{\text{B}}(\omega)] \quad (12.90)$$

The first observation we can make concerns the thermally averaged generic Schwinger term C. Since it is the same for the vector and the axial-vector correlators, by (12.80), the factor $D_{\text{S}}(\mathbf{p})$ must be the same as well. Equating them reproduces the first finite-temperature sum rule (12.82).

The essence of the argument of Das, Mathur, and Okubo is that spontaneous chiral symmetry breaking is a low-energy phenomenon. At very high energy it must disappear, at least in the limit that quark masses are zero and chiral symmetry is exact. Thus the difference between the vector and axial-vector propagators should go to zero at very high energy,

$$\lim_{p^0 \to \infty,\ \mathbf{p}\ \text{fixed}} \left[D_V^{\mu\nu}(p^0, \mathbf{p}) - D_A^{\mu\nu}(p^0, \mathbf{p}) \right] = 0 \quad (12.91)$$

If we do this for the time–time or time–space components of the propagators, that is, for the D_{L}, we again reproduce the first finite-temperature sum rule. Expanding to the next order in $1/p_0^2$ we obtain a finite-temperature generalization of the second zero-temperature sum rule, which is

$$\text{II-L} \qquad \int_0^{\infty} d\omega\, \omega \left[\rho_V^{\text{L}}(\omega, \mathbf{p}) - \rho_A^{\text{L}}(\omega, \mathbf{p}) \right] = 0 \quad (12.92)$$

Like the first, this sum rule involves only the longitudinal spectral densities, and so we call it II-L. Also like the first, it reduces to the original Weinberg sum rule as the temperature and/or chemical potential go to zero.

Next we consider the space–space components of the propagators. Examination of the D_{L}' in the infinite-energy limit gives us the sum rule II-L and nothing new. Examination of the D_{T} in the infinite-energy limit gives us another sum rule, which we call II-T because it involves the transverse spectral densities,

$$\text{II-T} \qquad \int_0^\infty d\omega\, \omega \left[\rho_V^{\mathrm{T}}(\omega, \mathbf{p}) - \rho_A^{\mathrm{T}}(\omega, \mathbf{p})\right] = 0 \qquad (12.93)$$

The finite-temperature sum rules II-L and II-T should become degenerate at $\mathbf{p} = \mathbf{0}$ because there ought not to be any difference between longitudinal and transverse excitations at rest. The sum rule II-T then reduces to the original second sum rule in the vacuum.

We want to emphasize that the sum rules derived in this section, I, II-L, and II-T, must be satisfied for every value of the momentum. Furthermore, our derivation is more general than QCD; any theory that satisfies the assumptions we have made must obey these sum rules.

Low-temperature behavior

As we are taking the zero-quark-mass limit here the pion is massless below any critical temperature for chiral symmetry restoration and/or deconfinement, and thus at parametrically low temperatures the heat bath is dominated by pions. In [20] the so-called Dey–Eletsky–Ioffe mixing theorem was proven, which says that, to order T^2, there is no change in the masses of vector and axial-vector mesons. What does change are the couplings to the currents. The finite-temperature correlators can be described by a mixing between the vector and axial-vector $T = 0$ correlators with a temperature-dependent coefficient:

$$D_V^{\mu\nu}(p, T) = (1 - \epsilon)D_V^{\mu\nu}(p, 0) + \epsilon D_A^{\mu\nu}(p, 0) \qquad (12.94)$$
$$D_A^{\mu\nu}(p, T) = (1 - \epsilon)D_A^{\mu\nu}(p, 0) + \epsilon D_V^{\mu\nu}(p, 0) \qquad (12.95)$$

These are valid to first order in $\epsilon \equiv T^2/6f_\pi^2$. This implies the same mixing of the spectral densities, namely,

$$\rho_V(p^0, \mathbf{p}, T) = (1 - \epsilon)\rho_V(s, 0) + \epsilon \rho_A(s, 0) \qquad (12.96)$$
$$\rho_A(p^0, \mathbf{p}, T) = (1 - \epsilon)\rho_A(s, 0) + \epsilon \rho_V(s, 0) \qquad (12.97)$$

with the appropriate longitudinal and transverse subscripts. The temperature dependence of the pion decay coupling was thus proven to be $f_\pi^2(T) = (1 - \epsilon)f_\pi^2$ for small T, consistent with the prediction of chiral perturbation theory [21]. Therefore, the finite-temperature sum rules I (12.82), II-L (12.92), and II-T (12.93), reduce to the original zero-temperature sum rules but with both sides of (12.62) and (12.63) multiplied by the factor $1 - 2\epsilon$. This satisfies the Dey–Eletsky–Ioffe mixing theorem.

The approach to chiral-symmetry restoration

Chiral transformations are rotations of the quark field with γ_5, and they may or may not have the $SU(N_f)$ (isospin) generators. The corresponding $U(1)_A$ and $SU(N_f)_A$ generators have different fates in QCD; the former is explicitly violated by the anomaly, the latter is broken spontaneously at low temperature and is restored at some critical temperature T_c, provided that the quark mass is strictly zero, as is assumed for the purposes of the current discussion. The ρ and a_1 currents are both unchanged by the $U(1)_A$ transformation but are mixed under $SU(N_f)_A$. Therefore, if this symmetry is restored at high temperatures then there should be no difference between the vector and the axial-vector correlators. In this section we speculate on exactly how this difference goes to zero with increasing temperature. Generally, one may suggest many different scenarios. Let us discuss the following three.

The simplest scenario is that the T-dependence factorizes. It means that the vector and axial-vector spectral densities mix, without changing their shape, as in the low-temperature limit considered in the previous section, only with a more general function $\epsilon(T)$. When the mixing becomes maximal, $\epsilon = 1/2$, chiral symmetry is restored. It is interesting to see the temperature at which this occurs using the lowest-order formula, $\epsilon = T^2/6f_\pi^2$. This estimate gives $T_{\text{complete mixing}} = \sqrt{3}f_\pi \approx 164$ MeV, which is indeed roughly equal to the expected critical temperature T_c.

The second scenario assumes that the ρ and a_1 mesons retain their identities and dominate the correlation function. However, their parameters change with temperature. In particular, the masses may move towards each other [22] or go to zero [23]. At T_c they become degenerate, and chiral symmetry is restored.

It is instructive then to look at the sum rules. Let us assume that vector meson dominance is a good approximation for the spectral densities and not worry about the continuum contribution for the time being. We focus on zero momentum for the sake of simplicity. When a pole mass is defined at finite temperature, it is usually defined as the energy of the excitation at zero momentum.

The vector spectral density is (note that there is no difference between the longitudinal and transverse cases at zero momentum)

$$\text{sign}(\omega)\,\rho_V(\omega) = -\frac{1}{\pi}\frac{m_\rho^4}{g_\rho^2}\,\text{Im}\,\frac{1}{\omega^2 - m_\rho^2 - \Pi_R^\rho(\omega) - i\Pi_I^\rho(\omega)} \qquad (12.98)$$

where Π_R^ρ and Π_I^ρ are the real and imaginary parts of the ρ self-energy at temperature T. In the narrow-width approximation this becomes

$$\text{sign}(\omega)\,\rho_V(\omega) = \frac{m_\rho^4}{g_\rho^2}\delta\big(\omega^2 - m_\rho^2 - \Pi_R^\rho(\omega)\big) \qquad (12.99)$$

The pole mass is determined self-consistently from $m_\rho^2(T) = m_\rho^2 + \Pi_R^\rho(m_\rho(T))$. Then the spectral density can be rewritten as

$$\text{sign}(\omega)\,\rho_V(\omega) = Z_\rho(T)\frac{m_\rho^4}{g_\rho^2}\delta\big(\omega^2 - m_\rho^2(T)\big) \qquad (12.100)$$

where the temperature-dependent residue is

$$Z_\rho^{-1}(T) = \left|1 - \frac{d}{d\omega^2}\Pi_R^\rho(\omega)\right| \qquad (12.101)$$

The normalization is $Z_\rho(0) = 1$. Similarly

$$\text{sign}(\omega)\,\rho_A(\omega) = Z_a(T)\frac{m_{a_1}^4}{g_a^2}\delta\big(\omega^2 - m_{a_1}^2(T)\big) + Z_\pi(T)f_\pi^2\omega^2\delta\big(\omega^2\big) \qquad (12.102)$$

Substituting these spectral densities into the finite-temperature sum rules I, II-L, and II-T tells us that the ρ and a_1 residues are equal:

$$Z_\rho(T) = Z_a(T) \qquad (12.103)$$

and that the pion residue is

$$Z_\pi(T) = 2Z_\rho(T)\left(\frac{m_\rho^2}{m_\rho^2(T)} - \frac{m_\rho^2}{m_{a_1}^2(T)}\right) \qquad (12.104)$$

We expect that $m_{a_1}^2(T) - m_\rho^2(T) \to 0$ as the temperature increases. Three types of behavior can be distinguished: both the ρ and the a_1 masses decrease with T, both masses increase with T, or the ρ mass increases while the a_1 mass decreases with T. The sum rules do not appear to rule out any of these possibilities. In any case, the result is that $Z_\pi(T) \to 0$ unless $Z_\rho(T) \to \infty$, which seems rather unphysical.

As distinct from the previous two scenarios, it may be that particles are not well defined as we approach a chiral-symmetry-restoring phase transition. That is, the imaginary part of the self-energy may become larger with increasing temperature. This broadening would also decrease the maximum peak value of the spectral density. Picturesquely, the vector and axial-vector mesons melt away in a very broad distribution of strength in the spectral densities.

Concluding this section, we say once more that the sum rules by themselves cannot of course tell which scenario is preferable. However, they can be used to restrict significantly the parametrization of the spectral densities at nonzero temperature.

12.4 Linear and nonlinear σ models

The $O(N)$ model as a quantum field theory in $d + 1$ dimensions [24] is a
basis or prototype for many interesting physical systems. The bosonic field
$\boldsymbol{\Phi}$ has N components. When the Lagrangian is such that the vacuum state
exhibits spontaneous symmetry breaking, it is known as a sigma model.
This is the case of interest to us here. In $d = 3$ space dimensions the linear
sigma model has the potential

$$\tfrac{1}{4}\lambda \left(\boldsymbol{\Phi}^2 - f_\pi^2 \right)^2$$

where λ is a positive coupling constant and f_π is the pion decay constant.
The model is renormalizable. In the limit $\lambda \to \infty$ the potential goes over
to a delta-function constraint on the length of the field vector and is then
known as a nonlinear sigma model.

When $N = 4$ one has a model for the low-energy dynamics of quantum chromodynamics (QCD). More explicitly, it is essentially the unique
description of the dynamics of very soft pions. This is basically due to
the isomorphism between the groups O(4) and SU(2) $\times$ SU(2), the latter
being the appropriate group for two flavors of massless quarks in QCD.
The linear sigma model, including the nucleon, goes back to the work of
Gell-Mann and Levy [25]. This subject has a vast literature.

As we have seen earlier in this chapter, much work has been done
on chiral perturbation theory that starts with the nonlinear sigma model
and adds higher-order, nonrenormalizable, terms to the Lagrangian; these
are determined by the dimensionality of the coefficients or field derivatives [26]. The goal is to construct an effective Lagrangian that describes
the low-energy properties of QCD to the desired accuracy. This whole
program owes a considerable amount to the classic works of Weinberg
[27, 28]

Finally, the standard model of the electroweak interactions, due to
Weinberg, Salam, and Glashow, has an SU(2) doublet scalar Higgs field
responsible for spontaneous symmetry breaking. If one neglects spin-1
gauge fields then the Higgs sector is also an O(4) field theory.

Since both linear and nonlinear σ models are prototypical field theories
in many respects, one expects that much insight on the nature of the
chiral-restoring phase transition, for example, can be had by studying
those at finite temperature.

12.4.1 Linear σ model at finite temperature

The linear σ model Lagrangian is

$$\mathcal{L} = \tfrac{1}{2}(\partial_\mu \boldsymbol{\Phi})^2 - \tfrac{1}{4}\lambda\big(\boldsymbol{\Phi}^2 - f_\pi^2 \big)^2 \tag{12.105}$$

where λ is a positive coupling constant. The bosonic field $\boldsymbol{\Phi}$ has N components. Rather arbitrarily, we define the first $N-1$ components to represent a pion field $\boldsymbol{\pi}$ and the last, Nth, component to represent the sigma field. Since the $O(N)$ symmetry is broken to an $O(N-1)$ symmetry at low temperatures, we immediately allow for a sigma condensate v whose value is temperature-dependent and yet to be determined. We write

$$
\begin{aligned}
\Phi_i(\mathbf{x},t) &= \pi_i(\mathbf{x},t) \qquad i = 1,\ldots,N-1 \\
\Phi_N(\mathbf{x},t) &= v + \sigma(\mathbf{x},t)
\end{aligned}
\tag{12.106}
$$

In terms of these fields the Lagrangian is

$$
\mathcal{L} = \tfrac{1}{2}\left(\partial_\mu \boldsymbol{\pi}\right)^2 + \tfrac{1}{2}\left(\partial_\mu \sigma\right)^2 - \tfrac{1}{4}\lambda\left(v^2 - f_\pi^2 + 2v\sigma + \sigma^2 + \boldsymbol{\pi}^2\right)^2 \tag{12.107}
$$

The action at finite temperature is obtained by rotating to imaginary time, $\tau = it$, and integrating τ from 0 to $\beta = 1/T$. The action is defined as

$$
\begin{aligned}
S = -&\tfrac{1}{4}\lambda\left(f_\pi^2 - v^2\right)^2 \beta V \\
+ &\int_0^\beta d\tau \int_V d^3x \Big\{ \tfrac{1}{2}\left[(\partial_\mu \boldsymbol{\pi})^2 - \bar{m}_\pi^2 \boldsymbol{\pi}^2 + (\partial_\mu \sigma)^2 - \bar{m}_\sigma^2 \sigma^2\right] \\
&+ \tfrac{1}{2}\lambda v\left(v^2 - f_\pi^2\right)\sigma - \lambda v\sigma(\boldsymbol{\pi}^2 + \sigma^2) - \tfrac{1}{4}\lambda(\sigma^2 + \boldsymbol{\pi}^2)^2 \Big\}
\end{aligned}
\tag{12.108}
$$

where the effective masses are

$$
\begin{aligned}
\bar{m}_\pi^2 &= \lambda\left(v^2 - f_\pi^2\right) \\
\bar{m}_\sigma^2 &= \lambda\left(3v^2 - f_\pi^2\right)
\end{aligned}
\tag{12.109}
$$

At any temperature v is chosen such that $\langle\sigma\rangle = 0$. This eliminates any one-particle reducible (1PR) diagrams in perturbation theory, leaving only one-particle irreducible (1PI) diagrams.

At zero temperature the potential is minimized when $v = f_\pi$. The pion is massless and the σ particle has a mass of $\sqrt{2\lambda}f_\pi$. The Goldstone theorem is satisfied.

Lin and Serot [29] argued that the σ meson should not be identified with the attractive s-wave interaction in the $\pi - \pi$ interaction, which is responsible for nuclear attraction. Rather, they argue that the σ meson should have a mass which is at least 1 GeV if not more. This means that λ is of order 50 or greater.

The simplest approximation at finite temperature is the mean field approximation. One allows for v to be temperature dependent; hence the effective masses are temperature dependent as well. However, interactions among the particles or collective excitations are neglected. The pressure includes only the contribution of the condensate and of the thermal motion of the independently moving particles. Thus

$$P = \frac{T}{V} \ln Z = -\frac{\lambda}{4} \left(f_\pi^2 - v^2 \right)^2 + P_0(T, m_\sigma) + (N-1)P_0(T, m_\pi) \tag{12.110}$$

The pressure of a free relativistic boson gas can be written in two ways:

$$P_0 = -T \int \frac{d^3 p}{(2\pi)^3} \ln(1 - e^{-\beta\omega}) = \int \frac{d^3 p}{(2\pi)^3} \frac{p^2}{3\omega} \frac{1}{e^{\beta\omega} - 1} \tag{12.111}$$

As pointed out earlier, this is a relatively simple but surprisingly powerful first approximation, which allows one to gain much insight into the behavior of relativistic quantum field theories at high temperature.

One expects that, as the temperature is raised, thermal fluctuations will tend to disorder the condensate field v, and at sufficiently high temperature it may even disappear. If there is a second-order phase transition then the correlation length should go to infinity, which is equivalent to the effective σ mass going to zero. With such an expectation one may expand the free-boson gas pressure about zero mass to obtain

$$P_0(T, m) = \frac{\pi^2}{90}T^4 - \frac{m^2 T^2}{24} + \frac{m^3 T}{12\pi} + \cdots \tag{12.112}$$

Since the masses are proportional to the square root of λ it is generally inconsistent to retain the cubic term in m because there exist loop diagrams which are not included in the mean field approximation but which contribute to the same order in λ. Therefore we take

$$P(T, v) = N\frac{\pi^2}{90}T^4 + \frac{\lambda}{2}v^2 \left(f_\pi^2 - \frac{N+2}{12}T^2 \right) - \frac{\lambda}{4}v^4 \tag{12.113}$$

where the pion and σ masses have been expressed in terms of λ, v, and f_π. Maximizing the pressure with respect to v gives

$$v^2 = f_\pi^2 - \frac{N+2}{12}T^2 \tag{12.114}$$

This result is easily understood. Going back to (12.108), we can differentiate $\ln Z$ with respect to v with the result that

$$v^2 = f_\pi^2 - 3\langle \sigma^2 \rangle - \langle \pi^2 \rangle \tag{12.115}$$

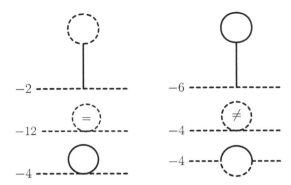

Fig. 12.5. The diagrams contributing to the one-loop pion self-energy, in the linear σ model. The broken lines represent the pion whereas the solid lines represent the σ. The overall sign and combinatoric factors are shown. In the contributions involving the pion four-point vertex, the signs $=$ and $\neq$ stand for cases where the pion loop and the external field have the same, or different, quantum numbers.

as long as we choose $\langle \sigma \rangle = 0$. For any free bosonic field ϕ with mass m,

$$\langle \phi^2 \rangle = \int \frac{d^3 p}{(2\pi)^3} \frac{1}{\omega} \frac{1}{e^{\beta \omega} - 1} \tag{12.116}$$

where $\omega = \sqrt{p^2 + m^2}$. In the limit where the temperature is greater than the mass, $\langle \phi^2 \rangle \to T^2/12$. This yields (12.114) directly.

The condensate goes to zero at a critical temperature given by

$$T_c^2 = \frac{12}{N+2} f_\pi^2 \tag{12.117}$$

Above this temperature thermal fluctuations are too large to allow a nonzero condensate. It is a straightforward exercise to show that the pressure and its first derivative are continuous at T_c but that the second derivative is discontinuous. This is therefore a second-order phase transition.

There are two major problems with the mean field approximation as described. The first is that the pion has a negative mass-squared at every temperature greater than zero. Not only is the Goldstone theorem not satisfied, but there are tachyons as well! The sigma particle also gets a negative mass-squared at temperatures above $\sqrt{8/(N+2)}\, f_\pi < T_c$. Recalling the analysis in Section 7.3, this violation of basic physical principles is resolved by recognizing that the finite-temperature corrections to the squared masses are proportional to λT^2 and that one-loop self-energy corrections, not included in the mean field analysis, are of the same order. This can be understood from the following analysis.

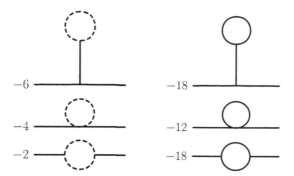

Fig. 12.6. The diagrams contributing to the one-loop σ self-energy.

At high temperatures, when the masses can be neglected in the loops, the mean field result is obtained by combining (12.110) and (12.114):

$$\bar{m}_\pi^2 = -\frac{N+2}{12}\lambda T^2$$

$$\bar{m}_\sigma^2 = 2\lambda f_\pi^2 - \frac{N+2}{4}\lambda T^2$$

(12.118)

The full one-loop self-energies for pions and the σ meson are shown in Figures 12.5 and 12.6. If one chooses $\langle\sigma\rangle = 0$ then there are no 1PR diagrams and the tadpoles should not be included; they are already included in the temperature dependence of v. One may check this by fixing $v = f_\pi$ and then computing the tadpole contributions to the effective masses. One gets precisely (12.118). The diagrams involving the four-point vertices contribute an amount $(N+2)\lambda T^2/12$ to both the pion and σ meson self-energies. When evaluated in the high-temperature approximation and at low frequency and momentum the 1PI diagrams involving three-point vertices may be neglected. (This follows from power counting. These diagrams involve two propagators instead of one, and so are only logarithmically divergent in the ultraviolet in the vacuum. The other diagrams are quadratically divergent, which leads to a T^2 behavior at finite temperature.) When all contributions of order λT^2 are included, the pole positions of the pion and σ meson propagators move, with the result that below T_c

$$m_\pi^2 = \bar{m}_\pi^2 + \Pi_\pi = 0$$

$$m_\sigma^2 = \bar{m}_\sigma^2 + \Pi_\sigma = 2\lambda f_\pi^2\left(1 - \frac{T^2}{T_c^2}\right)$$

(12.119)

and above T_c

$$m_\pi^2 = m_\sigma^2 = m_\Phi^2 = -\lambda f_\pi^2 + \Pi_\Phi = \frac{N+2}{12}\lambda\left(T^2 - T_c^2\right)$$

(12.120)

The Goldstone theorem is satisfied, there are no tachyons, and restoration of the full symmetry of the Lagrangian above T_c is evident.

It must be recognized that the results (12.118)–(12.120) are valid to order λ and cannot be extrapolated to $\lambda \to \infty$. At low temperature, where pions scatter from each other sequentially and there is essentially no propagation off mass shell between scatterings because of the low particle density, one may take the point of view that λ is a parameter to be adjusted to fit π–π scattering data and it does not matter how large λ is. This point of view cannot be taken at high temperature, where the pion number density is large, for then multiple scatterings will occur and they cannot be factorized into independent scatterings. This means that multiloop self-energy diagrams will be important at high temperature if λ is not perturbatively small.

The second major problem is that long-wavelength fluctuations very near the phase transition cannot be treated with perturbation theory because the self-interacting boson fields become massless just at the transition. Although this is a well-known problem in the statistical mechanics of second-order phase transitions, exactly how it affects the critical temperature is not known for the linear σ model in $3 + 1$ dimensions. The result presented here must be accepted for what it is: a one-loop estimate of the critical temperature.

12.4.2 Nonlinear σ model at finite temperature

The nonlinear σ model may be defined by the Lagrangian

$$\mathcal{L} = \tfrac{1}{2} \left(\partial_\mu \mathbf{\Phi} \right)^2 \tag{12.121}$$

together with the constraint

$$f_\pi^2 = \mathbf{\Phi}^2(\mathbf{x}, t) \tag{12.122}$$

The partition function is

$$Z = \int [d\mathbf{\Phi}] \, \delta\!\left(f_\pi^2 - \mathbf{\Phi}^2\right) \exp\left(\int_0^\beta d\tau \int d^3x \, \mathcal{L} \right) \tag{12.123}$$

Because the length of the chiral field is fixed and cannot be changed by thermal fluctuations it is often said that on the one hand chiral symmetry-breaking is built into this model and therefore there can be no chiral-symmetry-restoring phase transition. On the other hand, the linear sigma model does undergo a symmetry-restoring phase transition. Taking the quartic coupling constant λ to infinity essentially constrains the length of the chiral field to be f_π, just as in the nonlinear model. The critical temperature, however, is independent of λ at least in the mean field approximation. So it would seem that the phase transition survives. If this is true then one ought to be able to derive it entirely within the context of the nonlinear model. That is what we shall do, although it involves a lot more effort than the treatment of the linear model in the mean field approximation. Since the only parameter in the model is f_π and we are interested in temperatures comparable with it, we cannot make an expansion in powers of T/f_π. The only other parameter is N, the number of field components. This suggests an expansion in $1/N$.

We begin by representing the field-constraining delta function by an integral,

$$Z = \int [d\mathbf{\Phi}] \, [db'] \exp \left\{ \int_0^\beta d\tau \int d^3x \left[\mathcal{L} + ib' \left(\mathbf{\Phi}^2 - f_\pi^2 \right) \right] \right\} \quad (12.124)$$

As with the linear model, we define the first $N-1$ components of $\mathbf{\Phi}$ to be the pion field and the last component to be the sigma field. We allow for a zero-frequency and zero-momentum condensate of the sigma field, referred to as v. Following Polyakov [30] we also separate out explicitly the zero-frequency and zero-momentum mode of the auxiliary field b'. Integrating over all the other modes will give us an effective action involving the constant part of the fields. We will then minimize the free energy with respect to these constant parts, which gives us a saddle point approximation. Integrating over fluctuations about the saddle point is a finite-volume correction and of no consequence in the thermodynamic limit. The Fourier expansions are

$$\Phi_i(\mathbf{x}, \tau) = \pi_i(\mathbf{x}, \tau) = \sqrt{\frac{\beta}{V}} \sum_n \sum_\mathbf{p} e^{i(\mathbf{x} \cdot \mathbf{p} + \omega_n \tau)} \, \tilde{\pi}_i(\mathbf{p}, n)$$

$$\Phi_N(\mathbf{x}, \tau) = v + \sigma(\mathbf{x}, \tau) = v + \sqrt{\frac{\beta}{V}} \sum_n \sum_\mathbf{p} e^{i(\mathbf{x} \cdot \mathbf{p} + \omega_n \tau)} \, \tilde{\sigma}(\mathbf{p}, n)$$

$$b'(\mathbf{x}, \tau) = \tfrac{1}{2}im^2 + b(\mathbf{x}, \tau) = \tfrac{1}{2}im^2 + T\sqrt{\frac{\beta}{V}} \sum_n \sum_\mathbf{p} e^{i(\mathbf{x} \cdot \mathbf{p} + \nu_n \tau)} \, \tilde{b}(\mathbf{p}, n)$$

$$(12.125)$$

One must remember to exclude the zero-frequency and zero-momentum mode from the summations. The field $\mathbf{\Phi}$ must be periodic in imaginary time for the usual reasons, but there is no such requirement on the auxiliary field b, hence we must have $\omega_n = 2\pi nT$ and $\nu_n = \pi nT$. Since the field b has dimensions of inverse length squared we have inserted another factor of T so as to make its Fourier amplitude dimensionless, as is the case for the other fields. The action then becomes

$$S = \int_0^\beta d\tau \int_V d^3x \left\{ \tfrac{1}{2} \left[(\partial_\mu \boldsymbol{\pi})^2 - m^2 \boldsymbol{\pi}^2 + (\partial_\mu \sigma)^2 - m^2 \sigma^2 \right] \right.$$
$$\left. - ib \left(2v\sigma + \boldsymbol{\pi}^2 + \sigma^2 \right) \right\} + \tfrac{1}{2} m^2 \left(f_\pi^2 - v^2 \right) \beta V \qquad (12.126)$$

Note that terms linear in the fields integrate to zero because $\langle \pi_i \rangle = \langle \sigma \rangle = \langle b \rangle = 0$.

An effective action is derived by expanding $\exp(S)$ in powers of b and integrating over the pion and σ fields. The term linear in b vanishes on account of $\tilde{b}(\mathbf{0}, 0) \propto \langle b \rangle = 0$. The term proportional to b^2 is nonzero and is exponentiated, thus summing a whole series of contributions. The term proportional to b^3 is also nonzero and it, too, may be exponentiated, summing an infinite series of higher-order terms left out of the order-b^2 exponentiation. After making the scaling $b \to b/\sqrt{2N}$ the effective action becomes

$$S_{\text{eff}} = -\tfrac{1}{2} \sum_n \sum_{\mathbf{p}} \left(\omega_n^2 + p^2 + m^2 \right) \left[\tilde{\boldsymbol{\pi}}(\mathbf{p}, n) \cdot \tilde{\boldsymbol{\pi}}(-\mathbf{p}, -n) + \tilde{\sigma}(\mathbf{p}, n) \tilde{\sigma}(-\mathbf{p}, -n) \right]$$

$$- \tfrac{1}{2} \sum_n \sum_{\mathbf{p}} \left(\Pi(p, \omega_n, T, m) + \frac{2}{N} \frac{v^2}{\omega_n^2 + p^2 + m^2} \right) \tilde{b}(\mathbf{p}, 2n) \tilde{b}(-\mathbf{p}, -2n)$$

$$+ \tfrac{1}{2} m^2 \left(f_\pi^2 - v^2 \right) \beta V + \mathrm{O}\left(\frac{\tilde{b}^3}{\sqrt{N}} \right) \qquad (12.127)$$

Note that only even Matsubara frequencies contribute in the b-field: $\nu_n = 2\pi n T$. This may have been anticipated. There appears the one-loop function

$$\Pi(p, \omega_n, T, m) = T \sum_l \int \frac{d^3k}{(2\pi)^3} \frac{1}{(\omega_n - \omega_l)^2 + (\mathbf{p} - \mathbf{k})^2 + m^2} \frac{1}{\omega_l^2 + k^2 + m^2} \qquad (12.128)$$

The effective action is an infinite series in b. The coefficients are frequency and momentum dependent, arising from one-loop diagrams. In addition, each successive term is suppressed by $1/\sqrt{N}$ compared with the previous one. This is the large-N expansion.

The propagators for the π and σ fields are of the usual form,

$$\mathcal{D}_0^{-1}(p, \omega_n, m) = \omega_n^2 + p^2 + m^2 \qquad (12.129)$$

with an effective mass m yet to be determined. The propagator for the b-field is more complicated:

$$\mathcal{D}_b^{-1}(p, \omega_n, m) = \Pi(p, \omega_n, T, m) + \frac{2}{N} \frac{v^2}{\omega_n^2 + p^2 + m^2} \qquad (12.130)$$

The value of the condensate v is not yet determined, either.

Keeping only the terms up to order b^2 in S_{eff} (the rest vanish in the limit $N \to \infty$) allows us to obtain an explicit expression for the partition function and the pressure; this includes the next-to-leading order terms in N:

$$P = \frac{T}{V} \ln Z = \tfrac{1}{2} m^2 \left(f_\pi^2 - v^2 \right)$$

$$- \tfrac{1}{2} N T \sum_n \int \frac{d^3 p}{(2\pi)^3} \ln \left[\beta^2 \left(\omega_n^2 + p^2 + m^2 \right) \right]$$

$$- \tfrac{1}{2} T \sum_n \int \frac{d^3 p}{(2\pi)^3} \ln \left(\Pi(p, \omega_n, T, m) + \frac{2}{N} \frac{v^2}{\omega_n^2 + p^2 + m^2} \right)$$

$$(12.131)$$

The second term in the argument of the last logarithm should and will be set to zero at this order. It may be needed at higher order in the large-N expansion to regulate infrared divergences.

The pressure is extremized with respect to the mass parameter m. Therefore $\partial P / \partial m^2 = 0$. From the initial expression for Z this is seen to be equivalent to the thermal average of the constraint:

$$f_\pi^2 = \langle \mathbf{\Phi}^2 \rangle = v^2 + \langle \boldsymbol{\pi}^2 \rangle + \langle \sigma^2 \rangle \qquad (12.132)$$

If an approximation to the exact partition function is made, such as the large-N expansion, this constraint should still be satisfied. It may, in fact, single out a preferred value of m.

To leading order in N we may neglect the term involving Π entirely. The pressure is then

$$P = \tfrac{1}{2} m^2 \left(f_\pi^2 - v^2 \right) + N\, P_0(T, m) \qquad (12.133)$$

The pressure must be a maximum with respect to variations in the condensate v. This means that

$$\frac{\partial P}{\partial v} = -m^2 v = 0 \qquad (12.134)$$

which is equivalent to the condition $\langle \sigma \rangle = 0$. There are two possibilities.

1 $m = 0$ There exist massless particles, or Goldstone bosons, and the value of the condensate is determined by the thermally averaged constraint. This is the symmetry-broken phase.

2 $v = 0$ The thermally averaged constraint is satisfied by a nonzero temperature-dependent mass. There are no Goldstone bosons. This is the symmetry-restored phase.

Evidently there is a chiral-symmetry-restoring phase transition!

In the leading order of the large-N approximation the particles are represented by free fields with a potentially temperature-dependent mass m. Again, we may use

$$\frac{\partial P_0(T, m)}{\partial m^2} = \langle \phi^2 \rangle = \int \frac{d^3 p}{(2\pi)^3} \frac{1}{\omega} \frac{1}{e^{\beta\omega} - 1} \qquad (12.135)$$

with $\omega = \sqrt{p^2 + m^2}$. Thus extremizing the pressure with respect to m^2 is equivalent to satisfying the thermally averaged constraint

$$f_\pi^2 = v^2 + \langle \boldsymbol{\pi}^2 \rangle + \langle \sigma^2 \rangle \qquad (12.136)$$

Note however that the pion and σ fields have the same mass and therefore $\langle \boldsymbol{\pi}^2 \rangle = (N - 1)\langle \sigma^2 \rangle$. Consider now the two different phases.

In the asymmetric, symmetry-broken, phase the mass is zero. The above constraint is satisfied by a temperature-dependent condensate:

$$v^2(T) = f_\pi^2 - \frac{N T^2}{12} \qquad (12.137)$$

This condensate goes to zero at a critical temperature

$$T_c^2 = \frac{12}{N} f_\pi^2 \qquad \text{(leading-}N\text{ approximation)} \qquad (12.138)$$

Exactly at T_c the thermally averaged constraint is satisfied by the fluctuations of N massless degrees of freedom without the help of a condensate.

In the symmetric phase the condensate is zero. The constraint is satisfied by thermal fluctuations alone:

$$f_\pi^2 = N \int \frac{d^3 p}{(2\pi)^3} \frac{1}{\omega} \frac{1}{e^{\beta\omega} - 1} \qquad (12.139)$$

Thermal fluctuations decrease with increasing mass at fixed temperature. The constraint is only satisfied by massless excitations at one temperature, namely, T_c. At temperatures $T > T_c$ the mass must be greater than zero. Near the critical temperature the mass should be small, and the fluctuations may be expanded about $m = 0$ as

$$f_\pi^2 = NT^2 \left[\frac{1}{12} - \frac{m}{4\pi T} - \frac{m^2}{8\pi^2 T^2} \ln\left(\frac{m}{4\pi T}\right) - \frac{m^2}{16\pi^2 T^2} + \cdots \right] \qquad (12.140)$$

As T approaches T_c from above, the mass approaches zero as follows:

$$m(T) = \frac{\pi}{3T} \left(T^2 - T_c^2\right) + \cdots \qquad (12.141)$$

This is a second-order phase transition since there is no possibility of metastable supercooled or superheated states.

The mass must grow faster than the temperature at very high temperatures in order to keep the field fluctuations fixed and equal to f_π^2.

Asymptotically the particles move nonrelativistically. This allows us to compute the fluctuations analytically. We get

$$f_\pi^2 = N \left(\frac{T}{2\pi}\right)^{3/2} \sqrt{m}\, e^{-m/T} \tag{12.142}$$

This is a transcendental equation for $m(T)$. It can also be written as

$$m = T \ln\left(\frac{NT}{2\pi f_\pi} \sqrt{\frac{mT}{2\pi f_\pi^2}}\right) \tag{12.143}$$

Roughly, the solution behaves as follows:

$$m \sim T \ln\left(\frac{T^2}{T_c^2}\right) \tag{12.144}$$

It is rather amusing that, at the leading order of the large-N approximation, the elementary excitations are massless below T_c, become massive above T_c, and at asymptotically high temperatures move nonrelativistically.

The result to first order of the large-N expansion provides good insight into the nature of the two-phase structure of the nonlinear σ model, but it is not quite satisfactory for two reasons. First, it predicts N massless Goldstone bosons in the broken-symmetry phase when in fact we know there ought to be only $N-1$. Second, the square of the critical temperature is $12f_\pi^2/N$ whereas it is $12f_\pi^2/(N+2)$ in the linear σ model in the mean field approximation; we expect them to be the same in the limit $\lambda \to \infty$. Both these problems can be rectified by inclusion of the next-to-leading-order term in N, which gives the contribution of the b-field.

It is natural to expect that the b-field will contribute essentially one negative degree of freedom to the T^4 term in the pressure so as to give $N-1$ Goldstone bosons in the low-temperature phase. Therefore we move one of the N degrees of freedom and put it together with the b-field contribution as

$$P = \tfrac{1}{2}m^2\left(f_\pi^2 - v^2\right) - \tfrac{1}{2}(N-1)\, T \sum_n \int \frac{d^3p}{(2\pi)^3}\, \ln\left[\beta^2\left(\omega_n^2 + p^2 + m^2\right)\right]$$

$$-\tfrac{1}{2}T \sum_n \int \frac{d^3p}{(2\pi)^3}\, \ln\left[\beta^2\left(\omega_n^2 + p^2 + m^2\right)\Pi\right] \tag{12.145}$$

The function $\Pi(p, \omega_n, T, m)$ can be reduced to a one-dimensional integral:

$$\Pi = \frac{1}{8\pi^2 p} \int_0^\infty \frac{dk\, k}{\omega} \ln\left(\frac{k^2 + pk + \Lambda^2}{k^2 - pk + \Lambda^2}\right) \frac{1}{e^{\beta\omega} - 1} \tag{12.146}$$

where

$$\Lambda^2 = \Lambda^2(p, \omega_n, m) = \frac{(\omega_n^2 + p^2)^2 + 4m^2\omega_n^2}{4(\omega_n^2 + p^2)} \tag{12.147}$$

but unfortunately Π cannot be simplified any further. In any case, to the order in N to which we are working, the pressure is

$$P = \tfrac{1}{2}m^2\left(f_\pi^2 - v^2\right) + (N-1)P_0(T, m) + P_1(T, m) \tag{12.148}$$

The pressure can be thought of, in the low-temperature phase, as due to $N-1$ Goldstone bosons with an interaction term P_1.

Because of the logarithm, the main contribution to the interaction pressure will come when Π is very small compared to unity. This corresponds to very large values of the parameter Λ; in other words, to very high momentum, Matsubara frequency, or mass. In this limit,

$$\Pi \to \frac{1}{4\pi^2\Lambda^2}\int_0^\infty \frac{dk\,k^2}{\omega}\frac{1}{e^{\beta\omega}-1} = \frac{T^2}{2\pi^2\Lambda^2}h_3\left(\frac{m}{T}\right) \tag{12.149}$$

This may be considered as a high-energy approximation, and we shall henceforth refer to it as such. Then

$$P_1 = \tfrac{1}{2}T\sum_n\int\frac{d^3p}{(2\pi)^3}\ln\left[\beta^2\left(\omega_n^2 + p^2 + m^2\right)\Pi\right]$$

$$\approx -\tfrac{1}{2}T\sum_n\int\frac{d^3p}{(2\pi)^3}\ln\left(\frac{h_3}{\pi^2}\frac{(\omega_n^2 + p^2)(\omega_n^2 + p^2 + m^2)}{(\omega_n^2 + \omega_+^2)(\omega_n^2 + \omega_-^2)}\right) \tag{12.150}$$

with dispersion relations

$$\omega_\pm^2 = p^2 + 2m^2 \pm 2m\sqrt{p^2 + m^2} \tag{12.151}$$

The interaction pressure can now be determined in the usual way to be

$$P_1 = -T\int\frac{d^3p}{(2\pi)^3}\left[\ln(1 - e^{-\beta p}) + \ln(1 - e^{-\beta\omega(p)})\right.$$

$$\left. - \ln(1 - e^{-\beta\omega_+(p)}) - \ln(1 - e^{-\beta\omega_-(p)})\right] \tag{12.152}$$

Note that $h_3(m/T)$ has no effect within this approximation. Note also that in the broken-symmetry phase where $m = 0$ the contribution of the b-field cancels one of the massless degrees of freedom to give $N-1$ Goldstone bosons.

Now we are prepared to examine the behavior of the system near the critical temperature with the inclusion of next-to-leading terms in N. We make an expansion in m/T as before. The pressure is, up to and including order m^3,

$$P = (N-1)\frac{\pi^2}{90}T^4 - \frac{N+2}{24}m^2T^2 + \frac{1}{2}m^2\left(f_\pi^2 - v^2\right) + \frac{N}{12\pi}m^3T$$

(12.153)

In the high-temperature phase, where $v = 0$, maximization with respect to m yields

$$f_\pi^2 = T^2\left(\frac{N+2}{12} - \frac{N}{4\pi}\frac{m}{T}\right)$$

(12.154)

This gives the same critical temperature as in the mean field treatment of the linear σ model.

$$T_c^2 = \frac{12}{N+2}f_\pi^2 \qquad \text{(sub-leading-N approximation)}$$

(12.155)

The mass approaches zero from above as follows:

$$m(T) = \frac{\pi(N+2)}{3NT}\left(T^2 - T_c^2\right)$$

(12.156)

In the results obtained immediately above, an approximation for Π to which we have referred as a high-energy approximation has been used. Relaxing this approximation can be done, albeit at the cost of a numerical calculation. Of course, one should also go beyond the mean field approximation in the linear model.

12.4.3 Finite-temperature behavior of f_π

Consideration of correlation functions at finite temperature is more involved than at zero temperature. Lorentz invariance is not manifest because there is a preferred frame of reference, the frame in which the matter is at rest. Thus spectral densities and other functions may depend on energy and momentum separately and not just on their invariant s. Also, the number of Lorentz tensors is greater because there is a new vector available, namely, the vector $u_\mu = (1, 0, 0, 0)$ that specifies the rest frame of the matter.

In the usual fashion one may construct a Green's function for the axial-vector current $\mathcal{A}_a^\mu$:

$$G_{ab}^{\mu\nu}(z, \mathbf{q}) = \int_{-\infty}^{\infty}\frac{d\omega}{\omega - z}\,\rho_{ab}^{\mu\nu}(\omega, \mathbf{q})$$

(12.157)

where the spectral density tensor is

$$\rho_{ab}^{\mu\nu}(\omega, \mathbf{q}) = \frac{1}{Z} \sum_{m,n} (2\pi)^3 \delta(\omega - E_m + E_n)\delta(\mathbf{q} - \mathbf{p_m} + \mathbf{p_n})$$
$$\times \left(e^{-E_n/T} - e^{-E_m/T} \right) \langle n|\mathcal{A}_a^\mu(0)|m\rangle\langle m|\mathcal{A}_b^\nu(0)|n\rangle \quad (12.158)$$

The summation is over a complete set of energy eigenstates.

Owing to current conservation the spectral density tensor can be decomposed into longitudinal and transverse terms:

$$\rho_{ab}^{\mu\nu}(q) = \delta_{ab} \left[\rho_A^{\mathrm{L}}(q)P_{\mathrm{L}}^{\mu\nu} + \rho_A^{\mathrm{T}}(q)P_{\mathrm{T}}^{\mu\nu} \right] \quad (12.159)$$

In general the spectral densities depend on q^0 and $\mathbf{q}$ separately as well as on the temperature. In the vacuum we can always go to the rest frame of a massive particle and in that frame there can be no difference between longitudinal and transverse polarizations, so that $\rho_{\mathrm{L}} = \rho_{\mathrm{T}} = \rho$. We also observe that $P_{\mathrm{L}}^{\mu\nu} + P_{\mathrm{T}}^{\mu\nu} = -(g^{\mu\nu} - q^\mu q^\nu/q^2)$. The pion, being a massless Goldstone boson, is special. It contributes to the longitudinal axial spectral density and not to the transverse one. In vacuum,

$$\rho^{\mu\nu}(q) = \left(\frac{q^\mu q^\nu}{q^2} - g^{\mu\nu} \right) \rho_A(q^2) + f_\pi^2\delta(q^2)q^\mu q^\nu \quad (12.160)$$

This may be taken to be the definition of the pion decay constant at zero temperature. In fact, one can write the pion's contribution as

$$f_\pi^2\delta(q^2)q^\mu q^\nu = f_\pi^2 q^2\delta(q^2)P_{\mathrm{L}}^{\mu\nu} \quad (12.161)$$

This cannot be taken as the definition of the pion decay constant at finite temperature because the contribution of the pion to the longitudinal spectral density cannot be assumed to be a delta function in q^2. In general, as mentioned previously, the pion's dispersion relation will be more complicated and will develop a width at nonzero momentum. This smears out the delta function into something like a relativistic Breit–Wigner distribution. Fortunately, the Goldstone theorem [31] requires that there be a zero-frequency excitation when the momentum is zero (see Chapter 7). This implies that the width must go to zero at $\mathbf{q} = \mathbf{0}$, which results in a delta function at zero frequency. Explicit calculations support this assertion [32, 33]. Therefore it would seem to make sense to define

$$f_\pi^2(T) \equiv 2 \lim_{\epsilon \to 0} \int_0^\epsilon \frac{dq_0^2}{q_0^2} \rho_A^{\mathrm{L}}(q_0, \mathbf{q} = \mathbf{0}) \quad (12.162)$$

Physically this means that the pion decay constant at finite temperature measures the strength of the coupling of the Goldstone boson to the longitudinal part of the retarded axial-vector response function in the limit of zero momentum.

We shall first study the pion's contribution to the spectral density at temperatures small compared with f_π. We shall study both the nonlinear and linear σ models. At low temperatures the σ meson's contribution as a material degree of freedom is frozen out and one might expect the same dynamics to be operative in both models; in other words, one might expect the result to be the same and so independent of λ.

The nonlinear σ model

The nonlinear σ model was defined at the beginning of subsection 12.4.2. One can make a nonlinear redefinition of the field without changing the physical content of the theory. Various redefinitions may be found in the literature. First we will list the most common ones and then we will compute $f_\pi(T)$ for each of them, thereby illustrating that one always gets the same result. It is interesting to see how this comes about; it is also reassuring that it does.

A convenient way to express the sigma and pion fields that explicitly contains the constraint is

$$\begin{aligned}
\sigma &= f_\pi \, \cos(\phi/f_\pi) \\
\boldsymbol{\pi} &= f_\pi \hat{\boldsymbol{\phi}} \, \sin(\phi/f_\pi)
\end{aligned} \tag{12.163}$$

where $\phi = |\boldsymbol{\phi}|$ and $\hat{\boldsymbol{\phi}} = \boldsymbol{\phi}/\phi$. The Lagrangian may then be expressed in terms of the fields of choice:

$$\begin{aligned}
\mathcal{L} &= \tfrac{1}{2} \partial_\mu \boldsymbol{\pi} \cdot \partial^\mu \boldsymbol{\pi} + \tfrac{1}{2} \partial_\mu \sigma \, \partial^\mu \sigma \\
&= \tfrac{1}{2} \partial_\mu \boldsymbol{\pi} \cdot \partial^\mu \boldsymbol{\pi} + \tfrac{1}{2} \frac{(\boldsymbol{\pi} \cdot \partial_\mu \boldsymbol{\pi})(\boldsymbol{\pi} \cdot \partial^\mu \boldsymbol{\pi})}{f_\pi^2 - \pi^2} \\
&= \tfrac{1}{2} \frac{f_\pi^2}{\phi^2} \sin^2\left(\frac{\phi}{f_\pi}\right) \partial_\mu \boldsymbol{\phi} \cdot \partial^\mu \boldsymbol{\phi} + \tfrac{1}{2} \left[1 - \frac{f_\pi^2}{\phi^2} \sin^2\left(\frac{\phi}{f_\pi}\right) \right] \partial_\mu \partial^\mu \phi
\end{aligned} \tag{12.164}$$

Another representation to consider is due to Weinberg [27], who makes the definition

$$\mathbf{p} = 2 \frac{f_\pi^2}{\pi^2} \left(1 - \sqrt{1 - \frac{\pi^2}{f_\pi^2}} \right) \boldsymbol{\pi} \tag{12.165}$$

or inversely

$$\boldsymbol{\pi} = \frac{\mathbf{p}}{1 + p^2/4f_\pi^2} \tag{12.166}$$

In terms of Weinberg's field definition the Lagrangian is very compact:

$$\mathcal{L} = \frac{1}{2} \frac{\partial_\mu \mathbf{p} \cdot \partial^\mu \mathbf{p}}{(1 + p^2/4f_\pi^2)^2} \tag{12.167}$$

The $(\sigma, \boldsymbol{\pi})$ representation is cumbersome because of the constraint, although it can be handled by the Lagrange multiplier method of subsection 12.4.2. However, it is inconvenient for exposing the physical particle content and for doing perturbation theory in terms of physical particles. Among the three physical representations we choose to work with here, it is interesting to note the range of allowed values of the fields. The magnitude of the **p**-field can range from zero to infinity, the magnitude of the $\boldsymbol{\pi}$-field can range from 0 to f_π, and the magnitude of the $\boldsymbol{\phi}$-field can range from 0 to πf_π. This distinction is important when dealing with nonperturbative large-amplitude motion; whether it makes any difference in low orders of perturbation theory is not known.

The first step in our quest to extract the temperature dependence of f_π from the theory is to obtain the form of the axial-vector current in terms of the chosen fields. Starting from

$$\mathcal{A}_\mu = -\sigma\,\partial_\mu\boldsymbol{\pi} + \boldsymbol{\pi}\,\partial_\mu\sigma \tag{12.168}$$

one directly computes

$$
\begin{aligned}
\mathcal{A}_\mu &= -\sigma\left(\partial_\mu\boldsymbol{\pi} + \frac{\boldsymbol{\pi}\,(\boldsymbol{\pi}\cdot\partial_\mu\boldsymbol{\pi})}{f_\pi^2 - \pi^2}\right) \\
&= -\frac{f_\pi^2}{2\phi}\sin\left(\frac{2\phi}{f_\pi}\right)\partial_\mu\boldsymbol{\phi} - f_\pi\hat{\boldsymbol{\phi}}\left[1 - \frac{f_\pi}{2\phi}\sin\left(\frac{2\phi}{f_\pi}\right)\right]\hat{\boldsymbol{\phi}}\cdot\partial_\mu\boldsymbol{\phi} \\
&= -\frac{1}{f_\pi}\frac{1}{(1 + p^2/4f_\pi^2)^2}\left[\left(f_\pi^2 - \frac{1}{4}p^2\right)\partial_\mu\mathbf{p} + \frac{1}{2}\mathbf{p}\,(\mathbf{p}\cdot\partial_\mu\mathbf{p})\right]
\end{aligned}
\tag{12.169}
$$

Every form of the axial-vector current is an odd function of the pion field.

Obviously it is not possible to compute the axial-vector correlation function exactly. We will restrict our attention to low temperatures. Roughly speaking, a loop expansion of the correlation function is an expansion in powers of T^2/f_π^2, each additional loop contributing one more such factor. To one-loop order, we need the axial-vector current to third order in the pion field:

$$
\begin{aligned}
\mathcal{A}_\mu &= -f_\pi\,\partial_\mu\boldsymbol{\pi} + \frac{\pi^2}{2f_\pi}\,\partial_\mu\boldsymbol{\pi} - \frac{1}{f_\pi}\,\boldsymbol{\pi}\,(\boldsymbol{\pi}\cdot\partial_\mu\boldsymbol{\pi}) \\
&= -f_\pi\,\partial_\mu\boldsymbol{\phi} + \frac{2\phi^2}{3f_\pi}\,\partial_\mu\boldsymbol{\phi} - \frac{2}{3f_\pi}\,\boldsymbol{\phi}\,(\boldsymbol{\phi}\cdot\partial_\mu\boldsymbol{\phi}) \\
&= -f_\pi\,\partial_\mu\mathbf{p} + \frac{3p^2}{4f_\pi}\,\partial_\mu\mathbf{p} - \frac{1}{2f_\pi}\,\mathbf{p}\,(\mathbf{p}\cdot\partial_\mu\mathbf{p}) \tag{12.170}
\end{aligned}
$$

We will also need the Lagrangian to fourth order in the pion field:

$$\mathcal{L}_4 = \frac{1}{2f_\pi^2} \left(\boldsymbol{\pi} \cdot \partial_\mu \boldsymbol{\pi}\right) \left(\boldsymbol{\pi} \cdot \partial^\mu \boldsymbol{\pi}\right)$$

$$= \frac{1}{6f_\pi^2} \left[\left(\boldsymbol{\phi} \cdot \partial_\mu \boldsymbol{\phi}\right) \left(\boldsymbol{\phi} \cdot \partial^\mu \boldsymbol{\phi}\right) - \phi^2 \, \partial_\mu \boldsymbol{\phi} \cdot \partial^\mu \boldsymbol{\phi}\right]$$

$$= -\frac{1}{4f_\pi^2} \, p^2 \, \partial_\mu \mathbf{p} \cdot \partial^\mu \mathbf{p}$$

$$(12.171)$$

The correlation function $\langle \mathcal{A}_\mu^i(x) \mathcal{A}_\nu^j(y) \rangle$ will have a zero-loop contribution from the π–π correlation function $\langle \partial_\mu \pi^i(x) \partial_\nu \pi^j(y) \rangle$, a one-loop self-energy correction to the same π–π correlation function, and a one-loop contribution from the correlation function involving four pions, $\langle \partial_\mu \pi^i(x) \pi^j(y)\pi^k(y)\partial_\nu \pi^l(y) \rangle$.

The contribution of the bare-pion propagator $\mathcal{D}_0$ to the longitudinal spectral density is easily found to be

$$\rho_A^{\mathrm{L}}(q_0, \mathbf{q}) = f_\pi^2 \, q^2 \, \delta\left(q^2\right) \qquad (12.172)$$

At zero temperature this is just the definition of the pion decay constant.

The one-loop pion self-energy may be computed by standard diagrammatic or functional integral techniques. The results are:

$$\Pi_{\boldsymbol{\pi}}(q) = -\frac{T^2}{12f_\pi^2} \, q^2$$

$$\Pi_{\mathbf{p}}(q) = (N-1) \frac{T^2}{24f_\pi^2} \, q^2 \qquad (12.173)$$

$$\Pi_{\boldsymbol{\phi}}(q) = \frac{1}{3} \Pi_{\boldsymbol{\pi}}(q) + \frac{2}{3} \Pi_{\mathbf{p}}(q)$$

These are rather dependent on the definition of the pion field! Nevertheless, it is worth noting that the Goldstone theorem is satisfied on account of the fact that the self-energy is always proportional to q^2.

The final contribution comes from the correlation function for a pion at point x with three pions at point y. Again, standard diagrammatic or functional integral techniques may be used. To express the answers, we gather together the contributions from the bare propagator, from the one-loop self-energy, and from this correlation function and quote the coefficient of the term $f_\pi^2 q^2 \delta\left(q^2\right)$ in the longitudinal part of the axial-vector spectral density:

$$\boldsymbol{\pi}: \left(1 - \frac{T^2}{12f_\pi^2}\right) - (N-3)\frac{T^2}{12f_\pi^2}$$

$$\mathbf{p}: \left(1 + (N-1)\frac{T^2}{24f_\pi^2}\right) - \left(N - \frac{5}{3}\right)\frac{T^2}{8f_\pi^2} \qquad (12.174)$$

$$\phi: \left(1 + (N-2)\frac{T^2}{36f_\pi^2}\right) - (N-2)\frac{T^2}{9f_\pi^2}$$

In all three cases the results are the same and amount to a temperature dependence of

$$f_\pi^2(T) = f_\pi^2\left(1 - \frac{N-2}{12}\frac{T^2}{f_\pi^2}\right) \qquad (12.175)$$

This agrees with the analysis of Gasser and Leutwyler [21] for the only case where they can be compared, $N_{\mathrm{f}}^2 = N = 4$.

The linear σ model

It is now not surprising to discover that the linear σ model gives the same result for $f_\pi(T)$ at low temperatures as the nonlinear sigma model. The reason is that the σ meson is very heavy at low temperatures and cannot contribute materially in the way that the pions do. However, the way in which the linear σ model works out is very different.

Let us go back to the axial-vector current before shifting the sigma field:

$$\mathcal{A}_\mu = -\sigma\,\partial_\mu\boldsymbol{\pi} + \boldsymbol{\pi}\,\partial_\mu\sigma \qquad (12.176)$$

After making the shift $\sigma \to v + \sigma$ the current takes the form

$$\mathcal{A}_\mu = -v\,\partial_\mu\boldsymbol{\pi} - \sigma\,\partial_\mu\boldsymbol{\pi} + \boldsymbol{\pi}\,\partial_\mu\sigma \qquad (12.177)$$

By maximizing the pressure (which is equivalent to minimizing the effective potential) with respect to v at each temperature we effectively sum all tadpole diagrams, leaving only 1PI diagrams in any subsequent perturbative treatment. If this is done, one's inclination is to identify $v(T)$ with $f_\pi(T)$. This is wrong; $f_\pi(T)$ has additional contributions, as we shall now see.

The first contribution to $f_\pi^2(T)$ does come from $v^2(T)$ since it involves the cross term of $\partial_\mu\pi^a(x)$ with $\partial_\nu\pi^a(y)$. Following the analysis of subsection 12.4.1, but at low temperature rather than high, we simply leave out the contribution of the heavy σ meson. This gives

$$P(T, v) = (N - 1)\frac{\pi^2}{90}T^4 + \frac{\lambda}{2}v^2\left(f_\pi^2 - \frac{N - 1}{12}T^2\right) - \frac{\lambda}{4}v^4 \qquad (12.178)$$

Maximizing with respect to v gives

$$v^2(T) = f_\pi^2 - \frac{N - 1}{12}T^2 \qquad (12.179)$$

There is another, nonlocal, contribution to the vertex, corresponding to the emission and absorption of a virtual σ meson. One might think that it would be suppressed by the large σ mass, $m_\sigma^2 = 2\lambda f_\pi^2$, but in fact this is compensated by the coupling constant λ in the extra vertex. Evaluation of this diagram gives a contribution to $f_\pi^2(T)$ of $T^2/6f_\pi^2$.

Finally there is a contribution coming from the dressed pion propagator analogous to that in the nonlinear σ model. The full one-loop 1PI pion self-energy diagrams have been shown already in Figure 12.5. We know that the sum of the momentum-independent terms is zero on account of Goldstone's theorem. We just need the contribution that is quadratic in the energy and momentum of the pion. This can arise only from the so-called exchange diagram involving two $\sigma\pi\pi$ vertices. In imaginary time (Euclidean space) it is

$$\Pi_{\text{ex}}(\omega_n, \mathbf{q}) = -4\lambda^2 f_\pi^2 T \sum_l \int \frac{d^3k}{(2\pi)^3} \frac{1}{\omega_l^2 + \mathbf{k}^2} \frac{1}{(\omega_l + \omega_n)^2 + (\mathbf{k} + \mathbf{q})^2 + m_\sigma^2} \qquad (12.180)$$

Since $T \ll m_\sigma$ it is easy to extract the part that is quadratic in the momentum. Analytically continuing to Minkowski space ($\omega_n \to iq_0$), it is $q^2T^2/12f_\pi^2$.

The residue of the pion pole in the axial-vector correlation function can now be obtained by adding the vacuum contribution, the pion self-energy correction, and the tadpole and nonlocal vertex corrections as follows:

$$\left(1 - \frac{1}{12}\frac{T^2}{f_\pi^2}\right) - \frac{N - 1}{12}\frac{T^2}{f_\pi^2} + \frac{1}{6}\frac{T^2}{f_\pi^2}$$

The final result,

$$f_\pi^2(T) = f_\pi^2\left(1 - \frac{N - 2}{12}\frac{T^2}{f_\pi^2}\right) \qquad (12.181)$$

is identical to that of the nonlinear σ model. We remark that this cannot be used to compute the critical temperature since it was obtained under the condition that $T \ll f_\pi$.

The approach to chiral-symmetry restoration

Calculation of $f_\pi(T)$ as $T \to T_c$ is more involved than in the low-temperature limit. It was done for the nonlinear model by Jeon and Kapusta [34]. Here we just quote the result:

$$f_\pi^2(T) = f_\pi^2 - \frac{N+2}{12}T^2 \tag{12.182}$$

It goes to zero at the correct critical temperature. Notice that the coefficient of the T^2 term is different from that in the low-temperature limit. A relatively simple Padé approximation may be used to extrapolate smoothly from low temperatures to the critical temperature:

$$\frac{f_\pi^2(T)}{f_\pi^2} \approx \frac{1 - \dfrac{T^2}{T_c^2}}{1 - \dfrac{4}{(N+2)}\dfrac{T^2}{T_c^2}\left(1 - \dfrac{T^2}{T_c^2}\right)} \tag{12.183}$$

12.4.4 Finite-temperature scalar condensate

The scalar condensate is defined as $|\langle\mathbf{\Phi}\rangle|$. Our convention has been to allow the last, Nth, component of the field to condense, and to refer to this as either v, if the field is shifted, or $\langle\sigma\rangle$ if the field is not shifted. In this section we use the latter convention.

It is interesting to ask what happens to this condensate as a function of temperature in the nonlinear model. The constraint as an operator equation is $f_\pi^2 = \mathbf{\Phi}^2$ and as a thermal average is $f_\pi^2 = \langle\mathbf{\Phi}^2\rangle$; it is not $f_\pi = |\langle\mathbf{\Phi}\rangle|$. The condensate can indeed change with temperature. In fact we can quite easily compute it to two-loop order. Before doing so, we first discuss the connection of this condensate with the quark condensate $\langle\bar{\psi}\psi\rangle$.

In two-flavor QCD one often associates the sigma and pion fields with certain bilinear forms of the quark fields:

$$\bar{\psi}\psi \sim \sigma$$
$$i\bar{\psi}\gamma_5\boldsymbol{\tau}\psi \sim \boldsymbol{\pi}$$

This association is made because the quark bilinear forms transform in the same way under $\mathrm{SU}(2) \times \mathrm{SU}(2)$ as the corresponding meson fields. The dimensions do not match, so there must be some dimensional coefficient relating them; this coefficient could even be a function of the group invariant $\sigma^2 + \boldsymbol{\pi}^2 \sim \left(\bar{\psi}\psi\right)^2 - \left(\bar{\psi}\gamma_5\boldsymbol{\tau}\psi\right)^2$. Does this particular combination of four-quark condensates change with temperature? The temperature dependence of the four-quark condensates at low temperatures was first calculated in [35] with the help of the fluctuation–dissipation theorem. The contribution of pions alone was later discussed in [36] using soft-pion techniques. From [35, 36] one can state the two condensates separately:

$$\langle (\bar{\psi}\psi)^2 \rangle = \left(1 - \frac{T^2}{4f_\pi^2}\right) \langle 0|(\bar{\psi}\psi)^2|0\rangle - \frac{T^2}{12f_\pi^2} \langle 0|(\bar{\psi}\gamma_5\boldsymbol{\tau}\psi)^2|0\rangle \quad (12.184)$$

and

$$\langle (\bar{\psi}\gamma_5\boldsymbol{\tau}\psi)^2 \rangle = \left(1 - \frac{T^2}{12f_\pi^2}\right) \langle 0|(\bar{\psi}\gamma_5\boldsymbol{\tau}\psi)^2|0\rangle - \frac{T^2}{4f_\pi^2} \langle 0|(\bar{\psi}\psi)^2|0\rangle \quad (12.185)$$

Therefore there is no correction to this group invariant to order T^2/f_π^2 inclusive:

$$\langle (\bar{\psi}\psi)^2 - (\bar{\psi}\gamma_5\boldsymbol{\tau}\psi)^2 \rangle = \langle 0|(\bar{\psi}\psi)^2 - (\bar{\psi}\gamma_5\boldsymbol{\tau}\psi)^2|0\rangle \quad (12.186)$$

This result is consistent with our analysis of the nonlinear σ model in previous sections.

Now let us return to the business of computing the temperature dependence of the scalar condensate to one- and two-loop order. In terms of the three representations used in the discussion of the nonlinear σ model in Section 12.4.3 the σ field is given by

$$\frac{\sigma}{f_\pi} = \sqrt{1 - \frac{\boldsymbol{\pi}^2}{f_\pi^2}} = 1 - \frac{\boldsymbol{\pi}^2}{2f_\pi^2} - \frac{(\boldsymbol{\pi}^2)^2}{8f_\pi^4} + \cdots$$

$$= \left(1 - \frac{\mathbf{p}^2}{2f_\pi^2} + \frac{(\mathbf{p}^2)^2}{16f_\pi^4}\right)^{1/2} \left(1 + \frac{\mathbf{p}^2}{4f_\pi^2}\right)^{-1} = 1 - \frac{\mathbf{p}^2}{2f_\pi^2} + \frac{(\mathbf{p}^2)^2}{8f_\pi^4} + \cdots$$

$$= \cos\left(\frac{\phi}{f_\pi}\right) = 1 - \frac{\phi^2}{2f_\pi^2} + \frac{(\phi^2)^2}{24f_\pi^4} + \cdots \quad (12.187)$$

To second order in the pion field all three representations are the same. Using the free-field expression for the thermal average of the field squared we get

$$\frac{\langle \sigma \rangle}{f_\pi} = 1 - \frac{N-1}{2}\left(\frac{T^2}{12f_\pi^2}\right) + \cdots \quad (12.188)$$

For $N = 4$, the only value for that we can quantitatively compare with QCD, this agrees with the result of Gasser and Leutwyler [21].

The coefficient of the term that is fourth order in the pion field differs in sign and magnitude among the three representations. It would be a miracle if the thermal average of $\sqrt{1 - \boldsymbol{\pi}^2/f_\pi^2}$, $\cos(\phi/f_\pi)$, and the Weinberg expression were all the same! But regarding the order $(T^2/12f_\pi^2)^2$ we must recognize that the term that is second order in the pion field gets modified owing to a one-loop self-energy. This was computed for each representation in Section 12.4.3 and the results were listed in (12.173). The term that is fourth order in the pion field can be evaluated using free fields. The result is

$$\langle (\phi^2)^2 \rangle = (N^2 - 1) \left(\frac{T^2}{12} \right)^2 \tag{12.189}$$

and is obviously representation independent. The contributions for each representation are

$$\boldsymbol{\pi}: \ 1 - \frac{N-1}{2} \left(\frac{T^2}{12 f_\pi^2} \right) \left[1 - \left(\frac{T^2}{12 f_\pi^2} \right) \right] - \frac{N^2-1}{8} \left(\frac{T^2}{12 f_\pi^2} \right)^2$$

$$\mathbf{p}: \ 1 - \frac{N-1}{2} \left(\frac{T^2}{12 f_\pi^2} \right) \left[1 + \frac{N-1}{2} \left(\frac{T^2}{12 f_\pi^2} \right) \right] + \frac{N^2-1}{8} \left(\frac{T^2}{12 f_\pi^2} \right)^2$$

$$\phi: \ 1 - \frac{N-1}{2} \left(\frac{T^2}{12 f_\pi^2} \right) \left[1 + \frac{N-2}{3} \left(\frac{T^2}{12 f_\pi^2} \right) \right] + \frac{N^2-1}{24} \left(\frac{T^2}{12 f_\pi^2} \right)^2$$

$$\tag{12.190}$$

where the second term in each line comes from the square of the pion field and the last term comes from the pion field in fourth order. The sum of all terms is identical in all three representations; it is

$$\frac{\langle \sigma \rangle}{f_\pi} = 1 - (N-1) \left(\frac{T^2}{24 f_\pi^2} \right) - \frac{(N-1)(N-3)}{2} \left(\frac{T^2}{24 f_\pi^2} \right)^2 + \cdots \tag{12.191}$$

The miracle happens. It is a consequence of the fact that physical quantities must be independent of field redefinition. What is more, for $N = 4$ it agrees with the previously obtained result of Gasser and Leutwyler. However, we emphasize once more that this expression should not be used to infer a critical temperature because it has been derived under the assumption that the temperature is small compared with f_π.

A calculation of the scalar condensate in the nonlinear model near the critical temperature was made by Jeon and Kapusta [34]. The result is exactly the same as in the linear model, (12.114), namely

$$\langle \sigma \rangle^2 = v^2(T) = f_\pi^2 - \frac{N+2}{12} T^2 \tag{12.192}$$

This expression has corrections of order $v^2(T)/N$ and T^2/N in the large-N expansion.

12.5 Exercises

12.1 Use the exponential representation of the pseudoscalar fields (just above (12.6)) in the leading-order chiral Lagrangian $\mathcal{L}_2$ to calculate the four-pion interaction.

12.2 Use the four-pion interaction calculated in the first exercise to calculate the two-loop contribution to the pressure of a pion gas. Compare with (12.23).

12.3 Use the chiral Lagrangian $\mathcal{L}_2$ to compute the π–π scattering amplitude. Use it to calculate the pion self-energy as in Section 12.2. Compare your result with (12.173).

12.4 Read the paper by Dey, Eletsky, and Ioffe and rederive the mixing rule for vector and axial-vector correlators at finite temperature.

12.5 Derive (12.116).

12.6 Construct a Padé approximation for $\langle \sigma \rangle = v(T)$ to extrapolate from $T \ll f_\pi$ to $T \to T_{\mathrm{c}}$.

12.7 Do the linear and nonlinear σ models satisfy the Weinberg sum rules at finite temperature? Explain your answer.

12.8 How are conditions (12.103) and (12.104) modified if the ρ and a_1 spectral densities are taken to be relativistic Breit–Wigner distributions with momentum-independent but temperature-dependent widths instead of delta functions?

References

1. Kapusta, J. I. and Shuryak, E. V., *Phys. Rev. D* **49**, 4694 (1994).
2. Bochkarev, A., and Kapusta, J., *Phys. Rev. D* **54**, 4066 (1996).
3. Adler, S. L., *Phys. Rev.* **177**, 2426 (1969); Bell, J. S., and Jackiw, R., *Nuovo Cimento* **60A**, 47 (1967).
4. Gell-Mann, M., Oakes, R. J., and Renner, B., *Phys. Rev.* **175**, 2195 (1968).
5. Coleman, S., Wess, J., and Zumino, B., *Phys. Rev.* **177**, 2239 (1969); Callan, C. G., Coleman, S., Wess, J., and Zumino, B., *idem*, 2247.
6. Gasser, J., and Leutwyler, H., *Ann. Phys. (NY)* **158**, 142 (1984); *Nucl. Phys.* **B250**, 465 (1985).
7. Leutwyler, H., *Ann. Phys. (NY)* **235**, 165 (1994).
8. Gerber, P., and Leutwyler, H., *Nucl. Phys.* **B321**, 387 (1989).
9. Jeon, S., and Ellis, P. J., *Phys. Rev. D* **58**, 045013 (1998).
10. Sakurai, J. J. (1969). *Currents and Mesons* (University of Chicago Press, Chicago).
11. Collins, P. D. B. (1977). *An Introduction to Regge Theory and High Energy Physics* (Cambridge University Press, Cambridge).
12. Eletsky, V. L., Belkacem, M., Ellis, P. J., and Kapusta, J. I., *Phys. Rev. C* **64**, 035202 (2001).
13. Martell, A. T., and Ellis, P. J., *Phys. Rev. C* **69**, 065206 (2004).
14. Rapp, R., and Wambach, J., *Adv. Nucl. Phys.*, **25**, 1 (2000).

15. Rapp, R., and Gale, C., *Phys. Rev. C* **60**, 024903 (1999).
16. Post, M., Leupold, S., and Mosel, U., *Nucl. Phys.* **A741**, 81 (2004).
17. Weinberg, S., *Phys. Rev. Lett.* **18**, 507 (1967).
18. Shifman, M. A., Vainshtein, A. I., and Zakharov, V. I., *Nucl. Phys.* **B147** 385, 448, 519 (1979).
19. Das, T., Mathur, V. S., and Okubo, S., *Phys. Rev. Lett.* **18**, 761 (1967).
20. Dey, M., Eletsky, V. L., and Ioffe, B. L., *Phys. Lett.* **B252**, 620 (1990); Eletsky, V. L., and Ioffe, B. L., *Phys. Rev. D* **47**, 3083 (1993).
21. Gasser, J., and Leutwyler, H. *Phys. Lett.* **B184**, 83 (1987); *Nucl. Phys.* **B307**, 763 (1988).
22. Song, C., *Phys. Rev. D* **48**, 1375 (1993).
23. Brown, G., and Rho, M., *Phys. Rev. Lett.* **66**, 2720 (1991).
24. Wilson, K. G., *Phys. Rev. D* **7**, 2911 (1973).
25. Gell-Mann, M., and Levy, M., *Nuovo Cimento* **16**, 705 (1960).
26. Donoghue, J. F., Golowich, E., and Holstein, B. R. (1992). *Dynamics of the Standard Model* (Cambridge University Press, Cambridge).
27. Weinberg, S., *Phys. Rev.* **166**, 1568 (1968).
28. Weinberg, S., *Physica* **96A**, 327 (1979).
29. Lin, W., and Serot, B. D., *Nucl. Phys.* **A512**, 637 (1990).
30. Polyakov, A. M. (1987). *Gauge Fields and Strings* (Harwood, Chur).
31. Goldstone, J., *Nuovo Cimento* **19**, 154 (1961).
32. Goity, J., and Leutwyler, H., *Phys. Lett.* **B228**, 517 (1989).
33. Schenk, A., *Nucl. Phys.* **A363**, 97 (1991); *Phys. Rev. D* **47**, 5138 (1993).
34. Jeon, S., and Kapusta, J., *Phys. Rev. D* **54**, 6475 (1996).
35. Bochkarev, A., and Shaposhnikov, M., *Nucl. Phys.* **B268**, 220 (1986).
36. Eletsky, V. L., *Phys. Lett.* **B299**, 111 (1993).

Bibliography

Some reviews of chiral perturbation theory

Ecker, G., *Prog. Part. Nucl. Phys.* **35**, 1 (1995).
Pich, A., *Rep. Prog. Phys.* **58**, 563 (1995).
Scherer, S., *Adv. Nucl. Phys.* **27**, 277 (2003).

The operator product expansion and its application to QCD

Wilson, K. G., *Phys. Rev.* **179**, 1499 (1969).
Wilson, K. G., and Zimmermann, W., *Comm. Math. Phys.* **24**, 87 (1972).
Gross, D. J., and Wilczek, F., *Phys. Rev. D* **8**, 3635 (1973); *ibid* **9**, 980 (1974).
Georgi, H., and Politzer, H. D., *Phys. Rev. D* **9**, 416 (1974).

13

Nucleation theory

The dynamics of first-order phase transitions has fascinated scientists at least since the time of Maxwell and Van der Waals. Much work on the classical theory of the nucleation of gases and liquids was carried out in the early part of the 1900s, culminating in the theory of Becker and Döring [1]. There were and still are many important applications, such as cloud and bubble chambers, the freezing of liquids, and precipitation in the atmosphere. The modern theory of nucleation was pioneered by Langer [2]. Langer's theory is based in a more fundamental way on the microscopic interactions of atoms and molecules. It can also be applied close to a critical point. Nucleation theory has been extended to relativistic quantum field theory by Coleman and Callan [3] for zero temperature and by Affleck [4] and Linde [5] for finite temperature. A coarse-grained relativistic field theory description was developed by Csernai and Kapusta [6] for finite temperature and extended to finite density by Venugopalan and Vischer [7]. Langer's results are recovered in the nonrelativistic limit. Applications here are to elementary particle phase transitions in the early universe, heavy ion collisions, and even the nucleation of black holes.

The goal of nucleation theory is to compute the probability that a bubble or droplet of the A-phase appears in a system initially in the B-phase near the critical temperature. Homogeneous nucleation theory applies when the system is pure; inhomogeneous nucleation theory applies when impurities cause the formation of bubbles or droplets. For the applications we have in mind, namely the early universe and very-high-energy nuclear collisions, it seems that homogeneous nucleation theory is appropriate. In the everyday world it is usually the opposite; dust or ions in the atmosphere are much more efficient in producing precipitation. Nucleation theory is applicable for first-order phase transitions when the matter is not dramatically supercooled or superheated. If substantial supercooling or superheating is present, or if the phase transition is second-order, then

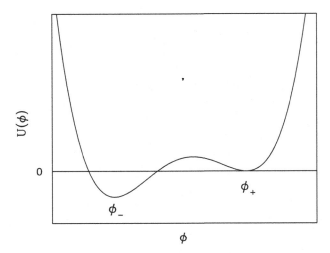

Fig. 13.1.   A potential with two nondegenerate minima.

the relevant dynamics is spinodal decomposition. In this chapter we concern ourselves only with homogeneous nucleation theory.

13.1 Quantum nucleation

A relativistic quantum field theory approach has been worked out by Coleman and Callan [3] for nucleation from one vacuum to another. This is essentially a straightforward extension of the semiclassical formula for tunneling through a barrier in quantum mechanics, generalizing from one degree of freedom to many degrees of freedom and then to an infinite number – a field theory. This approach will be illustrated for a single scalar field.

The Lagrangian is

$$\mathcal{L} = \tfrac{1}{2}\partial_\mu\phi\partial^\mu\phi - U(\phi) \qquad (13.1)$$

Suppose that U has a local minimum at ϕ_+ and a global minimum at ϕ_-, with $U(\phi_-) < U(\phi_+)$, as illustrated in Figure 13.1. If the system is at ϕ_+, the false vacuum, it can tunnel through the barrier to enter the region near the true vacuum, ϕ_-. In nonrelativistic quantum mechanics, the tunneling probability amplitude is dominated by the exponential of minus the action of a trajectory which goes from one side of the barrier to the other. The probability itself is proportional to the exponential of minus the action for a trajectory which begins near ϕ_+, goes through the barrier, and returns to its starting point (on account of time reversal). In the path integral approach to quantum mechanics, this corresponds to the motion of a point particle in imaginary time, as opposed to real time, or equivalently to the motion of a particle in the inverted potential. The process of starting near the false vacuum, going through the barrier, and returning to the starting point was called a "bounce" by Coleman.

Although we are interested in the vacuum tunneling rate we can still use the formalism of finite-temperature field theory, taking the zero temperature, or $\beta \to \infty$, limit in the end. In Euclidean space the classical equation of motion is

$$\frac{\partial^2 \phi}{\partial \tau^2} + \nabla^2 \phi = U'(\phi) \tag{13.2}$$

The boundary conditions we impose are

$$\phi(\mathbf{x}, 0) = \phi(\mathbf{x}, \beta) = \phi_+ \tag{13.3}$$

$$\lim_{|\mathbf{x}| \to \infty} \phi(\mathbf{x}, \tau) = \phi_+ \tag{13.4}$$

$$\frac{\partial \phi}{\partial \tau}(\mathbf{x}, \tau_0) = 0 \tag{13.5}$$

The first of these means that the bounce begins and ends at the false vacuum. The second means that the bounce is localized, being surrounded by false vacuum. The third means that the field has zero velocity at the time τ_0, the time at which the field penetrates the barrier: $U(\phi(\mathbf{x}, \tau_0)) = U(\phi_+)$. Solutions to the classical field equation will be dominant in the classical ($\hbar \to 0$) limit since they have minimal values of the action.

One should expect that the vacuum tunneling solution with the smallest action has O(4) invariance, from the symmetry of the problem. The bounce solution, referred to as $\bar{\phi}$, depends only on the variable $\rho = \sqrt{\tau^2 + \mathbf{x}^2}$. Rather than taking $0 < \tau < \beta$ one may just as well take $-\beta/2 < \tau < \beta/2$. Then the equation of motion simplifies to

$$\frac{d^2 \bar{\phi}}{d^2 \rho} + \frac{3}{\rho} \frac{d\bar{\phi}}{d\rho} = U'(\bar{\phi}) \tag{13.6}$$

The boundary conditions are

$$\lim_{\rho \to \infty} \bar{\phi}(\rho) = \phi_+ \tag{13.7}$$

$$\frac{\partial \bar{\phi}}{\partial \rho}(0) = 0 \tag{13.8}$$

The last of these is needed to avoid a singularity at the origin. The action is then computed from

$$S = 2\pi^2 \int_0^\infty d\rho \, \rho^2 \left[\frac{1}{2} \left(\frac{d\phi}{d\rho} \right)^2 + U(\phi) \right] \tag{13.9}$$

Let us refer to S evaluated with the bounce solution $\bar{\phi}$ as S_B. Since the bounce is a solution to the equation of motion and is localized, it will have finite action.

At zero temperature the system would sit at ϕ_-, where the potential energy is a minimum. Of course there will be quantum corrections to the energy density. The most important of these will arise from fluctuations about ϕ_-. From Chapter 2 we know that we can express the quadratic fluctuations to the partition function around this minimum as

$$N \left\{ \det\left[-\partial_\tau^2 - \nabla^2 + U''(\phi_-)\right] \right\}^{-1/2} \tag{13.10}$$

where N is a normalization constant. The bounce solution, together with the quadratic fluctuations about it, will contribute

$$N\beta V \exp(-S_B) \left\{ \det\left[-\partial_\tau^2 - \nabla^2 + U''(\bar{\phi})\right] \right\}^{-1/2} \tag{13.11}$$

where N is the same normalization. This expression neglects complications due to any zero eigenvalues when performing the functional integral. The factor of spacetime volume βV arises from integration over the position of the center of the bounce: it may be centered anywhere, not necessarily at $\tau = 0$, $\mathbf{x} = \mathbf{0}$ as assumed above. The vacuum energy density is computed in the limit $\beta \to \infty$ from the formula $E_0 = -\partial \ln Z_0 / \partial \beta$. In this semiclassical approximation,

$$\ln Z_0 = \ln \left\{ N \left[-\partial_\tau^2 - \nabla^2 + U''(\phi_-)\right]^{-1/2} \right\}$$
$$+ \ln \left\{ 1 + \frac{\det[-\partial_\tau^2 - \nabla^2 + U''(\bar{\phi})]^{-1/2}}{\det[-\partial_\tau^2 - \nabla^2 + U''(\phi_-)]^{-1/2}} \exp(-S_B) \right\} \tag{13.12}$$

Notice that the normalization N drops out from the second logarithm.

The operator $-\partial_\tau^2 - \nabla^2 + U''(\bar{\phi})$ has four zero eigenvalues owing to the invariance of the bounce solution under translation of its center. Thus, if $\bar{\phi}$ is a solution to the classical equation of motion then so are the $\phi_\mu = b\partial_\mu \bar{\phi}$, where b is a constant. The normalization b can be determined as follows. First, since $\bar{\phi}$ is a solution to the classical equation of motion, the action is stationary under general variations, in particular under the infinitesimal scale transformation

$$\delta\bar{\phi} = x_\nu \partial_\nu \bar{\phi} \tag{13.13}$$

Evaluating the action with $\bar{\phi} + \delta\bar{\phi}$ and setting the first-order variation of it to zero, we get

$$\int d^4x (\partial_\mu \bar{\phi})(\partial_\mu \bar{\phi}) = 4 \int d^4x \, \mathcal{L}(\bar{\phi}) = 4S_B \tag{13.14}$$

Requiring that each ϕ_μ be normalized to unity determines the factor $b = S_{\mathrm{B}}^{-1/2}$.

The functions $\partial_\mu \bar{\phi}$ all have one node, hence none of them represents the lowest state. There must be at least one mode with a negative eigenvalue. Owing to the power $1/2$ in (13.12), the bounce solution contributes an imaginary part to the vacuum energy density. This means that the bounce solution is actually a saddle point of the action, not a local minimum.

If we somehow prepare the system in a state at or near ϕ_+ then it will decay by quantum tunneling, and this is reflected in the imaginary part of the energy. If the bounce solution is left out of the sum over states in the partition function then the energy density is real, as it must be if we explicitly sum over the energy eigenstates of the Hamiltonian. If the bounce solution is kept, its contribution should be isolated and identified as an instability of a state that does not belong to the spectrum of the Hamiltonian.

Putting Planck's constant back into our formulae for the moment, we realize that in the semiclassical limit the bounce solution is exponentially suppressed via the factor $\exp(-S_{\mathrm{B}}/\hbar)$. To lowest order in this small quantity, the imaginary part of the energy density is

$$I = \left(\frac{S_{\mathrm{B}}}{2\pi}\right)^2 \left| \frac{\det'[-\partial_\tau^2 - \nabla^2 + U''(\bar{\phi})]}{\det[-\partial_\tau^2 - \nabla^2 + U''(\phi_-)]} \right|^{-1/2} \exp(-S_{\mathrm{B}}) \qquad (13.15)$$

where the prime means that the four zero eigenvalues are omitted from the determinant. The first factor arises from the integration over the four zero modes. The factor involving the ratio of determinants has the dimension $1/\mathrm{length}^4$ since four eigenvalues are deleted from one of the operators, yielding an I with the proper dimensions of the number of tunnelings per unit time per unit volume. The exponential is the dominant factor in the tunneling, and is analogous to the Boltzmann factor in producing a critical-sized droplet in the classical nucleation rate.

Generally the classical equation of motion must be solved numerically to obtain the bounce solution, which is then used to compute the bounce action and the tower of eigenvalues of the fluctuation operator. However, in some circumstances one can make a thin-wall approximation to obtain the bounce solution, the action, and the negative eigenvalue. For example, consider the potential

$$U(\phi) = \lambda \left(\phi^2 - a^2\right)^2 + \frac{\epsilon}{2a}(\phi - a) \qquad (13.16)$$

where ϵ is a small quantity that represents the breaking of the reflection symmetry of the potential. To lowest order in this quantity, $\phi_\pm = \pm a$. The bounce solution has the behavior that it equals $-a$ for $\rho \ll R$ and a

for $\rho \gg R$, and crosses zero at $\rho = R$. This defines the four-dimensional radius of the bounce R. The approximate solution is

$$\bar\phi(\rho) = \begin{cases} -a & \rho \ll R \\ a\tanh\left(\dfrac{\rho - R}{2\xi}\right) & \rho \approx R \\ a & \rho \gg R \end{cases} \tag{13.17}$$

Here $\xi \equiv 1/\sqrt{8\lambda}a$ is the correlation length. This ought to be a good approximation when the wall thickness, characterized by ξ, is much less than the radius R. Substitution into the action yields

$$S_{\mathrm{B}} = -\frac{\pi^2\epsilon}{2}R^4 + \frac{\pi^2}{12\lambda\xi^3}R^3 \tag{13.18}$$

which displays the competition between the four-dimensional volume energy and the three-dimensional surface energy. The radius is determined by minimization:

$$R_{\mathrm{B}} = \frac{1}{8\lambda\epsilon\xi^3} \tag{13.19}$$

For self-consistency, we must therefore require $\epsilon \ll 1/(8\lambda\xi^4)$. The resulting action is

$$S_{\mathrm{B}} = \frac{\pi^2}{6}\epsilon R_{\mathrm{B}}^4 \tag{13.20}$$

The semiclassical calculation ought to be valid when $S_{\mathrm{B}} \gg 1$. A detailed calculation proves that there is one and only one negative eigenvalue, which is $-3/(2R_{\mathrm{B}}^2)$.

13.2 Classical nucleation

The classical theory of nucleation culminated in the work of Becker and Döring [1]; it was nicely reviewed by McDonald [8]. This theory was developed to describe the nucleation of a liquid droplet in a dilute yet supersaturated vapor.

The classical expression for the nucleation of a droplet of dense liquid in a dilute gas is

$$I = a(i_*)\left(\frac{|\Delta E''(i_*)|}{2\pi T}\right)^{1/2} n_1 \exp\left(\frac{-\Delta E(i_*)}{T}\right) \tag{13.21}$$

where $\Delta E(i_*)$ is the formation energy of a critical sized droplet consisting of i_* molecules, a prime denotes differentiation with respect to the number of molecules i, T is the temperature, n_1 is the density of single molecules, and $a(i_*)$ is the accretion rate of single molecules on a critical droplet. Usually the accretion rate is taken to be

$$a(i_*) = \tfrac{1}{2} n_1 \bar{v} 4\pi R_*^2 s \qquad (13.22)$$

which is the flux of particles ($\bar{v}$ is the mean speed of gas molecules) striking the surface of the critical droplet times a "sticking fraction" s less than unity. The first term in the nucleation rate is a dynamical factor influencing the growth rate, the second term characterizes fluctuations about the critical droplet, and the product of the third and fourth terms gives the quasi-equilibrium number density of critical-sized droplets. The energy is measured with respect to the gas molecules, so that $\Delta E(1) = 0$.

To extend the classical expression to the nucleation of a droplet in a somewhat denser gas, the first thing to do is to multiply the Boltzmann factor by the number of states available to the hot droplet:

$$e^{-\Delta E/T} \rightarrow e^{-\Delta E/T} e^{\Delta S} \qquad (13.23)$$

Owing to the thermodynamic identities $S = -dF/dT$ and $F = E - TS$, this modifies the Boltzmann factor to $e^{-\Delta F/T}$.

The size of the droplet can be characterized not by the number of molecules it contains but by its radius. Then integration over quadratic fluctuations about the mean size will give the prefactor

$$\left(\frac{|\Delta F''(R_*)|}{2\pi T} \right)^{1/2} \qquad (13.24)$$

The accretion rate must be multiplied by the increase in radius per particle absorbed to compensate for this change of variable. Upon absorption of one more particle, the droplet free energy changes by

$$\delta \Delta F = \Delta F'(R_*) \delta R + \tfrac{1}{2} \Delta F''(R_*) (\delta R)^2 \qquad (13.25)$$

The derivatives are evaluated at R_*, where the first derivative vanishes. The (Gibbs) free energy added by one gas molecule is just minus the pressure of the gas molecules divided by their number density. Therefore the accretion rate is multiplied by the factor

$$\delta R = \left(-\frac{P_1}{n_1 \Delta F''(R_*)} \right)^{1/2} \qquad (13.26)$$

Putting everything together we arrive at

$$I = 2\pi s \bar{v} R_*^2 n_1^2 \left(\frac{P_1}{n_1 \pi T} \right)^{1/2} \exp\left(\frac{-\Delta F_*}{T} \right) \qquad (13.27)$$

Generalizing to different species of molecules we write

$$I = 2\pi R_*^2 n_1 \exp\left(\frac{-\Delta F_*}{T}\right) \sum_j s_j \bar{v}_j n_j \left(\frac{P_j}{n_j \pi T}\right)^{1/2} \qquad (13.28)$$

where P_j is the partial pressure of the jth species, n_j is their density, etc. The quasi-equilibrium density of critical droplets is normalized to the density of the lightest species of particles, n_1. Note especially the appearance of R_*^2 in the prefactor. This arises from the fact that the absorption rate is proportional to the surface area. In contrast, when the growth rate is dominated by dissipation, as will be the case in Sections 13.3 and 13.4, the prefactor has only one power of R_*.

13.3 Nonrelativistic thermal nucleation

The theory of nucleation developed by Langer [2] starts with the introduction of a set of variables η_i, $i = 1, \ldots, N$, that describe N collective degrees of freedom of the system. We introduce a distribution function $\rho(\{\eta\}, t)$ that is a probability density for the configurations $\{\eta\}$ as a function of time t. We assume that $\rho(\{\eta\}, t)$ satisfies a continuity equation of the form

$$\frac{\partial \rho}{\partial t} = \partial_t \rho = -\sum_{i=1}^N \frac{\partial J_i}{\partial \eta_i} \qquad (13.29)$$

where the probability current is given by

$$J_i = -\sum_{j=1}^N \mathcal{M}_{ij}\left(\frac{\partial F}{\partial \eta_j}\rho + T\frac{\partial \rho}{\partial \eta_j}\right) \qquad (13.30)$$

Here $\mathcal{M}$ is a generalized mobility matrix and $F\{\eta\}$ is a coarse-grained free energy. Both of these quantities will be discussed in more detail below. Note that (13.29)–(13.30) can be derived via standard statistical techniques by adding a suitable Langevin force to the Hamiltonian equations of motion

$$\partial_t \eta_i = -\sum_{j=1}^N A_{ij}\frac{\partial F}{\partial \eta_j} \qquad (13.31)$$

where A is an antisymmetric matrix with entries 0 or 1.

The choice of variables η_i will depend on the problem. Generally one chooses the smallest set that describes the system to sufficient accuracy yet allows for a tractable analysis. The equilibrium configurations, for which $\partial_t \rho = 0$, have a probability distribution of the form

$$\rho_{\text{eq}}\{\eta\} \propto \exp\left(\frac{-F\{\eta\}}{T}\right) \tag{13.32}$$

Such configurations represent either the initial metastable point in the η-space denoted by $\{\eta_0\}$, or the final state. The phase transition starts from a metastable point $\{\eta_0\}$ and moves to the vicinity of a stable point, a point where F has its minimum. In this process the system is likely to pass a saddle point. The configuration at the saddle point, $\{\bar{\eta}\}$, is close to $\{\eta_0\}$ except for the presence of one critical-sized droplet of the new phase. At the saddle point we assume stationary flow, $\partial_t \rho = 0$, and calculate the current across this saddle. The rate of probability flow, $\{\bar{\eta}\}$, determines the droplet-formation rate in the system. This rate is

$$I = I_0 \exp\left(\frac{-\Delta F}{T}\right) \tag{13.33}$$

It gives the number of critical-sized droplets created in unit volume in unit time. The activation energy ΔF is given by

$$\Delta F = F\{\bar{\eta}\} - F\{\eta_0\} \tag{13.34}$$

The prefactor I_0 in (13.33) is the product of two terms:

$$I_0 = \frac{\kappa}{2\pi}\Omega_0 \tag{13.35}$$

Here the dynamical prefactor is κ (with dimension inverse time) and the statistical prefactor is Ω_0 (with dimension inverse volume). Langer showed that the statistical prefactor can be written as

$$\Omega_0 = \mathcal{V}\left(\frac{2\pi T}{|\bar{\lambda}_1|}\right)^{1/2} \prod_{\alpha=\alpha_0+2}^{N} \left(\frac{2\pi T}{\bar{\lambda}_\alpha}\right)^{1/2} \prod_{\alpha=1}^{N} \left(\frac{\lambda_\alpha^{(0)}}{2\pi T}\right)^{1/2} \tag{13.36}$$

Here $\mathcal{V}$ is the volume of η-space available for the flux of probability flow and $\{\bar{\eta}\}$ and $\{\eta_0\}$ are the eigenvalues of the matrix

$$\frac{\partial^2 F\{\eta\}}{\partial \eta_i \partial \eta_j}$$

evaluated at the points $\{\bar{\eta}\}$ and $\{\eta_0\}$. We will evaluate Ω_0 in the next section.

Since $\{\eta_0\}$ is a minimum of F, all the $\lambda_\alpha^{(0)}$ must be positive. Because $\{\bar{\eta}\}$ resides at the highest point along the path of lowest energy leading away from $\{\eta_0\}$, there is only one eigenvalue $\bar{\lambda}_\alpha$ that is negative. This is the eigenvalue denoted by $\bar{\lambda}_1$ in (13.36). If $F\{\eta\}$ has translational symmetry in three-space then there will be at least three other eigenvalues $\bar{\lambda}_\alpha$, which are zero. These correspond to the three independent translations of the position of the symmetry-breaking fluctuation (bubble or droplet) described by $\{\bar{\eta}\}$. The product of $\bar{\lambda}$'s appearing in (13.36) starts with $\alpha = \alpha_0 + 2$, where α_0 is the total number of symmetries of F which are broken by $\{\bar{\eta}\}$. The integration over these α_0 degrees of freedom defines the factor $\mathcal{V}$ in (13.36).

The dynamical factor κ is the exponential growth rate of the unstable mode $\{\bar{\eta}\}$. To compute κ, we linearize (13.31) about $\eta_i = \bar{\eta}_i$:

$$\partial_t \nu_i = -\sum_{j,l=1}^{N} A_{ij} \frac{\partial^2 F}{\partial \bar{\eta}_j \partial \bar{\eta}_l} \nu_l \qquad (13.37)$$

where $\nu_i = \eta_i - \bar{\eta}_i$. Then, setting $\nu \propto \mathrm{e}^{\kappa t}$, we identify κ as the positive eigenvalue of the matrix

$$-\sum_{j=1}^{N} A_{ij} \frac{\partial^2 F}{\partial \bar{\eta}_j \partial \bar{\eta}_l} \qquad (13.38)$$

In the nucleation problem the instability described by κ is the initial growth rate of a bubble or droplet that has just exceeded the critical size.

The dynamical prefactor has been calculated by Langer and Turski [9, 10] and by Kawasaki [11] for a liquid–gas phase transition near the critical point, where the gas is not dilute, to be

$$\kappa = \frac{2\lambda \sigma T}{\ell^2 n_\ell^2 R_*^3} \qquad (13.39)$$

This involves the thermal conductivity λ, the surface free energy σ, the latent heat per molecule ℓ, and the density of molecules in the liquid phase n_ℓ. The interesting physics in this expression is the appearance of the thermal conductivity. In order for the droplet to grow beyond the critical size, latent heat must be conducted away from the surface into the gas. For a relativistic system of particles or quantum fields that has no net conserved charge, such as baryon number, the thermal conductivity vanishes. The reason is that there is no rest frame defined by the baryon density to refer to heat transport. Hence this formula obviously cannot be applied to such systems.

13.4 Relativistic thermal nucleation

The relativistic quantum field theory approach for nucleation from one vacuum to another as worked out in Section 13.1 was extended by Affleck [4] and Linde [5] to finite temperature. In the limit where thermal fluctuations dominate quantum fluctuations the rate is

$$I = \frac{\omega_-}{\pi} \left(\frac{S_3}{2\pi T} \right)^{3/2} \left\{ \frac{\det'[-\nabla^2 + U''(\bar{\phi}, T)]}{\det[-\nabla^2 + U''(0, T)]} \right\}^{-1/2} \exp\left(\frac{-S_3}{T} \right) \quad (13.40)$$

where S_3 is the three-dimensional action associated with the formation of a critical-sized bubble or droplet. This follows from the assumption that the radius of the bubble is much larger than the inverse temperature β. It is assumed that the bounce solution depends on three-dimensional r instead of four-dimensional ρ, namely, $\bar{\phi}(\rho) \rightarrow \bar{\phi}(r)$. Integration over τ in the action just produces an overall factor $\beta = 1/T$. The factor ω_- is the frequency of the unstable mode. The ratio of determinants is almost never evaluated because it would have to be done numerically. Usually dimensional analysis is invoked to approximate this pre-exponential factor by T^4 or by T_c^4, so that

$$I \approx T^4 \, e^{-S_3/T} \qquad \text{or} \qquad I = T_c^4 \, e^{-S_3/T} \quad (13.41)$$

The expression (13.40) is very similar to the nucleation rate given by Langer for nonrelativistic systems, which itself is a generalization from the classical nucleation rate. It is our goal here to derive an expression that is fully relativistic, has Langer's rate formula as a nonrelativistic limit, and is expressed in terms of physically measurable observables such as surface energy, latent heat, transport coefficients, and so on. This involves the use of collective coordinates and coarse-graining.

The model of nucleation adopted here will be defined by the choice of the statistical variables, η_i, and the corresponding coarse-grained free energy $F\{\eta\}$. The conventional formulation of classical many-body statistical mechanics in terms of particle positions and momenta is not very convenient for the present purpose. Nucleation is characterized by semi-macroscopic fluctuations involving large numbers of particles. Therefore hydrodynamic-type collective variables are more appropriate to describe the formation of bubbles or droplets.

Hydrodynamics can be derived from microscopic kinetic theory by a coarse-graining or cellular method. That is, one divides up the macroscopic system into semimacroscopic cells of a given volume and assigns specific densities and flows to each of these cells. The free energy computed by performing a partition sum subject to the cellular constraints is the coarse-grained F that we are talking about. There is no problem, in principle, in summing over the cellular densities and flows to obtain the true equilibrium free energy. Moreover, as long as each cell comes to local thermal equilibrium rapidly compared with the times required for the hydrodynamic processes that one wants to consider, then one can use the coarse-grained F for computing nonequilibrium properties of the system.

The question that arises at this point is, what is a suitable size for the coarse-graining cells? In order for the hydrodynamic description to make sense, the cell volume must be much larger than the average volume per molecule. However, the cells cannot have linear dimensions appreciably larger than a correlation length. If the cells are chosen to be too large, phase separation will occur within single cells and the interesting details of the condensation mechanism will be lost in the process of taking cellular averages. To put this another way, we expect F as a function of the average energy density ϵ to be a nonconvex function with distinct minima corresponding to the two phases. But, if the cell size is large enough for well-defined phase separation to occur within a cell then F must approach its convex envelope and cannot possibly have the above property. We conclude that the cell size can be neither much larger nor much smaller than a correlation length.

13.4.1 Relativistic fluid dynamics

The equations of motion of relativistic fluid dynamics, $\partial_\mu T^{\nu\mu} = 0$, can be given in terms of $E \equiv T^{00}$ and $M^i = T^{0i}$, that is, $E = (\epsilon + Pv^2)\gamma^2$ and $\mathbf{M} = (\epsilon + P)\gamma^2 \mathbf{v}$, where ϵ is the energy density and P is the pressure; see Section 6.9. The low-speed limit of relativistic fluid dynamics ($\gamma^2 \approx 1$ and $Pv^2 \ll \epsilon$, but P not assumed small compared to ϵ) is given by

$$\partial_t \epsilon = -\nabla \cdot \mathbf{M} \tag{13.42}$$

and

$$\partial_t \mathbf{M} = -\nabla \cdot \left(\frac{1}{w} \mathbf{M} \otimes \mathbf{M} \right) - \nabla P \tag{13.43}$$

Here $w = \epsilon + P$ is the enthalpy density, and we have assumed that the relativistic energy density is $E = (\epsilon + Pv^2)\gamma^2 \approx \epsilon$ and that the relativistic momentum density is $\mathbf{M} = w\gamma^2 \mathbf{v} \approx w\mathbf{v}$. The low-speed limit of relativistic fluid dynamics finds applications not only in cosmology and astrophysics but also in terrestrial environments dominated by radiation processes, such as nuclear detonations, high-energy shock waves, and rocket engines.

With the above-mentioned restrictions in mind we will try to find a suitable form for the coarse-grained free energy F. This is not a trivial problem. We choose as our basic variables the local energy density and momentum density fields, $\epsilon(\mathbf{x}, t)$ and $\mathbf{M}(\mathbf{x}, t)$. The free energy F must consist of a kinetic energy F_K and an interaction term F_I. The kinetic term is simply

$$F_{\mathrm{K}}(\epsilon, \mathbf{M}) = \frac{1}{2} \int d^3x \ w\mathbf{v}^2 = \int d^3x \ \frac{\mathbf{M}^2}{2w} \tag{13.44}$$

We shall assume that F_{I} is a functional of ϵ only and that it can be written in the form

$$F_I[\epsilon(\mathbf{x})] = \int d^3x \left(\frac{1}{2}K(\nabla\epsilon)^2 + f(\epsilon) \right) \tag{13.45}$$

where $f(\epsilon)$ is the Helmholtz free energy density and $\frac{1}{2}K(\nabla\epsilon)^2$ is the usual gradient energy. The quantity K is a constant to be determined. Note that in this discussion we assume that the temperature T is constant.

Using the above F with the mobility matrix

$$
\begin{aligned}
\mathcal{M}_{ij} &= \partial_j(M_i) + (M_i)\partial_j - \frac{M_j}{2w}(\partial_i w) \\
\mathcal{M}_{i0} &= -\partial_i \epsilon \\
\mathcal{M}_{00} &= 0 \\
\mathcal{M}_{0i} &= (\partial_i w) + w\partial_i
\end{aligned}
\tag{13.46}
$$

the equations of motion for ϵ and $\mathbf{M}$ are obtained as the low-speed limit of relativistic fluid dynamics. The equation for energy conservation is

$$\partial_t \epsilon = -(\nabla w) \cdot \frac{\delta F_{\mathrm{K}}}{\delta \mathbf{M}(\mathbf{x})} - w\nabla \cdot \frac{\delta F_{\mathrm{K}}}{\delta \mathbf{M}(\mathbf{x})} = -\nabla \cdot \mathbf{M}(\mathbf{x}) \tag{13.47}$$

and the equation for momentum conservation, the Euler equation, is

$$
\begin{aligned}
\partial_t \mathbf{M} &= -\left[\nabla\mathbf{M} + \mathbf{M}\,\nabla - \frac{\mathbf{M}}{2w}\nabla w \right] \cdot \frac{\delta F_{\mathrm{K}}}{\delta \mathbf{M}(\mathbf{x})} + \frac{\delta F}{\delta \epsilon(\mathbf{x})}\nabla\epsilon \\
&= -\nabla \cdot \left(\frac{1}{w}\mathbf{M} \otimes \mathbf{M} \right) - K(\nabla^2\epsilon)\nabla\epsilon + \frac{\partial f}{\partial \epsilon}\nabla\epsilon
\end{aligned}
\tag{13.48}
$$

In the limit where we have a uniform system in equilibrium it is clear, from (13.43) and (13.48), that we must identify the last term on the right-hand side with the gradient of the pressure,

$$\frac{\partial f}{\partial \epsilon}\nabla\epsilon = \nabla f \longrightarrow -\nabla P \tag{13.49}$$

Note that when $\epsilon(\mathbf{x})$ is varying so slowly that the gradient energy can be neglected, (13.45) is consistent with

$$. \ f(\epsilon) = \epsilon - Ts = -P \tag{13.50}$$

13.4.2 Parametrization of the free energy

Imagine having two phases in equilibrium with each other at temperature T and, furthermore, that there is an interface separating them. This interface cannot be perfectly sharp. It must have a finite thickness of the order of a correlation length. In a local-density picture the energy density ϵ should vary smoothly from one phase to the other. Since first-order phase transitions have a latent heat, this means that we need to know the free energy density $f(\epsilon)$ for values of the energy density ranging between one phase and the other. To be specific, in what follows the low-temperature low-energy-density phase will be denoted by the subscript L, and the high-temperature high-energy-density phase will be denoted by the subscript H. In addition to the need to know $f(\epsilon)$ for $\epsilon_L < \epsilon < \epsilon_H$ we will also encounter situations where we need to know $f(\epsilon)$ for a range of values about ϵ_L and ϵ_H. Statistical fluctuations about local thermal equilibrium would require such knowledge, for example.

For a range of temperatures about T_c, $f(\epsilon)$ should have minima located at $\epsilon_L(T)$ and $\epsilon_H(T)$. There should also be a barrier between these two minima located at some $\epsilon_0(T)$. We require that

$$f(\epsilon_L(T)) = - P_L(T)$$
$$f(\epsilon_H(T)) = - P_H(T) \tag{13.51}$$

Therefore, at fixed T we shall parametrize $f(\epsilon)$ by a fourth-order polynomial in ϵ. Owing to the pinning of the two local minima shown above, $f(\epsilon)$ will have its global minimum at $\epsilon_H(T)$ when $T > T_c$ and its global minimum at $\epsilon_L(T)$ when $T < T_c$. At the critical temperature the two minima of $f(\epsilon)$ are equal. Our parametrization is

$$f(\epsilon) = f_0 + \frac{f_0''(\epsilon - \epsilon_0)^2}{2} - \frac{(\epsilon_L + \epsilon_H - 2\epsilon_0)f_0''}{3(\epsilon_L - \epsilon_0)(\epsilon_H - \epsilon_0)}(\epsilon - \epsilon_0)^3$$
$$+ \frac{f_0''}{4(\epsilon_L - \epsilon_0)(\epsilon_H - \epsilon_0)}(\epsilon - \epsilon_0)^4 \tag{13.52}$$

where $\epsilon_L(T)$, $\epsilon_H(T)$, $P_L(T)$ and $P_H(T)$ are specified functions of T and f_0'' is the curvature of f at the top of the barrier located at ϵ_0 ($f_0'' < 0$). Let us define $\Delta\epsilon \equiv \epsilon_H - \epsilon_L > 0$ and $\Delta P \equiv P_L - P_H$. In terms of these variables,

$$\epsilon_0 = \frac{\epsilon_L + \epsilon_H}{2} + \frac{f_0''(\Delta\epsilon)^3}{12\Delta P} \pm \left[\left(\frac{f_0''(\Delta\epsilon)^3}{12\Delta P} \right)^2 + \frac{(\Delta\epsilon)^2}{4} \right]^{1/2} \tag{13.53}$$

where $+$ ($-$) corresponds to $\Delta P > 0$ ($\Delta P < 0$) and

$$f_0 = -P_H + \frac{f_0''}{12} \frac{(\epsilon_H - \epsilon_0)^2(\epsilon_H - 2\epsilon_L + \epsilon_0)}{\epsilon_L - \epsilon_0} \tag{13.54}$$

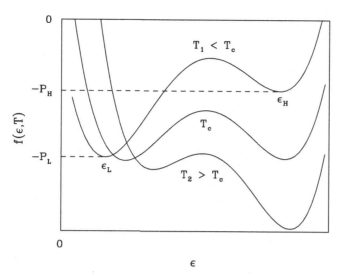

Fig. 13.2. Free-energy functional for extrapolating states away from equilibrium.

The first derivative of f is

$$f'(\epsilon) = \frac{\partial f}{\partial \epsilon} = \frac{f_0''(\epsilon - \epsilon_0)(\epsilon - \epsilon_L)(\epsilon - \epsilon_H)}{(\epsilon_L - \epsilon_0)(\epsilon_H - \epsilon_0)} \qquad (13.55)$$

Thus, if the location of the two minima and their depths are given for fixed T then only one free parameter, f_0'', remains. In particular, this parameter determines the barrier height, position, and curvature at all energy densities:

$$f''(\epsilon) = \frac{f_0''}{(\epsilon_L - \epsilon_0)(\epsilon_H - \epsilon_0)}$$
$$\times \left[(\epsilon - \epsilon_0)(\epsilon - \epsilon_L) + (\epsilon - \epsilon_0)(\epsilon - \epsilon_H) + (\epsilon - \epsilon_L)(\epsilon - \epsilon_H)\right] \quad (13.56)$$

See Figure 13.2 for illustrations of $f(\epsilon)$ when T is greater than, equal to, or less than T_c. Unless we can extract this free-energy function from the Lagrangian in a more fundamental way we shall be content to use this parametrization in the following analyses.

13.4.3 Surface profile

We restrict ourselves to the case of idealized bubbles or droplets. That is, we consider only the limit in which the nucleating fluctuation described by $\{\bar{\eta}\}$ is, indeed, a well-defined sphere of the L-phase with radius R large compared with the interface thickness or the correlation length ξ (to be defined below). In principle we need not make this restriction in the present theory. As we shall see, however, it is the appropriate one in the cases of interest here. By going to this limit we can do all our calculations analytically instead of having to resort to numerical methods.

The stationary point $\{\bar{\eta}\}$ is given by $\mathbf{v}(\mathbf{x}) = 0$ and $\epsilon(\mathbf{x}) = \bar{\epsilon}(\mathbf{x})$, where $\bar{\epsilon}$ satisfies

$$\frac{\delta F_{\mathrm{I}}}{\delta \bar{\epsilon}(r)} = -K\nabla^2\bar{\epsilon} + \frac{\partial f}{\partial \bar{\epsilon}} = 0 \qquad (13.57)$$

Given a spherical bubble of L-phase surrounded by H-phase at $T < T_{\mathrm{c}}$ the energy density $\bar{\epsilon}$ depends only on the distance r from the center of the bubble. Deep inside the bubble the energy density will be ϵ_{L}; far away from the bubble the energy density will be ϵ_{H}. The energy density profile $\bar{\epsilon}(r)$ then describes a smooth transition from one phase to the other. As discussed above, we will assume that the surface is located at a distance R from the center that is much greater than the surface thickness.

Using our parametrization of $f(\epsilon)$ the static profile equation becomes

$$-K\left(\frac{d^2}{dr^2} + \frac{2}{r}\frac{d}{dr}\right)\bar{\epsilon} + f_0''\frac{(\bar{\epsilon} - \epsilon_0)(\bar{\epsilon} - \epsilon_{\mathrm{L}})(\bar{\epsilon} - \epsilon_{\mathrm{H}})}{(\epsilon_{\mathrm{L}} - \epsilon_0)(\epsilon_{\mathrm{H}} - \epsilon_0)} = 0 \qquad (13.58)$$

We introduce a correlation length defined at the top of the barrier by $\xi_0^2 \equiv -K/f_0''$. Then

$$\frac{d^2\bar{\epsilon}}{dr^2} + \frac{2}{r}\frac{d\bar{\epsilon}}{dr} + \frac{(\bar{\epsilon} - \epsilon_0)(\bar{\epsilon} - \epsilon_{\mathrm{L}})(\bar{\epsilon} - \epsilon_{\mathrm{H}})}{\xi_0^2(\epsilon_{\mathrm{L}} - \epsilon_0)(\epsilon_{\mathrm{H}} - \epsilon_0)} = 0 \qquad (13.59)$$

Let us find the behavior of the solution in each of three regions.

(i) In the interior of the bubble $\bar{\epsilon} = \epsilon_{\mathrm{L}} + g_1(r)$; $g_1(r)$ is a small deviation from the equilibrium L-phase energy density satisfying

$$\frac{d^2 g_1}{dr^2} + \frac{2}{r}\frac{dg_1}{dr} - \xi_{\mathrm{L}}^{-2}g_1 = 0 \qquad (13.60)$$

where

$$\xi_{\mathrm{L}}^2 = \xi_0^2\frac{\epsilon_{\mathrm{H}} - \epsilon_0}{\Delta\epsilon} \qquad (13.61)$$

defines the correlation length in the L-phase. The solution of this equation is

$$g_1(r) = \frac{A_1}{r}\sinh\left(\frac{r}{\xi_{\mathrm{L}}}\right) + \frac{B_1}{r}\cosh\left(\frac{r}{\xi_{\mathrm{L}}}\right) \qquad (13.62)$$

From the requirement that the solution be finite at the origin we get $B_1 = 0$. In order to match onto the interface region, A_1 must be very small, proportional to $\mathrm{e}^{-R/\xi_{\mathrm{L}}}$. Then $\bar{\epsilon}(r) \approx \epsilon_{\mathrm{L}}$ throughout most of the interior.

(ii) Near R we can write $\bar{\epsilon} = \epsilon_0 + g_2(r)$. Linearizing in g_2 leads to

$$\frac{d^2 g_2}{dr^2} + \frac{2}{r}\frac{dg_2}{dr} + \xi_0^{-2}g_2 = 0 \tag{13.63}$$

The general solution is

$$g_2(r) = \frac{A_2}{r}\sin\left(\frac{r}{\xi_0}\right) + \frac{B_2}{r}\cos\left(\frac{r}{\xi_0}\right) \tag{13.64}$$

We require that $g_2(R) = 0$, which is equivalent to defining the location of the surface by the equation $\bar{\epsilon}(R) = \epsilon_0$. Thus the solution for $\bar{\epsilon}$ in the vicinity of the bubble's surface is

$$\bar{\epsilon} = \epsilon_0 + \frac{A_2}{r}\sin\left(\frac{r-R}{\xi_0}\right) \tag{13.65}$$

and $A_2 > 0$.

(iii) The exterior solution has the same functional form as in the interior except that $g_3(r \to \infty) = 0$ is required by the boundary condition. The exterior solution is therefore

$$\bar{\epsilon} = \epsilon_H - \frac{A_3}{r}e^{-r/\xi_H} \tag{13.66}$$

where

$$\xi_H^2 = \xi_0^2 \frac{\epsilon_0 - \epsilon_L}{\Delta\epsilon} \tag{13.67}$$

defines the correlation length in the H-phase and $A_3 > 0$.

At the critical temperature $f(\epsilon_L) = f(\epsilon_H)$. Then the free energy becomes symmetric, $\epsilon_0 = (\epsilon_L + \epsilon_H)/2$, and $\xi_H^2 = \xi_L^2 = \xi_0^2/2$. In this case the interfacial profile has a nice analytical solution in the planar $(R \to \infty)$ limit:

$$\bar{\epsilon}(x) = \frac{1}{2}\left[\epsilon_L + \epsilon_H + \Delta\epsilon \tanh\left(\frac{x}{2\xi_H}\right)\right] \tag{13.68}$$

Here the surface is located at $x = 0$ with L-phase on the left and H-phase on the right.

Suppose that an L-phase bubble has formed in the H-phase at $T < T_c$ because of statistical fluctuations. The change in free energy of the system is

$$\Delta F = \frac{4\pi}{3}(f_L - f_H)R^3 + 4\pi R^2 \sigma \tag{13.69}$$

where σ is the surface free energy. For baryon free matter,

$$\Delta F = \frac{4\pi}{3}[P_H(T) - P_L(T)]R^3 + 4\pi R^2 \sigma \tag{13.70}$$

The hadronic droplet is stationary if $\partial_R \Delta F = 0$, which leads to Laplace's formula

$$P_{\mathrm{L}}(T) - P_{\mathrm{H}}(T) = \frac{2\sigma}{R(T)} \tag{13.71}$$

Thus the activation energy, in our approximation, is

$$\Delta F = \frac{4}{3}\pi\sigma R^2 \tag{13.72}$$

The surface free energy can be calculated from our parametrization of F_{I}. For a planar interface or for a sphere whose radius is much greater than its surface thickness the formula was given by Cahn and Hilliard [12]:

$$\sigma = K \int_{-\infty}^{\infty} dx \left(\frac{d\bar{\epsilon}}{dx}\right)^2 \tag{13.73}$$

Inserting the solution for the planar interface at $\mathrm{T_c}$, this integral takes the form

$$\sigma = K \left(\frac{\Delta\epsilon}{2}\right)^2 \frac{1}{2\xi_{\mathrm{H}}} \int_{-\infty}^{\infty} dz \frac{1}{\cosh^4 z} = \frac{K(\Delta\epsilon)^2}{6\xi_{\mathrm{H}}} \tag{13.74}$$

The correlation length and the surface free energy determine the parameters $-f_0''$ and K in the coarse-grained free energy. In principle these parameters are temperature dependent. Their temperature dependence is, however, generally difficult to obtain.

13.4.4 The prefactor

The prefactor is a product of two terms: the statistical prefactor and the dynamical prefactor. The statistical prefactor, Ω_0, is a measure of both the available phase space as the system goes over the saddle and of statistical fluctuations at the saddle relative to the equilibrium states. The dynamical prefactor, κ, is the exponential growth rate of the bubble or droplet at the saddle point. This is the more difficult to calculate. We shall evaluate it using techniques exactly analogous to those employed by Turski and Langer [9, 10].

The general expression for the statistical prefactor was given in (13.36). To evaluate it, we first consider the eigenvalues of the matrix of second derivatives of F, the λ_α. The $\lambda_\alpha^{(0)}$ are eigenvalues of the operator

$$\left.\frac{\delta^2 F_{\mathrm{I}}}{\delta\epsilon(\mathbf{x})\delta\epsilon(\mathbf{x}')}\right|_{\epsilon=\epsilon_{\mathrm{H}}} = \left(-K\nabla^2 + \frac{\partial^2 f}{\partial\epsilon_{\mathrm{H}}^2}\right)\delta(\mathbf{x} - \mathbf{x}') \tag{13.75}$$

Here by $\partial^2 f / \partial \epsilon_H^2$ we mean the second derivative of f with respect to ϵ at fixed temperature evaluated in the equilibrium H-phase. This is a measure of the fluctuations in the system and cannot be determined from knowledge of the equation of state alone. Since the right-hand side of (13.75) depends on $\mathbf{x}$ only through ∇^2, its eigenfunctions are plane waves, with wave vectors $\mathbf{q}$ and eigenvalues

$$\lambda_{\mathbf{q}}^{(0)} = K\mathbf{q}^2 + \frac{\partial^2 f}{\partial \epsilon_H^2} \qquad (13.76)$$

There is also a set of eigenvalues, formally to be included among the $\lambda_\alpha^{(0)}$, which come from the kinetic term F_K. In Langer and Turski [10] it was concluded that these eigenvalues are spurious; that is, they do not describe physically relevant fluctuations but only bulk motion of the system. Hence they do not appear in the final formula for any nucleation quantity.

At the saddle point, $\epsilon(\mathbf{x}) = \bar{\epsilon}(r)$, the operator

$$\left. \frac{\delta^2 F_I}{\delta \epsilon(\mathbf{x}) \delta \epsilon(\mathbf{x}')} \right|_{\epsilon = \bar{\epsilon}(r)} = \left(-K\nabla^2 + \frac{\partial^2 f}{\partial \bar{\epsilon}^2} \right) \delta(\mathbf{x} - \mathbf{x}') \qquad (13.77)$$

is no longer translationally invariant because of the r-dependence of $\bar{\epsilon}$. As was discussed by Langer [13], the resulting spherically symmetric Schrödinger-like eigenvalue equation has an s-wave ground state with a radial eigenfunction proportional to $d\bar{\epsilon}/dr$ and a negative eigenvalue

$$\bar{\lambda}_1 \approx -\frac{2K}{R^2} \qquad (13.78)$$

This eigenstate is associated with the instability of the critical bubble against uniform expansion or contraction. The next states are the three p-waves, with eigenvalues $\bar{\lambda} = 0$, which occur because of the broken translational symmetry. Then there are higher-order partial waves with positive $\bar{\lambda}$ corresponding to volume-conserving deformations of the shape of the droplet. Finally, there is a continuum of nonlocalized eigenfunctions starting at $\bar{\lambda} = \partial^2 f / \partial \epsilon_H^2$. These eigenfunctions are similar to the states associated with the $\lambda^{(0)}$ in that they describe fluctuations in the bulk plasma but here these fluctuations are perturbed by the presence of the bubble. As before, the eigenvalues associated with the kinetic part of F are spurious and can be disregarded.

We can recognize the products over α in (13.36) as representing fluctuation corrections to the mean field excess free energy of the bubble. If we were to evaluate ΔF using measured values of the surface energy and thermodynamic potential, it would be inconsistent to include fluctuation corrections to ΔF in the prefactor Ω_0. Strictly speaking, the

nucleation formula used here requires that ΔF be first evaluated at the
stationary point obtained from (13.57), and then corrected by the fluctuation terms in Ω_0. But this procedure would imply that the radius
of the critical droplet is determined by the expression for σ given in
(13.73), which is not necessarily the same as the experimental surface
free energy because of the fluctuation corrections. What we shall do,
instead, is to delete the explicit fluctuation terms in Ω_0 and interpret
σ everywhere as the true surface energy; we shall make a similar assumption concerning other thermodynamic quantities that appear. Possibly
this procedure can be justified by going beyond the Gaussian approximations for η-space integrations which were used in deriving (13.36); that is,
by constructing a renormalized perturbation expansion in the neighbor-
hood of $\{\bar{\eta}\}$. If this program can be carried out, we might also be able
to compute systematically curvature corrections to the surface energy.
These corrections will be omitted here, and we shall focus our attention
on other ingredients of the nucleation formula, particularly the dynamical
prefactor.

Note that there are $\alpha_0 + 1 = 4$ more terms in the product over the $\lambda_\beta^{(0)}$
than in the product over the $\bar{\lambda}_\alpha$ in (13.36). This means that the logarithm
of the combined products is not precisely a free-energy difference. To see
what is happening here, it is useful to think in terms of a one-to-one
pairing between the $\lambda_\beta^{(0)}$ and the $\bar{\lambda}_\alpha$. At the top of the spectra (large positive $\lambda^{(0)}$ and $\bar{\lambda}$) both kinds of eigenvalue correspond to short-wavelength
fluctuations that extend throughout the volume of the system V. We can
pair these eigenvalues so that their contributions cancel each other to
the extent that the droplet volume is negligible compared with the total
volume of the system. At the bottom of the continuum a finite set of
$\bar{\lambda}$ values, which correspond to localized deformations of the bubble, fall
appreciably below their associated $\lambda^{(0)}$ values. Thus, by pairing the λ's as
described, the correction to ΔF remains of order R^3 in the limit $V \to \infty$,
as it must. This procedure leaves four unpaired $\lambda^{(0)}$'s at the bottom of the
spectrum that are not accounted for by the revised ΔF. Specifically, we
have

$$\lim_{V \to \infty} \prod_{\beta=1}^{4} \left(\frac{\lambda_\beta^{(0)}}{2\pi T} \right)^{1/2} = \left(\frac{1}{2\pi T} \frac{\partial^2 f}{\partial \epsilon_{\mathrm{H}}^2} \right)^2 \qquad (13.79)$$

remaining as the sole explicit contribution from the complicated products
over the α.

Having written down the value for $\bar{\lambda}_1$, we need only evaluate the factor
$\mathcal{V}$ to complete the calculation of Ω_0. The formula for $\mathcal{V}$ was given by

Langer [2, 13]:

$$\mathcal{V} = V \left[\frac{1}{3} \int dr (\nabla \bar{\epsilon})^2 \right]^{3/2} = V \left[\frac{4\pi R^2 \sigma}{3K} \right]^{3/2} \tag{13.80}$$

Here we have made use of the fact that $d\bar{\epsilon}/dr$ is appreciable only in a narrow region near $r = R$, where R is the radius of the bubble.

The resulting expression for Ω_0 is

$$\Omega_0 = V \left(\frac{4\pi R^2 \sigma}{3K} \right)^{3/2} \left(\frac{\pi T R^2}{K} \right)^{1/2} \left(\frac{1}{2\pi T} \frac{\partial^2 f}{\partial \epsilon_H^2} \right)^2 \tag{13.81}$$

Identifying the correlation length ξ_H in the H-phase by

$$\frac{1}{K} \frac{\partial^2 f}{\partial \epsilon_H^2} = \frac{1}{\xi_H^2} \tag{13.82}$$

we can write (13.81) in the form

$$\Omega_0 = \frac{2}{3\sqrt{3}} \left(\frac{\sigma}{T} \right)^{3/2} \left(\frac{R}{\xi_H} \right)^4 V \tag{13.83}$$

If one considers the nucleation rate to be per unit volume then the volume V should be divided out of the above expression. Usually we do mean the rate per unit volume and so Ω_0 will not include the factor V in subsequent discussion.

The dynamical prefactor κ should be obtained as the positive eigenvalue of the matrix given in (13.38). Using the mobility matrix and the fact that the bubble solution is spherically symmetric and satisfies (13.57), one finds that $\kappa = 0$. This means that the bubble does not grow. The reason was discussed by Langer and Turski [10]. In order for a bubble (or droplet) to grow, latent heat must be transported away from the surface region: for the nonrelativistic systems they were considering, they discovered that heat conduction was necessary to allow for growth. This eventually led to (13.39), which says that κ is proportional to the thermal conductivity λ. It is clear that to get our bubble to grow we must include the effects of dissipation in the dynamics.

We now want to determine the equations of motion of dissipative fluid dynamics (Section 6.9) for small deviations about the stationary configuration $\epsilon(\mathbf{x}, t) = \bar{\epsilon}(r), \mathbf{v}(\mathbf{x}, t) = 0$. To that end we write $\epsilon = \bar{\epsilon}(r) + \nu(\mathbf{x}, t)$ and $\mathbf{v} = \mathbf{v}(\mathbf{x}, t)$ and linearize the full equations of motion, including the gradient term F_K, in terms of ν and $\mathbf{v}$:

$$\partial_t \nu = -\nabla \cdot \mathbf{M} = -\nabla \cdot (\bar{w}\mathbf{v}) \tag{13.84}$$

$$\partial_t (\bar{w}\mathbf{v}) = \nabla \bar{\epsilon} \left[-K \nabla^2 \nu + f'' \nu \right] + \nabla \left[(\zeta + 4\eta/3) \nabla \cdot \mathbf{v} \right] \tag{13.85}$$

Hereafter when we write f, f', or f'' we intend that they be evaluated at the stationary configuration, so that they are complicated functions of r.

To determine κ we look for radial perturbations of the form

$$\nu(\mathbf{x}, t) = \nu(r) e^{\kappa t} \tag{13.86}$$

$$\mathbf{v}(\mathbf{x}, t) = v(r) \hat{r} e^{\kappa t} \tag{13.87}$$

These radial deviations are governed by the equations of motion

$$\kappa \nu(r) = -\frac{1}{r^2} \frac{d}{dr} \left[r^2 \bar{w} v(r) \right] \tag{13.88}$$

and

$$\kappa \bar{w} v(r) = -\frac{d\bar{\epsilon}}{dr} \left[-K \left(\frac{d^2}{dr^2} + \frac{2}{r} \frac{d}{dr} \right) + f'' \right] \nu(r)$$
$$+ \frac{d}{dr} \left\{ \left(\zeta + \frac{4\eta}{3} \right) \frac{1}{r^2} \frac{d}{dr} \left[r^2 v(r) \right] \right\} \tag{13.89}$$

Eliminating $\nu(r)$ using the first equation we obtain a linear third-order differential equation for the velocity profile:

$$\kappa^2 \bar{w} v(r) = -\frac{d\bar{\epsilon}}{dr} \left[K \left(\frac{d^2}{dr^2} + \frac{2}{r} \frac{d}{dr} \right) - f'' \right] \left\{ \frac{1}{r^2} \frac{d}{dr} \left[r^2 \bar{w} v(r) \right] \right\}$$
$$+ \frac{d}{dr} \left\{ \kappa \left(\zeta + \frac{4\eta}{3} \right) \frac{1}{r^2} \frac{d}{dr} \left[r^2 v(r) \right] \right\} \tag{13.90}$$

Self-consistent solutions of this equation, together with the boundary conditions, should provide us with the allowed values of κ. Unfortunately, it is not a trivial equation to solve. Therefore we will first analyze the behavior of the solution in three regions: the interior of the bubble, the exterior of the bubble, and the surface region. We first note a constraint that follows from (13.88) and the conditions that $v(r)$ vanishes at the origin and at infinity, namely

$$\int_0^\infty dr \, 4\pi r^2 \nu(r) = 0 \tag{13.91}$$

In the interior region, from the origin to within a few correlation lengths of the surface, recall that $\bar{\epsilon} \approx$ constant. Then the first term on the right-hand side of (13.90) vanishes, and the equation for $v(r)$ reduces to

$$r^2 v'' + 2rv' - (a_L^2 r^2 + 2)v = 0 \tag{13.92}$$

where $a_L^2 = \kappa w_L (\zeta_L + 4\eta_L/3)^{-1}$. The general solution of this differential equation is

$$v(r) = A \left(\frac{a_L}{r} - \frac{1}{r^2} \right) e^{a_L r} + B \left(\frac{a_L}{r} + \frac{1}{r^2} \right) e^{-a_L r} \tag{13.93}$$

where A and B are constants. We must require that v and v' vanish at $r = 0$. Consequently both A and B are zero, so that the velocity vanishes in the interior of the bubble. This is true to the extent that $\bar{\epsilon} = \text{constant}$ in this region.

In the exterior region, far outside the surface, the energy and enthalpy densities approach their equilibrium values in the bulk H-phase, $\bar{\epsilon} \to \epsilon_H$ and $\bar{w} \to w_H$. Then the first term on the right-hand side of (13.90) can again be neglected as a first approximation. The solution with the correct large-r behavior is

$$v(r) = C \left(\frac{a_H}{r} + \frac{1}{r^2} \right) e^{-a_H r} \tag{13.94}$$

where C is a constant and $a_H^2 = \kappa w_H (\zeta_H + 4\eta_H/3)^{-1}$.

In the region of the surface, $r \approx R$, the stationary configuration $\bar{\epsilon}(r)$ is varying rapidly and $d\bar{\epsilon}/dr$ is nonzero. Therefore, unlike in the deep interior or the exterior of the bubble, the first term on the right-hand side of (13.90) cannot be dropped. In fact, as we shall see, κ is proportional to the viscosity, which we assume to be very small. Then the other two terms in the equation are of second order in the viscosity, and we shall ignore them. Thus, to good approximation, in the surface region $\nu(r)$ satisfies

$$\left(-K\nabla^2 + f'' \right) \nu(r) = 0 \tag{13.95}$$

Given that $\bar{\epsilon}(r)$ satisfies (13.57) and that $\nu(r)$ must go to zero at the origin and at infinity, the solution to the above equation is

$$\nu(r) \sim \frac{d\bar{\epsilon}}{dr} \tag{13.96}$$

Together with (13.88) this implies that in the surface region

$$v(r) = \frac{D}{r^2 \bar{w}(r)} \int_0^r dr' r'^2 \frac{d\bar{\epsilon}}{dr'} \tag{13.97}$$

where D is a constant. For distances r which exceed the bubble radius R by more than a few correlation lengths but which are less than $2R$, (13.97) can be integrated to give

$$v(r) \approx \frac{D \Delta\epsilon}{w_H} \frac{R^2}{r^2} \tag{13.98}$$

Remember that, as always, we are assuming weak to moderate supercooling, so that $R \gg \xi$.

It is necessary to distinguish between the actual radius of the bubble, R, and the radius of the bubble in the stationary or metastable configuration, R_*, determined by Laplace's formula. If the stationary bubble is perturbed only slightly then the energy-density profile will change by only a minute amount. The transport of heat away from the surface will be a very slow process because of the assumed smallness of the viscosity. As the bubble slowly begins to expand, the energy-density profile will not change much, but the profile moves out a small distance dR in a time dt. The energy flux density (the energy per unit area per unit time) that must be transported outwards is $\Delta w\, dR/dt$. Here we do not distinguish between the difference in energy densities and the difference in enthalpy densities of the two bulk phases because the pressure difference is small compared with the energy-density differences; we shall refer to them interchangeably as the latent heat. This energy flux must be balanced by that due to dissipation, which is $-(\zeta + 4\eta/3)v\, dv/dr$. We will evaluate the flow velocity just outside the surface of the bubble. According to (13.98) the derivative is $dv/dr \approx -2v/R$. Therefore energy balance gives us the relation

$$\Delta w \frac{dR}{dt} = 2\left(\zeta_{\mathrm{H}} + \frac{4\eta_{\mathrm{H}}}{3}\right)\frac{v^2}{R} \qquad (13.99)$$

The outward momentum flux density (the momentum per unit area per unit time) is $\Delta w\, v^2$. (This neglects a small contribution from viscous terms that can be considered to be a higher-order effect.) The momentum flux density must be equated to the force per unit area, which comes from the Laplace formula

$$\Delta w\, v^2 = 2\sigma\left(\frac{1}{R_*} - \frac{1}{R}\right) \qquad (13.100)$$

Again, the velocity is to be evaluated just outside the surface.

Using both energy and momentum conservation we can eliminate the velocity and solve for dR/dt:

$$\frac{dR}{dt} = \frac{4(\zeta_{\mathrm{H}} + 4\eta_{\mathrm{H}}/3)\sigma(R - R_*)}{(\Delta w)^2\, R^2\, R_*} \qquad (13.101)$$

This is a differential equation for $R(t)$, from which we can read off the value of κ. It is

$$\kappa = \frac{4\sigma(\zeta_{\mathrm{H}} + 4\eta_{\mathrm{H}}/3)}{(\Delta w)^2\, R_*^3} \qquad (13.102)$$

This may be considered the principal result of this section.

Putting it all together gives the nucleation rate

$$I = \frac{4}{\pi}\left(\frac{\sigma}{3T}\right)^{3/2} \frac{\sigma(\zeta_{\mathrm{H}} + 4\eta_{\mathrm{H}}/3)R_*}{\xi_{\mathrm{H}}^4(\Delta w)^2} \mathrm{e}^{-\Delta F/T} \qquad (13.103)$$

where $\Delta F = 4\pi\sigma R_*^2/3$ and R_* is given by the Laplace formula (13.71). This is the probability per unit volume per unit time of nucleating an L-phase bubble out of the H-phase. If one considers nucleating an H-phase droplet in the L-phase instead, one just needs to evaluate the correlation length and the viscosities in the L-phase rather than the H-phase. At the critical temperature, $R_* \to \infty$, and the rate vanishes because of the exponential. The system must supercool a minute amount at least in order that the rate attain a finite value. Note that at the critical temperature the pre-exponential factor is linearly divergent in R_*, which is qualitatively unlike the simple dimensionless estimate of a constant I_0.

Venugopalan and Vischer [7] extended the calculation of κ to incorporate a net baryon number and therefore the effect of thermal conduction. The result is

$$\kappa = \frac{2\sigma[\lambda_H T + 2(\zeta_H + 4\eta_H/3)]}{(\Delta w)^2 R_*^3} \qquad (13.104)$$

This is proportional to a linear combination of the three dissipation coefficients. It reduces to the expression derived above when thermal conduction can be neglected and to the expression of Langer and Turski in the nonrelativistic limit and when shear and bulk viscosities are small.

This completes our calculation and analysis of the thermal nucleation rate for systems with zero or negligibly small baryon number. In a subsequent chapter we shall use it in a set of rate equations for the time evolution of phase transitions in the early universe and in ultrarelativistic nuclear collisions.

13.5 Black hole nucleation

In a beautiful and original work Gross, Perry, and Yaffe [14] calculated the nucleation rate for black holes in a thermal bath of gravitons. Their result is

$$I = 1.752\, T \left(\frac{M_0}{T}\right)^{212/45} \left(\frac{m_P}{4\pi}\right)^3 \exp\left(\frac{-m_P^2}{16\pi^2 T^2}\right) \qquad (13.105)$$

where $m_P \equiv G^{-1/2}$ is the Planck mass and G is Newton's constant. The quantity M_0 is a regulator mass, undetermined in the pure Einstein theory but supposed to be of the order of m_P in a more complete quantum theory of gravitation. Physically the reason for this instability of flat space is that statistical fluctuations will produce small black holes. According to Hawking [15] the effective temperature of a black hole is $m_P^2/8\pi M$ where M is its mass. If the mass is too large then the black hole temperature will be smaller than that of its surroundings and it will accrete matter. If the mass is too small, the black hole temperature will be greater than its surroundings and it will evaporate and eventually explode. The critical mass for this unstable equilibrium is $M_* = m_P^2/8\pi T$.

The calculation of the nucleation rate by Gross, Perry, and Yaffe is based upon small fluctuations about a Schwarzschild instanton in a path-integral formulation of Einstein's theory. There is one negative eigenvalue, which gives rise to the instability of flat space. The calculation is at the same time elegant and lengthy. However, the main features of the result can be obtained from the classical theory of nucleation [16].

Consider a volume V with gravitons at temperature T. The probability that a fluctuation will produce a black hole of critical mass is $\exp(-\Delta F)$, where ΔF is the change in free energy of the system with T and V held fixed. Now $\Delta F = F_* - F_{\mathrm{g}}$, where F_* is the free energy of the black hole and F_{g} is the free energy of the thermal gravitons displaced by the black hole. The black hole free energy F_* is related to M_* by

$$M_* = F_* - T\frac{dF_*}{dT} \tag{13.106}$$

or

$$F_* = \frac{M_*}{2} = \frac{m_{\mathrm{P}}^2}{16\pi T} \tag{13.107}$$

whereas F_{g} is given by

$$F_{\mathrm{g}} = -\frac{\pi^2}{45}T^4\frac{4\pi}{3}r^3 \tag{13.108}$$

where r is of the order of or slightly greater than the Scharzschild radius. Thus F_{g}/T is of the order of 10^{-2} to 10^{-3} and will be neglected.

Knowing the probability for one statistical fluctuation, we can estimate the density for fluctuations to occur. Consider quantum density fluctuations on the smallest scale possible, namely, the Planck wavelength $\lambda_{\mathrm{P}} = 2\pi/m_{\mathrm{P}}$. Imagine a cube with fluctuations spaced $\lambda_{\mathrm{P}}/2$ apart. The quantum density of fluctuations necessary to produce a black hole of critical mass is then estimated to be

$$n_* = \left(\frac{m_{\mathrm{P}}}{\pi}\right)^3\exp\left(\frac{-m_{\mathrm{P}}^2}{16\pi T^2}\right) \tag{13.109}$$

The rate of change of n_* can be calculated as

$$\frac{dn_*}{dt} = \frac{1}{T}\left|\frac{dM_*}{dt}\right|n_* = \frac{m_{\mathrm{P}}^2}{8\pi T^3}n_* \tag{13.110}$$

The rate of increase in the black hole mass may be estimated by the rate at which gravitons cross the Schwarzschild radius R_{S}:

$$\frac{dM_*}{dt} = 2\times 4\pi R_{\mathrm{S}}^2\int_{\mathrm{hemisphere}}\frac{d^3p}{(2\pi)^3}\frac{p}{\exp(p/T)-1} = \frac{\pi}{120}T^2 \tag{13.111}$$

Putting everything together we get

$$I = \frac{8\pi}{15} T \left(\frac{m_P}{4\pi}\right)^3 \exp\left(\frac{-m_P^2}{16\pi T^2}\right) \qquad (13.112)$$

Comparing (13.105) and (13.112) we see that in the former there still remains a factor $(M_0/T)^{212/45}$ to interpret. The origin of this term is a quantum correction to the free energy of the black hole [17]:

$$\frac{F_*^{\text{quantum}}}{T} = -\frac{106}{45} \chi \ln\left(\frac{M_0}{T}\right) \qquad (13.113)$$

The factor χ is a topological invariant of the space, being 2 for the Schwarzschild metric and 0 for flat space. The final formula derived heuristically is

$$I = \frac{8\pi}{15} T \left(\frac{M_0}{T}\right)^{212/45} \left(\frac{m_P}{4\pi}\right)^3 \exp\left(\frac{-m_P^2}{16\pi T^2}\right) \qquad (13.114)$$

It is remarkable that not only the functional dependence on T and m_P is reproduced, but also the absolute normalization is very close. This is more than could reasonably be expected.

13.6 Exercises

13.1 Show that the functions $\phi_\mu = b\partial_\mu\bar{\phi}$ are solutions to the equation of motion given in Section 13.1 and that they have one node.

13.2 Write down the false vacuum decay rate including explicitly Planck's constant.

13.3 Make a numerical estimate of the nucleation rate for a critical-sized water droplet in an atmosphere that is oversaturated by 10% at 10 degrees C.

13.4 Derive (13.40) along the same lines used to derive the vacuum decay rate.

13.5 Derive (13.88)–(13.90) and from them patch together an approximate solution for $v(r)$ valid from $r = 0$ to $r = \infty$.

13.6 Calculate the black hole formation rate with the inclusion of N_f massless spin-1/2 fermions and N_b massless spin-0 bosons.

References

1. Becker, R., and Döring, W., *Ann. Phys. (NY)* **24**, 719 (1935).
2. Langer, J. S., *Ann. Phys. (NY)* **54**, 258 (1969).
3. Coleman, S., *Phys. Rev. D* **15**, 2929 (1977); **16**, 1248(E) (1977); Callan, C. G., and Coleman, S., *ibid.* **16**, 1762 (1977).
4. Affleck, I., *Phys. Rev. Lett.* **46**, 388 (1981).

5. Linde, A. D., *Nucl. Phys.* **B216**, 421 (1983). There was an overall factor of $\omega_-/\pi T$ missing in the expression for I in this paper; see equation (13.40) of the present text.

6. Csernai, L. P., and Kapusta, J. I., *Phys. Rev. D* **46**, 1379 (1992).

7. Venugopalan, R., and Vischer, A. P., *Phys. Rev. E* **49**, 5849 (1994).

8. McDonald, J. E., *Amer. J. Phys.* **30**, 870 (1962); **31**, 31 (1963).

9. Turski, L. A., and Langer, J. S., *Phys. Rev. A* **22**, 2189 (1980).

10. Langer, J. S., and Turski, L. A., *Phys. Rev. A* **8**, 3230 (1973).

11. Kawasaki, K., *J. Stat. Phys.* **12**, 365 (1975).

12. Cahn, J. W., and Hilliard, J. E., *J. Chem. Phys.* **28**, 258 (1958); **31**, 688 (1959).

13. Langer, J. S., *Ann. Phys. (NY)* **41**, 108 (1967).

14. Gross, D. J., Perry, M. J., and Yaffe, L. G., *Phys. Rev. D* **25**, 330 (1982).

15. Hawking, S. W., *Nature* **248**, 30 (1974).

16. Kapusta, J. I., *Phys. Rev. D* **30**, 831 (1984).

17. Hawking, S. W. (1979). In *General Relativity: An Einstein Centenary Survey*, ed. Hawking, S. W., and Israel, W. (Cambridge University Press, Cambridge).

Bibliography

A nice review of nucleation theory in nonrelativistic and condensed matter systems is the following:

Gunton, J. D., San Miguel, M., and Sahni, P. S. (1983). In *Phase Transitions and Critical Phenomena*, eds. Domb, C., and Lebowitz, J. L. (Academic Press, London) Vol. 8.

14

Heavy ion collisions

The only practical way of creating and studying hot and dense strongly interacting matter in the laboratory is by colliding heavy nuclei at high energies. Some of the pioneering studies have used nuclear emulsion data of highly energetic cosmic ray events. However, a serious handicap there is the lack of control over the physical beam characteristics. For a few decades now, there has existed a vibrant experimental program seeking to explore the physics of nuclear collisions in different energy regimes and with different combinations of beam and target nuclei. The pioneering experiments at the Lawrence Berkeley National Laboratory (Berkeley, USA) have been followed by several other experimental ventures. It is impossible to enumerate all the facilities, but some important efforts at the high end of the energy spectrum have been pursued at the GSI (Darmstadt, Germany), CERN (Geneva, Switzerland), and at Brookhaven National Laboratory (Upton, USA). The Relativistic Heavy Ion Collider (RHIC) is located at BNL, and the Large Hadron Collider (LHC) has a heavy ion program expected to begin at CERN around 2007. A healthy experimental program in high energy nuclear collisions requires a basis in nucleon–nucleon and nucleon–nucleus collisions. These in fact constitute a crucial category of control experiments for the more complex nucleus–nucleus events. The study of strongly interacting matter at high temperature and density enjoys an active and fruitful collaboration between the experimental and theoretical communities.

In relativistic nuclear collisions, multiple scatterings involving both the primary constituents (the original nucleons) and the secondary particles (mostly created pions) can, in principle, drive the system towards a state of local thermodynamic equilibrium. The reason for this originates in the phenomenology of hadronic collisions. From those studies it is known that, at energies relevant for the applications considered in this chapter,

317

the large inelastic part of the nucleon–nucleon cross section will cause considerable energy loss of the colliding constituents. This energy loss ultimately translates into the creation of a large number of light mesons, mostly appearing in the central rapidity region, which is midway between the projectile and target fragmentation regions. The identity of the primordial fields first materializing at mid-rapidity (partons or composites) is not completely clear but should depend on the initial energy density. The key issue, however, is the following: because of the large particle multiplicities involved, the relativistic collisions of heavy nuclei will create zones of short mean free paths. This condition will pave the way to the statistical treatment of heavy ion collisions that we shall discuss in this chapter. We have seen that QCD predicts a transition from hot hadronic matter to quark–gluon plasma, provided that the energy density is large enough. We also will review some of the probes that have been proposed to study hot and dense systems and to determine whether a new state of matter has been created.

14.1 Bjorken model

Fermi was the first to apply statistical techniques to hadronic particle production in p–p collisions [1]. Shortly thereafter, the first application of relativistic hydrodynamics to a strongly–interacting system was made by Landau [2]. The power, elegance, and simplicity of hydrodynamics is essentially contained in the statement that the entire system can be described by a few macroscopic thermodynamic fields. The conditions necessary for this to be so are that any modification of the state of the system is reflected instantaneously in the fields. Quantitatively, this statement identifies any relaxation time as shorter than any other time scale in the system under scrutiny. Local thermal equilibrium is therefore assumed. We also assume that the net baryon number and electric charge are zero. Not only does this simplify the analysis but it is a very good approximation in high energy collisions because of the large number of particles produced.

We have already seen, in Chapter 6, that the energy–momentum tensor may be written as

$$T^{\mu\nu} = -Pg^{\mu\nu} + (\epsilon + P)u^\mu u^\nu \qquad (14.1)$$

where P is the pressure, ϵ is the energy density, and $u^\mu = (\gamma, \gamma\mathbf{v})$ is the local flow velocity relative to some fixed reference frame. In a frame in which the fluid is locally at rest, $u^\mu = (1, 0, 0, 0)$, $T^{00} = \epsilon$, $T^{ij} = P\delta_{ij}$, and $T^{i0} = 0$. The conservation of energy and momentum is expressed as

$$\partial_\mu T^{\mu\nu} = 0 \qquad (14.2)$$

This vector equation, (14.2), represents a set of four scalar equations. However, there are five unknown quantities: the three independent components of the flow velocity u^μ (the normalization condition $u^2 = 1$ defines three independent and one dependent component), the energy density, and the pressure. To close this system, another equation must be supplied, and this is the equation of state. This set of equations can always be solved numerically. However, their solution is a complicated task in three spatial dimensions unless simplifying assumptions are placed on the symmetry of the system. There is a wide body of literature devoted to the techniques used in numerical simulations using relativistic hydrodynamics.

Insight can be gained by considering some simple limits. Motivated by empirical observations, Bjorken [3] was led to explore the consequences of the existence of a central plateau structure in the inclusive particle production as a function of the spacetime rapidity y, defined as

$$y = \frac{1}{2} \ln \left(\frac{t + z}{t - z} \right) \tag{14.3}$$

where the z-axis is oriented along the beam direction. Theoretically, the existence of this plateau implies that the initial conditions, viewed at the same proper time after the beginning of the nuclear collision, are invariant with respect to Lorentz transformations along the longitudinal (or beam) direction.

Another assumption of the Bjorken scenario is that essentially all the baryon number is carried by the receding Lorentz-contracted nuclei that have just collided. The produced particles then occupy the central rapidity region and the high multiplicity will ensure rapid thermalization followed by hydrodynamic evolution. At this point it is appropriate to note that this approach is really a conceptual idealization. In actual practice, the manifest success of the hydrodynamic model in relativistic nuclear collisions at RHIC energies suggests a very early thermalization, even though the microscopic mechanisms that would drive it currently remain unclear.

In keeping with Bjorken's line of thought, we shall be interested in the early stages of the hydrodynamic development of the central collision of high-energy nuclei. There the flow can be assumed one dimensional, owing largely to the initial symmetry of the colliding system. At slightly later times, larger than those associated with the size of the nucleus ($t > 1.2 A^{1/3}$ fm/c), the rarefaction wave coming in from the nuclear surface will be fully formed and a three-dimensional expansion will set in. Therefore, the early solution will be independent of the rapidity, and nothing in the time evolution will spoil this symmetry. One may write the general solutions as $\epsilon(\tau)$, $P(\tau)$, $T(\tau)$, $u^\mu(\tau)$, with proper time $\tau = \sqrt{t^2 - z^2}$. Solving for t and z in terms of τ and y yields

$$t = \tau \cosh y \qquad z = \tau \sinh y \tag{14.4}$$

Then

$$u^\mu = \frac{dx^\mu}{d\tau} = (\cosh y, 0, 0, \sinh \tau) \tag{14.5}$$

and indeed, $u^\mu u_\mu = 1$. One may then write

$$u^\mu \frac{\partial \tau}{\partial x^\mu} = u^0 \frac{\partial \tau}{\partial t} + u^3 \frac{\partial \tau}{\partial z} = \cosh^2 y - \sinh^2 y \equiv 1$$

and

$$\frac{\partial \tau}{\partial x^\mu} = \frac{x_\mu}{\tau} \tag{14.6}$$

The equation for the conservation of energy and momentum is

$$\partial_\mu T^{\mu\nu} = \frac{\partial T^{\mu\nu}}{\partial x^\mu} = \frac{\partial(\epsilon + P)}{\partial \tau} \frac{\partial \tau}{\partial x^\mu} u^\mu u^\nu + (\epsilon + P) \frac{\partial u^\mu}{\partial x^\mu} u^\nu$$

$$+ (\epsilon + P) u^\mu \frac{\partial u^\nu}{\partial x^\mu} - g^{\mu\nu} \frac{\partial P}{\partial \tau} \frac{\partial \tau}{\partial x^\mu} = 0 \tag{14.7}$$

With the help of (14.6), this reduces to

$$\frac{\partial \epsilon}{\partial \tau} + \frac{\epsilon + P}{\tau} = 0 \tag{14.8}$$

Defining an entropy density $s = S/V = (\epsilon + P)/T$ and using the facts that $u^\mu \partial/\partial x^\mu = d/d\tau$ and that at constant volume $d\epsilon = T ds$, we may rewrite the above equation as

$$\frac{ds}{d\tau} + \frac{s}{\tau} = 0 \tag{14.9}$$

the solution of which clearly satisfies

$$\frac{s(\tau)}{s(\tau_0)} = \frac{\tau_0}{\tau} \tag{14.10}$$

Also implied by (14.9) is

$$\frac{\partial (s u^\mu)}{\partial x^\mu} = \partial_\mu s^\mu = 0 \tag{14.11}$$

Entropy is therefore a conserved quantity. Furthermore, since a volume element in this geometry is $dV = d^2 x_\perp \tau dy$, (14.10) also means that the entropy per unit rapidity, dS/dy, is a constant with respect to proper time.

Let us now study the time evolution implicit in the formalism we have just written down. We start by considering the case of a first-order phase transition. Our pragmatic approach will be to describe the quark–gluon plasma as a noninteracting gas of eight massless gluons and two flavors (u, d) of massless quarks. Note that massive strange quarks could also be included self-consistently. A bag constant B [4] is used to simulate the effect of confinement in the hadron phase, which is described as a noninteracting gas of massless pions. Thus the pressure, energy density, and entropy density in each of the two phases are

$$P_q = 37aT^4 - B \qquad \epsilon_q = 111aT^4 + B \qquad s_q = 148aT^3$$
$$P_h = 3aT^4 \qquad \epsilon_h = 9aT^4 \qquad s_h = 12aT^3 \tag{14.12}$$

where $a = \pi^2/90$. The critical temperature is determined by pressure balance to be

$$T_c = \left(\frac{B}{34a}\right)^{1/4} \tag{14.13}$$

Thus B may be eliminated in favor of T_c. The latent heat necessary to liberate the color degrees of freedom is $4B$.

We may write

$$\frac{d\epsilon}{d\tau} = \frac{d\epsilon}{dP}\frac{dP}{dT}\frac{dT}{d\tau} = -\frac{sT}{\tau} \tag{14.14}$$

where (14.8) has been used. Note that $dP = sdT$ at constant volume. The sound velocity is

$$v_s^2 = \frac{dP}{d\epsilon} \tag{14.15}$$

Putting all this together,

$$\frac{1}{T}\frac{dT}{d\tau} = -\frac{v_s^2}{\tau} \tag{14.16}$$

which yields

$$T = T_0 \left(\frac{\tau_0}{\tau}\right)^{v_s^2} \tag{14.17}$$

For the equations of state in (14.12), $v_s^2 = 1/3$ except at T_c. At T_c it is necessary to specify in addition the volume fraction f of the quark–gluon phase. The entropy density is

$$s(f, T_c) = s_q(T_c)f + s_h(T_c)(1 - f) \tag{14.18}$$

and similarly for the energy density.

We assume now that the nucleus–nucleus collision produces a quark–gluon plasma with initial entropy density $s_0 > s_q(T_c)$. The temperature evolves according to (14.17) until T drops to T_c. This occurs in the proper time interval $\tau_0 < \tau \leq \tau_1 = (T_0/T_c)^3 \tau_0$. Assuming that the nucleation of the hadron phase is fast, the system then enters the mixed phase. In the mixed phase the entropy density decreases, not by decreasing T but by converting quark–gluon plasma to hadron matter at lower entropy density but still at T_c. The fraction $f(\tau)$ is easily derived to be

$$f(\tau) = \frac{1}{r-1} \left(r\frac{\tau_1}{\tau} - 1 \right) \tag{14.19}$$

where $r = 37/3$ is the ratio of number of degrees of freedom in the two phases. Thus $1 > f > 0$ for $\tau_1 < \tau < \tau_2 = r\tau_1$. The mixed phase terminates at τ_2 whereupon the temperature begins to fall again according to

$$T(\tau) = T_c \left(\frac{\tau_2}{\tau} \right)^{1/3} \tag{14.20}$$

for $\tau > \tau_2$. The expansion continues until the pions can no longer maintain thermal contact. One can take this as a final breakup temperature T_f, also called the freezeout temperature. In totally dynamical simulations of the nuclear collision this sharp cutoff is avoided.

In the case $s_q(T_c) > s_0 > s_h(T_c)$ we assume that the matter is initially formed in the mixed phase, with volume fraction f_0 determined by

$$s = s_q(T_c)f_0 + s_h(T_c)(1 - f_0) \tag{14.21}$$

It follows that

$$f(\tau) = \frac{1}{r-1} \left\{ [1 + (r - 1)f_0] \frac{\tau_0}{\tau} - 1 \right\} \tag{14.22}$$

The system evolves in the mixed phase until $\tau_2 = [1 + (r - 1)f_0]\tau_0$. The evolution then follows (14.17) in the hadron phase.

Let us now suppose that the equation of state leads to a second-order phase transition. We parametrize the effective number of massless bosonic degrees of freedom in each of the two phases as

$$N_h(T) = 3 + be^{(T-T_c)/d} \qquad N_q(T) = 37 - ce^{(T-T_c)/d} \tag{14.23}$$

It is straightforward to verify that this leads to a second-order phase transition (P and s continuous but ds/dT discontinuous) provided that $b + c = 34$, $b > 0$, $b \neq 17$. Consistently with our discussion of the Weinberg sum rules in Chapter 12, let us require the ρ and a_1 mesons to become effectively massless at T_c. Then $b = 18$. Setting $c = 16$ produces 21 massless bosonic degrees of freedom at T_c, corresponding to the up and down quarks. The missing 16 degrees of freedom correspond to the eight massless gluons, which may not be readily available at T_c. The entropy is $4aT^3N(T)$, and the evolution can easily be charted using (14.10). The parameter d controls the degree-of-freedom conversion rate. For the sake of illustration we choose $d = 0.034\,T_c$.

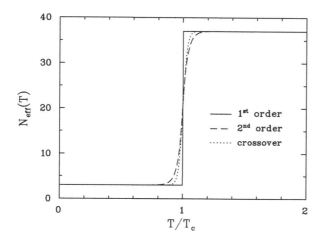

Fig. 14.1. The number of degrees of freedom as a function of the temperature for equations of state producing phase transitions of the first order (solid line) and of the second order (broken line). The effect of a rapid crossover (dotted line) is also shown.

A rapid-crossover scenario is produced by the parametrization

$$N(T) = 20 + 17 \tanh\left(\frac{T - T_{\mathrm{c}}}{d}\right) \qquad (14.24)$$

The transitions from one set of degrees of freedom to another are shown in Figure 14.1, for the different schemes we have considered: a first-order phase transition, a second-order transition, and a rapid crossover. Similarly, the temperature evolution associated with each of these is shown in Figure 14.2.

Another powerful feature of the Bjorken model is the particle production. Since the entropy density of a gas of massless pions is proportional to the pion number density, it follows that the entropy can be determined by measuring the charged-particle multiplicity N_{ch}. Considering a head-on collision of equal-mass nuclei, one finds approximately

$$\frac{dN_{\mathrm{ch}}}{dy} = \left(f_0 + \frac{1 - f_0}{r}\right) 3\pi R^2 \tau_0 T_0^3 \qquad (14.25)$$

where R is the nuclear radius, $f_0 = 0$ if $T_0 < T_{\mathrm{c}}$, $0 \le f_0 \le 1$ if $T_0 = T_{\mathrm{c}}$, and $f_0 = 1$ if $T_0 > T_{\mathrm{c}}$. Those arguments are not significantly altered even if the rather large latent heat is shrunk to zero so that the first-order phase transition turns into a second-order one, or even if there is no proper phase transition at all. The essential requirement is that the number of degrees of freedom should increase by a factor r in a small temperature interval $\Delta T \approx d$. The conservation of entropy density enables one to relate

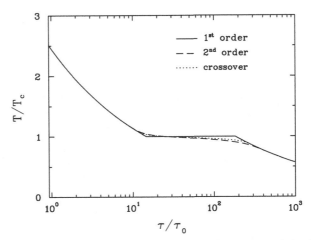

Fig. 14.2. The temperature evolution in the Bjorken model. Note that the plateau in T starts and ends at $\tau = \tau_1$ and τ_2, respectively. See the text for details.

measurements in the final state to parameters that determine the initial conditions for thermal equilibrium and hydrodynamic flow.

In this section, we have used a simple dynamical model for ultrarelativistic nucleus–nucleus collisions, and simple parametrizations of the equation of state to give a flavor of this branch of high-energy nuclear physics. For more sophisticated discussions, the reader is referred to the literature cited at the end of the chapter.

14.2 The statistical model of particle production

As mentioned previously, Fermi's seminal paper was instrumental to the development of statistical techniques for particle production in strongly interacting systems [1]. Fermi's original application was to proton–proton collisions. Our discussion will concentrate on nucleus–nucleus collisions, where the applicability of the model is arguably maximal, but the statistical model has even been applied in the case of e^+e^- collisions [5].

If we assume that the approach to equilibrium can be modeled by a transport equation of the Boltzmann type for the phase-space density $f(x,p)$, we may write

$$\left(\frac{p^\mu}{m}\frac{\partial}{\partial x^\mu} + F^\mu \frac{\partial}{\partial p^\mu}\right) f(x,p) = C[f] \qquad (14.26)$$

where F^μ is a generalized force term and $C[f]$ is a collision term that ensures entropy growth. At equilibrium, detailed balance makes the right-hand side of this equation vanish, and thermal distributions functions are recovered. In fact, in high-energy nuclear collisions a statistical approach is natural, as the high multiplicity will provide a physical environment appropriate for realization of equipartition. Specifically, $\lambda \sim 1/\sigma n$, where σ is a total cross-section, λ is a mean free path, and $n \sim \int d^3 p\, f(x,p)$. Thus as the multiplicities increase, the mean free path will decrease. Furthermore, the relevance of statistical arguments should improve in high-temperature environments, owing to the same arguments.

The fundamental quantity that regulates the thermal composition of particle species is the partition function. We will work in the grand canonical ensemble. We have already encountered this quantity in Chapter 1; it is given by $Z = \operatorname{Tr} \hat{\rho}$, where $\hat{\rho}$, the statistical density matrix, is given by (1.1). In a system that we are modeling as a gas of relativistic hadrons (stable and unstable), the quantum numbers we choose to be conserved are electric charge, baryon number, and strangeness. The grand canonical partition function can then be written as a sum of partition functions for individual hadrons and resonances:

$$\ln Z(V, T, \mu_Q, \mu_B, \mu_S) = \sum_i \ln Z_i(V, T, \mu_Q, \mu_B, \mu_S) \qquad (14.27)$$

where

$$\ln Z_i(V, T, \mu_Q, \mu_B, \mu_S) = \pm (2s_i + 1) \frac{V}{2\pi^2} \int_0^\infty dp\, p^2 \ln\left[1 \pm \lambda_i \exp(-\beta \omega_i)\right] \qquad (14.28)$$

The $+$ or $-$ sign is for fermions or bosons, $2s_i + 1$ is the spin degeneracy factor, $\omega_i = \sqrt{p^2 + m_i^2}$, $\beta = 1/T$, and the fugacity is

$$\lambda_i(T, \mu_Q, \mu_B, \mu_S) = \exp\left[\beta(\mu_Q Q_i + \mu_B B_i + \mu_S S_i)\right] \qquad (14.29)$$

The coordinate-space density of species i is then

$$n_i(T, \mu_Q, \mu_B, \mu_S) = \frac{N_i}{V} = (2s_i + 1) \frac{T}{2\pi^2} \sum_{\ell=1}^\infty \frac{(\pm 1)^{\ell+1}}{\ell} \lambda_i^\ell m_i^2 K_2(\ell \beta m_i) \qquad (14.30)$$

where $K_2(x)$ is a modified Bessel function. In actual comparisons with experiment, it is especially important to account for resonances decaying into lighter hadrons; then we get a net number

$$N_i^{\text{net}}(T, \mu) = N_i(T, \mu) + \sum_k N_k(T, \mu) B_{k \to i + X} \qquad (14.31)$$

where $B_{k \to i + X}$ is the branching ratio for the decay $k \to i + X$. At high temperatures (around and above the pion mass) the yield of the light mesons is indeed dominated by feed-down from the higher-lying resonances.

In practical applications to measured particle numbers and, especially, ratios, the temperature T and the baryon chemical potential μ_B are the two main parameters of the model. Note that there is no quantum number associated with the conservation of meson number, unlike baryons. Also, overall strangeness conservation fixes μ_S. Note that this would actually be rigorously true if measurements covered all the phase space, so that all fragments were measured. For measurements performed at mid-rapidity, however, the strangeness entering one region in rapidity is approximately canceled by that leaving. Therefore, even in experiments with a limited phase-space coverage, the strangeness chemical potential can be taken to vanish. In addition, charge conservation requirements have a small influence at RHIC energies and above. Finally, the volume V drops out in analyses of particle number ratios. It can actually be fixed by measuring the total pion multiplicity and requiring agreement between the theoretical expression and the empirical value.

Putting these ingredients together, one may further assume chemical equilibrium and thus verify how far this assumption will hold. Chemical equilibrium implies that if $c \rightleftharpoons a + b$ then $\mu_c = \mu_a + \mu_b$. Therefore, the chemical potential of a given resonance is fixed by its decay systematics and can be written in terms of μ_B. In the final analysis, decay cascades (where several generations of particle decays contribute) are also included. Also of practical concern is whether to use only data at mid-rapidity or data that is integrated over the full phase space. A popular and pragmatic choice is to restrict the analysis to a slice at mid-rapidity centered at zero with a total width of 2 units of rapidity [6]. From CERN experiments, the ratios of particle abundances were fitted at fixed-target bombarding energies of 40 and 158 GeV per nucleon, for collisions of Pb on Pb. At RHIC energies ($\sqrt{s} = 130$ and 200 GeV in the nucleon–nucleon center-of-mass frame), Au + Au collisions were analyzed. At 40 GeV per nucleon, 11 particle ratios were included in the fit while that number was 24 at 158 GeV per nucleon. The lower RHIC energy included 13 species, while the higher energy included five particle ratios; these numbers are continuously updated as the experimental analyses continue. Weak-decay systematics are extremely important: those species that are unstable against the weak interaction will eventually decay and their products will be measured by the experimental detectors. The goodness of fit was evaluated via the minimization of

$$\chi^2 = \sum_i \frac{\left(R_i^{\text{expt}} - R_i^{\text{model}}\right)^2}{\sigma_i^2} \tag{14.32}$$

where R_i is the fraction of particles of species i in the total number of particles of all species and σ_i is its experimental uncertainty. The set of (T, μ_B) values that minimize the above relation is plotted in Figure 14.3. The values of χ^2 attained are about 1 per degree of freedom [6].

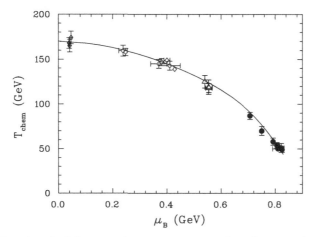

Fig. 14.3. The chemical freezeout temperature against baryon chemical potential as extracted from several fits to measured particle ratios at different energies. The solid line is a curve for which the average energy per hadron $\langle E \rangle / \langle N \rangle = 1$ GeV [7]. The data have been collected from experiments performed at the GSI [8], the AGS at BNL [9], CERN [10, 6], and RHIC [11].

There have been efforts [12] to improve the fits to hadron-yield ratios by invoking a departure from chemical equilibrium and looking for evidence of this deviation in the data. For example, the density of pions is parametrized by generalizing the thermal distribution function to

$$\frac{N_\pi}{V} = 3 \int \frac{d^3p}{(2\pi)^3} \frac{1}{\gamma_\pi^{-1} e^{\omega_\pi/T} - 1} \tag{14.33}$$

where γ_π is a parameter that regulates the absolute chemical equilibrium and is therefore unity in that limit. Values of $\gamma_\pi \neq 1$ would constitute, in this interpretation, a signature of nonequilibrium. We will not pursue this further here, but it is a topic of current investigation.

The fitted values from Figure 14.3 can be reconciled with a global picture that emerged from years of heavy ion phenomenology at CERN's SPS, which we now very briefly summarize. The intuitive picture is as follows. The nuclear system is first heated and compressed. This is followed by a phase of decompression where both the temperature and the density drop. Note here the use of the word temperature, which stems again from years of phenomenological analysis. Two freezeout temperatures may be identified. As the hot, interacting system cools, it eventually breaks apart and its constituents begin free-streaming towards the detectors to be measured individually. A criterion for this to happen is that the mean free path, as defined earlier by the inverse of the product of density and cross section, becomes comparable with the spatial dimensions of the system:

$$\lambda \sim \frac{1}{n\sigma} \sim R \qquad (14.34)$$

The particle population will be dominated by pions, as they are the lightest species. Some insight on the behavior of the system may then be had by considering chiral perturbation theory. For temperatures below the pion mass, the elastic cross section essentially saturates the total cross section: the inelastic channels manifest themselves at higher powers of the chiral expansion [13]. This means that number-changing interactions will cease before purely elastic interactions do, as the system expands and cools. Another way of thinking about this is related to the fact that inelastic reactions have energy thresholds, whereas elastic interactions do not. Therefore, there will exist a region where $T_{\mathrm{kin}} < T < T_{\mathrm{chem}}$. Here T_{kin} is the kinetic freezeout temperature (where transverse-momentum spectra cease to evolve) and T_{chem} is the temperature below which the particle numbers do not change.

The fact that the curve corresponding to an average energy per particle of 1 GeV traces the path laid out by the thermal-model fit is very suggestive of a critical phenomenon. However, numerical simulations of relativistic nuclear collisions have correlated the energy per particle value of 1 GeV with the onset of inelastic thresholds [14], at least at beam energies that correspond to those spanned in the experimental fits shown in Figure 14.3. It is very suggestive that the low-μ_B chemical freezeout temperatures found in the thermal analysis of experimental nucleus–nucleus data are consistent with the critical temperature extracted from the lattice simulation of QCD, as mentioned in Section 10.5. This would be the case if the chemical composition of the hadrons being measured were established during the hadronization of the quark–gluon plasma. Note also the similarity between Figures 14.3 and 10.9. Although very suggestive, these connections remain the source of much current research.

14.3 The emission of electromagnetic radiation

In theoretical studies of hot and dense strongly interacting systems, electromagnetic radiation constitutes a class of penetrating probes. This is essentially a reflection of the near absence of final-state interactions for photons (real and virtual) that are produced in relativistic nuclear collisions. More quantitatively, at scales relevant for hadronic phenomenology, $\alpha/\alpha_{\mathrm{s}} \sim 0.002 \ll 1$. This means that electromagnetic radiation, once created, will leave the system unscathed. In line with the rest of this chapter, we assume that nuclear collisions at high energies form a thermalized system. As mentioned previously, this assertion receives considerable empirical support. We will now proceed to derive the rate of emission of electromagnetic radiation from a thermal, strongly interacting, medium. As in any hadronic collision, there will always be emission of electromagnetic radiation from the very first interactions, those involving cold matter. In nucleon–nucleus and in nucleus–nucleus events, this primordial photon and lepton pair emission is usually treated (up to aspects like the Cronin effect, which we do not discuss here) as an additive superposition of nucleon–nucleon contributions, calculated using the techniques of perturbative QCD. The details of this are outside the scope of this book.

Consider generic hadronic states $|i\rangle$ and $|f\rangle$ and a transition between them that involves the absorption or emission of a photon with four-momentum $k^\mu = (\omega, \mathbf{k})$ and polarization ϵ^μ. To make things more definite we shall concentrate on the case of real photons here, and extend our analysis to lepton pair production later. The transition rate between the two states is

$$R_{fi} = \frac{|S_{fi}|^2}{tV} \tag{14.35}$$

tV being the proper four-volume. To leading order in the interaction Hamiltonian (or equivalently, in the one-photon approximation), the S-matrix element is

$$S_{fi} = \langle f| \int d^4 x\, \hat{J}_\mu(x) A^\mu(x) |i\rangle \tag{14.36}$$

$\hat{J}_\mu(x)$ being the hadronic electromagnetic current operator. Considering a free vector field

$$A^\mu(x) = \frac{\epsilon^\mu}{\sqrt{2\omega V}} \left(e^{ik\cdot x} + e^{-ik\cdot x} \right) \tag{14.37}$$

and, invoking translation invariance for the matrix element

$$\langle f|\hat{J}_\mu(x)|i\rangle = e^{i(p_f - p_i)\cdot x} \langle f|\hat{J}_\mu(0)|i\rangle$$

one may write

$$R_{fi} = -\frac{g^{\mu\nu}}{2\omega V}(2\pi)^4 \left[\delta(p_i + k - p_f) + \delta(p_i - k - p_f) \right]$$
$$\times \langle f|\hat{J}_\mu(0)|i\rangle \langle i|\hat{J}_\nu(0)|f\rangle \tag{14.38}$$

One delta function corresponds to the absorption process and the other to emission. The differential thermal emission rate is obtained by keeping the appropriate delta function, summing over final states, and averaging over initial states with a Boltzmann weight $e^{-\beta \hat{K}_i}/Z$, where $Z = \sum_i e^{-\beta \hat{K}_i}$, and $\hat{K} = \hat{H} - \mu\hat{N}$:

$$\frac{d^3R}{d^3k} = -\frac{g^{\mu\nu}}{2\omega V}\frac{V}{(2\pi)^3}\frac{1}{Z}\sum_i e^{-\beta\hat{K}_i}\sum_f (2\pi)^4\delta(p_i - p_f - k)$$

$$\times \langle j|\hat{J}_\mu(0)|i\rangle\langle i|\hat{J}_\nu(0)|f\rangle \tag{14.39}$$

Defining, as in Section 6.2, spectral functions associated respectively with absorption and emission,

$$f_{\mu\nu}^\pm(k) = \pm\frac{1}{Z}\sum_{i,f} e^{-\beta\hat{K}_i}(2\pi)^3\delta(p_i - p_f \pm k)$$

$$\times \langle f|\hat{J}_\mu(0)|i\rangle\langle i|\hat{J}_\nu(0)|f\rangle \tag{14.40}$$

one may use the identity relating them, $f_{\mu\nu}^+(k) = -e^{\beta\omega}f_{\mu\nu}^-(k)$, to write

$$\omega\frac{d^3R}{d^3k} = \frac{g^{\mu\nu}}{(2\pi)^3}\,\pi f_{\mu\nu}^- \tag{14.41}$$

Note that the symbol used here (f) is different from that used in Chapter 6 (ρ), to make clear the fact that here the correlation functions involve the current operator. One relates the current–current correlators to those involving the fields through the equation of motion $\partial^\mu\partial_\mu A_\nu(x) = J_\nu(x)$, written here in the Feynman gauge. Doing this, and using the fact that the spectral density $\rho_{\mu\nu}^n(k)$ is proportional to the imaginary part of the retarded propagator, (6.33), one obtains

$$\omega\frac{d^3R}{d^3k} = -\frac{g^{\mu\nu}}{(2\pi)^3}\,\mathrm{Im}\,\Pi_{\mu\nu}'^R(\omega, \mathbf{k}) \tag{14.42}$$

Here the finite-temperature retarded improper self-energy, $\Pi_{\mu\nu}'^R$, is defined through the appropriate Schwinger–Dyson equation, $D = D^0 - D^0\Pi'D^0$. Therefore, to leading order in the electromagnetic interaction but to all orders in the strong interaction,

$$\omega\frac{d^3R}{d^3k} = -\frac{g^{\mu\nu}}{(2\pi)^3}\,\mathrm{Im}\,\Pi_{\mu\nu}^R(\omega, \mathbf{k}) \tag{14.43}$$

where $\Pi_{\mu\nu}^R$ is the finite-temperature retarded photon self-energy.

Repeating this derivation, but for a virtual photon that converts to a lepton pair, we are led to

$$E_+E_-\frac{d^6R}{d^3p_+ d^3p_-} = \frac{2e^2}{(2\pi)^6}\frac{1}{k^4}\left[p_+^\mu p_-^\nu + p_+^\nu p_-^\mu - g^{\mu\nu}\left(p_+ \cdot p_- + m_\ell^2\right)\right]$$

$$\times \Pi_{\mu\nu}^R(\omega, \mathbf{k})\frac{1}{e^{\beta\omega} - 1} \tag{14.44}$$

where the invariant mass of the virtual photon is $M^2 = k^2 = (p_+ + p_-)^2$, p_+ and p_- are the momenta of the lepton pair components, and m_ℓ is the lepton mass.

14.4 Photon production in high-energy heavy ion collisions

The formation and observation of quark–gluon plasma in ultrarelativistic collisions between heavy nuclei is an important goal of modern nuclear physics. Among the proposed probes of the plasma are the directly produced real photons [15–21]. Microscopically, these could come from the annihilation process $q\bar{q} \to g\gamma$ and from the QCD Compton process $qg \to q\gamma$, $\bar{q}g \to \bar{q}\gamma$. These photons interact only electromagnetically, unlike pions, and so their mean free paths are typically much larger than the transverse size of the region of hot matter created in any nuclear collision. As a result, high-energy photons produced in the interior of the plasma usually pass through the surrounding matter without interacting, carrying information directly from wherever they were formed to the detector. This makes them an interesting object of study to both theorists and experimenters.

Here we concern ourselves with the following questions. What is the spectral emissivity of quark–gluon plasma? What is the spectral emissivity of hot hadronic matter? How do they compare at the same temperature? These are important questions. Suppose we put hadron gas in one box and quark–gluon plasma in another and maintain them at the same temperature T. Can we tell which box contains the quark–gluon plasma by looking through a small window and measuring the photon spectrum? If we wait long enough the answer is clearly no: even if we do not put any photons into the boxes at the beginning, the matter will eventually come to equilibrium under the electromagnetic interactions; to a good approximation the final photon distribution will be just the Planck distribution at temperature T. Fortunately, in conditions more appropriate to a nuclear collision the answer is yes. A closer analog to a nuclear collision is to make the boxes smaller than the photon mean free path and to make the walls transparent to photons, so that the photons always escape and the photon distribution stays far from equilibrium. The spectral emissivity then directly reflects the dynamics of real photon-producing reactions in the matter, which may be different for the two phases. The thermal production rates in the two phases are important in another sense. Suppose that quark–gluon plasma is formed in a collision. It will expand and eventually hadronize in a first- or second-order phase transition or rapid crossover. The hadrons themselves may maintain local thermal equilibrium for a while, also producing photons. The total yield is a sum of the yields from both phases. To make the method clear, we shall mainly

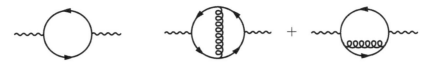

Fig. 14.4.   One- and two-loop contributions to the photon self-energy in QCD.

concentrate here on the radiation from the partonic phase of QCD, and follow the treatment in [21].

In an expansion in diagram topologies, the one- and two-loop contributions to $\Pi_{\mu\nu}$ are shown in Figure 14.4. The imaginary part is obtained by cutting the diagrams. Cutting the one-loop diagram gives zero when the photon is on the mass shell since $q\bar{q} \to \gamma$ has no phase space. Certain cuts of the two-loop diagrams give order-g^2 corrections to the nonexistent reaction $q\bar{q} \to \gamma$, while other cuts correspond to the reactions $q\bar{q} \to g\gamma$, $qg \to q\gamma$ and $\bar{q}g \to \bar{q}\gamma$. Let $\mathcal{M}_i$ represent the amplitude for one of these. The contribution to the rate in relativistic kinetic theory for a photon-producing reaction $1 + 2 \to 3 + \gamma$ is

$$R_i = \mathcal{N} \int \frac{d^3p_1}{2E_1(2\pi)^3} \frac{d^3p_2}{2E_2(2\pi)^3} f_1(E_1) f_2(E_2) (2\pi)^4 \delta(p_1^\mu + p_2^\mu - p_3^\mu - p^\mu)$$

$$\times |\mathcal{M}_i|^2 \frac{d^3p_3}{2E_3(2\pi)^3} \frac{d^3p}{2E(2\pi)^3} [1 \pm f_3(E_3)] \qquad (14.45)$$

where $\mathcal{N}$ is a degeneracy factor, the f's are the Fermi–Dirac or Bose–Einstein distribution functions as appropriate, and there is either a Bose-enhancement or a Pauli-suppression of the strongly interacting particle in the final state. (Another example of the connection between the imaginary part of the finite-temperature retarded self-energy and relativistic kinetic theory can be found in Section 16.6.)

This rate can be simplified. Define $s = (p_1 + p_2)^2$ and $t = (p_1 - p)^2$. Insert integrations over s and t with a delta function for each of these identities. This is a natural thing to do because the invariant amplitude depends only on these two variables. Converting the total rate to a differential one, all but four of the integrations can be done without approximation:

$$E\frac{d^3R_i}{d^3p} = \frac{\mathcal{N}}{(2\pi)^7} \frac{1}{16E} \int ds\, dt\ |\mathcal{M}_i(s,t)|^2 \int dE_1\, dE_2\, f_1(E_1) f_2(E_2)$$

$$\times [1 \pm f_3(E_1 + E_2 - E)]\theta(E_1 + E_2 - E)(aE_1^2 + bE_1 + c)^{-1/2}$$

$$(14.46)$$

where

$$a = -(s + t)^2$$
$$b = 2(s + t)(Es - E_2 t) \tag{14.47}$$
$$c = st(s + t) - (Es + E_2 t)^2$$

At present we are interested in the case where the photon energy is large, $E_1 + E_2 > E \gg T$. In this limit it is a good approximation to make the replacement

$$f_1(E_1) f_2(E_2) \rightarrow e^{-(E_1 + E_2)/T} \tag{14.48}$$

Even though E_1 or E_2 separately need not be large, phase space is unfavorable for it. This approximation can be checked numerically (see Exercise 14.4). Then the integrals over E_1 and E_2 can be done, with the relatively simple result

$$E \frac{d^3 R_i}{d^3 p} = \frac{\mathcal{N}}{(2\pi)^6} \frac{T}{32E} e^{-E/T} \int \frac{ds}{s} \ln(1 \pm e^{-s/4ET})^{\pm 1} \int dt \, |\mathcal{M}_i(s, t)|^2 \tag{14.49}$$

The upper sign is to be taken when particle 3 is a fermion, the lower sign when it is a boson.

For massless particles the amplitude is related to the differential cross section by

$$\frac{d\sigma}{dt} = \frac{|\mathcal{M}|^2}{16\pi s^2} \tag{14.50}$$

For the annihilation diagram,

$$\frac{d\sigma}{dt} = \frac{8\pi\alpha\alpha_s}{9s^2} \frac{u^2 + t^2}{ut} \tag{14.51}$$

where u and t are Mandelstam variables, and $\mathcal{N} = 20$ when summing over the up and down quarks. For the Compton reaction,

$$\frac{d\sigma}{dt} = \frac{-\pi\alpha\alpha_s}{3s^2} \frac{u^2 + s^2}{us} \tag{14.52}$$

and $\mathcal{N} = 320/3$. The integral over t just gives the total cross section. But the total cross section involving the exchange of a massless particle is infinite: the differential cross sections have a pole at t and/or $u = 0$. Many-body effects are necessary to screen this divergence. We will show how this works. For now we delete the region of phase space causing the divergence. We integrate over

$$-s + k_c^2 \leq t \leq -k_c^2$$
$$2k_c^2 \leq s < \infty \tag{14.53}$$

where k_c is an infrared cutoff and $T^2 \gg k_c^2 > 0$. This way of regulating the divergence treats u and t symmetrically and maintains the identity $s + t + u = 0$ that is appropriate for all massless particles.

In the limit $k_c^2 \to 0$ we find

$$E \frac{d^3 R}{d^3 p}^{\text{Compton}} = \frac{5}{9} \frac{\alpha \alpha_s}{6\pi^2} T^2 e^{-E/T} \left[\ln \left(\frac{4ET}{k_c^2} \right) + C_{\text{F}} \right] \qquad (14.54)$$

$$E \frac{d^3 R}{d^3 p}^{\text{annihilation}} = \frac{5}{9} \frac{\alpha \alpha_s}{3\pi^2} T^2 e^{-E/T} \left[\ln \left(\frac{4ET}{k_c^2} \right) + C_{\text{B}} \right] \qquad (14.55)$$

where

$$C_{\text{F}} = \frac{1}{2} - \gamma_{\text{E}} + \frac{12}{\pi^2} \sum_{n=2}^{\infty} \frac{(-1)^n}{n^2} \ln n$$

$$= 0.0460 \ldots \qquad (14.56)$$

$$C_{\text{B}} = -1 - \gamma_{\text{E}} - \frac{6}{\pi^2} \sum_{n=2}^{\infty} \frac{1}{n^2} \ln n$$

$$= -2.1472 \ldots \qquad (14.57)$$

and γ_{E} is Euler's constant. These expressions use the full Fermi–Dirac or Bose–Einstein distribution functions in the final state. Although $E \gg T$, it is not necessarily the case that $E_3 \gg T$. Taking this into account, one gets slightly different results if one uses the Boltzmann distribution in the final state instead:

$$E \frac{d^3 R}{d^3 p}^{\text{Compton}} = \frac{5}{9} \frac{2\alpha \alpha_s}{\pi^4} T^2 e^{-E/T} \left[\ln \left(\frac{4ET}{k_{c^2}} \right) + \frac{1}{2} - \gamma_{\text{E}} \right] \quad (14.58)$$

$$E \frac{d^3 R}{d^3 p}^{\text{annihilation}} = \frac{5}{9} \frac{2\alpha \alpha_s}{\pi^4} T^2 e^{-E/T} \left[\ln \left(\frac{4ET}{k_c^2} \right) - 1 - \gamma_{\text{E}} \right] \quad (14.59)$$

Corrections to these formulae vanish in the limit $k_c \to 0$.

The essential factors in these rates are easy to understand. There is a factor $5/9$ from the sum of the squares of the electric charges of the u and d quarks, a factor $\alpha \alpha_s$ coming from the topological structure of the diagrams, a factor T^2 from phase space, which gives the overall dimension to the rate, the ubiquitous Boltzmann factor $e^{-E/T}$ for photons of energy E, and a logarithm due to the infrared behavior.

The infrared divergence in the photon production rate discussed above is caused by a diverging differential cross section when the momentum transfer goes to zero. Often, long-range forces can be screened by

Fig. 14.5. HTL-corrected photon self-energy, in QCD.

many-body effects at finite temperature. In fact, we have already seen concrete examples of this mechanism in Chapter 9. From the hard thermal loops (HTL) analysis, we know that a propagator must be dressed if the momentum flowing through it is soft, on a scale set by the temperature T. For the present application we would begin by replacing the bare propagators and vertices in the one-loop diagram of Figure 14.4 by effective propagators and vertices. The reason is that the propagation of soft momenta is connected with infrared divergences in loops; if we do not dress these propagators we get infinite answers, so the corrections due to the dressing of the propagators are also infinite and therefore necessary. Thus, the results with soft propagators dressed are really the lowest-order finite results. In our case it is necessary to dress one of the quark propagators because our results diverge otherwise. It is not necessary to dress both, nor is it necessary to dress either of the vertices, because these produce only finite corrections that are of higher order in g. We are thus led to evaluate the diagram shown in Figure 14.5. Some insight can be gained by expanding the diagram as a power series in g^2. The zeroth-order term reproduces the one-loop diagram of Figure 14.4. The order-g^2 term reproduces one of the two-loop diagrams of Figure 14.4, with the recognition that the quark self-energy is not the exact one-loop self-energy but is approximated by its high-temperature limit. Clearly this is a summation of an infinite set of diagrams that is purposely designed to regulate infrared problems of the type encountered here.

Starting with Figure 14.5, and summing over u and d quarks, we find

$$\Pi^{\mu\nu}(p) = -6 \times \frac{5}{9} e^2 T \sum_{k_0} \int \frac{d^3 k}{(2\pi)^3} \, \mathrm{Tr} \left[\mathcal{G}^*(k) \gamma^\mu \mathcal{G}(p-k) \gamma^\nu \right] \tag{14.60}$$

where

$$\mathcal{G}^*(k) = \mathcal{G}^*_+(k) \frac{\gamma_0 - \mathbf{k} \cdot \boldsymbol{\gamma}}{2} + \mathcal{G}^*_-(k) \frac{\gamma_0 + \mathbf{k} \cdot \boldsymbol{\gamma}}{2} \tag{14.61}$$

is the dressed propagator for a quark with four-momentum k, already encountered in Section 9.4, and

$$\mathcal{G}(q) = g_+(q) \frac{\gamma_0 - \mathbf{q} \cdot \boldsymbol{\gamma}}{2} + g_-(q) \frac{\gamma_0 + \mathbf{q} \cdot \boldsymbol{\gamma}}{2} \tag{14.62}$$

is the bare propagator for a quark with four-momentum $q = p - k$. The propagator $\mathcal{G}^*(k)$ was defined in (9.37), and

$$g_\pm(q) = (-q_0 \pm \mathbf{q})^{-1} \tag{14.63}$$

Using these expressions for the quark propagators, and evaluating the traces, we obtain

$$\Pi_\mu^{\mathrm{R},\mu}(p) = \frac{20}{3}e^2 T \sum_{k_0} \int \frac{d^3k}{(2\pi)^3} \left\{ \mathcal{G}_+^*(k) \left[g_+(q) \left(1 - \mathbf{k} \cdot \mathbf{q}\right) + g_-(q) \left(1 + \mathbf{k} \cdot \mathbf{q}\right) \right] \right.$$

$$\left. + \mathcal{G}_-^*(k) \left[g_+(q) \left(1 + \mathbf{k} \cdot \mathbf{q}\right) + g_-(q) \left(1 - \mathbf{k} \cdot \mathbf{q}\right) \right] \right\} \tag{14.64}$$

That the self-energy is retarded means that p_0 has a small positive imaginary part, as is appropriate in linear response analysis.

We then follow Braaten, Pisarski, and Yuan [22] in computing the imaginary part in the following elegant way:

$$\mathrm{Im}\, T \sum_{k_0} F_1(k_0)\, F_2(p_0 - k_0)$$

$$= \frac{1}{2i}\, \mathrm{Disc}\, T \sum_{k_0} F_1(k_0)\, F_2(p_0 - k_0)$$

$$= \pi(1 - e^{E/T}) \int_{-\infty}^{+\infty} d\omega \int_{-\infty}^{+\infty} d\omega'\, N_\mathrm{F}(\omega)\, N_\mathrm{F}(\omega')$$

$$\times \delta(E - \omega - \omega')\, \rho_1(\omega)\, \rho_2(\omega') \tag{14.65}$$

Here N_F is the Fermi–Dirac occupation number and ρ_1 and ρ_2 are the spectral densities for the two chosen functions F_1 and F_2. Specifically, these are related by

$$F(k_0) = \int_{-\infty}^{+\infty} \frac{d\omega}{\omega - k_0 - i\epsilon}\, \rho(\omega) \tag{14.66}$$

We need the spectral density functions $\rho_\pm^*$ and $r_\pm$ for the dressed and bare propagators, respectively. The latter can be obtained in a straightforward fashion, and the former were given in Chapter 9. Putting this information together we obtain

$$\mathrm{Im}\, \Pi_\mu^{\mathrm{R},\mu} = -\frac{20\pi}{3}e^2 (e^{E/T} - 1) \int \frac{d^3k}{(2\pi)^3} \int_{-\infty}^{+\infty} d\omega \int_{-\infty}^{+\infty} d\omega'\, \delta(E - \omega - \omega')$$

$$\times\ N_\mathrm{F}(\omega) N_\mathrm{F}(\omega') \left[(1 + \mathbf{q} \cdot \mathbf{k})(\rho_+^* r_- + \rho_-^* r_+) \right.$$

$$\left. + (1 - \mathbf{q} \cdot \mathbf{k})(\rho_+^* r_+ + \rho_-^* r_-) \right] \tag{14.67}$$

with $r_\pm(\omega', \mathbf{q}) = \delta(\omega' \mp |\mathbf{q}|)$. In these expressions ρ_+^* and ρ_-^* (9.39) are evaluated at $(\omega, \mathbf{k})$.

In the kinetic theory calculation we were forced to put a cutoff k_c^2 on the four-momentum transfer t (and on u) to avoid an infrared divergence. This cutoff removes only the small region of phase space left out by (14.53). Anything else must necessarily be higher order in g. Inspection of Figure 14.4 shows that the exchanged quark must be dressed and must satisfy

$$-k_c^2 \leq \omega^2 - \mathbf{k}^2 \leq 0. \tag{14.68}$$

This means that the delta functions (representing poles) in the spectral densities do not contribute to this order, but only the functions $\beta_\pm$ (representing branch cuts); see (9.40).

The energy-conserving delta function, together with the mass-shell delta functions of $r_\pm$, can be used to evaluate the integral over ω' and the integral over the angle between $\mathbf{k}$ and $\mathbf{q}$ in (14.67). Then, making use of the inequalities $E \gg T$ and $0 \leq \mathbf{k}^2 - \omega^2 \leq k_c^2 \ll T^2$, we get

$$\mathrm{Im}\, \Pi_\mu^{\mathrm{R},\mu} = -\frac{5e^2}{6\pi} \left(\mathrm{e}^{E/T} - 1 \right) \mathrm{e}^{-E/T}$$
$$\times \int_0^{k_c} d|\mathbf{k}| \int_{-|\mathbf{k}|}^{|\mathbf{k}|} d\omega \left[(|\mathbf{k}| - \omega)\beta_+(\omega, \mathbf{k}) + (|\mathbf{k}| + \omega)\beta_-(\omega, \mathbf{k}) \right]$$
$$\tag{14.69}$$

The integral involving β_- is the same as the integral involving β_+, so we only need to determine the latter and multiply by 2. Furthermore it is convenient to make the change of variables $|\mathbf{k}| = \tau \cosh\eta$ and $\omega = \tau \sinh\eta$. Then we have for the above double integral

$$2 \int_{-\infty}^{+\infty} d\eta \int_0^{k_c} \tau \, d\tau \, (|\mathbf{k}| - \omega)\beta_+(\omega, \mathbf{k})$$
$$= \frac{m_q^2}{4} \int_{-\infty}^{+\infty} \frac{d\eta}{\cosh^2\eta} \left\{ \ln\left(\frac{(\Theta + y_c \cosh^2\eta)^2 + 1}{\Theta^2 + 1} \right) \right.$$
$$\left. -2\Theta \left[\tan^{-1}(\Theta + y_c \cosh^2\eta) - \tan^{-1}(\Theta) \right] \right\} \tag{14.70}$$

where

$$\Theta = \frac{2}{\pi} \frac{Q_0(\sinh\eta) - Q_1(\sinh\eta)}{1 - \tanh\eta} \tag{14.71}$$

and

$$y_c = \frac{2}{\pi} \frac{k_c^2}{m_q^2} \tag{14.72}$$

The quantities $Q_0(z)$ and $Q_1(z)$ are Legendre functions.

We still have some freedom in choosing the cutoff k_c. Since g is supposed to be perturbatively small for this whole analysis to make sense let us choose k_c to lie somewhere in the interval

$$m_q \ll k_c \ll T \tag{14.73}$$

Then we are allowed to take the limit $y_c \gg 1$ in (14.70). Doing so, and dropping terms that vanish in the limit $y_c \to \infty$, we find that the right-hand side becomes

$$m_q^2 \ln\left(\frac{k_c^2}{m_q^2}\right) + \frac{m_q^2}{4} \int_{-\infty}^{\infty} \frac{d\eta}{\cosh^2\eta} \left[\ln\left(\frac{4}{\pi^2} \frac{\cosh^4\eta}{\Theta^2+1}\right) - 2\Theta\left(\frac{\pi}{2} - \tan^{-1}\Theta\right)\right] \tag{14.74}$$

This remaining integral is a pure number and is evaluated as $-4\ln 2$.

Now we have all the items we need in order to write down the contribution to the rate coming from the infrared-sensitive (IR) part of phase space:

$$E\frac{d^3R^{\mathrm{IR}}}{d^3p} = \frac{5}{9}\frac{\alpha\alpha_s}{2\pi^2}T^2 e^{-E/T}\ln\left(\frac{k_c^2}{2m_q^2}\right) \tag{14.75}$$

where

$$2m_q^2 = \frac{g^2 T^2}{3} \tag{14.76}$$

Adding the contributions from both the hard momentum transfers, (14.54) and (14.55), and the soft momentum transfers, (14.75), we get the net rate

$$E\frac{d^3R}{d^3p} = \frac{5}{9}\frac{\alpha\alpha_s}{2\pi^2}T^2 e^{-E/T}\ln\left(\frac{2.912}{g^2}\frac{E}{T}\right) \tag{14.77}$$

This is independent of the cutoff k_c! The HTL resummation method works beautifully to screen the infrared divergence. (Inclusion of the exact Bose–Einstein and Fermi–Dirac distributions in the initial state instead of the Boltzmann limit (14.48) leads to a replacement of the numerical factor 2.912 in the logarithm by 3.739. See Exercise 14.4.)

It is apparent that our asymptotic formula breaks down when $E \leq g^2 T/2.9$ because the logarithm goes negative. For photon energies that are small on a scale set by the the temperature, a complete calculation should include bremsstrahlung processes. Also, the effective cutoff was determined under the assumption that the photon energy was large. If it is not, then all propagators and vertices in Figure 14.5 must be dressed.

The rate for photon emission described above was computed by taking the imaginary part of Figure 14.5. In a Feynman diagram representation, the HTL correction induces a thermal mass which screens the singularity that would appear when the intermediate-quark propagator goes on-shell. Moving on to a higher topology in the number of loops and taking the imaginary part gives contributions like those of Figure 14.6. These

Fig. 14.6. Two photon-producing processes that appear to be of higher order in α_s than the Compton and annihilation contributions.

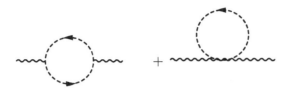

Fig. 14.7. The two Feynman diagrams that contribute to the ρ self-energy. The wavy lines are a neutral ρ, whereas the broken lines represent charged pions.

bremsstrahlung and pair-annihilation plus scattering contributions to the photon emission are superficially of higher order in α_s (they appear to be $\mathcal{O}(\alpha_s^2)$) than the ones we have discussed previously [23]. If the virtuality of the off-shell quark going into the vertex where the photon is being emitted is very small, there is an enhancement in the net thermal emission rate. This can be seen in the prefactor: $\alpha_s^2 T^2/m_q^2 \sim \alpha_s$. Those diagrams, naively of higher order in the strong coupling, contribute parametrically at the same order as the previous ones for low energy photons! The resolution of this apparent paradox was provided by a systematic identification of all processes contributing, to the leading order in α_s, to photon and lepton pair production [24].

For the evaluation of the emissivity of hot matter in the confined, hadronic sector, calculations have mainly followed the techniques outlined in this section. In particular, most practitioners have used relativistic kinetic theory and considered the contributing processes, such as $\pi\rho \to \pi\gamma$ and $\pi\pi \to \rho\gamma$, channel by channel. This closely parallels the first part of this section, where the annihilation and Compton contributions to the photon spectrum in hot QCD were considered. Many authors have contributed to this line of study. An early analysis was that given in reference [21]. A recent assessment of this issue can be found in [25].

14.5 Dilepton production

The calculation of dilepton radiation from a medium of strongly interacting partons follows steps very similar to those used for the calculation of real-photon emission. The first calculation using the HTL resummation technique was performed by Braaten, Pisarski, and Yuan [22]. The dilepton sector has also profited from a reappraisal of the electromagnetic emissivities, complete to leading order in α_s [26]. Instead of concentrating on the techniques that are appropriate for QCD again, we choose to consider radiation from a hot gas of mesons. This is more representative of conditions existing at temperatures below that of the phase transition, or just before the strongly interacting system freezes out. In a similar way, this discussion will illustrate the use of effective interactions to calculate the in-medium vector spectral density, as alluded to in Section 12.2. Conversely, we shall see that the methods in that chapter for inferring the spectral density from experimental data can be used to evaluate the emission of electromagnetic radiation.

This discussion closely follows that of Gale and Kapusta [27]. We start with the interaction between a vector meson and a conserved current. This is known to be renormalizable even if the vector meson is massive. For the case at hand, charged pions interact with a neutral ρ meson via the Lagrangian

$$\mathcal{L} = |D_\mu \Phi|^2 - m_\pi^2 |\Phi|^2 - \tfrac{1}{4}\rho_{\mu\nu}\rho^{\mu\nu} + \tfrac{1}{2}m_\rho^2\rho_\mu\rho^\mu \qquad (14.78)$$

where Φ is the complex charged pion field, $\rho_{\mu\nu} = \partial_\mu\rho_\nu - \partial_\nu\rho_\mu$ is the ρ field strength, and $D_\mu = \partial_\mu + ig_\rho\rho_\mu$ is the covariant derivative. The one-loop ρ self-energy in a gas of pions is represented by the two diagrams of Figure 14.7. In Euclidean space,

$$\Pi^{\mu\nu}(k) = -g_\rho^2 T \sum_n \int \frac{d^3p}{(2\pi)^3} \frac{(2p+k)^\mu(2p+k)^\nu}{(p^2+m_\pi^2)\left[(p+k)^2+m_\pi^2\right]}$$
$$+ 2\delta^{\mu\nu}g_\rho^2 T \sum_n \int \frac{d^3p}{(2\pi)^3} \frac{1}{p^2+m_\pi^2} \qquad (14.79)$$

Here, p_4 or $k_4 = 2\pi T \times$ an integer. The zero-temperature part of the self-energy may be evaluated using dimensional regularization. The vacuum part is then

$$\Pi^{\mu\nu}_{\text{vac}}(k) = (k^\mu k^\nu - k^2\delta^{\mu\nu})\frac{1}{3}\left(\frac{g_\rho}{4\pi}\right)^2$$
$$\times \left[\left(1+\frac{4m_\pi^2}{k^2}\right)^{3/2}\ln\left(\frac{\sqrt{1+4m_\pi^2/k^2}+1}{\sqrt{1+4m_\pi^2/k^2}-1}\right) - \frac{8m_\pi^2}{k^2} + C\right]$$
$$\qquad (14.80)$$

where C is a renormalization constant. The contribution from $T > 0$ is

$$\Pi^{44}_{\text{mat}}(k) = -\left(\frac{g_\rho^2}{2\pi}\right)\int_0^\infty \frac{dp\, p^2}{\omega}\frac{1}{e^{\beta\omega}-1}\left(\frac{4\omega^2-k_4^2}{2p|\mathbf{k}|}\ln a - 4 + \frac{2ik_4\omega}{p|\mathbf{k}|}\ln b\right)$$

(14.81)

$$\Pi^{4i}_{\text{mat}}(k) = -\frac{k^i k_4}{\mathbf{k}^2}\Pi^{44}_{\text{mat}}(k)$$

(14.82)

$$\Pi^{ij}_{\text{mat}}(k) = A\delta^{ij} + B\frac{k^i k^j}{\mathbf{k}^2}$$

(14.83)

The scalar functions A and B are given by

$$A = -\frac{1}{2}\left(\frac{g_\rho}{2\pi}\right)^2\int_0^\infty\frac{dp\, p^2}{\omega}\frac{1}{e^{\beta\omega}-1}\left(\frac{4(k_4^2-\mathbf{k}^2)}{\mathbf{k}^2} - \frac{2ik_4\omega(k_4^2+\mathbf{k}^2)}{p|\mathbf{k}|^3}\ln b\right.$$
$$\left.+ \frac{k_4^2(k_4^2-4\omega^2)+\mathbf{k}^2(\mathbf{k}^2+2k_4^2-4p^2)}{2p|\mathbf{k}|^3}\ln a\right)$$

(14.84)

$$B = -\frac{1}{2}\left(\frac{g_\rho}{2\pi}\right)^2\int_0^\infty\frac{dp\, p^2}{\omega}\frac{1}{e^{\beta\omega}-1}\left(\frac{4(\mathbf{k}^2-3k_4^2)}{\mathbf{k}^2} + \frac{2ik_4\omega(3k_4^2+\mathbf{k}^2)}{p|\mathbf{k}|^3}\ln b\right.$$
$$\left.+ \frac{3k_4^2(4\omega^2-k_4^2)+\mathbf{k}^2(4p^2-2k_4^2-\mathbf{k}^2)}{2p|\mathbf{k}|^3}\ln a\right)$$

(14.85)

with

$$a = \frac{\left(k_4^2+\mathbf{k}^2-2p|\mathbf{k}|\right)^2+4\omega^2 k_4^2}{\left(k_4^2+\mathbf{k}^2+2p|\mathbf{k}|\right)^2+4\omega^2 k_4^2}$$

$$b = \frac{\left(k_4^2+\mathbf{k}^2\right)^2-4\left(p|\mathbf{k}|+ik_4\omega\right)^2}{\left(k_4^2+\mathbf{k}^2\right)^2-4\left(p|\mathbf{k}|-ik_4\omega\right)^2}$$

(14.86)

and $\omega = \sqrt{p^2+m_\pi^2}$. Switching back to Minkowski space, we may write as in (5.46)

$$\Pi^{\mu\nu} = FP_{\text{L}}^{\mu\nu} + GP_{\text{T}}^{\mu\nu}$$

(14.87)

where $P_{\text{T/L}}^{\mu\nu}$ are the transverse and longitudinal projection operators. Using the relation between the self-energy and the full and bare propagators,

$$\Pi^{\mu\nu} = \left(\mathcal{D}^{-1}\right)^{\mu\nu} - \left(\mathcal{D}_0^{-1}\right)^{\mu\nu} \qquad (14.88)$$

and (14.87), we obtain

$$\mathcal{D}^{\mu\nu} = -\frac{P_{\mathrm{L}}^{\mu\nu}}{k^2 - m_\rho^2 - F} - \frac{P_{\mathrm{T}}^{\mu\nu}}{k^2 - m_\rho^2 - G} - \frac{k^\mu k^\nu}{m_\rho^2 k^2} \qquad (14.89)$$

For any linear response analysis and for lepton pair production rates we need the retarded ρ propagator. Therefore we will analytically continue the Matsubara frequency, $k_4 = 2\pi n T$, to $ik_4 = k_0 = E + i\epsilon$, where $\epsilon \to 0^+$. The scalar functions F and G acquire an imaginary part when a or b are negative. This happens when the variable of integration, p, lies in the interval

$$\left| E\sqrt{1 - 4m_\pi^2/M^2} - |\mathbf{k}| \right| \leq 2p \leq E\sqrt{1 - 4m_\pi^2/M^2} + |\mathbf{k}| \quad (14.90)$$

Here $M = \sqrt{k^2}$ is the invariant mass of the ρ and $E = \sqrt{M^2 + \mathbf{k}^2}$ is the total energy in the rest frame of the pion gas.

At zero temperature, dimensional regularization and renormalization yield equal longitudinal and transverse self-energies, which are finite:

$$
\begin{aligned}
F_{\mathrm{vac}} &= G_{\mathrm{vac}} \\
&= \frac{g_\rho^2}{48\pi^2} M^2 \Bigg\{ (1 - 4m_\pi^2/M^2)^{3/2} \\
&\quad \times \left(\ln\left| \frac{\sqrt{1 - 4m_\pi^2/M^2} + 1}{\sqrt{1 - 4m_\pi^2/M^2}\,1} \right| - i\pi\theta(M^2 - 4m_\pi^2) \right) \frac{8m_\pi^2}{M^2} + C \Bigg\}
\end{aligned}
$$

$$(14.91)$$

The bare and renormalized fields and masses are related by

$$\rho_\mu^{(0)} = \mathcal{Z}^{1/2} \rho_\mu \qquad \mathcal{Z}_0 = \left(m_\rho / m_\rho^{(0)} \right)^2 \qquad (14.92)$$

and the coupling constants are related by

$$\mathcal{Z}_0 g_\rho^{(0)} = \mathcal{Z}^{1/2} g_\rho \qquad (14.93)$$

We may choose $\mathcal{Z}_0 = \mathcal{Z}$ for convenience. Finally, for the physical mass to be m_ρ, we choose C in such a way that $\operatorname{Re} F_{\mathrm{vac}}(k^2 = m_\rho^2) = 0$.

Finally, at $T > 0$, $F = F_{\text{vac}} + F_{\text{mat}}$ and $G = G_{\text{vac}} + G_{\text{mat}}$, where

$$F_{\text{mat}} = \frac{g_\rho^2}{4\pi^2} \frac{M^2}{\mathbf{k}^2} \int_0^\infty \frac{dp\, p^2}{\omega} \frac{1}{e^{\beta\omega} - 1}$$

$$\times \left[\frac{4\omega^2 + E^2}{2p|\mathbf{k}|} (\ln|a| - i\pi\Delta) - 4 + \frac{2\omega E}{p|\mathbf{k}|} (\ln|b| + i\pi\Delta) \right] \qquad (14.94)$$

$$G_{\text{mat}} = \frac{g_\rho^2}{4\pi^2} \int_0^\infty \frac{dp\, p^2}{\omega} \frac{1}{e^{\beta\omega} - 1} \left[\frac{2(E^2 + \mathbf{k}^2)}{\mathbf{k}^2} - \frac{E\omega M^2}{p|\mathbf{k}|^3} (\ln|b| + i\pi\Delta) \right.$$

$$\left. + \frac{\mathbf{k}^2(4p^2 - \mathbf{k}^2 + 2E^2) - E^2(E^2 + 4\omega^2)}{4p|\mathbf{k}|^3} (\ln|a| - i\pi\Delta) \right]$$

$$(14.95)$$

where a and b are given in (14.86) and

$$\Delta = \begin{cases} 1 & \text{if } \left| E\sqrt{1 - 4m_\pi^2/M^2} - |\mathbf{k}| \right| \leq 2p \leq E\sqrt{1 - 4m_\pi^2/M^2} + |\mathbf{k}| \\ 0 & \text{otherwise} \end{cases}$$

$$(14.96)$$

We have shown (14.44) that the dilepton emission rate is related to the imaginary part of the retarded photon self-energy, at finite temperature. The vector meson dominance model (VMD) states that the hadronic electromagnetic current operator is given by the current–field identity

$$J_\mu = -\frac{e}{g_\rho} m_\rho^2 \rho_\mu - \frac{e}{g_\omega} m_\omega^2 \omega_\mu - \frac{e}{g_\phi} m_\phi^2 \phi_\mu \qquad (14.97)$$

The VMD is nonperturbative in the strong interaction and has had an impressive phenomenological success [29]. See also Exercise 14.7. We have encountered VMD before, in Section 12.2. The current–field identity turns the current–current correlation function into a field–field correlation function. Therefore to order e^2 but to all orders in the strong coupling, the dilepton emission rate can be written in terms of the in-medium vector spectral density, which is itself calculated with the effective hadronic Lagrangian:

$$E_+ E_- \frac{d^6 R}{d^3 p_+ d^3 p_-}$$

$$= \frac{2}{(2\pi)^6} \frac{e^4}{g_\rho^2} \frac{m_\rho^4}{M^4} \left(p_+^\mu p_-^\nu + p_+^\nu p_-^\mu - g^{\mu\nu} p_+ \cdot p_- \right) \text{Im}\, \mathcal{D}_{\mu\nu}^{\text{R}}(\omega, \mathbf{k}) \frac{1}{e^{\beta\omega} - 1}$$

$$(14.98)$$

To make the longitudinal and transverse contributions manifest, we may use $k^\mu = p_+^\mu + p_-^\mu$ and $q^\mu = p_+^\mu - p_-^\mu$ to write the expression in terms of

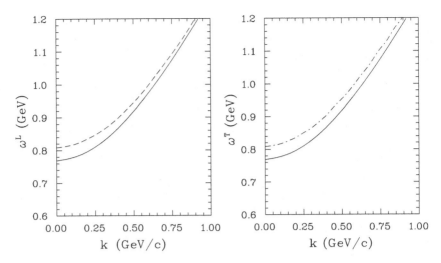

Fig. 14.8.   The dispersion relations for a ρ meson in its longitudinal and transverse polarization states. The curves are for $T = 0$ (lower) and 150 MeV (upper).

the real and imaginary parts of $F = F_R + iF_I$, and $G = G_R + iG_I$:

$$
\begin{aligned}
E_+ E_- & \frac{d^6 R}{d^3 p_+ d^3 p_-} \\
= & \frac{1}{(2\pi)^6} \frac{e^4}{g_\rho^2} \frac{m_\rho^4}{M^4} \left\{ \left[\mathbf{q}^2 - \left(\mathbf{q} \cdot \hat{\mathbf{k}} \right)^2 \right] \frac{-F_I}{\left(M^2 - m_\rho^2 - F_R \right)^2 + F_I^2} \right. \\
& \left. + \left[2M^2 - \mathbf{q}^2 + \left(\mathbf{q} \cdot \hat{\mathbf{k}} \right)^2 \right] \frac{-G_I}{\left(M^2 - m_\rho^2 - G_R \right)^2 + G_I^2} \right\} \frac{1}{e^{\beta\omega} - 1} \quad (14.99)
\end{aligned}
$$

This treatment may be generalized and extended to other mesons [30, 31]. This is necessary for a realistic treatment including chiral symmetry.

Finally, the effects of the interactions on the ρ meson may be quantified further by considering the longitudinal and transverse dispersion relations, which are found by locating the poles in the ρ propagator. They are generated by obtaining the self-consistent solutions of

$$
\begin{aligned}
(\omega^2)^L &= \mathbf{k}^2 + m_\rho^2 + F_R(\omega^L, |\mathbf{k}|, T) \\
(\omega^2)^T &= \mathbf{k}^2 + m_\rho^2 + G_R(\omega^T, |\mathbf{k}|, T)
\end{aligned}
\quad (14.100)
$$

The longitudinal and transverse dispersion relations are plotted in Figure 14.8. Observe that the in-medium energy asymptotically goes over to the free energy, with increasing momentum. This behavior is characteristic of a many-body effect.

14.6 *J/ψ* suppression

In the search for the quark–gluon plasma, the experimental signature of this new state of matter that enjoys the most popularity is that associated with the suppression of the J/ψ vector meson. The main ingredients of this simple and elegant idea [32] can briefly be summarized as follows. As the temperature increases, so will the effect of color Debye screening, which will ultimately cause the dissociation of the charmonium bound states. Suppression of the J/ψ was predicted before its experimental observation!

In nonrelativistic charmonium models, the interaction potential is most simply modeled as

$$V(r) = \sigma r - \frac{\alpha_{\text{eff}}}{r} \tag{14.101}$$

where σ is the string tension and α_{eff} is an effective Coulombic interaction coupling. The energy of the lowest bound state can be roughly estimated in a semiclassical approximation [32]. We start by writing

$$E(r) = 2m + \frac{1}{mr^2} + V(r) \tag{14.102}$$

where m is the c quark rest mass. The second term is obtained by invoking the uncertainty relation to write the kinetic term involving the reduced mass in coordinate space. The lowest bound state is found by minimizing the energy with respect to r. Taking $\alpha_{\text{eff}} \simeq 1/2$, $m \simeq 1.5$ GeV, and $\sigma = 0.19$ GeV^2 one obtains $r_{J/\psi} \simeq 0.3$ fm. This value is in qualitative agreement with that obtained through more sophisticated approaches and also confirms that, at $T = 0$, the size of the J/ψ is largely set by the confining part of the potential.

Now consider the high-temperature plasma phase. If the transition is first order, this is tantamount to choosing $T > T_{\text{c}}$. Since the quark–antiquark pair is heavy, it makes sense to use a static potential for their mutual interaction. We have discussed this already in Chapters 8 and 10. At leading order in the coupling, the interaction is modeled by one-gluon exchange, and at small momenta the gluon propagator develops an electric mass related to $\Pi_{00}(k)$. In pure SU(N) gauge theory, one calculates the real part of the finite-temperature one-loop gluon self-energy.

The Debye-screened color Coulomb potential is

$$V(r) = -\frac{N^2 - 1}{2N} \frac{g^2}{4\pi r} \exp(-m_{\text{el}} r) = -\frac{\alpha_{\text{s}}^{\text{eff}}}{r} \exp(-m_{\text{el}} r) \tag{14.103}$$

with $m_{\mathrm{el}}^2 = Ng^2T^2/3$. The coupling $\alpha_{\mathrm{s}}^{\mathrm{eff}}$ obtained above T_{c} is generally different from that used in the zero-temperature potential (14.101). Using [33]

$$g^2(T) = \frac{24\pi^2}{11N\ln\left(19.2T/\Lambda_{\overline{\mathrm{MS}}}\right)} \tag{14.104}$$

one may get an estimate for $T \sim \Lambda_{\overline{\mathrm{MS}}}$, which implies that $\alpha_{\mathrm{s}}^{\mathrm{eff}} \simeq 0.3$.

An interesting phenomenon, revealed by keeping the first powers in the momentum expansion of Π_{00}, is that of Friedel oscillations in QCD [33]. To see this, it is useful to recall that $F = -\Pi_{00} = F_{\mathrm{vac}} + F_{\mathrm{mat}}$. A low-momentum expansion for $F_{\mathrm{mat}}(0,\mathbf{k})$ has been performed in the temporal axial gauge and is [34]

$$\begin{aligned}
F_{\mathrm{mat}}(0,\mathbf{k}) = {}& \frac{1}{3}g^2NT^2 - \frac{1}{4}g^2NT|\mathbf{k}| \\
& - \frac{11}{48\pi^2}g^2N\mathbf{k}^2\left[\ln\left(\frac{\mathbf{k}^2}{T^2}\right) + \frac{2}{33} + 2(\gamma_{\mathrm{E}} - \ln 4\pi)\right]
\end{aligned} \tag{14.105}$$

The first term in this expansion is the electric mass, which is gauge invariant. The second term, linear in $\mathbf{k}$, is also gauge invariant. It is the same in the temporal-axial, Coulomb, and all covariant gauges. The reason is that this term modifies the plasmon effect in the thermodynamic potential; see Section 8.3. The next term has the same coefficient as that of the vacuum term, as it must in order to produce a temperature-dependent coupling constant. Keeping the terms that are subleading in the low-momentum expansion produces

$$V(r) = -\frac{N^2-1}{2N}\frac{g^2(T)}{2\pi^2 r}\int_0^\infty dz\,\frac{z\sin zx}{z^2 - 2tz + 1} \tag{14.106}$$

where $x = m_{\mathrm{el}}r$ and $t = 3m_{\mathrm{el}}/8T$. A contour integration puts the integral into the form

$$V(r) = -\frac{N^2-1}{2N}\frac{g^2(T)}{4\pi r}S(x,t) \tag{14.107}$$

The dimensionless screening function is

$$\begin{aligned}
S(x,t) = {}& 2\left(\cos tx + \frac{t}{\sqrt{1-t^2}}\sin tx\right)\exp\left(-x^2\sqrt{1-t^2}\right) \\
& - \frac{4t}{\pi}\int_0^\infty dy\,\frac{y^2\exp(-xy)}{(1-y^2)^2 + 4t^2y^2}
\end{aligned} \tag{14.108}$$

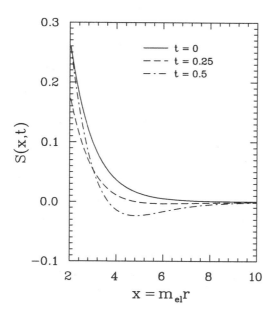

Fig. 14.9. The screening function $S(x, t)$ for different values of $t = 3m_{\text{el}}/8T$.

$S(x, t)$ has the following asymptotic expansion. For fixed x and small t (the high-temperature limit), $S \to e^{-x}$, the Debye-screening result. For fixed t and $x \to \infty$ (the long-distance limit), $S \to 8t/\pi x^3$. At very large distances the potential is repulsive and falls as a power, not as an exponential:

$$V(r) \to \frac{9}{4\pi^3} \left[\frac{N^2 - 1}{N^2} \right] \frac{1}{T^3 r^4}$$

The screening function is derived under the assumption that $x > 1$; that is, the low-momentum expansion of $F(0, \mathbf{k})$ has been used. This expression cannot be written in terms of elementary functions. The integration in (14.108) must be done numerically, and the results are plotted in Figure 14.9. We see that in general the inclusion of the momentum dependence of the gluon self-energy increases the screening for $1 < x < 3$ (or between one and three Debye lengths) but decreases the screening at greater distances. In fact, for large distances there is a slight antiscreening: the potential is repulsive instead of attractive.

Going back to the low-momentum expansion, we keep only the leading term. Inserting (14.103) into (14.102) and minimizing produces a value for $r_{J/\psi}$. All the temperature dependence is now contained in the value of m_{el}. After minimization, the algebraic equation to be solved is

$$\frac{2m_{el}}{m\alpha_{eff}} = x(x+1)e^{-x} \tag{14.109}$$

where $x = rm_{el}$. Notice that the left-hand side increases linearly with temperature (the variation of α_{eff} with T is only logarithmic). The right-hand side has a maximum at 1.62. At that point we have $(m_{el})_{max} = 0.81 m\alpha_s^{eff}$. Extracting the logarithmic dependence, and using the definition of the electric mass, yields the equation

$$T = \frac{0.81}{3\pi} mg(T) \tag{14.110}$$

Using $\Lambda_{\overline{MS}} = 220$ MeV, the above turns into a nonlinear equation for the maximum temperature at which the J/ψ exists. Solving it, one obtains $T_{max} \simeq 200$ MeV. Should the J/ψ disappear because of the mechanism discussed here, the higher-lying excitations of the charmonium bound states will already have dissolved. Remember that the J/ψ is an $n = 1$, $\ell = 0$ state, whereas the lesser-bound states are $\psi'(n = 2, \ell = 0)$ and $\chi_c(n = 2, \ell = 1)$.

The whole analysis in terms of potential models can only give a general idea of the dissociation pheomenon. First, the heavy quark potential should be determined directly using lattice QCD simulations at finite temperature. The nonperturbative studies could allow the study of the evolution of the gap between the charmonium bound-state mass and the open charm threshold, among other things. It has recently become possible to study directly the finite-temperature charmonium spectral density on the lattice [35]. All such studies are currently based on quenched lattices: they do not include quark loops, thermal or otherwise. This obvious shortcoming will have to be addressed in order to extract any quantitative result. Furthermore, one needs to reconstruct the thermal spectral densities from the thermal correlators: recent progress on this has been made possible by the use of Bayesian techniques in lattice analysis [36]. This topic is one for specialists. See Chapter 10 for the basic notions of QCD lattice gauge theory.

Any analysis of the charmonium spectrum in nuclear collisions will be incomplete unless supplemented by knowledge of what happens to those states in cold nuclear matter and in hot hadronic systems. These are all environments that are liable of influencing the measured J/ψ yields as well as those of the higher-lying states. There is at present considerable uncertainty, because the J/ψ sits at an energy scale that is not high enough for perturbative QCD to be totally reliable and because there is no direct experimental data on J/ψ–hadron cross sections. However, the yields of charmonium bound states as a function of the muon pair

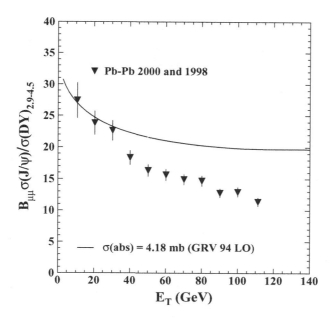

Fig. 14.10. The average of two data sets showing the J/ψ to Drell–Yan ratio (multiplied by the branching ratio into a dimuon pair) as a function of the transverse energy, in the invariant mass region 2.9 GeV/c^2 < M < 4.5 GeV/c^2. The solid line shows the effect of nuclear absorption with an absorption cross section that is extracted from proton–nucleus data. This plot is from [37], with kind permission of Springer Science and Business Media.

transverse energy (and hence of event-centrality) in proton–nucleus collisions can reveal the features that are germane to absorption in cold nuclear matter. In fact, proton–nucleus analyses provide an important class of control experiments, as plasma formation is not expected to occur there. Drell–Yan muon pairs serve as a background estimator, as they constitute the dominant source of continuum dileptons at the invariant masses of interest here. The ratios of cross sections for proton–nucleus collisions are then fitted to a Glauber prescription of normal nuclear matter absorption; this procedure leads to a value $\sigma_{\mathrm{abs}} = 4.18 \pm 0.35$ mb [37]. The extra absorption in nucleus–nucleus events, shown in Figure 14.10, is then deemed anomalous. Owing to the large multiplicities that are common in heavy ion collision environments, the interaction of the newly formed J/ψ with this hot hadronic matter also has to be considered. At the present time, many hadronic approaches claim to reproduce the anomalous SPS J/ψ absorption data with various degrees of success, thereby making the arguments claiming a a new state of matter considerably less compelling. This topic is still under investigation and will continue to be so in experimental measurements at RHIC and at the LHC.

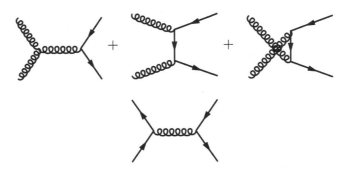

Fig. 14.11.   Tree-level diagrams for the processes $gg \to s\bar{s}$, and $q\bar{q} \to s\bar{s}$.

14.7 Strangeness production

Another signature of the presence of a nascent quark–gluon plasma created in high-energy nuclear collisions is that of strangeness production [38]. Strange quarks and antiquarks are absent in cold nuclear matter. They are found only in the parton distribution functions of the sea quarks, probed by deep inelastic scattering experiments. As a consequence their abundances are typical of those of quantum fluctuations. In a hot partonic system, however, the situation is different. High initial temperatures, greater than the strange quark mass, imply an abundance comparable with that of the lighter up and down quarks. The loss of confinement suggests comparable rates of production of up, down, and strange quarks.

Starting with no strange quarks (or antiquarks), estimates for the production of $s\bar{s}$ pairs can be obtained from lowest-order perturbative QCD. The contributing channels are those of gluon fusion and light $q\bar{q}$ annihilation. The relevant Feynman diagrams are shown in Figure 14.11. The invariant matrix elements have been calculated by several groups of workers to leading order in the strong coupling constant [39]. Labeling the processes in Figure 14.11 by a, b, c and d, respectively (going from left to right, starting from the top), the squared matrix elements summed over initial color, spin, and flavor states are

$$
\begin{aligned}
\sum |\mathcal{M}_a|^2 &= 16 \times 6(\pi\alpha_{\mathrm{s}})^2 \frac{(m^2 - t)(m^2 - u)}{3s^2} \\[2mm]
\sum |\mathcal{M}_b|^2 &= 16 \times 6(\pi\alpha_{\mathrm{s}})^2 \frac{2}{27} \frac{(m^2 - t)(m^2 - u) - 2m^2(m^2 + t)}{(m^2 - t)^2} \\[2mm]
\sum |\mathcal{M}_c|^2 &= 16 \times 6(\pi\alpha_{\mathrm{s}})^2 \frac{2}{27} \frac{(m^2 - t)(m^2 - u) - 2m^2(m^2 + u)}{(m^2 - u^2)^2} \\[2mm]
\sum |\mathcal{M}_d|^2 &= N_{\mathrm{f}}\, 6^2 (\pi\alpha_{\mathrm{s}})^2 \frac{16}{81} \frac{(m^2 - t)^2 + (m^2 - u)^2 + 2m^2 s}{s^2}
\end{aligned}
\tag{14.111}
$$

the interference terms being

$$\sum \mathcal{M}_a \mathcal{M}_b^* = 16 \times 6 \, (\pi \alpha_s)^2 \frac{(m^2 - t)(m^2 - u) + m^2(u - t)}{12s(m^2 - t)}$$

$$\sum \mathcal{M}_a \mathcal{M}_c^* = 16 \times 6 \, (\pi \alpha_s)^2 \frac{(m^2 - t)(m^2 - u) + m^2(u - t)}{12s(m^2 - u)} \quad (14.112)$$

$$\sum \mathcal{M}_b \mathcal{M}_c^* = 16 \times 6 \, (\pi \alpha_s)^2 \frac{m^2(s - 4m^2)}{108(m^2 - u)(m^2 - t)}$$

Here s, t, and u are the usual Mandelstam variables, N_f is the number of fermion flavors, and m is the strange quark mass. In the equations above, the numerical prefactors correspond to products of the degeneracy factors (spin × color) for the gluons (2×8) and quarks (2×3). For the processes under consideration a scale appropriate for the evaluation of the strong coupling yields $\alpha_s(s)$.

Given the above, the cross sections averaged over initial states are evaluated to be

$$\bar{\sigma}_{gg \to s\bar{s}} = \frac{2\pi \alpha_s^2}{3s} \left[\left(1 + \frac{4m^2}{s} + \frac{m^4}{s^2} \right) \tanh^{-1} w(s) - \left(\frac{7}{8} + \frac{31}{8} \frac{m^2}{s} \right) w(s) \right]$$

$$\bar{\sigma}_{q\bar{q} \to s\bar{s}} = \frac{8\pi \alpha_s^2}{27s} \left(1 + \frac{2m^2}{s} \right) w(s) \quad (14.113)$$

where $w(s) = \sqrt{1 - 4m^2/s}$. The rate for pair production can then be calculated using the usual formalism of relativistic kinetic theory. Quite generally, one may write a rate for the reaction $a_1 + a_2 \to X$, in the independent-particle limit, as

$$R(a_1 + a_2 \to X) = \frac{1}{1 + \delta_{a_1, a_2}} \int \frac{d^3 k_1}{(2\pi)^3} f(\mathbf{k}_1) \frac{d^3 k_2}{(2\pi)^3} f(\mathbf{k}_2) \sigma(a_1 + a_2 \to X) \, v_{\text{rel}}$$

$$(14.114)$$

with

$$v_{\text{rel}} = \frac{(k_1 \cdot k_2)^2 - m_a^4}{E_1 E_2} \quad (14.115)$$

In the case where the initial-state fields are massless, $v_{\text{rel}} = s/(2E_1 E_2)$ with $s = (k_1 + k_2)^2$.

The invariant rate (the number of reactions per unit time per unit volume) is then

$$R = \frac{d^4 N}{dt d^3 x} = \frac{1}{2} \int_{4m^2}^{\infty} ds \, s \, \delta(s - (k_1 + k_2)^2) \int \frac{d^3 k_1}{(2\pi)^3 E_1} \int \frac{d^3 k_2}{(2\pi)^3 E_2}$$

$$\times \left(\frac{1}{2} f_g(\mathbf{k}_1) f_g(\mathbf{k}_2) \bar{\sigma}_{gg \to s\bar{s}}(s) + N_f \, f_q(\mathbf{k}_1) f_{\bar{q}}(\mathbf{k}_2) \bar{\sigma}_{q\bar{q} \to s\bar{s}}(s) \right)$$

$$(14.116)$$

Note that the distribution functions here contain the appropriate degeneracy factor,

$$f_g(\mathbf{k}) = 16 \, \frac{1}{e^{\beta|\mathbf{k}|} - 1}$$

so that the gluon density is

$$n_g = \frac{N_g}{V} = \int \frac{d^3k}{(2\pi)^3} f_g$$

and similarly for the quarks and antiquarks.

Inserting the appropriate Bose–Einstein or Fermi–Dirac distribution functions, the net rate may be computed numerically. Doing this with the quark chemical potential set to zero, one finds that the gluon contribution dominates the contribution with the $q\bar{q}$ initial state. For the gluon fusion rate, expanding the Bose–Einstein distribution functions for $T \ll m$ yields

$$R_g = \frac{4T}{\pi^4} \int_{4m^2}^{\infty} ds \, s^{3/2} \bar{\sigma}_{gg \to s\bar{s}} \sum_{k,\ell=1} \frac{1}{\sqrt{k\ell}} K_1\left(\frac{\sqrt{k\ell s}}{T}\right)$$

$$\simeq \frac{7}{6\pi^2} \alpha_s^2 m T^3 e^{-2m/T} \left(1 + \frac{51}{14}\frac{T}{m} + \cdots\right) \qquad (14.117)$$

One may divide out the temperature dependence and plot a dimensionless rate, $R/\alpha_s T^4$, against m/T, where m is the strange quark mass. A parametrization of these results over the temperature range considered here is perhaps useful for modeling purposes. An excellent parametrization for the range of m/T plotted is provided by $R/\alpha_s^2 T^4 = (a + bx^2)\exp(-cx)$, with $x = m/T$, $a = 0.937$, $b = 0.958$, and $c = 2.715$. The fit is shown, together with the result of the numerical rate calculation, in Figure 14.12.

When the density of $s\bar{s}$ pairs increases, their annihilation will start to deplete the population of strange quarks. This depletion rate will be proportional to the square of the strange quark density. Then the rate equation for a static (nonexpanding) system is

$$\frac{dn_s(t)}{dt} = R\left[1 - \left(\frac{n_s(t)}{n_s^{eq}}\right)^2\right] \qquad (14.118)$$

For small departures from equilibrium, such that $n(t) = n^{eq} + \delta n(t)$ where $|\delta n(t)| \ll n^{eq}$, we may linearize (14.118):

$$\frac{d\,\delta n_s(t)}{dt} = -\frac{\delta n_s(t)}{\tau_{eq}} \qquad \tau_{eq} = \frac{n_s^{eq}}{2R} \qquad (14.119)$$

Therefore, a large rate means a short equilibration time.

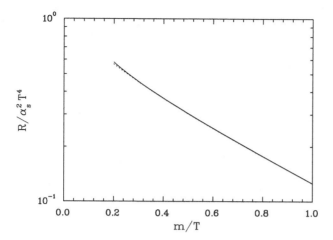

Fig. 14.12.   The dimensionless rate for emission of $s\bar{s}$ pairs from a sum of all tree-level partonic processes with u and d quarks, as a function of the ratio of the strange quark mass to the temperature (solid line). Here $\alpha_{\rm s} = 0.6$ and $m = 150$ MeV. The parametrization discussed in the text corresponds to the dotted line.

We now consider the fate of the strange hadrons in the portion of the system's spacetime trajectory that is in the confined sector. Let us assume that the system has zero net baryon number and that the species present are light pseudoscalars only. Anticipating the effect of high temperatures, we approximate the distribution functions to be of the Boltzmann type with vanishing chemical potentials. Then all integrals but one can be performed in (14.114), to yield

$$R(a_1 + a_2 \to X) = \frac{T^6}{16\pi^4} \int_{z_0}^{\infty} \sigma(E) z^2 (z^2 - 4z_a^2) K_1(z)\, dz \quad (14.120)$$

where $z = E/T$, E is the center-of-mass energy, and $z_a = m_a/T$. For the annihilation process, $z_0 = 2z_a$. If $a_1 + a_2 \to b + c$ and $2m_a < m_b + m_c$ then $z_0 = (m_b + m_c)/T$. The reader is invited to verify that this expression agrees with the leading term ($k = \ell = 1$) in (14.117).

There is not much data on strangeness production in mesonic annihilation. Some estimates exist of the cross section for the process $\pi^+\pi^- \to K^+K^-$ from measurements of $\pi^-p \to K^+K^-n$ [40]. These estimates find that the cross section is roughly constant as a function of energy, with a mean value of $\sigma_0 = 5/3$ mb. With a total of three isospin channels, the total cross section is thus $3\sigma_0 = 5$ mb. Then

$$R(\pi\pi \to K\bar{K})$$
$$= \frac{3\sigma_0 T^6}{16\pi^4} \left[z_0^2 (z_0^2 - 4z_a^2 + 8) K_0(z_0) + 4z_0(z_0^2 - 2z_a^2 + 4) K_1(z_0) \right] \quad (14.121)$$

In this case, $z_0 = 2m_K/T$ and $z_a = m_\pi/T$.

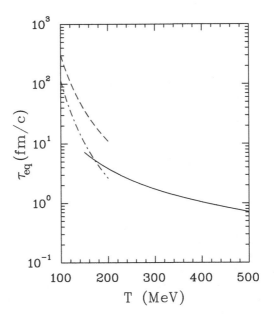

Fig. 14.13. The time constant τ_{eq} for chemical equilibration as a function of temperature for different processes: solid curve, τ_{eq} for the partonic reactions $gg \to s\bar{s}$ and $q\bar{q} \to s\bar{s}$ (with $q = u$, d); broken curve, τ_{eq} for $\pi\pi \to K\bar{K}$; broken and dotted curve, the equilibration time for the $K\bar{K}$ annihilation processes. The parameters used here are $\alpha_{\mathrm{s}} = 0.6$ and m (strange quark mass) = 150 MeV.

Another useful reaction for evaluating the population of strangeness-carrying hadrons is $K^+K^- \to$ nonstrange hadrons. Its magnitude may be estimated from that of $p\bar{p} \to$ charged hadrons [41]:

$$\sigma(p\bar{p} \to \text{charged hadrons}) = A' + \frac{B'}{\sqrt{(E/m_p)^2 - 4}} \qquad (14.122)$$

with $A' = 38.25$ mb and $B' = 36$ mb. Since K^+K^- has four valence quarks whereas $p\bar{p}$ has six, a simple estimate may then be obtained by multiplying A' and B' by $(2/3)^2$ and replacing m_p by m_K. Finally, bear in mind that a K^- can equally annihilate on a K^0 or a K^+. Putting all this together we arrive at

$$R(K\bar{K} \text{ annihilation}) = 2(R_A + R_B)$$

where

$$R_A = \frac{AT^6}{8\pi^4} z_0^3 K_3(z_0)$$

$$R_B = \frac{BT^6}{64\pi^3} z_0(3 + 3z_0 + z_0^2)e^{-z_0}$$

(14.123)

with $z_0 = 2m_K/T$, $A = 17$ mb, and $B = 16$ mb. It is revealing to plot the relevant strangeness-equilibration time constants, as evaluated from (14.119). The rates were integrated numerically in the Boltzmann limit for the distribution functions and were used to obtain the various relaxation times. These are shown in Figure 14.13. Of course, not all processes operate over the complete temperature range shown there.

This figure is revealing in many aspects. First, note the strong temperature-dependence. Second, this plot shows why strangeness-enhancement is considered a promising probe for the formation of the quark–gluon plasma. The smallness of τ_{eq} for the partonic contributions indicates that gluons and light quarks will reach equilibrium during the early stages of the plasma phase. However, note that the time constant for $s\bar{s}$ in the plasma phase is within a factor 2 of that for $K\bar{K}$ annihilation in the hadron phase for the interesting temperature interval of $150 < T < 250$ MeV. This suggests that strangeness production and annihilation in the hadronic phase will be comparable in magnitude with that in the plasma phase. This also means that the actual usefulness of this observable will depend on the details of the evolution scenario. Finally, the relationship between the rates for $\pi\pi \to K\bar{K}$ and for $K\bar{K}$ annihilation is as it should be. If the latter pairs could only annihilate into a pair of pions, the two rates would be equal by detailed balance. However, two kaons may annihilate into a many-pion (more than two) final state, and this will increase the net rate and decrease the related time constant.

In order to calculate how the strangeness density evolves in time, the spacetime evolution is needed. An increase in volume will cause a proportionate decrease in density even in the absence of interactions. This means that (14.118) needs to be supplemented by a dilution term:

$$\frac{dn_s(t)}{dt} = R\left[1 - \left(\frac{n_s(t)}{n_s^{eq}}\right)^2\right] - \frac{n_s(t)}{V(t)}\frac{dV(t)}{dt}$$

(14.124)

In the Bjorken model, the volume grows linearly with time because the entropy density drops inversely with time, (14.10). Therefore in this case

$$\frac{dn_s(t)}{dt} = R(T(t))\left[1 - \left(\frac{n_s(t)}{n_s^{eq}}\right)^2\right] - \frac{n_s(t)}{t}$$

(14.125)

The initial condition, $n_s(t_0)$, needs to be chosen. Three possibilities include: (i) no strange quarks; (ii) strange quarks in chemical equilibrium; (iii) a strange quark abundance determined by proton–proton collisions at the same energy. If the equilibration rate is high enough compared with the expansion rate, strange quarks will come to equilibrium no matter what the initial condition.

The equation above is valid in the plasma phase. Once the plasma begins to convert into hadrons, the rate equation for kaons is

$$\frac{dn_{K^-}(t)}{dt} = R_h(T_c) \left(1 - \left(\frac{n_{K^-}(t)}{n_{K^-}^{\text{eq}}(T_c)} \right)^2 \right)$$

$$- \frac{n_{K^-}(t)}{\hat{f}(t)\,t} \frac{d}{dt}\left[\hat{f}(t)\,t\right] + \frac{1}{2}\frac{n_s(t)}{\hat{f}(t)}\frac{d\hat{f}(t)}{dt} \qquad (14.126)$$

where $\hat{f}(t) = 1 - f(t)$ is the volume fraction in the plasma phase and R_h is the rate in the hadron phase. The dilution term is slightly different from its previous version. The last term given the gain from strangeness conversion into the hadron phase from the plasma phase. The factor $1/2$ accounts for the fact that a given s quark is equally likely to end up in a $\bar{K}^0$.

The strangeness content as a function of time is given by integration of the differential equations presented in this section. It is clear how to adapt this rate equation to the evolution in the purely hadronic phase. When comparing with actual measurements, perhaps a more realistic estimate will need to include multistrange baryons and strange antibaryons, as well as to consider the effect of different and more sophisticated spacetime evolution scenarios. The interested reader is invited to consult the research literature for the current status of strangeness as a probe of heavy ion collisions.

14.8 Exercises

14.1   Construct the pressure and energy density as functions of temperature for the three equations of state presented in Section 14.1.

14.2   A simple way to model the effect of the transition from one-dimensional to three-dimensional expansion is to replace the formula $s(\tau) = s(\tau_0)\tau_0/\tau$ in the Bjorken model by $s(\tau) = [s(\tau_0)\tau_0 R^2]/[\tau(\tau^2 + R^2)]$, where R is the nuclear radius. This takes into account the time delay for the rarefaction wave from the surface to reach the center of the hot matter. Calculate the temperature as a function of proper time for the three equations of state of Section 14.1, and plot the result similarly to Figure 14.2.

14.3 Derive the expression for the photon production rate in relativistic kinetic theory (14.46).

14.4 Consider the rates for photon emission through the Compton and annihilation processes, (14.54) and (14.55). These were evaluated assuming that the initial-state distribution functions could be approximated by their Maxwell–Boltzmann form. Show that, keeping the quantum distribution functions in the initial state, the rates from the Compton and annihilation processes become [42]

$$E\frac{d^3 R^{\text{Compton}}}{d^3 p} = \frac{5}{9}\frac{\alpha\alpha_{\text{s}}}{4\pi^2}\frac{1}{e^{E/T}+1}T^2\left[\ln\left(\frac{4ET}{k_{\text{c}}^2}\right) + C_{\text{F}}'\right]$$

where

$$C_{\text{F}}' = -\gamma_{\text{E}} + \frac{1}{2} - \frac{8}{\pi^2}\sum_{n=0}^{\infty}\ln\frac{(2n+1)}{(2n+1)^2}$$

and

$$E\frac{d^3 R^{\text{annihilation}}}{d^3 p} = \frac{5}{9}\frac{\alpha\alpha_{\text{s}}}{4\pi^2}\frac{1}{e^{E/T}+1}T^2\left[\ln\left(\frac{4ET}{k_{\text{c}}^2}\right) + C_{\text{B}}'\right]$$

where

$$C_{\text{B}}' = -\gamma_{\text{E}} - 1 - \frac{8}{\pi^2}\sum_{n=0}^{\infty}\ln\frac{(2n+1)}{(2n+1)^2}$$

14.5 Prove (14.65).

14.6 Derive the expressions (14.94) and (14.95) for the scalar functions F and G at finite temperature.

14.7 Show that the pion electromagnetic form factor in the vector meson dominance model (VMD) is predicted to be

$$F_\pi(M) = \frac{m_\rho^2 + F_{\text{vac}}(0)}{m_\rho^2 + F_{\text{vac}}(M) - M^2}$$

and show that this reproduces the Gounaris–Sakurai formula [28].

14.8 Construct the grand canonical partition function for a gas of hadrons containing all light mesons, baryons, and resonances, up to a mass of 2 GeV. Use the *Particle Data Table* [43]. For several combinations of temperature and chemical potential (say $T = 100$, 150, and 200 MeV; $\mu_B = 250$ and 550 MeV), evaluate the density of positively charged pions (including the resonance-decay contribution) divided by that of the thermal pions.

14.9   Calculate numerically the rate in (14.116), and show that the contribution from quark–antiquark annihilation is negligible with respect to the gluon fusion rate. You should plot the two rates for $100 < T < 300$ MeV.

14.10   Assuming a first-order phase transition, obtain the behavior of the strangeness density as a function of time in the Bjorken model. Do the calculation for two initial temperatures, $T_0 = 250$ and 500 MeV. Plot $n_s(t)/n_s^{eq}$ as a function of time, and compare with the results shown in [44].

References

1. Fermi, E., *Phys. Rev.* **81**, 83 (1951).
2. Landau, L. D., *Izv. Akad. Nauk SSSR, ser. fiz.* **17**, 51 (1953).
3. Bjorken, J. D., *Phys. Rev. D* **27**, 140 (1983).
4. Chodos, A., Jaffe, R. L., Johnson, K., Thorn, C. B., and Weisskopf, V. F., *Phys. Rev. D* **9**, 3471 (1974).
5. Becattini, F., *Z. Phys. C* **69**, 485 (1996).
6. Braun-Munzinger, P., Redlich, K., and Stachel, J. (2003). In *Quark–Gluon Plasma 3* (World Scientific, Singapore).
7. Cleymans, J., and Redlich, K., *Phys. Rev. Lett.* **81**, 5284 (1998); *Phys. Rev. C* **60**, 054908 (1999).
8. Cleymans, J., Oeschler, H., and Redlich, K., *Phys. Rev. C* **59**, 1663 (1999); Averbeck, R., Holzmann, R., Metag, V., and Simon, R. S., *Phys. Rev. C* **67**, 024903 (2003); Becattini, F., Cleymans, J., Keranen, A., Suhonen, E., and Redlich, K., *Phys. Rev. C* **64**, 024901 (2001).
9. Braun-Munzinger, P., Stachel, J., Wessels, J. P., and Xu, N., *Phys. Lett.* **B344**, 43 (1995); **B365**, 1 (1996); Braun-Munzinger, P., Heppe, I., and Stachel, J., *Phys. Lett.* **B465**, 15 (1999); Becattini, F., Gazdzicki, M., Keranen, A., Manninen, J., and Stock, R., *Phys. Rev. C* **69**, 024905 (2004).
10. Redlich, K., and Tounsi, A., *Eur. Phys. J. C* **24**, 589 (2002).
11. Braun-Munzinger, P., Magestro, D., Redlich, K., and Stachel, J., *Phys. Lett.* **B518**, 41 (2001); Cleymans, J., Kampfer, B., Kaneta, M., Wheaton, S., and Xu, N., *Phys. Rev. C* **71**, 054901 (2005); Broniowski, W., and Florkowski, W., *Phys. Rev. C* **65**, 064905 (2002).
12. Letessier, J., and Rafelski, J. (2002). *Hadrons and Quark–Gluon Plasma* (Cambridge University Press, Cambridge).
13. Gerber, P., Leutwyler, H., and Goity, J. L., *Phys. Lett.* **B246**, 513 (1990).
14. Bleicher, M., and Aichelin, J., *Phys. Lett.* **B530**, 81 (2002): Bravina, L., *et al., Nucl. Phys.* **A698**, 383 (2002); *Phys. Rev. C* **66**, 014906 (2002).
15. Shuryak, E. V., *Sov. J. Nucl. Phys.* **28**, 408 (1978).
16. Kajantie, K., and Miettinen, H. I., *Z. Physik* **C9**, 341 (1981).
17. Halzen, F., and Liu, H. C., *Phys. Rev. D* **25**, 1842 (1982).
18. Sinha, B., *Phys. Lett.* **B128**, 91 (1983).
19. Hwa, R. C., and Kajantie, K., *Phys. Rev. D* **32**, 1109 (1985).

20. Staadt, G., Greiner, W., and Rafelski, J., *Phys. Rev. D* **33**, 66 (1986).
21. Kapusta, J., Lichard, P., and Seibert, D., *Phys. Rev. D* **44**, 2774 (1991).
22. Braaten, E., Pisarski, R. D., and Yuan, T. C., *Phys. Rev. Lett.* **64**, 2242 (1990).
23. Aurenche, P., Gélis, F., Kobes, R., and Petitgirard, E., *Phys. Rev. D* **54**, 5274 (1996); Aurenche, P., Gélis, F., Kobes, R., and Zaraket, H., *Phys. Rev. D* **58**, 085003 (1998).
24. Arnold, P., Moore, G. D., and Yaffe, L. G., *JHEP*, **0112**, 9 (2001); **0206**, 30 (2002).
25. Turbide, S., Rapp, R., and Gale, C., *Phys. Rev. C* **69**, 014903 (2004).
26. Aurenche, P., Gélis, F., Moore, G. D., and Zaraket, H., *JHEP* **0212**, 6 (2002).
27. Gale, C., and Kapusta, J. I., *Nucl. Phys.* **B357**, 65 (1991).
28. Gounaris, G. J. and Sakurai, J. J., *Phys. Rev. Lett.* **21**, 244 (1968).
29. Sakurai, J. J. (1969). *Currents and Mesons* (University of Chicago Press, Chicago).
30. Rapp, R., and Wambach, J., *Adv. Nucl. Phys.* **25**, 1 (2000).
31. Rapp, R., and Gale, C., *Phys. Rev. C* **60**, 024903 (1999).
32. Matsui, T., and Satz, H., *Phys. Lett.* **B178**, 416 (1986).
33. Gale, C., and Kapusta, J., *Phys. Lett.* **B198**, 89 (1987).
34. Kajantie, K., and Kapusta, J., *Ann. Phys. (NY)* **160**, 477 (1985).
35. Datta, S., Karsch, F., Petreczky, P., and Wetzorke, I., *Phys. Rev. D* **69**, 094507 (2004).
36. Asakawa, M., Hatsuda, T., and Nakahara, Y., *Prog. Part. Nucl. Phys.* **46**, 459 (2001).
37. Alessandro, B., *et al.*, A new measurement of J/ψ suppression in Pb-Pb collisions at 158 GeV per nucleon, *Eur. J. Phys. C* **39**, 335 (2005).
38. Rafelski, J., and Müller, B., *Phys. Rev. Lett.* **48**, 1066 (1982); **56**, 2334 (1986); Koch, P., Müller, B., and Rafelski, J., *Phys. Rep.* **142**, 167 (1986).
39. Georgi, H. M., *et al.*, *Ann. Phys. (NY)*, **114**, 273 (1978); Combridge, B. L., *Nucl. Phys.* **B151**, 429 (1979); Matsui, T., Svetitsky, B., and McLerran, L. D., *Phys. Rev. D* **34**, 783 (1986); **34**, 2047 (1986); **37**, 844 (1988).
40. Grayer, G., *et al.*, (1973). In *AIP Conf. Proc. 13* (AIP, New York).
41. Hamilton, R. P., *et al.*, *Phys. Rev. Lett.* **44**, 1182 (1980).
42. Wong, C.-Y. (1994). *Introduction to High-Energy Heavy Ion Collisions* (World Scientific, Singapore).
43. Eidelman, S., *et al.*, *Phys. Lett.* **B592**, 1 (2004).
44. Kapusta, J., and Mekjian, A., *Phys. Rev. D* **33**, 1304 (1986).

Bibliography

A reprint volume containing many of the pioneering papers on relativistic heavy ion collisions

Kapusta, J., Müller, B., and Rafelski, J. (2003). *Quark–Gluon Plasma: Theoretical Foundations* (Elsevier, Amsterdam).

Some books on the physics of relativistic heavy ion collisions

Csernai, L. P. (1994). *Introduction to Relativistic Heavy Ion Collisions* (Wiley and Sons, New York).

Wong, C.-Y. (1994). *Introduction to High-Energy Heavy Ion Collisions* (World Scientific, Singapore).

Letessier, J., and Rafelski, J., (2002). *Hadrons and Quark–Gluon Plasma* (Cambridge University Press, Cambridge).

Hwa, R., and Wang, X.-N. (eds.), (2003). *Quark–Gluon Plasma 3* (World Scientific, Singapore).

Shuryhak, E. V. (2004). *The QCD Vacuum, Hadrons and the Superdense Matter* (World Scientific, New Jersey).

A series of international conferences on the subject of relativistic heavy ion collisions:

Quark Matter 2002: The 16th Int. Conf. on Ultra-Relativistic Nucleus–Nucleus Collisions, Gutbrod, H., Aichelin, G., and Werner, K., (eds.), *Nucl. Phys.* **A715**, 3 (2003).

Quark Matter 2004: The 17th Int. Conf. on Ultra-Relativistic Nucleus–Nucleus Collisions, Ritter, H.-G. and Wang, X.-N. (eds.), *J. Phys. G* **30**, S633 (2004).

15

Weak interactions

In the 1960s and early 1970s a theory was developed by Glashow [1], Weinberg [2], Salam [3], 't Hooft [4], and others that unified the weak and electromagnetic interactions. This theory is presently in accord with all experimental information. It is not our purpose here to go into a detailed exposition of the model or the history of weak interaction physics. Rather, we want to show that the spontaneously broken gauge symmetry that is the cornerstone of the theory can be restored in a phase transition at a critical temperature of order 100 GeV. The existence and order of this transition depend on details that we shall discuss in this chapter.

15.1 Glashow–Weinberg–Salam model

We begin with a theory involving bosons only. The essence of the model can be found without the inclusion of fermions: they will be added later. The Lagrangian is

$$\mathcal{L} = (D_\mu \Phi)^\dagger (D^\mu \Phi) + c^2 \Phi^\dagger \Phi - \lambda \left(\Phi^\dagger \Phi \right)^2$$
$$- \tfrac{1}{4} g^{\mu\nu} g_{\mu\nu} - \tfrac{1}{4} f_a^{\mu\nu} f_{\mu\nu}^a \tag{15.1}$$

This Lagrangian has an SU(2) × U(1) symmetry. There is an SU(2) gauge field A_μ^a and a U(1) gauge field B_μ. The field strengths are

$$f_{\mu\nu}^a = \partial_\mu A_\nu^a - \partial_\nu A_\mu^a - g\epsilon^{abc} A_\mu^b A_\nu^c \tag{15.2}$$
$$g_{\mu\nu} = \partial_\mu B_\nu - \partial_\nu B_\mu \tag{15.3}$$

There is a covariant derivative

$$D_\mu = \partial_\mu + \tfrac{1}{2} ig A_\mu^a \tau^a + \tfrac{1}{2} ig' B_\mu \tag{15.4}$$

which acts on a complex SU(2) field

$$\Phi = \frac{1}{\sqrt{2}}\begin{pmatrix} \phi_1 + i\phi_2 \\ \phi_3 + i\phi_4 \end{pmatrix} \tag{15.5}$$

Note that, according to (15.1), A_μ^a and B_μ are massless spin-1 bosons and, if $c^2 > 0$, Φ is a tachyon. Thus we should expect spontaneous symmetry breaking. Altogether, there are apparently 12 spin degrees of freedom.

Owing to the gauge symmetry we may choose, without loss of generality, the vacuum expectation value

$$\langle \Phi \rangle = \frac{1}{\sqrt{2}}\begin{pmatrix} 0 \\ v \end{pmatrix} \tag{15.6}$$

where v is a real constant. Then, for arbitrary Φ, we write

$$\Phi = \frac{1}{\sqrt{2}}U^{-1}(\boldsymbol{\zeta})\begin{pmatrix} 0 \\ v + \eta \end{pmatrix} \tag{15.7}$$

where $\boldsymbol{\zeta}(\mathbf{x},t)$ and $\eta(\mathbf{x},t)$ are the independent fields and

$$U(\boldsymbol{\zeta}) = \exp\left(\frac{-i\boldsymbol{\zeta}\cdot\boldsymbol{\tau}}{2v}\right) \tag{15.8}$$

This is the so-called unitary, or U, gauge. It is a useful gauge since it makes the particle content of the theory manifest.

Now let

$$\Phi \to \Phi' = U(\boldsymbol{\zeta})\Phi = \frac{1}{\sqrt{2}}\begin{pmatrix} 0 \\ v + \eta \end{pmatrix} \tag{15.9}$$

This is just a particular SU(2) gauge transformation, such that

$$B_\mu \to B_\mu \ ,$$
$$\boldsymbol{\tau}\cdot\boldsymbol{A}_\mu \to \boldsymbol{\tau}\cdot\boldsymbol{A}_\mu' = U(\boldsymbol{\zeta})\left(\boldsymbol{\tau}\cdot\boldsymbol{A}_\mu - \frac{i}{g}U^{-1}(\boldsymbol{\zeta})\partial_\mu U(\boldsymbol{\zeta})\right)U^{-1}(\boldsymbol{\zeta}) \tag{15.10}$$

After some algebra, the Lagrangian is expressed in terms of the independent fields as

$$\begin{aligned}
\mathcal{L} &= \tfrac{1}{2}(\partial_\mu\eta)(\partial^\mu\eta) + \tfrac{1}{2}c^2(v+\eta)^2 - \tfrac{1}{4}\lambda(v+\eta)^4 \\
&\quad + \tfrac{1}{4}\Phi'^\dagger(g'B_\mu + g\boldsymbol{\tau}\cdot\boldsymbol{A}_\mu)(g'B^\mu + g\boldsymbol{\tau}\cdot\boldsymbol{A}^\mu)\Phi' \\
&\quad - \tfrac{1}{4}g^{\mu\nu}g_{\mu\nu} - \tfrac{1}{4}f_a^{\mu\nu}f_{\mu\nu}^a
\end{aligned} \tag{15.11}$$

This can be written as the sum of a classical part, $\mathcal{L}_{\text{cl}}$, a part quadratic in the fields, $\mathcal{L}_{\text{quad}}$, and a part giving rise to interactions that is cubic and quartic in the fields, $\mathcal{L}_{\mathrm{I}}$;

$$\mathcal{L}_{\mathrm{cl}} = \tfrac{1}{2}c^2 v^2 - \tfrac{1}{4}\lambda v^4 \tag{15.12}$$

$$
\begin{aligned}
\mathcal{L}_{\mathrm{quad}} = {} & \tfrac{1}{2}(\partial_\mu \eta)^2 - \tfrac{1}{2}(3\lambda v^2 - c^2)\eta^2 \\
& - \tfrac{1}{4}(\partial_\mu B_\nu - \partial_\nu B_\mu)^2 - \tfrac{1}{4}(\partial_\mu A_\nu^c - \partial_\nu A_\mu^c)^2 \\
& + \tfrac{1}{8}v^2 \left[(g' B_\mu - g A_\mu^3)^2 + g^2 (A_\mu^1)^2 + g^2 (A_\mu^2)^2 \right]
\end{aligned} \tag{15.13}
$$

We define new fields

$$
\begin{aligned}
W_\mu^\pm &= \left(A_\mu^1 \pm i A_\mu^2 \right) / \sqrt{2} \\
Z_\mu &= \left(g' B_\mu - g A_\mu^3 \right) / \sqrt{g^2 + g'^2} \\
A_\mu &= \left(g B_\mu + g' A_\mu^3 \right) / \sqrt{g^2 + g'^2}
\end{aligned} \tag{15.14}
$$

The masses are

$$
\begin{aligned}
m_\eta^2 &= 3\lambda v^2 - c^2 \\
m_A &= 0 \\
m_W &= \tfrac{1}{2}g v \\
m_Z &= \tfrac{1}{2}\sqrt{g^2 + g'^2}\, v
\end{aligned} \tag{15.15}
$$

The tachyon is avoided as long as $v^2 \geq c^2/3\lambda$. In fact, from (15.12) we see that the classical minimum occurs at $v^2 = v_0^2 = c^2/\lambda$, so that indeed the model shows spontaneous symmetry breaking.

After addition of the fermions, it becomes possible to identify the fields and parameters described above: A_μ is the photon, $W^\pm$ and Z are the weak interaction bosons, and η is the as yet unobserved Higgs boson. Since all these are massive except for the photon, the total number of spin degrees of freedom is 12, the same as before, since the $W^\pm$ and Z each acquirie one degree of freedom from the Φ field. The electric charge is

$$e = \frac{gg'}{\sqrt{g^2 + g'^2}} \tag{15.16}$$

and the Weinberg angle is defined by

$$\tan\theta_{\mathrm{W}} = \frac{g'}{g} \tag{15.17}$$

Experimentally, it is found that $e = 0.3028\ldots$ and $\sin^2\theta_{\mathrm{W}} = 0.226 \pm 0.004$. This leads to $g = 0.637$ and $g' = 0.344$. It turns out that the vacuum field v_0 is related directly to the Fermi constant: $v_0^2 = (\sqrt{2}G_{\mathrm{F}})^{-1} = (246\,\mathrm{GeV})^2$. The predicted masses of the gauge bosons in the tree approximation are then $m_W = 78.4$ GeV and $m_Z = 89.0$ GeV. (Radiative corrections increase these by several GeV.) These are consistent with observation. Only the combination c^2/λ is known, so there remains one undetermined parameter. It may be taken to be the Higgs mass. The theoretical bounds are currently $130 < m_\eta < 190$ GeV [5]. The lower bound comes from the requirement that the standard model vacuum be stable. The upper bound comes from the requirement that λ be small enough that perturbation theory can be used. It should remain valid up to a supposed grand unification scale $\Lambda_{\mathrm{GUT}} \sim 10^{16}$ GeV.

The fermions are included according to the following scheme. The quark mass eigenstates are not eigenstates of the weak interactions. The matrix connecting the different sets of eigenstates is the Cabibbo–Kobayashi–Maskawa (CKM) matrix. By convention the charge 2/3 quarks (u, c, t) are unmixed. The CKM matrix, U_{CKM}, is unitary and relates the eigenstates of the charge $-1/3$ quarks (d, s, b) as

$$\begin{pmatrix} d' \\ s' \\ b' \end{pmatrix} = U_{\mathrm{CKM}} \begin{pmatrix} d \\ s \\ b \end{pmatrix} \tag{15.18}$$

The elements of U_{CKM} will not be needed in the subsequent discussion.

The fermions are then grouped into left-handed SU(2) doublets and right-handed SU(2) singlets. For example, the electron and its neutrino form the doublet

$$L = \begin{pmatrix} \nu_e \\ e^- \end{pmatrix}_L \tag{15.19}$$

where $e_L^- = \frac{1}{2}(1 - \gamma_5)e^-$, and a singlet $R = \frac{1}{2}(1 + \gamma_5)e^-$. These are coupled to the gauge bosons via

$$\bar{R}\left(i\slashed{\partial} - g'\slashed{B}\right)R + \bar{L}\left(i\slashed{\partial} - \tfrac{1}{2}g'\slashed{B} + \tfrac{1}{2}g\slashed{A}^a\tau^a\right)L \tag{15.20}$$

The other leptons and quarks are included in an analogous way. The coupling to γ, $W^\pm$, and Z can be written compactly as

$$e\bar{\psi}\gamma^\mu \left\{ QA_\mu + \frac{1}{2^{3/2}\sin\theta_{\mathrm{W}}}(1-\gamma_5)\left(T^+W_\mu^+ + T^-W_\mu^-\right) \right.$$
$$\left. + \frac{1}{\sin\theta_{\mathrm{W}}\cos\theta_{\mathrm{W}}}\left[\tfrac{1}{2}(1-\gamma_5)T_3 - Q\sin^2\theta_{\mathrm{W}}\right]Z_\mu \right\}\psi \tag{15.21}$$

where ψ is one of the following doublets,

$$\begin{pmatrix} u \\ d' \end{pmatrix} \quad \begin{pmatrix} c \\ s' \end{pmatrix} \quad \begin{pmatrix} t \\ b' \end{pmatrix} \quad \begin{pmatrix} \nu^e \\ e^- \end{pmatrix} \quad \begin{pmatrix} \nu^\mu \\ \mu^- \end{pmatrix} \quad \begin{pmatrix} \nu^\tau \\ \tau^- \end{pmatrix} \tag{15.22}$$

and where Q is the electric charge operator, T_3 is the third component of the weak SU(2) spin (with eigenvalue 1/2 for $\nu_e, \nu_\mu, \nu_\tau,\ u,\ c,\ t$, and eigenvalue $-1/2$ for e^-, μ^-, τ^-, d, s, b), and $T^\pm$ is the raising or lowering operator, which acts on the left-handed particles. The weak hypercharge Y is determined by $Q = T_3 + \frac{1}{2}Y$.

In order to retain the left-handed SU(2) symmetry for the fermions it is not possible to add a mass term in the usual form $-\bar{\psi}M\psi$. The allowable term for the electron, for example, is of the Yukawa form

$$-f_e \left(\bar{R}\Phi^\dagger L + \bar{L}\Phi R \right) \qquad (15.23)$$

After using (15.9) we obtain the electron mass as $m_e = \frac{1}{2}f_e v^0$ and zero neutrino mass. A similar situation prevails for the other fermions. Thus all quarks and leptons receive their masses on account of spontaneous symmetry breaking. In the vacuum, a quark or lepton mass is therefore $\sim f_i G_F^{-1/2}$ where f_i is a dimensionless coupling constant. For all but the t quark the f_i are very small since $G_F^{-1/2} = 293$ GeV.

15.2 Symmetry restoration in mean field approximation

The existence of phase transitions in the early universe has been a question that has preoccupied a generation of cosmologists. Early on, Kirzhnits [6] found that the symmetry between the weak and electromagnetic interactions would be restored at high temperatures. This result was soon complemented by similar works by Weinberg [7] Dolan and Jackiw [8], and Kirzhnits and Linde [9]. Some consequences of this phase transition will be discussed in Chapter 16. In the sections that follow, the stage will be set for the theoretical investigation of the electroweak phase transition, its existence, and its order.

The Glashow–Weinberg–Salam model is relatively easy to study at finite temperatures in the mean field approximation. At high temperature, $T > 50$ GeV, the fermion masses can be ignored except for that of the top quark. For simplicity, we shall ignore that for the moment as well. First, we shall use the U-gauge and show that it leads to an erroneous result, at least in the mean field approximation. This can be corrected in a covariant gauge.

The U-gauge has the advantage of displaying immediately the physical degrees of freedom. From (15.12) and (15.15) we can write the pressure as

$$\begin{aligned} P_{\mathrm{MF}} = &-\tfrac{1}{4}c^4/\lambda + \tfrac{1}{2}c^2 v^2 - \tfrac{1}{4}\lambda v^4 \\ &+ 6P_0(m_W) + 3P_0(m_Z) + 2P_0(0) + P_0(m_\eta) + \tfrac{7}{8}\pi^2 T^4 \end{aligned}$$

$$(15.24)$$

The last term is the contribution from three generations of massless quarks and leptons. The previous four terms are the boson contributions, with

$$P_0(m) = \int \frac{d^3k}{(2\pi)^3} \frac{k^2}{3\omega} \frac{1}{e^{\beta\omega}-1} \sim \frac{\pi^2}{90}T^4 - \frac{m^2}{24}T^2 \tag{15.25}$$

where the high-temperature limit is given. Using this limit we obtain

$$P_{\mathrm{MF}} = \left(\tfrac{7}{8} + \tfrac{2}{15}\right)\pi^2 T^4 + \tfrac{1}{2}v^2\left[c^2 - \tfrac{1}{4}T^2\left(\lambda + \tfrac{3}{4}g^2 + \tfrac{1}{4}g'^2\right)\right]$$
$$-\tfrac{1}{4}\lambda v^4 - \tfrac{1}{4}c^4/\lambda \tag{15.26}$$

Maximizing the pressure with respect to the mean field v gives the temperature dependence $v^2(T) = \left[c^2 - \tfrac{1}{4}T^2\left(\lambda + \tfrac{3}{4}g^2 + \tfrac{1}{4}g'^2\right)\right]/\lambda$ if $T^2 < 4c^2/\left(\lambda + \tfrac{3}{4}g^2 + \tfrac{1}{4}g'^2\right)$ and $v(T) = 0$ otherwise. This would indicate restoration of the gauge symmetry that was spontaneously broken at $T = 0$.

However, the result (15.26) is wrong. The reason can be traced to the U-gauge itself. Although it makes the physical particle content of the theory manifest, it is not, in practice, a renormalizable gauge. This follows from the poor ultraviolet behavior of the massive vector meson propagators, which is $p^\mu p^\nu/m^2 p^2$ instead of $1/p^2$. The implication for finite temperature is serious since T effectively acts as a physical high-momentum cutoff. Another way to see the difficulty is to consider the transformation (15.8) in the high-temperature phase, where v is supposed to vanish.

A more appropriate gauge for our purpose is the R-gauge, suitably generalized from its first application to the Abelian Higgs model, given in Section 7.4. Now we take as the independent fields η and ζ, defined by

$$\Phi = \frac{1}{\sqrt{2}}\begin{pmatrix} 0 \\ v+\eta \end{pmatrix} + \frac{i\zeta \cdot \tau}{\sqrt{2}v}\begin{pmatrix} 0 \\ v \end{pmatrix} = \frac{1}{\sqrt{2}}\begin{pmatrix} \zeta_2 + i\zeta_1 \\ v + \eta - i\zeta_3 \end{pmatrix} \tag{15.27}$$

which is suggested by (15.7) and (15.8). We choose the SU(2) gauge-fixing function to be

$$F^a = \partial^\mu A_\mu^a - \tfrac{1}{2}\rho g v \zeta^a - f^a(\mathbf{x}, \tau) \tag{15.28}$$

and the U(1) gauge-fixing function to be

$$F = \partial^\mu B_\mu + \tfrac{1}{2}\rho g' v \zeta^3 - f(\mathbf{x}, \tau) \tag{15.29}$$

The gauge-fixing delta functions $\delta(F)$ and $\delta(F^a)$ in the functional integral expression for Z are multiplied by

$$\exp\left\{-\frac{1}{2\rho}\int d^3x\, d\tau (f_a^2 + f^2)\right\}$$

and integration over $f^a(\mathbf{x}, \tau)$ and $f(\mathbf{x}, \tau)$ is carried out. The result is to add to the Lagrangian the gauge-fixing terms

$$-\frac{1}{2\rho}\left(\partial^\mu A_\mu^a - \tfrac{1}{2}\rho g v \zeta^a\right)^2 - \frac{1}{2\rho}\left(\partial^\mu B_\mu + \tfrac{1}{2}\rho g' v \zeta^3\right)^2 \qquad (15.30)$$

The cross terms in (15.30) between the gauge fields and ζ are ρ-independent. They combine with the cross terms from $(D_\mu \Phi)^\dagger (D^\mu \Phi)$ to produce total divergences that integrate to zero. Thus one advantage of using (15.28) and (15.29) is that there is no mixing between the fields. The terms $-(\partial^\mu A_\mu^a)^2/2\rho - (\partial^\mu B_\mu)^2/2\rho$ are familiar from the covariant gauge. The last terms in (15.30), when combined with the quadratic terms in $c^2 \Phi^\dagger \Phi - \lambda(\Phi^\dagger \Phi)^2$, yield the masses

$$\begin{aligned}
m_\eta^2 &= 3\lambda v^2 - c^2 \\
m_{\zeta_1}^2 = m_{\zeta_2}^2 &= \lambda v^2 - c^2 + \tfrac{1}{4}\rho g v^2 \\
m_{\zeta_3}^2 &= \lambda v^2 - c^2 + \tfrac{1}{4}\rho(g^2 + g'^2)v^2
\end{aligned} \qquad (15.31)$$

The fact that the ζ masses are gauge or ρ-dependent suggests that these do not represent physical particles.

The determinants must be analyzed. They are $\det(\partial F/\partial \alpha)$ and $\det(\partial F^a/\partial \alpha^b)$, where the infinitesimal gauge transformations are parametrized by $\alpha(\mathbf{x}, \tau)$ and $\alpha^b(\mathbf{x}, \tau)$. With the help of (8.11), we find

$$\begin{aligned}
\frac{\partial F^a}{\partial \alpha^b} &= -\partial^2 \delta^{ab} - \tfrac{1}{4}\rho g^2 v^2 \delta^{ab} + \text{linear terms} \\
\frac{\partial F}{\partial \alpha} &= -\partial^2 - \tfrac{1}{4}\rho g'^2 v^2 + \text{linear terms}
\end{aligned} \qquad (15.32)$$

where "linear terms" indicates terms that are linear in A_μ^a, ζ, and/or η. The determinants can be written as functional integrals over the ghost fields C_a and C. The ghost masses can be read off directly from (15.32):

$$\begin{aligned}
m_{C_a}^2 &= \tfrac{1}{4}\rho g^2 v^2 \\
m_C^2 &= \tfrac{1}{4}\rho g'^2 v^2
\end{aligned} \qquad (15.33)$$

The propagators for the W and Z fields are

$$D^{\mu\nu} = \frac{g^{\mu\nu} - p^\mu p^\nu/m^2}{p^2 - m^2} + \frac{p^\mu p^\nu/m^2}{p^2 - \rho m^2} \qquad (15.34)$$

where $m^2 = m_Z^2$ or m_W^2. The first term is the usual propagator for a massive vector boson. The second term looks like the propagator for an unphysical longitudinally propagating particle.

Now we are ready to put together this strange zoo of real and fictitious particles. Again, in the mean field approximation at high temperature we have

$$P_{\mathrm{MF}} = -\tfrac{1}{4}c^4/\lambda + \tfrac{1}{2}c^2v^2 - \tfrac{1}{4}\lambda v^4 + \left(\tfrac{7}{8} + \tfrac{2}{15}\right)\pi^2 T^4$$
$$- \tfrac{1}{24}T^2 \left[3m_Z^2 + 6m_W^2 + m_\eta^2 + \left(m_{\zeta_1}^2 + m_{\zeta_2}^2 + m_{\zeta_3}^2 + \rho m_Z^2\right.\right.$$
$$\left.\left. + 2\rho m_W^2 - 6m_{C_a}^2 - 2m_C^2\right)\right]$$

(15.35)

The quantity in the second parentheses is all that distinguishes the U-gauge from the R-gauge. These mass-squared terms add up to $3(\lambda v^2 - c^2)$. The pressure is thus

$$P_{\mathrm{MF}} = \tfrac{121}{120}\pi^2 T^4 + \tfrac{1}{2}v^2\left[c^2 - \tfrac{1}{4}T^2\left(2\lambda + \tfrac{3}{4}g^2 + \tfrac{1}{4}g'^2\right)\right]$$
$$-\tfrac{1}{4}\lambda v^4 - \tfrac{1}{4}c^4/\lambda + \tfrac{1}{6}c^2 T^2$$

(15.36)

This should be compared with (15.26). Note that all ρ-dependence has vanished: a nice check on the calculation.

Minimizing P_{MF} with respect to v we obtain

$$v^2(T) = \begin{cases} (c^2/\lambda)\left(1 - T^2/T_{\mathrm{c}}^2\right) & T \leq T_{\mathrm{c}} \\ 0 & T \geq T_{\mathrm{c}} \end{cases}$$

(15.37)

$$P_{\mathrm{MF}} = \begin{cases} \tfrac{121}{120}\pi^2 T^4 + \tfrac{1}{4}(c^4/\lambda)\left(1 - T^2/T_{\mathrm{c}}^2\right)^2 + \tfrac{1}{6}c^2 T^2 - \tfrac{1}{4}c^4/\lambda & T \leq T_{\mathrm{c}} \\ \tfrac{121}{120}\pi^2 T^4 + \tfrac{1}{6}c^2 T^2 - \tfrac{1}{4}c^4/\lambda & T \geq T_{\mathrm{c}} \end{cases}$$

(15.38)

and

$$T_{\mathrm{c}}^2 = \frac{4c^2}{2\lambda + \tfrac{3}{4}g^2 + \tfrac{1}{4}g'^2}$$

(15.39)

This yields a second-order symmetry-restoring phase transition at T_{c} since $\partial P/\partial T$ is continuous but $\partial^2 P/\partial^2 T$ is not. If we take the zero-temperature Higgs mass to be 120 GeV then $c = 84.9$ GeV and $\lambda = 0.119$. The critical temperature is $T_{\mathrm{c}} = 225$ GeV. The effective potential is plotted in Figure 15.1 as a function of v for several values of the temperature, including the critical value. Here the effective potential is $\Omega_{\mathrm{MF}}^{\mathrm{eff}}(v) \equiv P_{\mathrm{MF}}(0, T) - P_{\mathrm{MF}}(v, T)$; in the literature it is also written as V_{eff}. Minimizing the effective potential is equivalent to maximizing the pressure.

All the previously discussed difficulties associated with spontaneous symmetry breaking and nonabelian gauge theories at finite temperature arise in the Glashow–Weinberg–Salam model as well. For example, at sufficiently high temperature the Higgs-mass-squared of (15.15) becomes

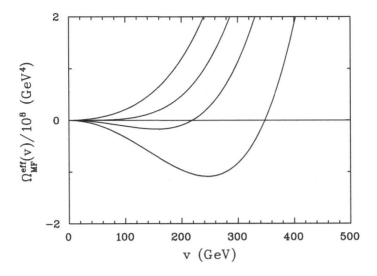

Fig. 15.1. The effective potential in the mean field approximation, as described in the text. The curves shown correspond to potentials calculated at $T = 0$, $T = 175\,\mathrm{GeV}$, $T = T_\mathrm{c} = 225\,\mathrm{GeV}$, and $T = 275\,\mathrm{GeV}$, from bottom to top, respectively.

negative, and loop self-energy corrections are necessary to cure it. In the high-temperature phase the mean field masses of the gauge fields are zero. Thus the same infrared problems will arise as in QCD. The contributions of exchange and ring diagrams to the pressure may be computed.

15.3 Symmetry restoration in perturbation theory

The applicability of finite-temperature perturbation expansions in the electroweak theory will now be more closely examined. Consider a scalar field theory, $\lambda\phi^4$, like that discussed elsewhere in this book. At each order in a loop expansion there will be terms of the form

$$T \sum_n \int \frac{d^3p}{(2\pi)^3} f(\omega_n, \mathbf{p}) \qquad (15.40)$$

where $f(\omega_n, \mathbf{p})$ is a functional of propagators and vertices. The tadpole diagram is a simple example, namely

$$T \sum_n \int \frac{d^3p}{(2\pi)^3} \frac{1}{\omega_n^2 + \omega^2} \qquad (15.41)$$

with $\omega = \sqrt{\mathbf{p}^2 + m^2}$. Clearly this will be dominated by the Matsubara zero mode. Omitting the integral over momentum, this fact is expressed as

$$T \sum_n \frac{1}{\omega_n^2 + \omega^2} \sim \frac{T}{\omega^2} \tag{15.42}$$

What might constitute a dimensionless loop-expansion parameter? The argument above suggests that when the vertices contribute an overall constant λ then the expansion parameter that controls the convergence is $\lambda T/\omega \sim \lambda T/m_{\text{eff}}$ for bosons (where m_{eff} is some soft scale in our theory) and $\lambda T/T = \lambda$ for fermions. For bosons the perturbation expansion could be ill defined if $m_{\text{eff}} < \lambda T$, and then non-perturbative techniques would be required. What happens in the standard model is more complicated because of the inclusion of the gauge bosons. In what follows we study the electroweak theory with the inclusion of the ring diagrams, that are known to be important for long wavelengths.

In the Glashow–Weinberg–Salam model, the gauge boson mass term is of the form

$$(A_\mu^a, B_\mu) \, M^2 \begin{pmatrix} A_a^\mu \\ B^\mu \end{pmatrix}$$

with a non-diagonal mass matrix

$$M^2(v) = \frac{v^2}{4} \begin{pmatrix} g^2 & 0 & 0 & 0 \\ 0 & g^2 & 0 & 0 \\ 0 & 0 & g^2 & -gg' \\ 0 & 0 & -gg' & g'^2 \end{pmatrix} \tag{15.43}$$

The customary procedure is to define the physical fields $W_\mu^\pm$, Z_μ, and A_μ as linear combinations of the A_μ^a and B_μ fields, such that the physical masses are $m_W^2(v) = g^2 v^2/4$, $m_Z^2(v) = (g^2 + g'^2)v^2/4$, and $m_A^2(v) = 0$. For this application, we now assume that only the top quark Yukawa coupling, f_t, is nonzero. The shift in the Higgs field generates a mass through the term $\mathcal{L}_{\text{Yukawa}} = f_t \bar{t} t v/\sqrt{2}$.

The one-loop contribution to the thermodynamic potential is split into zero-temperature and finite-temperature contributions. Following Carrington [10], one may write the contribution from the Higgs boson (ϕ), the gauge boson (gb), and the top quark (ψ) loops as

$$\Omega_1(v) = \Omega_1^{\text{vac}}(v) + \Omega_1^{\text{mat}}(v) \tag{15.44}$$

where

$$\begin{aligned} \Omega_1^{\text{vac}}(v) &= \Omega_{1,\phi}^{\text{vac}}(v) + \Omega_{1,gb}^{\text{vac}}(v) + \Omega_{1,\psi}^{\text{vac}}(v) \\ \Omega_1^{\text{mat}}(v) &= \Omega_{1,\phi}^{\text{mat}}(v) + \Omega_{1,gb}^{\text{mat}}(v) + \Omega_{1,\psi}^{\text{mat}}(v) \end{aligned} \tag{15.45}$$

The zero-temperature loops are regularized using a cut-off. Their contribution may be obtained from Section 7.3; for an arbitrary mass $m_x(v)$ it is

$$\frac{\Lambda_c^2}{32\pi^2} m_x^2(v) + \frac{m_x^4(v)}{64\pi^2} \left[\ln\left(\frac{m_x^2(v)}{\Lambda_c^2}\right) - \frac{1}{2} \right] \tag{15.46}$$

This procedure generates a correction to the tree-level zero-temperature effective potential, so that

$$\Omega^{\text{vac}}(v) = \Omega^{\text{tree}}(v) + \Omega_1^{\text{vac}}(v) \tag{15.47}$$

with

$$\Omega^{\text{tree}} = -\frac{1}{2}c^2 v^2 + \frac{1}{4}v^4$$

and

$$\Omega_1^{\text{vac}}(v)$$
$$= \frac{3}{32\pi^2}\lambda c^2 v^2 - \frac{v^4}{64\pi^2}\left(6\lambda^2 + \frac{3}{16}g^4 + \frac{3}{32}(g^2 + g'^2)^2 - \frac{3}{2}f^4\right)$$
$$+ \frac{1}{64\pi^2}\left[6m_W^4(v)\ln\left(\frac{\lambda v^2}{c^2}\right) + 3m_Z^4(v)\ln\left(\frac{\lambda v^2}{c^2}\right) - 12m_t^4(v)\ln\left(\frac{\lambda v^2}{c^2}\right)\right.$$
$$\left. + m_1^4(v)\ln\left(\frac{m_1^2(v)}{2c^2}\right) + 3m_2^4(v)\ln\left(\frac{m_2^2(v)}{2c^2}\right)\right] \tag{15.48}$$

The one-loop finite-temperature thermodynamic potential for bosons and fermions is just the negative of the pressure for the free particle of mass $m_x(v)$. There will be contributions to the ring diagrams from both gauge and Higgs bosons. The finite-temperature part of the one-loop potential will combine with the ring contribution to define a potential in terms of the shifted mass-squared. Therefore we need to evaluate the gauge boson and Higgs boson self-energies in the leading infrared limit. For the ith Higgs field,

$$\Pi_i(0) = \Pi_\phi^{(A_\mu^a)}(0) + \Pi_\phi^{(B_\mu)}(0) + \Pi_\phi^{(\phi)}(0) + \Pi_\phi^{(\psi)}(0) \tag{15.49}$$

where the individual contributions are

$$\Pi_\phi^{(A_\mu^a)}(0) = \frac{1}{8}g^2 T^2 \qquad \Pi_\phi^{(B_\mu)}(0) = \frac{1}{16}\left(g^2 + g^2\right)T^2$$
$$\Pi_\phi^{(\phi)}(0) = \frac{1}{2}\lambda T^2 \qquad \Pi_\phi^{(\psi)}(0) = \frac{1}{4}f_t T^2 \tag{15.50}$$

The ring contribution for the Higgs field is

$$\Omega_{\text{ring}}^{\text{mat}}(v) = -\frac{1}{2}T\sum_n \int \frac{d^3 q}{(2\pi)^3} \sum_{\ell=1}^{\infty} \frac{1}{\ell}\left(-\frac{1}{\omega_n^2 + \mathbf{q}^2 + m_i^2(v)}\Pi_i(0)\right)^\ell \tag{15.51}$$

which, combined with the finite-temperature part of the one-loop potential, gives

$$\Omega_\phi^{\mathrm{mat}}(v) = -P_0(\widetilde{m}_1) - 3P_0(\widetilde{m}_2) \tag{15.52}$$

where $P_0(m)$ is as in Section 15.2, $\widetilde{m}_i^2 = m_i^2(v) + \Pi_i(0)$, and the factor 3 is a degeneracy factor. For the gauge boson polarization tensors, as used in Section 5.4,

$$\Pi_{\mu\nu}(0) = \Pi_{\mathrm{T}}(0)P_{\mathrm{T}\,\mu\nu} + \Pi_{\mathrm{L}}(0)P_{\mathrm{L}\,\mu\nu} \tag{15.53}$$

In the static infrared limit $\Pi_{\mu\nu}^{AB}(0) = \Pi_{00}^{AB}(0)P_{\mathrm{L}\,\mu\nu}$, and Π_{00}^{AB} is approximately diagonal if the ratio of the gauge boson masses and the temperature is small:

$$\Pi_{00}(0) = \begin{bmatrix} \Pi_{00}^{(2)}(0) & 0 & 0 & 0 \\ 0 & \Pi_{00}^{(2)}(0) & 0 & 0 \\ 0 & 0 & \Pi_{00}^{(2)}(0) & 0 \\ 0 & 0 & 0 & \Pi_{00}^{(1)}(0) \end{bmatrix} \tag{15.54}$$

Here the superscripts (1) and (2) refer to the U(1) and SU(2) gauge bosons, respectively. One defines as $\Pi_{gb}^{(2)}(0)$, $\Pi_\phi^{(2)}(0)$, and $\Pi_\psi^{(2)}(0)$, the contribution to the SU(2) gauge boson polarization tensor from the gauge boson, Higgs boson, and t quark loops. One may use a similar notation for the polarization of the U(1) gauge boson. Then

$$\begin{aligned}
\Pi_{00}^{(1)}(0) &= \Pi_\phi^{(1)}(0) + \Pi_\psi^{(1)}(0) \\
\Pi_{00}^{(2)}(0) &= \Pi_{gb}^{(2)}(0) + \Pi_\phi^{(2)}(0) + \Pi_\psi^{(2)}(0)
\end{aligned} \tag{15.55}$$

where

$$\begin{aligned}
&\Pi_\phi^{(1)}(0) = \frac{1}{6}g'^2 T^2 \qquad \Pi_\psi^{(1)}(0) = \frac{5}{3}g'^2 T^2 \\
&\Pi_{gb}^{(2)}(0) = \frac{2}{3}g^2 T^2 \qquad \Pi_\phi^{(2)}(0) = \frac{1}{6}g^2 T^2 \qquad \Pi_\psi^{(2)}(0) = g^2 T^2
\end{aligned} \tag{15.56}$$

The rest of the calculation for the ring contribution to the gauge boson effective potential proceeds as in the case of the Higgs particle. In terms of the mass and self-energy matrices, it may be written as

$$\Omega_{\mathrm{ring}}^{gb}(v) = -\frac{T}{12\pi}\,\mathrm{Tr}\left\{[M^2(v) + \Pi_{00}(0)]^{3/2} - M^3(v)\right\} \tag{15.57}$$

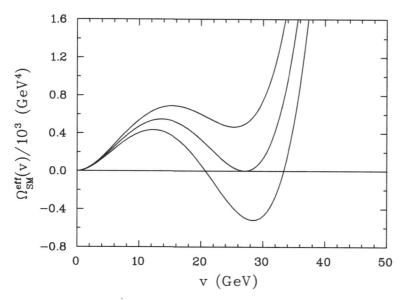

Fig. 15.2. The ring-improved effective potential that is relevant for the electroweak phase transition in the standard model. The physical parameters that enter this calculation are the masses of the Higgs particle and the top quark. The values here are $m_H = 120$ GeV, and $m_t = 175$ GeV. The curve at the critical temperature $T_c = 140.42$ GeV is enclosed by curves at $T = 140.40$ (lower) and 140.43 GeV (upper).

One may show that

$$\text{Tr}[M^2(v) + \Pi_{00}(0)]^{3/2} = 2a^{3/2} + \frac{1}{2\sqrt{2}} \left[(a+c) - \sqrt{(a-c)^2 + 4b^2} \right]^{3/2}$$

$$+ \left[(a+c) + \sqrt{(a-c)^2 + 4b^2} \right]^{3/2} \tag{15.58}$$

where $a = g^2 v^4/4 + \Pi_{00}^{(2)}(0)$, $b = -gg' v^2/4$, and $c = g'^2 v^2/4 + \Pi_{00}^{(1)}(0)$.

The final expression for the effective potential is obtained by adding to the zero-temperature parts the ring-improved finite-temperature expressions [10]. We remark that two-loop topologies have also been considered, along with their contribution (with resummations) to the effective potential [11, 12, 13].

With the methods described here, it has been shown that the standard model has a first-order phase transition, driven by the v^3 term [10, 14]. Using modern values of the physical parameters yields the effective potential shown in Figure 15.2. One observes that the perturbation approach appears to predict a very weak first-order phase transition with a critical temperature $T_c = 140.42$ GeV. However, this brings us to the core of the issue. Consider the behavior of the perturbative expansion in the standard model. To make the discussion specific, consider a temperature near T_c. A parameter can be associated with each loop in the expansion of the effective potential. For example, the expansion parameter for vector loops is $g^2 T_c / m_W(v_c)$ (generically writing g^2 for a linear combination of g^2 and g'^2), according to the analysis earlier in this section. We may write $g^2 T_c / m_W(v_c) \sim g T_c / v_c \sim \lambda / g^2$. This last value is $\sim m_H^2 / m_W^2$, evaluated at $T = 0$. The current experimental value of the mass of the W boson is 80.425 ± 0.033 GeV [15], and comes from direct measurements. The Higgs boson, at the time of this writing, is still a hypothetical particle. The bounds on its mass placed by self-consistent arguments have been reviewed earlier. Indirect experimental bounds for the standard model Higgs mass can also be obtained from precision electroweak measurements and from fits to measured top quark and $W^\pm$ masses. The global electroweak fits give a preferred value of 96^{+40}_{-38} GeV [15]. However, a recent high-precision measurement of the top quark mass raised the world average for m_t to 178.0 ± 4.3 GeV [16]. The impact on the best standard-model fit of the Higgs mass is that it is raised from 96 to 117 GeV. In line with arguments presented earlier, those numbers clearly cast doubt on the usefulness of a perturbative loop expansion in theoretical searches for an electroweak phase transition in the standard model. It is therefore important to consider lattice-based nonperturbative numerical approaches.

15.4 Symmetry restoration in lattice theory

As the quartic self-coupling λ becomes large, the accuracy of perturbative calculations decreases. For large enough λ, corresponding to a large Higgs mass, the order of the phase transition, and even its existence, cannot be determined using perturbation theory. As for QCD one might turn to numerical calculations of electroweak theory on a lattice. In general this is a more intensive numerical endeavor than in the QCD case for several reasons: there are two types of gauge field, there is a scalar doublet field, and there are three generations of fermion fields to deal with. Also, surprisingly, the weaker gauge coupling makes the simulations more demanding since it introduces a scale hierarchy that is very difficult to handle numerically.

Significant progress in finite-temperature lattice calculations of electroweak-like gauge theories has been realized in recent years with the help of the technique of dimensional reduction. Provided that we are interested in the computation of static quantities, we may generically write a four-dimensional boson or fermion field in terms of its Matsubara modes:

$$\phi(\tau, \mathbf{x}) = \sum_{n=-\infty}^{\infty} \exp(i2n\pi T\tau)\, \phi_n(\mathbf{x})$$
$$\psi(\tau, \mathbf{x}) = \sum_{n=-\infty}^{\infty} \exp(i(2n+1)\pi T\tau)\, \psi_n(\mathbf{x}) \tag{15.59}$$

The original four-dimensional theory is formally equivalent to a three-dimensional theory albeit with an infinite number of fields, each corresponding to a mode. The three-dimensional "masses" of bosons are $m_{\mathrm{B}} = 2\pi n T$ and those of fermions are $m_{\mathrm{F}} = (2n+1)\pi T$. If we are concerned with soft physics below some scale Λ, the heavier fields (on that scale) may be integrated out. This leaves an effective field theory where the parameters of the effective Lagrangian are functions of the temperature and of the scale Λ. This integration over heavy modes might be done perturbatively, and the expansion parameter would be $\Lambda/\pi T$. If the relevant scale is T or smaller then all fermionic modes, and all bosonic modes with $n \neq 0$, will have masses larger than πT and can be integrated out.

The effective three-dimensional action can be written as

$$S_{\mathrm{eff}} = bVT^3 + \int d^3x\, \mathcal{L}_{\mathrm{eff}} + \sum_n \frac{O_n}{T^n} \tag{15.60}$$

Here $\mathcal{L}_{\mathrm{eff}}$ is a three-dimensional effective Lagrangian with temperature-dependent parameters, b is some number that is related to the number of degrees of freedom, V is the volume, and the O_n represent the contribution from operators of dimension n. The latter will be suppressed by powers of the temperature but, in the high-T limit, the three-dimensional couplings contained therein will also be large. A typical way to rewrite the last term in the equation above is $\mathcal{O}(m_i^2(T)/T^2)$, the $m_i(T)$ being relevant mass scales for the problem at hand, such as inverse screening lengths, etc. The condition for omitting the last term in the effective action is tantamount to that controlling the convergence of the zero-temperature perturbative expansion, namely, $g^2 \ll 1$ where g is a dimensionless coupling constant. At first it would appear that little has been gained by formulating the problem in a reduced number of dimensions. However, the expansion parameter is different at zero and finite temperature. At finite T the perturbative expansion should prove useful if $g^2 T/\Lambda = g_3^2/\Lambda \ll 1$, where g_3 is the three-dimensional coupling. Therefore, at finite temperature it is entirely possible for the four-dimensional perturbation expansion to be unsuitable but for the dimensionally reduced theory to be applicable. For applications in the vicinity of a critical temperature, it turns out that the criterion of applicability of dimensional reduction is satisfied for

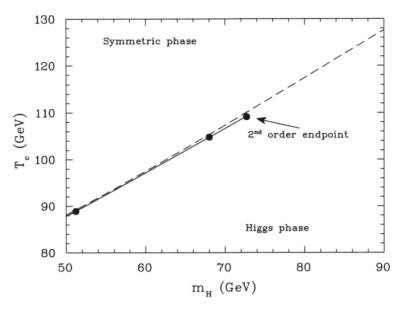

Fig. 15.3. A plot of the phase diagram of the standard model, investigated with lattice Monte Carlo techniques. The broken line is the perturbative (first-order) result; the solid line is a fit to the numerical results and shows a first-order transition with a second-order endpoint. The figure is adapted from Ref. [18].

electroweak theory but not for QCD, since the four-dimensional gauge coupling is not small at T_c.

Finite-temperature electroweak theory has been studied in lattice Monte Carlo simulations for a number of Higgs mass values by Kajantie *et al.* [17]. The effective Lagrangian they use is

$$\mathcal{L}_{\text{eff}} = \tfrac{1}{4} f_{ij}^a f_{ij}^a + (D_i \Phi)^\dagger (D_i \Phi) + m_3^2 \Phi^\dagger \Phi + \lambda_3 (\Phi^\dagger \Phi)^2 \qquad (15.61)$$

This is electroweak theory in three-dimensions without the U(1) gauge field, without fermions, and where the time component of the SU(2) gauge field has been integrated out. There are three parameters: g_3, that enters via the covariant derivative D_i, m_3, and λ_3. To lowest order (and ignoring Yukawa couplings and g') they are

$$g_3^2 = g^2 T$$
$$m_3^2 = \left(\tfrac{3}{16} g^2 + \tfrac{1}{2} \lambda \right) T^2 - c^2 \qquad (15.62)$$
$$\lambda_3 = \lambda T$$

where g, c, and λ are all parameters in the fundamental four-dimensional theory (15.1). These parameters have been computed with one-loop corrections too. This allows for a precise connection between physically measurable quantities such as the W, Z, and Higgs-boson masses and the thermodynamic properties of the electroweak theory.

The numerical results indicate that the theory has a first-order phase transition for small Higgs masses. The transition gets weaker as m_H grows and terminates around $m_H \sim 80$ GeV at a second-order endpoint. Those results, together with those obtained in perturbation theory, are summarized in Figure 15.3. It might be that these conclusions are modified by physics beyond the standard model; this is a topic still under investigation.

15.5 Exercises

15.1 Find explicitly the "linear terms" in (15.32).

15.2 Verify that (15.34) is the propagator for the W and Z bosons.

15.3 Express T_c in (15.39) in terms of the observable parameters e and θ_W and the zero-temperature Higgs mass. Assuming that perturbation theory is valid and using the quoted bounds on m_H, determine the allowable range for T_c.

15.4 Show that the term in the effective potential that is cubic in the vacuum expectation value of the scalar field is from the Matsubara zero-mode.

15.5 Derive the three-dimensional couplings in (15.62).

15.6 How is the temperature dependence of m_3^2 in (15.62) related to the critical temperature given by (15.39)?

References

1. Glashow, S., *Nucl. Phys.* **22**, 579 (1961).
2. Weinberg, S., *Phys. Rev. Lett.* **19**, 1264 (1967).
3. Salam, A., (1968). *Nobel Symposium on Elementary Particle Theory* (Wiley, New York).
4. 't Hooft, G., *Nucl. Phys.* **B35**, 167 (1971).
5. Cabibbo, N. *et al.*, *Nucl. Phys.* **B158**, 295 (1979); Hambye, T., and Riesselmann, K., *Phys. Rev. D* **55**, 7255 (1997); Isidori, G., Ridolfi, G., and Strumia, A., *Nucl. Phys.* **B609**, 387 (2001).
6. Kirzhnits, D. A., *JETP Lett.* **15**, 374 (1972).
7. Weinberg, S., *Phys. Rev. D* **9**, 3357 (1974).
8. Dolan, L., and Jackiw, R., *Phys. Rev. D* **9**, 3320 (1974).
9. Kirzhnits, D. A., and Linde, A. D., *Ann. Phys. (NY)* **101**, 195 (1976).
10. Carrington, M., *Phys. Rev. D* **45**, 2933 (1992).
11. Arnold, P., and Espinosa, O., *Phys. Rev. D* **47**, 3546 (1993).
12. Bagnasco, J. E., and Dine, M., *Phys. Lett.* **B303**, 308 (1993).
13. Fodor, Z., and Hebecker, A., *Nucl. Phys.* **B432**, 127 (1994).

14. Dine, M., Leigh, R. G., Huet, P., Lindé, A., and Lindé, D., *Phys. Rev. D* **46**, 550 (1992).
15. Eidelman S., *et al.*, *Phys. Lett.* **B592**, 1 (2004).
16. Abazov, V. M., *et al.*, *Nature* **429**, 638 (2004).
17. Kajantie, K., Laine, M., Rummukainen, K., and Shaposhnikov, M., *Phys. Rev. Lett.* **77**, 2887 (1996).
18. Laine, M., and Rummukainen, K., *Nucl. Phys.* **B73** (Proc. Suppl.), 180 (1999).

Bibliography

The standard SU(2) × U(1) electroweak model

Glashow, S., *Nucl. Phys.* **22**, 579 (1961).
Weinberg, S., *Phys. Rev. Lett.* **19**, 1264 (1967).
Salam, A. (1968). *Nobel Symposium on Elementary Particle Theory* (Wiley, New York).
't Hooft, G., *Nucl. Phys.* **B35**, 167 (1971).

Pioneering works on the electroweak phase transition

Kirzhnits, D. A., *JETP Lett.* **15**, 374 (1972).
Dolan, L., and Jackiw, R., *Phys. Rev. D* **9**, 3320 (1974).
Weinberg, S., *Phys. Rev. D* **9**, 3357 (1974).
Kirzhnits, D. A., and Linde, A. D., *Ann. Phys. (NY)* **101**, 195 (1976).

Dimensional reduction and its use in electroweak lattice calculations

Ginsparg, P., *Nucl. Phys.* **B170**, 388 (1980).
Appelquist, T., and Pisarski, R., *Phys. Rev. D* **23**, 2305 (1981).
Landsman, N. P., *Nucl. Phys.* **B322**, 498 (1989).
Farakos, K., Kajantie, K., Rummukainen, K., and Shaposhnikov, M. E., *Nucl. Phys.* **B425**, 67 (1994); **B442**, 317 (1995).
Kajantie, K., Laine, M., Rummukainen, K., and Shaposhnikov M. E., *Nucl. Phys.* **B458**, 90 (1996); **B466**, 189 (1996).

16

Astrophysics and cosmology

Finite-temperature field theory finds extensive applications in astrophysical environments and cosmology. This chapter is devoted to an introduction to these applications. A more comprehensive discussion could easily fill whole books.

The end product of the evolution of any star is a white dwarf star, a neutron star, or a black hole, depending on the initial mass of the star. A white dwarf star is held up against gravitational contraction by electron degeneracy pressure (Section 16.1) whereas a neutron star is held up by baryon degeneracy pressure and repulsive baryon interactions (Section 16.2). The sun will end its days by swelling up into a red giant and then collapsing to a white dwarf. Neutron stars are formed in the gravitational collapse of stars with initial mass in the range from about two to eight solar masses. The collapse is sudden and may be seen as a supernova. The resulting star is initially quite warm, perhaps 10 to 40 MeV in temperature, but cools rapidly by neutrino emission (Section 16.3). If the initial mass of the dying star is too great then it will end as a black hole.

There was some excitement when it was realized that a first-order QCD phase transition about one microsecond after the big bang could influence the abundances of the light isotopes such as deuterium, helium, and lithium. However, quantitative calculations now show that this is very unlikely (Section 16.4); in addition QCD, with its known set of quark masses, probably does not undergo a first-order phase transition, as we saw in Chapter 10.

Going further back in time, it seems quite likely that the final baryon and lepton numbers of the universe were determined at around the electroweak temperature scale of 100 GeV. Sphaleron transitions were the last phenomena that were able to change these numbers (Section 16.5). Baryogenesis and leptogenesis may have originated at some much earlier epoch, in the context of grand unified or supersymmetric theories. It may be that some very massive particles in such a theory preferentially decayed into baryons rather than antibaryons. The formation and decay rates of such particles are considered in Section 16.6.

16.1 White dwarf stars

A white dwarf is the end result of a star of about one solar mass after it has burned all its nuclear fuel. It is held up against gravitational collapse by the degeneracy pressure of electrons, although essentially all its mass is contributed by baryons. It is interesting to inquire to what extent the equation of state of the degenerate electron gas influences the structure of white dwarfs.

In a white dwarf, the pressure of the electrons dominates the pressure of the atomic nuclei while the mass density of the baryons dominates the total energy density. Therefore the energy density is approximately

$$\epsilon = \frac{m_{\mathrm{N}} n_e}{Y_e} \tag{16.1}$$

where m_{N} is the nucleon mass, n_e is the electron density, and Y_e is the number of electrons per baryon. For a star composed predominantly of helium $Y_e = 1/2$, while for a star composed predominantly of iron $Y_e = 26/56$. These values follow from the requirement of electrical neutrality. There are small corrections due to the binding energy of the atomic nuclei and to their average kinetic energy.

To determine the mass and structure of cold, nonrotating, spherically symmetric stars, we use the Tolman–Oppenheimer–Volkoff equation from general relativity,

$$r^2 \frac{dP}{dr} = -G(\epsilon + P)(\mathcal{M} + 4\pi r^3 P)\left(1 - \frac{2G\mathcal{M}}{r}\right)^{-1} \tag{16.2}$$

where

$$\mathcal{M}(r) = 4\pi \int_0^r \epsilon(r') r'^2 dr'$$

The function $\mathcal{M}(r)$ is the total mass contained within a sphere of radius r. We can neglect the pressure in comparison with the energy density. We can also neglect the general relativistic change in the metric. To an excellent approximation Newtonian gravitational physics applies.

It turns out that as the central density ϵ_c of the star increases, the mass increases at first while the radius decreases. As the central density is increased further, an asymptotic limit is reached for the stellar mass. White dwarfs with a mass greater than this "Chandrasekhar limit" cannot exist. To understand this limit we recognize that at very high density the electrons become ultrarelativistic. The electron pressure for noninteracting electrons is then

$$P_e = \frac{\mu_e^4}{12\pi^2} \tag{16.3}$$

and the density is

$$n_e = \frac{\partial P_e}{\partial \mu_e} = \frac{\mu_e^3}{3\pi^2} \tag{16.4}$$

Together with (16.1) this results in the equation of state

$$P = K\epsilon^{4/3} \tag{16.5}$$

where K is a constant. This has the form of a polytrope (pressure proportional to the energy density raised to a power). Newtonian gravitational physics then predicts the unique asymptotic mass

$$M_\infty = 4.555 \left(\frac{K}{G}\right)^{3/2} = 5.735 Y_e^2 M_{\text{sun}} \tag{16.6}$$

where the second equality expresses it in terms of the mass of the sun [1]. This mass is independent of the central density and the radius, which is given by

$$R = 3.891 \left(\frac{K}{G}\right)^{1/2} \epsilon_{\text{c}}^{-1/3} = 4.20 \left(\frac{M_{\text{sun}}}{\epsilon_{\text{c}}}\right)^{1/3} Y_e^{2/3} \tag{16.7}$$

The physical constants used above are: the average nucleon mass $m_{\text{N}} = 0.939$ GeV; Newton's constant $G = 6.707 \times 10^{-39}$ GeV^{-2}; the solar mass $M_{\text{sun}} = 1.989 \times 10^{30}$ kg; and the solar radius $R_{\text{sun}} = 6.961 \times 10^8$ km. For a white dwarf composed of helium $M_\infty = 1.43 M_{\text{sun}}$. The Chandrasekhar limit is one of the fundamental concepts in astrophysics.

The story is not complete. When the electron density becomes high enough, roughly when $\mu_e = 5m_e$, electrons are captured by protons to form neutrons (the neutrinos escape from the star). The electron-to-baryon ratio Y_e decreases, and so does the mass. As a function of increasing central density the mass goes up, reaches a maximum just below the Chandrasekhar limit, and then decreases. When the star mass falls with increasing central density the star is gravitationally unstable and collapses further.

It is clear that several other more minor effects have been left out of this analysis. Among these is the change in the equation of state of the electron gas owing to interactions among the electrons. Let us see how important these interactions are. From our previous studies we know that the first-order correction to the pressure in the limit $\mu_e \gg m_e, T$ is

$$P_e = \frac{\mu_e^4}{12\pi^2}\left(1 - \frac{3}{2}\frac{\alpha}{\pi}\right) \qquad (16.8)$$

and the correction to the density is

$$n_e = \frac{\partial P_e}{\partial \mu_e} = \frac{\mu_e^3}{3\pi^2}\left(1 - \frac{3}{2}\frac{\alpha}{\pi}\right) \qquad (16.9)$$

This means that the coefficient K is modified:

$$K \to K\left(1 - \frac{3}{2}\frac{\alpha}{\pi}\right)^{-1/3} \qquad (16.10)$$

This changes the Chandrasekhar limit by only 0.2%. So, after all this hard work we find that the perturbative corrections in an ultrarelativistic electron gas are probably impossible to discern by measuring white dwarf masses and radii.

16.2 Neutron stars

A neutron star consists of almost pure neutron matter with a central density greater than that in atomic nuclei. This represent the final state in the evolution of many stars. Owing to their high central density, neutron stars serve as distant laboratories for the study of dense, relativistic, strongly interacting systems. Their central cores may have some component of hyperon matter or quark matter. Much theoretical work has been published on this topic over the last forty years. Here we can just touch on some of the important issues by studying a few illustrative theories of cold dense baryonic matter.

To first approximation the star consists of pure neutron matter. However, neutrons undergo beta decay by the process $n \to p + e^- + \bar{\nu}_e$. This decay will continue until the density of protons and electrons is high enough for the Pauli exclusion principle to prevent any further decays; this happens when the chemical potentials satisfy $\mu_n = \mu_p + \mu_e$. The neutrinos escape from the star. In fact, neutrino radiation is an important mechanism for the cooling of a neutron star from its initial temperature of 10 to 40 MeV following its birth by supernova. The details of neutrino cooling are a fascinating, and complicated, story in themselves. The interested reader is referred to Section 16.3 and to the bibliography at the end of the chapter.

As the central density increases, so does the baryon chemical potential. Eventually it becomes high enough that hyperons can be produced and coexist in chemical equilibrium with the neutrons and protons. The lowest spin-1/2 baryon octet consists of $p, n, \Lambda, \Sigma^+, \Sigma^0, \Sigma^-, \Xi^0, \Xi^-$. If the baryon density is high enough, muons may appear too.

We will consider three different models for the equation of state. The first consists of relativistic but non-interacting neutrons. (It can be shown that the inclusion of noninteracting protons, whose abundance is determined by beta equilibrium with neutrons, does not modify the equation of state and therefore the structure of neutron stars by very much.) The second model consists of protons and neutrons in beta equilibrium, interacting via the exchange of σ, ω, and ρ mesons in the relativistic mean field approximation. The first two mesons heve been discussed already, in Chapter 11; the ρ meson is required here to reproduce the measured charge-symmetry energy of nuclear matter. The third model starts with the second and adds the six hyperons in the baryon octet. In addition, the vector meson ϕ is included, since it couples to the hyperons and represents vector repulsion among them.

All three models for the equation of state are based on the Lagrangian

$$
\begin{aligned}
\mathcal{L}_{\text{strong}} = &\sum_j \bar{\psi}_j (i\slashed{\partial} - m_j + g_{\sigma j}\sigma - g_{\omega j}\slashed{\omega} - g_{\phi j}\slashed{\phi} - g_{\rho j}\slashed{\rho}^a T_a)\psi_j \\
&+ \tfrac{1}{2}(\partial_\mu \sigma \partial^\mu \sigma - m_\sigma^2 \sigma^2) - \tfrac{1}{3}b m_N (g_\sigma \sigma)^3 - \tfrac{1}{4}c(g_\sigma \sigma)^4 \\
&- \tfrac{1}{4}\omega^{\mu\nu}\omega_{\mu\nu} + \tfrac{1}{2}m_\omega^2 \omega_\mu \omega^\mu - \tfrac{1}{4}\phi^{\mu\nu}\phi_{\mu\nu} + \tfrac{1}{2}m_\phi^2 \phi_\mu \phi^\mu \\
&- \tfrac{1}{4}\rho_a^{\mu\nu}\rho_{\mu\nu}^a + \tfrac{1}{2}m_\rho^2 \rho_\mu^a \rho_a^\mu
\end{aligned}
\tag{16.11}
$$

Here j runs over the spin-1/2 baryons in the octet and T^a is the isospin generator. The various models discussed above correspond to the inclusion or exclusion of some of the terms in $\mathcal{L}_{\text{strong}}$.

In the relativistic mean field approximation we allow the meson fields to acquire density-dependent average values; the nonzero ones are $\bar{\sigma}$, $\bar{\omega}_0$, $\bar{\phi}_0$, and $\bar{\rho}_0^3$. These are driven by the finite densities of particle number, baryon number, strangeness, and isospin asymmetry, respectively. From the Lagrangian, one can read off the effective baryon masses m_j^*,

$$
m_j^* = m_j - g_{\sigma j}\bar{\sigma}
\tag{16.12}
$$

and effective baryon chemical potentials μ_j^*,

$$
\mu_j^* = \mu_j - g_{\omega j}\bar{\omega}_0 - g_{\phi j}\bar{\phi}_0 - I_{3j}g_{\rho j}\bar{\rho}_0^3
\tag{16.13}
$$

where I_{3j} is the third component of the isospin of the jth baryon (1/2 for the proton, $-1/2$ for the neutron, etc.).

The particle densities are given in terms of the Fermi momenta by

$$
n_j = p_{Fj}^3/3\pi^2
\tag{16.14}
$$

The Fermi momenta, in turn, are related to the effective chemical potentials by

$$\mu_j^* = \sqrt{m_j^{*2} + p_{\mathrm{F}j}^2} \tag{16.15}$$

In a neutron star the matter is electrically neutral and in equilibrium under the strong, electromagnetic, and weak interactions. Chemical equilibrium among the baryons listed above, as well as the electrons and muons, implies the relations

$$
\begin{aligned}
\mu_p &= \mu_n - \mu_e & \mu_\Lambda &= \mu_n \\
\mu_{\Sigma^+} &= \mu_n - \mu_e & \mu_{\Sigma^0} &= \mu_n \\
\mu_{\Sigma^-} &= \mu_n + \mu_e & \mu_{\Xi^0} &= \mu_n \\
\mu_{\Xi^-} &= \mu_n + \mu_e &
\end{aligned}
\tag{16.16}
$$

where $\mu_e = \sqrt{m_e^2 + p_{\mathrm{F}e}^2}$, $n_e = p_{\mathrm{F}e}^3/3\pi^2$, and similarly for the muons. Electrical neutrality then requires

$$n_p + n_{\Sigma^+} = n_e + n_\mu + n_{\Sigma^-} + n_{\Xi^-} \tag{16.17}$$

The hyperons and muons will only appear when the baryon chemical potential μ_n is high enough to give them a nonvanishing Fermi momentum.

The total pressure and energy density are expressed in terms of the effective masses and chemical potentials as

$$
\begin{aligned}
P = &\sum_j P_{\mathrm{FG}}(\mu_j^*, m_j^*) + P_{\mathrm{FG}}(\mu_e, m_e) + P_{\mathrm{FG}}(\mu_\mu, m_\mu) \\
&- \tfrac{1}{2}m_\sigma^2\bar{\sigma}^2 - \tfrac{1}{3}bm_{\mathrm{N}}(g_\sigma\bar{\sigma})^3 - \tfrac{1}{4}c(g_\sigma\bar{\sigma})^4 \\
&+ \tfrac{1}{2}m_\omega^2\bar{\omega}_0^2 + \tfrac{1}{2}m_\phi^2\bar{\phi}_0^2 + \tfrac{1}{2}m_\rho^2(\bar{\rho}_0^3)^2
\end{aligned}
\tag{16.18}
$$

$$
\begin{aligned}
\epsilon = &\sum_j \epsilon_{\mathrm{FG}}(\mu_j^*, m_j^*) + \epsilon_{\mathrm{FG}}(\mu_e, m_e) + \epsilon_{\mathrm{FG}}(\mu_\mu, m_\mu) \\
&+ \tfrac{1}{2}m_\sigma^2\bar{\sigma}^2 + \tfrac{1}{3}bm_{\mathrm{N}}(g_\sigma\bar{\sigma})^3 + \tfrac{1}{4}c(g_\sigma\bar{\sigma})^4 \\
&+ \tfrac{1}{2}m_\omega^2\bar{\omega}_0^2 + \tfrac{1}{2}m_\phi^2\bar{\phi}_0^2 + \tfrac{1}{2}m_\rho^2(\bar{\rho}_0^3)^2
\end{aligned}
\tag{16.19}
$$

where P_{FG} and ϵ_{FG} are the Fermi-gas expressions with the quoted effective masses and chemical potentials

The values of the mean vector fields are determined in a transparent way:

$$m_\omega^2 \bar{\omega}_0 = \sum_j g_{\omega j} n_j$$

$$m_\phi^2 \bar{\phi}_0 = \sum_j g_{\phi j} n_j \qquad (16.20)$$

$$m_\rho^2 \bar{\rho}_0^3 = \sum_j I_{3j} g_{\rho j} n_j$$

The mean value of the scalar field must be determined numerically from the self-consistency condition

$$m_\sigma^2 \bar{\sigma} + b m_{\mathrm{N}} g_{\sigma N}^3 \bar{\sigma}^2 + c g_{\sigma N}^4 \bar{\sigma}^3 = \sum_j g_{\sigma j} n_j^{\mathrm{s}} \qquad (16.21)$$

where n_j^{s} is the scalar density of the jth baryon.

There are many parameters in $\mathcal{L}_{\mathrm{strong}}$. The masses are known. The coupling constants $g_{\omega N}$, $g_{\sigma N}$, b, and c were determined in Chapter 11 on the basis of the nuclear saturation density, binding energy, compressibility, and Landau mass. The ρ–nucleon coupling constant can be determined from the charge symmetry coefficient in the symmetry energy:

$$a_{\mathrm{sym}} = \left(\frac{g_{\rho N}}{m_\rho} \right)^2 \frac{p_{\mathrm{F}}^3}{12\pi^2} + \frac{p_{\mathrm{F}}^2}{6 m_{\mathrm{L}}} = 32.5 \text{ MeV} \qquad (16.22)$$

There is considerable uncertainty surrounding the coupling constants in the strange sector. Here we choose $g_{\phi N} = 0$ on the basis that the nucleons have no strange valence quarks while the ϕ meson is composed of $s\bar{s}$. A study of Λ hypernuclei by Rufa *et al.* [2] in the relativistic mean field approximation gives $g_{\sigma\Lambda} = 0.48 g_{\sigma N}$ and $g_{\omega\Lambda} = 0.56 g_{\omega N}$. A study by Keil, Hofmann, and Lenske [3] gives similar numbers, namely, $g_{\sigma\Lambda} = 0.49 g_{\sigma N}$ and $g_{\omega\Lambda} = 0.55 g_{\omega N}$. (For comparison, a study of low-energy nucleon–nucleon and hyperon–nucleon scattering by Maessen, Rijken, and de Swart [4] gives $g_{\sigma\Lambda} = 0.58 g_{\sigma N}$ and $g_{\omega\Lambda} = 0.66 g_{\omega N}$.) These two coupling constants are highly correlated, $g_{\sigma\Lambda}$ being somewhat smaller than $g_{\omega\Lambda}$. The reason is that the binding energy of a Λ hyperon in a nucleus or in nuclear matter depends mainly on the depth of the mean field potential, which is $g_{\omega\Lambda}\bar{\omega}_0 - g_{\sigma\Lambda}\bar{\sigma}_0$. Thus both coupling constants can be increased or decreased together to yield the same mean field potential. For the sake of illustration we shall use the values from Keil *et al.*; based on quark-counting we then estimate that $g_{\sigma\Sigma} = g_{\sigma\Xi} = 0.49 g_{\sigma N}$, $g_{\phi\Lambda} = g_{\omega\Lambda}$, $g_{\omega\Xi} = g_{\omega N}/3$, and $g_{\phi\Xi} = 2 g_{\phi\Lambda}$.

The equation of state for electrically neutral matter, P versus ϵ, is plotted in Figure 16.1. At *low* energy density the pressure of a gas of

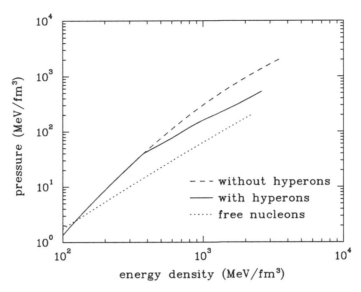

Fig. 16.1.   Equation of state for electrically neutral dense nuclear matter.

noninteracting nucleons (including electrons and muons) is greater than that of nuclear matter that takes into account interactions. The reason is that attractive interactions lower the pressure; in fact, for isospin-symmetric nuclear matter the pressure is zero at the saturation density of nuclear matter. At *high* energy density the situation is reversed; repulsive interactions involving vector mesons cause an increase in the pressure. When hyperons are included the pressure is reduced and the equation of state is said to be softened, on account of energy having been put into hyperon masses rather than into the kinetic energy of nucleons.

The star mass as a function of central energy density, for each of the three model equations of state, is plotted in Figure 16.2. These are obtained as solutions to the Tolman–Oppenheimer–Volkoff equation. The star mass at first increases with central density, reaches a maximum, and then decreases. The maximum mass represents the limit of stability. A star cannot be supported against gravitational collapse to a black hole by going beyond that limit. As can be seen by comparing Figures 16.1 and 16.2, a stiffer equation of state can support a higher maximum mass. A large number of neutron star masses have been measured in binary star systems. The most accurately measured ones tend to fall in the range between 1.4 and 1.5 solar masses. This proves observationally that nuclear interactions are crucial in supporting a neutron star from gravitational collapse; a gas of free neutrons, protons, electrons, and muons can only produce a star with maximum mass less than 0.7 solar mass.

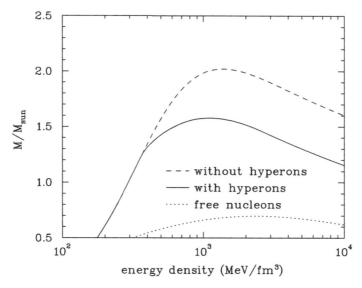

Fig. 16.2. Star mass as a function of central energy density for the three equations of state represented in Figure 16.1.

The chemical abundances of the baryons are very interesting. These are shown in Figure 16.3 for the model equation of state that includes hyperons. At low baryon density the matter is dominated by neutrons. Neutron decay is Pauli-blocked by a small admixture of protons and electrons. As the density goes up it is advantageous for more neutrons to be converted to protons and electrons. Eventually it becomes favorable for nucleons to be converted into hyperons. This is a general feature. However, the order of appearance of hyperons with density and their relative abundances depend sensitively on the numerical values of the coupling constants. Increasing the coupling to the scalar field decreases the effective mass, and decreasing the coupling to the vector fields increases the effective chemical potential, both of which work to favor the appearance of a given hyperon. Note, however, that the maximum-mass star only probes the equation of state up to an energy density of about $1\,\mathrm{GeV\,fm}^{-3}$ and a baryon density of about $0.9\,\mathrm{fm}^{-3} \approx 6n_0$, where n_0 is the nuclear saturation density.

Whether the central density in the most massive neutron stars is great enough to support a core of quark matter has been a topic of much study and debate over the last three decades; if so, the core may be a color superconductor, as described in Section 8.9. Unfortunately, it is very difficult to probe the deep interior of a cold neutron star. A neutron star is born in a supernova, however, and therefore has an initial temperature

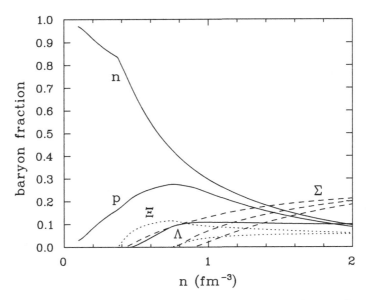

Fig. 16.3. Baryon chemical composition for the equation of state that includes hyperons. Note that the cusps correspond to particle production thresholds.

that may be as high as 40 MeV. The interior of the star cools by several mechanisms, including neutrino production. This is a topic to which we turn our attention now.

16.3 Neutrino emissivity

As mentioned in the previous section, neutron stars are born with a significant amount of thermal energy. A great deal of this is lost by neutrino emission. The microscopic processes are quite varied and complicated. The environments for these processes are usually separated into the outer crust and the inner core; the inner core may be nonsuperfluid or it may be superfluid and magnetized.

Two of the most important reactions in the crust are pair annihilation, $e^+e^- \to \nu\bar{\nu}$, and plasma decay, $\gamma \to \nu\bar{\nu}$. Pair annihilation is quite straightforward, but it was not until 1993 that a fully relativistic treatment of plasma decay (actually the decay of collective excitations of the plasma) was carried out, by Braaten and Segel [5]. One of the most important reactions in the crust is the direct Urca process, $n \to pe^-\bar{\nu}_e$, and the related reaction $pe^- \to n\nu_e$. (The process was named after a casino in Rio de Janeiro by Gamow and Schoenberg [6] who likened thermal energy to money and neutrinos to the casino that takes it away.) There is also a modified Urca process, in which a spectator nucleon N facilitates the process, namely, $nN \to pNe^-\bar{\nu}_e$ and $pNe^- \to nN\nu_e$. The nucleon N, or the neutron or proton for that matter, may be replaced by a hyperon, depending on the chemical conditions in the core. Then there is neutrino cooling by more exotic processes, such as pion condensation, kaon condensation, the Urca process for quarks, or color superconductivity. We shall consider some of these processes in this section. For a comprehensive survey the reader should consult the review by Yakovlev *et al.* [7].

16.3.1 Pair annihilation

When the temperature of the crust or the core reaches 100 keV or so, which is a significant fraction of the electron mass, there will be a significant number of electrons and positrons which can annihilate into neutrino–antineutrino pairs. The rate (number of reactions per unit time per unit volume) can be calculated directly from the cross section:

$$dR = \sigma(e^+e^- \to \nu_l\bar{\nu}_l)v_{\rm rel} \left(2\frac{d^3p_-}{(2\pi)^3}N_{\rm F}^-(p_-) \right) \left(2\frac{d^3p_+}{(2\pi)^3}N_{\rm F}^+(p_+) \right) \quad (16.23)$$

Here the subscript l specifies the neutrino flavor and $v_{\rm rel} = \sqrt{(p_+ \cdot p_-)^2 - m_e^4}/E_+E_-$; the quantities in large parentheses represent the thermal phase space for electrons and positrons, including the spin factor 2 (the Fermi–Dirac occupation numbers are the same as in (5.57)). This expression assumes that neutrinos escape so that there is no Pauli-blocking in the final state. Note that the cross section is proportional to the imaginary part of the forward scattering amplitude and to the square of the invariant amplitude, as discussed in Section 12.2. The same expression can be derived from the finite-temperature field theory rules using the standard model Lagrangian. For the present situation, where the temperature and chemical potential are smaller than the electroweak scale of 100 GeV, we might as well use the cross section as calculated in many texts on the standard model.

The neutrino emissivity Q is the energy radiated into neutrinos per unit time per unit volume. This involves multiplication of dR by the total energy $E_+ + E_-$ and integration over all phase space:

$$
\begin{aligned}
Q_{\rm pair} = \frac{G_{\rm F}^2}{3\pi} \int &\left(\frac{d^3p_-}{(2\pi)^3}N_{\rm F}^-(p_-) \right) \left(\frac{d^3p_+}{(2\pi)^3}N_{\rm F}^+(p_+) \right) (E_+ + E_-) \\
&\times \left\{ C_+^2 \left[m_e^4 + 3m_e^2(p_- \cdot p_+) + 2(p_- \cdot p_+)^2 \right] \right. \\
&\left. + 3m_e^2 C_-^2 \left[m_e^2 + (p_- \cdot p_+) \right] \right\}
\end{aligned}
\quad (16.24)
$$

The Fermi constant is denoted by $G_{\rm F}$. The quantities $C_\pm^2 = \sum_l (C_{Vl}^2 \pm C_{Al}^2)$ are sums over neutrino flavors of the vector and axial-vector constants. Electron neutrinos can be produced via charged or neutral current interactions, involving W and Z vector bosons, respectively, while muon and tau neutrinos can only be produced via neutral current interactions. Thus $C_{Ve} = 2\sin^2\theta_W + 1/2$, $C_{Ae} = 1/2$, $C_{V\mu} = C_{V\tau} = 2\sin^2\theta_W - 1/2$, $C_{A\mu} = C_{A\tau} = -1/2$, with $\sin^2\theta_W \approx 0.23$. The six-dimensional integral for Q_{pair} can be reduced to products of one-dimensional integrals. The latter cannot be found in closed form in general, but they can be evaluated numerically; simple parametrizations for them also exist (see [7]).

A particularly simple limit, although not the most relevant for the majority of periods of neutron star cooling, is the nondegenerate ($N_F \ll 1$) ultrarelativistic ($T \gg m_e$) limit;

$$Q_{\text{pair}} \to \frac{7\zeta(5)}{12\pi} C_+^2 G_F^2 T^9 \tag{16.25}$$

This illustrates how rapidly the cooling rate increases with temperature. In this limit, a ten-fold increase in T results in a billion-fold increase in the emissivity!

16.3.2 Plasma decay

We saw in Chapter 6 that the photon propagator at finite temperature has singularities corresponding to the propagation of transverse and longitudinal modes. Both modes have a finite energy at zero momentum. As a consequence, they will decay into a neutrino–antineutrino pair. This occurs via the coupling of the photon to a (virtual) e^+e^- pair, which then annihilates into neutrinos. The general expression for the emissivity is

$$Q_{\text{plasma}} = \int \frac{d^3k}{(2\pi)^3} \left[2N_B(\omega_T)\omega_T\Gamma_T(\omega_T) + N_B(\omega_L)\omega_L\Gamma_L(\omega_L) \right] \tag{16.26}$$

The N_B are the Bose-Einstein distributions, ω_T and ω_L are the energies of the transverse and longitudinal modes with momentum k, and Γ_T and Γ_L are the decay rates into a $\nu\bar\nu$ pair.

The complete one-loop analysis of the plasma decay rates at arbitrary temperature and chemical potential was carried out by Braaten and Segel [5]. The rates are expressed in terms of the photon longitudinal and transverse self-energies, F and G, and the residues of their poles, Z_L and Z_T. Specifically, $Z_L^{-1}(k_0, \mathbf{k}) = 1 - \partial F(k_0, \mathbf{k})/\partial k_0^2$ and

$Z_T^{-1}(k_0, \mathbf{k}) = 1 - \partial G(k_0, \mathbf{k})/\partial k_0^2$. We have

$$\Gamma_T(k_0, \mathbf{k}) = \frac{G_F^2}{48\pi^2 \alpha} Z_T(k_0, \mathbf{k}) \frac{k^2}{k_0} \left[C_V^2 G^2(k_0, \mathbf{k}) + C_A^2 \Pi_A^2(k_0, \mathbf{k}) \right]$$

$$\Gamma_L(k_0, \mathbf{k}) = \frac{G_F^2}{48\pi^2 \alpha} Z_L(k_0, \mathbf{k}) \frac{k^2}{k_0} C_V^2 F^2(k_0, \mathbf{k}) \tag{16.27}$$

The transverse rate also involves a new axial self-energy Π_A. To leading order in α it is given by

$$\Pi_A(k) = e^2 \frac{k^2}{|\mathbf{k}|} \int \frac{d^3 p}{(2\pi)^3 E} \left[N_F^-(E) - N_F^+(E) \right] \frac{k_0(p \cdot k) - Ek^2}{(p \cdot k)^2 - (k^2)^2/4} \tag{16.28}$$

where $E = \sqrt{\mathbf{p}^2 + m_e^2}$. To first order in α, the term $(k^2)^2/4$ in the denominator can be set to zero; it corresponds to an imaginary part arising from the production of electron–positron pairs. This is unphysical since it does not take into account the dispersion relation of electrons to the same order in α. The resulting expression for Π_A can be expressed as a one-dimensional integral that in general must be done numerically. When used to calculate the emissivity, all functions above are evaluated using the appropriate dispersion relation, either $k_0 = \omega_L(\mathbf{k})$ or $k_0 = \omega_T(\mathbf{k})$.

For neutron star cooling it is numerically efficient to have simple, accurate, analytic formulas for the emissivity. Nice formulas were derived by Braaten and Segel with this in mind. The following expressions were shown to be correct in the classical, degenerate, and relativistic limits for all momenta and correct at small momenta for all temperatures and densities; they were interpolated to an accuracy of order α in between these limits (in what follows $k = |\mathbf{k}|$):

$$\omega_T^2 = k^2 + \omega_P^2 \frac{3\omega_T^2}{2v_*^2 k^2} \left[1 - \frac{\omega_T^2 - v_*^2 k^2}{2v_* k \omega_T} \ln\left(\frac{\omega_T + v_* k}{\omega_T - v_* k} \right) \right] \qquad 0 \le k < \infty \tag{16.29}$$

$$\omega_L^2 = \omega_P^2 \frac{3\omega_L^2}{v_*^2 k^2} \left[\frac{\omega_L}{2v_* k} \ln\left(\frac{\omega_T + v_* k}{\omega_T - v_* k} \right) - 1 \right] \qquad 0 \le k < k_{\max} \tag{16.30}$$

$$k_{\max} = \sqrt{ \frac{3}{v_*^2} \left[\frac{1}{2v_*} \ln\left(\frac{1 + v_*}{1 - v_*} \right) - 1 \right] } \omega_P \tag{16.31}$$

$$v_*^2 \omega_P^2 = \frac{4\alpha}{3\pi} \int_0^\infty \frac{dp\, p^2}{E} \left[5 \left(\frac{p}{E} \right)^2 - 3 \left(\frac{p}{E} \right)^4 \right] N_F(E) \tag{16.32}$$

In these expressions ω_P is the plasma frequency, defined in Chapter 6. The variable v_* lies between 0 and 1. Since we start with two independent variables, T and μ, it is quite natural that the two independent variables ω_P and v_* appear in the result. The longitudinal and transverse energies must still be solved self-consistently from this set of equations.

When evaluated with the dispersion relations calculated above, the self-energies and residues are approximated to the same accuracy, as follows:

$$F = \omega_L^2 - k^2 \tag{16.33}$$

$$G = \omega_T^2 - k^2 \tag{16.34}$$

$$Z_T = \frac{2\omega_T^2(\omega_T^2 - v_*^2 k^2)}{3\omega_P^2\omega_T^2 + (\omega_T^2 + k^2)(\omega_T^2 - v_*^2 k^2) - 2\omega_T^2(\omega_T^2 - k^2)} \tag{16.35}$$

$$Z_L = \frac{2\omega_L^2(\omega_L^2 - v_*^2 k^2)}{(\omega_L^2 - k^2)[3\omega_P^2 - (\omega_L^2 - v_*^2 k^2)]} \tag{16.36}$$

$$\Pi_A = \omega_A k \frac{(\omega_T^2 - k^2)[3\omega_P^2 - 2(\omega_T^2 - k^2)]}{\omega_P^2(\omega_T^2 - v_*^2 k^2)} \tag{16.37}$$

One new frequency appears, which is

$$\omega_A = \frac{2\alpha}{3\pi} \int_0^\infty dp \left[3\left(\frac{p}{E}\right)^2 - 2\left(\frac{p}{E}\right)^4 \right] [N_F^-(E) - N_F^+(E)] \tag{16.38}$$

To calculate the emissivity, first the two dispersion relations must be solved numerically and inserted into the functions appearing in the integrand, and then the one-dimensional integral must be evaluated numerically. However, several limits can be evaluated analytically. Consider the high-temperature limit defined by $T \gg \omega_P$. It can be shown that the contribution of the longitudinal part is smaller than that of the transverse part by a factor of order ω_P^2/T^2, and the axial part is smaller by a factor of order ω_A^2/T^2. The transverse part can be evaluated by setting the factor $\omega_T^2 - k^2$ equal to $m_P^2 = G(k_0 = |\mathbf{k}|)$ (see Section 6.7) because the integral is dominated by $k \gg \omega_T$, and otherwise setting $\omega_T = k$. The emissivity is then given by

$$Q_{\text{plasma}} \to \frac{G_F^2}{24\pi^4\alpha} C_V^2 \zeta(3) m_P^6 T^3 \tag{16.39}$$

In the limit $T \gg |\mu_e|$ and $T \gg m_e$, $m_P^2 \propto \alpha T^2$. Then the emissivity goes as $\alpha^2 G_F^2 T^9$. The powers of the couplings follow from the lowest-order diagrams needed to make the process go, and the power of the temperature follows from dimensional analysis.

16.3.3 Direct Urca process for quarks

The analog of the direct Urca process for quarks is $d \to u + e^- + \bar{\nu}_e$ and $u + e^- \to d + \nu_e$. In beta equilibrium the chemical potentials are related by

$$\mu_d = \mu_s = \mu_u + \mu_e \tag{16.40}$$

If the particles are assumed to be massless, electrical neutrality is achieved without any electrons:

$$n_u = n_d = n_s = n$$
$$n_e = 0 \tag{16.41}$$

where n is the baryon density. At low temperatures the quark Urca process can only occur when all particles are near their Fermi surface; hence, there is very little phase space for the reactions to occur. In particular, if all particles are massless then energy and momentum conservation requires the up quark, down quark, and electron momenta all to be collinear. Giving the d quark a slightly greater mass than the u quark, say 7 MeV versus 5 MeV, does allow the decay to proceed, but very slowly. Iwamoto [8] showed that interactions among the quarks change the situation dramatically.

From Chapter 8 we know that the relation between the Fermi momentum, defined via the density, and the chemical potential is

$$\mu_q = \left(1 + \frac{2}{3\pi}\alpha_s\right) p_{Fq} \tag{16.42}$$

for quark flavors $q = u, d$. For relativistic electrons,

$$\mu_e \approx p_{Fe} \tag{16.43}$$

Therefore $p_{Fd} - p_{Fu} - p_{Fe} \approx -(2/3\pi)\alpha_s p_{Fe} < 0$. This opens up the phase space for the reactions and allows them to occur at a much higher rate. Knowing the decay rate for the down quark, and the cross section for the flavor-changing reaction, both of which could easily be calculated within the standard model, Iwamoto found their sum to be

$$Q_{\text{quark Urca}} = \frac{457}{630}G_F^2 \alpha_s \cos^2\theta_C \, p_{Fd}\, p_{Fu}\, p_{Fe}\, T^6 \tag{16.44}$$

where θ_C is the Cabibbo angle with $\cos^2\theta_C \approx 0.948$. The electron Fermi momentum would be zero if the strange quark mass were zero, but it is not. For the temperatures of interest, say 5 to 50 MeV, p_{Fe} is comparable to T, while p_{Fd} and p_{Fu} are definitely larger than T. The QCD coupling is in the range of 0.1 to 1.0. Therefore the quark Urca process provides quite a large emissivity.

There is also the direct Urca process in which the strange quark replaces the down quark. The current-quark value of the strange quark mass at the scales of relevance is around 105 to 150 MeV. This suppresses the reaction $u + e^- \to s + \nu_e$ but enhances the decay $s \to u + e^- + \bar{\nu}_e$. However, the latter is suppressed by the factor $\sin^2 \theta_\mathrm{C} \approx 0.052$ because it is a strangeness-changing process. Overall one finds that the direct Urca process with the strange quark is smaller than with the down quark.

If the electron Fermi momentum becomes too small then the modified quark Urca process $d + q \to u + q + e^- + \bar{\nu}_e$ and $u + q + e^- \to d + q + \nu_e$ dominates. This was calculated by Burrows [9].

16.4 Cosmological QCD phase transition

The main interest in a cosmological quark–gluon to hadron phase transition arises from its potential to influence the big bang nucleosynthesis. Whether QCD with its known set of parameters undergoes a first-order transition or something smoother is still not completely settled. Assuming that there is a first-order phase transition one needs nucleation theory to understand how the transition proceeds; this topic was discussed in Chapter 13. In this section we first discuss how it can be that nucleosynthesis is affected by a QCD phase transition, and then we analyze the dynamics of a first-order phase transition during the expanding early universe.

16.4.1 Inhomogeneous big bang nucleosynthesis

A cosmological first-order phase transition at $T \sim 160 - 180$ MeV could leave spatial inhomogeneities in the baryon-to-entropy ratio and in the ratio of protons and neutrons. If these inhomogeneities survive to $T \sim 0.1 - 1$ MeV then they could influence nucleosynthesis. This was first pointed out and analyzed by Witten [10], by Applegate, Hogan, and Scherrer [11], and by Alcock, Fuller, and Mathews [12]. In thermal and chemical equilibrium one might expect that the baryon density in the quark–gluon phase is higher than in the hadron phase. This is called the baryon density contrast. Assuming a critical temperature of $160 < T_\mathrm{c} < 180$ MeV, Kapusta and Olive [13] computed this baryon density contrast to be 1.5 to 2.5 when hadronic interactions were neglected and 5 to 7 when they were included. One would expect that the last regions of space to undergo the phase conversion would contain more baryons per unit volume than the first regions to phase-convert because of the lack of time for baryons to diffuse. After phase completion the neutrons will diffuse more rapidly than protons because they are electrically neutral and therefore do not Coulomb-scatter on electrons. This leads to isospin inhomogeneities, at least temporarily.

A detailed calculation of inhomogeneous nucleosynthesis with a comparison to the observed abundances of the light elements was performed

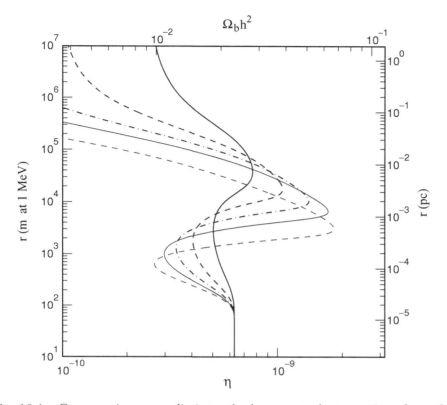

Fig. 16.4. Conservative upper limit to the baryon-to-photon ratio η from the ^{4}He abundance $Y_p \leq 0.248$ and the deuterium abundance $D/H \geq 1.5 \times 10^{-5}$. The three thicker curves are for volume fractions covered by the high-density regions of $1/(2\sqrt{2})$ (solid), $1/8$ (broken), and $1/(16\sqrt{2})$ (broken and dotted). The two thinner curves are for volume fractions $1/64$ (solid) and $1/256$ (broken). From [15].

by Kurki-Suonio *et al.* [14]. They considered baryon density contrasts ranging from 1 to 100 and matter-fractions in the high-density regions ranging from $1/64$ to $1/4$. The average separation of the high-density regions l was left as a free parameter, as was the average baryon-to-photon ratio of the universe. The differential diffusion of protons and neutrons was accounted for and then a standard nucleosynthesis code was run. By fitting the observed abundances of ^{4}He, D, ^{3}He, and ^{7}Li they concluded that the baryon-to-photon ratio must lie between 2×10^{-10} and 7×10^{-10} (or 20×10^{-10} if certain constraints on ^{7}Li were relaxed). They also concluded that $l < 150$ m at the time of nucleosynthesis, whereas at the completion of the QCD phase transition this upper limit would have been only about 1 m. A quantitative theoretical estimate of the latter scale is the purpose of the next subsection.

Recently the inhomogeneous nucleosynthesis calculation was redone, with technical improvements and updated estimates of the cosmic abundances of the relevant light elements, by Kainulainen, Kurki-Suonio, and Sihvola [15]. Their results are shown in Figure 16.4. The high-density matter was distributed in spheres. The inhomogeneities are ineffective in influencing nucleosynthesis unless the high-density regions are separated by more than about 150 m at $T = 1$ MeV.

16.4.2 Dynamics of the phase transition

The nucleation rate for a system of particles or fields that has negligible baryon number compared with the entropy was derived in Section 13.4. Here we mention only the essential details. The change in free energy due to the appearance of a bubble of hadronic matter in quark–gluon plasma is

$$\Delta F = \frac{4\pi}{3} r^3 \left[P_q(T) - P_h(T) \right] + 4\pi r^2 \sigma \tag{16.45}$$

where r is the radius. The critical-sized bubble has radius

$$r_* = \frac{2\sigma}{P_h(T) - P_q(T)} \tag{16.46}$$

which leads to

$$\Delta F_* = \frac{4\pi}{3} \sigma r_*^2 \tag{16.47}$$

The nucleation rate is

$$I = \frac{4}{\pi} \left(\frac{\sigma}{3T} \right)^{3/2} \frac{\sigma (3\zeta_q + 4\eta_q) r_*}{3(\Delta w)^2 \xi_q^4} e^{-\Delta F_*/T} \tag{16.48}$$

It is proportional to the shear viscosity η_q and the bulk viscosity ζ_q in the quark–gluon plasma and is inversely proportional to the square of the enthalpy ($w = \epsilon + P$) difference between the two phases.

For numerical purposes we use a simple bag-model-type equation of state with

$$P_q = (45.5 + 14.25) \frac{\pi^2}{90} T^4 - B$$
$$P_h = (5.5 + 14.25) \frac{\pi^2}{90} T^4 \tag{16.49}$$

The constant 45.5 approximates the effective number of degrees of freedom arising from massless gluons and up and down quarks and a strange quark mass comparable with the temperature. The constant 5.5 approximates the hadronic equation of state near T_c arising from a multitude of massive hadrons. The constant 14.25 arises from photons, neutrinos, electrons, and muons common to both phases. The bag constant B is chosen to give $T_c = 160$ MeV. For definiteness we take $\sigma = 50$ MeV/fm^2, $\xi_q = 0.7$ fm, and $\eta_q = 18T^3$ (see Section 9.6 and Baym *et al.* [16]).

Given the nucleation rate one would like to know the (volume) fraction of space $h(t)$ that has been converted from the quark–gluon plasma to hadronic gas at proper time t in the early universe. This requires kinetic equations that use the nucleation rate I as an input. Here we use a rate equation first proposed by Csernai and Kapusta [17]. The nucleation rate I is the probability of forming a bubble of critical size per unit time per unit volume. If the system cools to T_c at time t_c then at some later time t the fraction of space that has been converted to the hadronic phase is

$$h(t) = \int_{t_c}^{t} dt' \, I(T(t'))[1 - h(t')]V(t', t) \tag{16.50}$$

Here $V(t', t)$ is the volume of a hadronic bubble at time t that was nucleated at an earlier time t'; this takes into account bubble growth. The factor $1 - h(t')$ takes into account the fact that new bubbles can only be nucleated in the fraction of space not already occupied by the hadronic gas. This conservative approach neglects any spatial variation in the temperature. However, it does allow for completion of the transition without violating any of the fundamental laws of thermodynamics.

Next we need a dynamical equation that couples the time evolution of the temperature to the fraction of space converted to the hadronic phase. We use Einstein's equations as applied to the early universe, neglecting curvature. The evolution of the energy density is

$$\frac{d\epsilon}{dR} = -\frac{3w}{R} \tag{16.51}$$

where R is the scale factor at time t. This assumes kinetic but not phase equilibrium and is basically a statement of energy conservation. We express the energy density as

$$\epsilon = h\epsilon_h(T) + (1 - h)\epsilon_q(T) \tag{16.52}$$

where ϵ_h and ϵ_q are the energy densities in the two phases at the temperature T. There is a similar equation for the enthalpy w. The time dependence of the scale factor is determined by the equation of motion

$$\frac{1}{R}\frac{dR}{dt} = \sqrt{\frac{8\pi G\epsilon}{3}} \tag{16.53}$$

This expression can be used to relate the time to the scale factor using the normalization $R(t_c) = 1$.

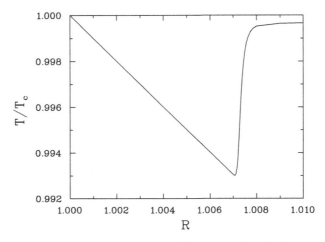

Fig. 16.5.   Temperature as a function of scale factor.

We also need to know how fast a bubble expands once it is created. This is a subtle issue since by definition a critical-sized bubble is metastable and will not grow without a perturbation. After applying a perturbation, a critical-sized bubble begins to grow. As the radius increases the surface curvature decreases, and an asymptotic interfacial velocity is approached. The asymptotic radial-growth velocity will be referred to as $v(T)$. The expected qualitative behavior of $v(T)$ is that the closer T is to T_c the more slowly the bubbles grow. At T_c there is no motivation for bubbles to grow at all since one phase is as good as the other. The bubble-growth velocity was studied by Miller and Pantano [18]. Their hydrodynamical results may be parametrized by the simple formula

$$v\gamma = 3\left(1 - \frac{T}{T_\mathrm{c}}\right)^{3/2} \tag{16.54}$$

which indeed has the expected behavior. A simple illustrative model for bubble growth is then

$$V(t',t) = \frac{4\pi}{3}\left[r_*(T(t')) + \int_{t'}^{t} dt''\, v(T(t''))\right]^3 \tag{16.55}$$

This expression can also be written in terms of R, R', R'' instead of t, t', t''.

We now have a complete set of coupled integro-differential equations, which must be solved numerically. These equation take into account bubble nucleation and growth, energy conservation, and Einstein's equations. They make no assumption about entropy conservation.

Figure 16.5 shows the temperature as a function of the scale factor. For practical purposes, nucleation begins near the bottom of the cooling

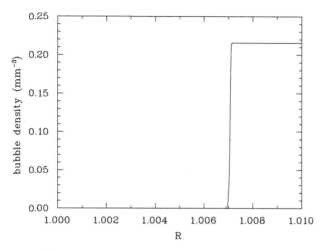

Fig. 16.6. Average bubble density as a function of scale factor.

line. Thereafter, the nucleation and growth of bubbles release latent heat that causes the temperature to rise. The increasing temperature shuts off nucleation, and the phase transition continues owing to the growth of already nucleated bubbles. The temperature can never quite reach T_c; if it did, bubble growth would cease and the transition would never complete. This is a result of the equations of motion and is not an imposition.

Figure 16.6 shows the average bubble density

$$n(R(t)) = \int_{t_c}^{t} dt'\, I(T(t'))[1 - h(t')] \qquad (16.56)$$

as a function of the scale factor. The bubble density rises rapidly just before R reaches 1.007 and reaches its asymptotic value just after 1.007.

Figure 16.7 shows the nucleation rate as a function of scale factor. The rate has a very sharp maximum between 1.0070 and 1.0071. The turn-on and turn-off of the nucleation rate corresponds precisely with the fall and rise of the temperature shown in Figure 16.5.

Figure 16.8 shows the fraction of space h that has made the conversion to the hadronic phase. When $h = 1$ the transition is complete and the temperature will begin to fall again. This occurs when $R \approx 1.4464$, to be compared with the value one would obtain from an ideal Maxwell construction, $R_{\mathrm{Maxwell}} = (239/79)^{1/3} = 1.446\,30. \dots$ In fact the whole curve $h(R)$ is very close to the ideal Maxwell construction, apart from its

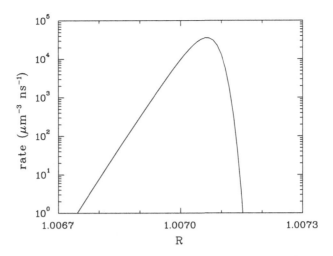

Fig. 16.7.   Nucleation rate as a function of scale factor.

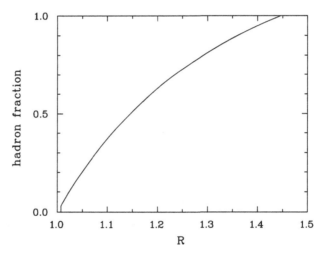

Fig. 16.8.   Volume fraction of space h occupied by the hadronic phase as a function of scale factor.

delayed start, apparent in the figure. The interested reader could work out the Maxwell formula from the equations given here.

Figure 16.9 shows the average bubble radius $\bar{r}$ as a function of scale factor, obtained from

$$\frac{4\pi}{3}\bar{r}^3 n = h \tag{16.57}$$

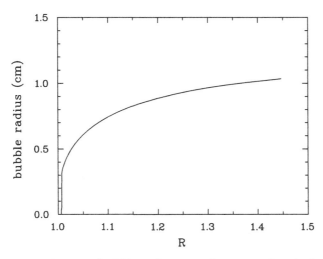

Fig. 16.9. Average bubble radius as a function of scale factor.

It grows with time and with the scale factor, of course. At the end of the
phase transition it is of order 1 cm. This is also the order of magnitude of
the distance between the final quark–gluon plasma regions. Unfortunately,
it is two orders of magnitude too small to affect nucleosynthesis. This
result is rather robust against reasonable variations in any of the input
parameters.

Nucleosynthesis is affected by remnant inhomogeneities in the baryon-
to-entropy ratio and in isospin if the high-baryon-density regions immediately following a QCD phase transition are separated by at least 1 m.
A set of dynamical equations can be written and solved for the evolution
of the universe through such a phase transition all the way to completion.
The evolution of the temperature and hadronic volume fraction as functions of time and scale factor are hardly different from the results of an
idealized Maxwell construction. The information not available in the latter construction is the length scale of the inhomogeneities, that is, bubble
sizes and so on. The characteristic distance between the last regions of
quark–gluon plasma seem to be of order 1 cm, too small to affect nucleosynthesis. However, qualifications and improvements can be made. For
example, when the fraction of space occupied by bubbles exceeds about
50%, interactions among the bubbles probably cannot be neglected. It
is unlikely, though, that further improvements in the dynamics would
qualitatively change the current picture of the transition. Indeed, crude
estimates of the effects of bubble fusion on the dynamics of the QCD
transition in heavy ion collisions indicate that the transition completes
only a little faster, and that the average bubble size is greater (Csernai
et al. [19]). At least this is in the right direction to be interesting.

16.5 Electroweak phase transition and baryogenesis

The standard model conserves baryon and lepton number at the classical level but not at the quantum level. This violation is always a possibility when the current is associated with a global symmetry rather than with a local gauge symmetry. Electric charge, for example, is conserved at both the classical and quantum levels. This phenomenon is called the Adler–Bell–Jackiw anomaly (Bell and Jackiw [20]; Adler [21]). In the standard model the divergence of the baryon current is

$$\partial_\mu J_B^\mu = \frac{N_{\text{fam}}}{64\pi^2} \epsilon^{\mu\nu\rho\sigma} \left(g^2 f_{\mu\nu}^a f_{\rho\sigma}^a + g'^2 g_{\mu\nu} g_{\rho\sigma} \right) \qquad (16.58)$$

Entering on the right-hand side are the field strength tensors for the SU(2) and U(1) gauge fields, in the same notation as in Chapter 15. There is a factor N_{fam} on the right-hand side equal to the number of quark families (the standard model has three). The divergence of the lepton current is exactly the same, so that if the numbers of families of quarks and leptons are the same, as in the standard model, the baryon number minus the lepton number, $B - L$, is conserved. Of course, baryon and lepton number will change only if the field configurations are such that the right-hand side does not vanish.

Gerard 't Hooft [22] showed that, indeed, the conservation of baryon number is violated by the instanton of the weak SU(2) group. (For instantons in QCD see Chapter 8.) The rate for baryon number violation is proportional to the factor $\exp(-16\pi^2/g^2) \approx 10^{-170}$. The probability of observing this effect is exceedingly small with any reasonable estimate of the prefactor. The proton lifetime, for example, has been estimated to be many orders of magnitude larger than the age of the universe. It would seem that this effect is merely a curiosity of quantum field theory.

However, Kuzmin, Rubakov, and Shaposhnikov [23] showed that this is not the case at high temperatures. The reason that baryon number can be violated at zero or low temperatures is that the weak instanton involves tunneling between inequivalent vacua with different baryon numbers. This tunneling is exponentially suppressed by the aforementioned factor. At high temperatures the transition can occur because of thermal fluctuations, and if the temperature is high enough the corresponding Boltzmann factor may not be nearly as small as the tunneling probability. Specifically, they calculated the free energy of a static classical field configuration involving the SU(2) gauge field and the Higgs field. The Boltzmann factor for the baryon-number-violating process is

$$\exp\left(\frac{-F_{\text{sphaleron}}}{T} \right) = \exp\left[\int_0^\beta d\tau \int d^3x \, \mathcal{L}_{\text{eff}}(A_i^a(\mathbf{x}), \Phi(\mathbf{x})) \right] \qquad (16.59)$$

The calculation is done at fixed temperature. Therefore the resummed effective Lagrangian derived in Sections 9.3 and 15.4 can be used. This is a beautiful example of the use of the effective resummed theory. To lowest order, this means that the coupling constant and the Higgs condensate become functions of temperature, $g(T)$, $v(T)$. Before describing the relevant classical solution to the field equations, let us understand the connection between baryon (and lepton) number violation and the Adler–Bell–Jackiw anomaly. Here we follow Klinkhamer and Manton [24], who coined the word *sphaleron* to refer to this and related classical solutions.

We compute the time rate of change of total baryon number as $dB/dt = \int d^3x \partial J_B^0/\partial t$. Let us assume that either the spatial baryon current $\mathbf{J}_B$ vanishes at spatial infinity or that it is periodic in a large box of volume V. In either case Gauss's theorem can be used to express the volume integral of the divergence of the spatial current in terms of a surface integral, which vanishes under the above assumptions. The change in the baryon number, relative to its value as $t \rightarrow -\infty$, is associated with the baryon number of the sphaleron,

$$B_{\text{sphaleron}} = \frac{N_{\text{fam}}g^2}{64\pi^2} \int_{-\infty}^{t} dt' \int d^3x \ \epsilon^{\mu\nu\rho\sigma} f_{\mu\nu}^a f_{\rho\sigma}^a \qquad (16.60)$$

The integrand can be expressed as the divergence of a current:

$$\begin{aligned} \partial_\mu K^\mu &= \frac{1}{2}\epsilon^{\mu\nu\rho\sigma} f_{\mu\nu}^a f_{\rho\sigma}^a \\ K^\mu &= \epsilon^{\mu\nu\rho\sigma} \left(f_{\nu\rho}^a A_\sigma^a - \frac{2}{3}\epsilon_{abc}A_\nu^a A_\rho^b A_\sigma^c \right) \end{aligned} \qquad (16.61)$$

This can be proven by using the classical equations of motion.

To proceed we must have time-dependent fields with finite energy at all times. Furthermore, we want these fields to evolve from the trivial vacuum, $A_\mu^a = 0$, at $t \rightarrow -\infty$ to the sphaleron configuration at time t. Moreover, we want A_μ^a to be a pure gauge field at spatial infinity such that $\mathbf{K} = 0$ there. Then we can write

$$B_{\text{sphaleron}} = \frac{N_{\text{fam}}g^2}{32\pi^2} \int d^3x \ K^0(\mathbf{x}, t) \qquad (16.62)$$

Whether this is nonzero depends on the field configuration. Notice that the sphaleron configuration we discussed earlier was time independent. In fact, to make the identification of baryon number with sphaleron, we first find a static configuration of fields and then make a gauge transformation to satisfy the conditions given above.

Define the dimensionless variable $\zeta = gvr$. The static sphaleron ansatz is

$$A^0 = 0$$
$$\mathbf{A} = v\frac{f(\zeta)}{\zeta}\hat{\mathbf{r}} \times \boldsymbol{\sigma} \tag{16.63}$$
$$\Phi = \frac{v}{\sqrt{2}}h(\zeta)\hat{\mathbf{r}} \cdot \boldsymbol{\sigma} \begin{pmatrix} 0 \\ 1 \end{pmatrix}$$

with boundary conditions $f(0) = h(0) = 0$, $f(\infty) = h(\infty) = 1$. The resulting equations of motion are

$$\zeta^2 f'' = 2f(1-f)(1-2f) - \frac{\zeta^2}{4}(1-f)h^2 \tag{16.64}$$

$$\zeta^2 h'' = -2\zeta h' + 2(1-f)^2 h - \frac{\lambda}{g^2}(1-h^2)h \tag{16.65}$$

These cannot be solved exactly in closed form, although analytic approximations can be found. Klinkhamer and Manton showed that the resulting free energy is $F_{\text{sphaleron}} = (4\pi v/g)F_0(\lambda/g^2)$. The factor F_0 varies smoothly from 1.566 at $\lambda = 0$ to 2.722 at $\lambda = \infty$, with $F_0(1) = 2.10$. The characteristic size of the sphaleron is found to be $1/gv$ simply from dimensional analysis. Note that the characteristic energy is $4\pi v/g \approx 5$ TeV when the parameters are those appropriate to the vacuum.

In order to compute the baryon number of the sphaleron we must make a gauge transformation. We choose the gauge transformation

$$U(\mathbf{x}) = \exp\left(\frac{i}{2}\Theta(r)\boldsymbol{\sigma} \cdot \mathbf{x}\right) \tag{16.66}$$

with a function $\Theta(r)$ that varies smoothly from 0 to π as r varies from 0 to ∞. The function should be chosen so that $\mathbf{A}$ goes to zero faster than $1/r$ as $r \to \infty$, so that $\mathbf{K}$ does not contribute to the integral yielding the baryon number. In particular

$$A_i^a = \frac{[1 - 2f(gvr)]\cos\Theta(r) - 1}{gr^2}\epsilon_{iab}x_b$$
$$+ \frac{[1 - 2f(gvr)]\sin\Theta(r)}{gr^2}\left(\delta_{ia}r^2 - x_i x_a\right) + \frac{1}{g}\frac{d\Theta}{dr}\frac{x_i x_a}{r^2} \tag{16.67}$$

By using this formula in K^0 it is easy to show that the baryon number of the sphaleron is $B_{\text{sphaleron}} = N_{\text{fam}}/2$. This is reasonable since the sphaleron interpolates between two sectors that differ by baryon number 1 for each family. The same holds true for lepton number.

The rate of sphaleron transitions involves primarily the Boltzmann factor, but for numerical purposes the prefactor is needed too. Calculation of the prefactor is analogous to that for the nucleation of bubbles in a first-order phase transition, as analyzed in Chapter 13. The first calculation was performed by Arnold and McLerran [25] who found that

$$\Gamma_{\text{sphaleron}} = \frac{\omega_-}{2\pi}(gv)^3 V \mathcal{N}_{\text{tran}} 8\pi^2 \mathcal{N}_{\text{rot}} \left(\frac{gT}{v}\right)^3 \kappa \exp\left(\frac{-F_{\text{sphaleron}}}{T}\right)$$
(16.68)

Here ω_- is the magnitude of the negative mode causing the instability. It was estimated to be of order gv. The volume of phase space associated with translational zero modes is $(gv)^3 V$, where the volume of the box or universe is V. The volume of rotation space, SO(3), is $8\pi^2$. The factors $\mathcal{N}_{\text{tran}}$ and $\mathcal{N}_{\text{rot}}$ relate to the normalization. They are given as integrals involving the functions f and h describing the sphaleron. Finally there is a determinantal factor κ (not to be confused with the quantity used in Chapter 13), that depends on the ratio λ/g^2. It is this last quantity that is very difficult to compute; this must be done numerically with great care. Carson *et al.* [26] found that $\mathcal{N}_{\text{tran}}$ is a smoothly increasing function, and $\mathcal{N}_{\text{rot}}$ a smoothly decreasing function, of λ/g^2. However, their product has the approximately constant value 90 for $0.1 < \lambda/g^2 < 10$. They found that ω_- is a slowly increasing function of the same ratio of couplings and differs from gv by only 30% as λ/g^2 varies by two orders of magnitude. They calculated κ for four different values of λ/g^2. Baacke and Junker [27] also calculated κ for seven values of λ/g^2. Their results are in approximate numerical agreement. It turns out that κ peaks at $\lambda/g^2 \approx 0.4$ and falls off rapidly for both smaller and larger values of λ/g^2. A simple parametrization that captures this feature is

$$\ln\kappa = \ln\kappa_{\text{max}} - 0.09\left(\frac{\lambda}{g^2} - 0.4\right)^2 - 0.13\left(\frac{g^2}{\lambda} - 2.5\right)^2$$
(16.69)
$$\ln\kappa_{\text{max}} = -3$$

If we now put everything together we find the rate per unit volume,

$$\frac{\Gamma_{\text{sphaleron}}}{V} = 56.3gv(g^2T)^3 \frac{\kappa(\lambda/g^2)}{\kappa_{\text{max}}} \exp\left(-\frac{4\pi v}{gT} F_0(\lambda/g^2)\right)$$
(16.70)

This depends on two scales, gv and g^2T, as well as on the ratio of the quartic and gauge couplings.

For what range of temperature is the sphaleron rate formula given above valid? It assumes that the baryon- and lepton-changing transitions are dominated by the sphaleron configuration and that higher excitations are unimportant. This means that on the one hand the argument of the exponential must be larger than unity, or $T < 4\pi v/g$. On the other hand, it assumes that gv provides an infrared cutoff smaller than the temperature,

$gv < T$. Therefore the expected range of validity is

$$gv < T < 4\pi v/g \qquad (16.71)$$

The values of v, λ, and g are those appropriate to T, not the zero-temperature values. Of these, v changes the most rapidly with T, as we saw in Sections 15.2 and 15.3. If we use the vacuum value $g = 0.637$ and 10% of the vacuum value of $v = 246$ GeV, the temperature range is 16 to 480 GeV. This is centered directly on the electroweak energy scale, which suggests that the baryon and lepton numbers of the universe were essentially determined when the universe had that range of temperatures.

To relate the sphaleron rate to the baryon-number-changing rate we follow Arnold and McLerran [25]. Suppose that the universe has different sectors of baryon and lepton number and a sphaleron appears. It is associated with baryon and lepton numbers equal to $N_{\text{fam}}/2$. The change in free energy of the universe when a sphaleron is formed now involves the extra term $(\Delta N_B \mu_B + \Delta N_L \mu_L)/T$; $\Delta N_B = \Delta N_L = \pm N_{\text{fam}}/2$, the sign being determined by whether the sphaleron increases or decreases the baryon and lepton numbers. The difference in the forward and backward rates involves the factor

$$e^{(\mu_B + \mu_L)N_{\text{fam}}/2T} - e^{-(\mu_B + \mu_L)N_{\text{fam}}/2T} \approx (\mu_B + \mu_L)N_{\text{fam}}/T \qquad (16.72)$$

where the last approximate equality follows because the chemical potentials are extremely small (the observed baryon-to-photon ratio is about 10^{-9}). Furthermore, the sphaleron facilitates the transition between two sectors that differ by a baryon number value equal to the number of families N_{fam}. Therefore the baryon-changing rate is

$$\frac{dN_B}{dt} = -N_{\text{fam}}^2 \frac{\mu_B + \mu_L}{T} \Gamma_{\text{sphaleron}} \qquad (16.73)$$

We need to relate the baryon number to the chemical potentials. We allow for a third chemical potential μ_{E} associated with electric charge. Taking three families of fermions, calculating the electric charge density and setting it to zero, and solving for the chemical potentials we find that $\mu_{\text{E}} = (3\mu_L - \mu_B)/8$. Then the densities are

$$n_B = \frac{5\mu_B + \mu_L}{8} T^2$$
$$n_L = \frac{9\mu_L + \mu_B}{8} T^2 \qquad (16.74)$$

If we further assume that the baryon and lepton numbers of the universe are equal we get $\mu_L = \mu_B/2$ and finally $n_B = (11/16)\mu_B T^2$. Putting this into the baryon-changing rate we finally get

$$\frac{1}{N_B}\frac{dN_B}{dt} = -1100\frac{\kappa(\lambda/g^2)}{\kappa_{max}}g^7 v \exp\left(-\frac{4\pi v}{gT}F_0(\lambda/g^2)\right) \qquad (16.75)$$

The absolute baryon number is decreased by sphalerons no matter whether it starts out positive or negative.

The characteristic time for the relaxation of baryon and lepton numbers to their equilibrium value of 0 is just given by the previous equation. This should be compared with the expansion rate of the universe. According to Einstein's equations the scale factor of the universe evolves according to (16.53). For an equation of state corresponding to $N_{dof} \approx 100$ massless bosonic degrees of freedom the characteristic expansion time scale is found from

$$\frac{1}{R}\frac{dR}{dt} = 1.66\sqrt{N_{dof}}\frac{T^2}{m_{Planck}} \qquad (16.76)$$

where $m_{Planck} = G^{-1/2} = 1.22 \times 10^{19}$ GeV. The baryon-number-changing rate is greater than the expansion rate of the universe for temperatures greater than T_*, that is determined approximately by

$$T_* \ln\left(\frac{v m_{Planck}}{T_*^2}\right) = \frac{4\pi v F_0}{g} \qquad (16.77)$$

The solution to this equation is approximately given by $T_* = v(T_*)$. Within a factor 2 we can estimate T_* to be about 100 GeV, the electroweak scale, that is within the range of validity of the sphaleron approximation to the baryon-changing rate. We would expect the net baryon and lepton numbers of the universe to be determined somewhere around T_*.

For some range of temperatures above the regime of validity of the sphaleron calculation the baryon- and lepton-number-changing reactions are not expected to be suppressed. When $T > 4\pi v/g$ there is no longer a barrier to these reactions. On dimensional grounds the rate per unit volume is then expected to be $Ag^{10}\ln(1/g^2)T^4$, where A is a constant [28, 29]. This involves a factor $(g^2T)^3$, arising from the spatial volume associated with the scale g^2T, and a factor $g^4\ln(1/g^2)T$ arising from the relaxation time. Since the rate per unit volume grows as T^4 and the particle density grows approximately as T^3, the rate per particle grows as T. This should be compared with the T^2 growth of the expansion rate of the universe. Therefore baryon- and lepton-number-changing processes will be predominant for $T_{**} > T > T_*$; it is left as an exercise for the reader to estimate T_{**}.

One can ask a different question. Is it possible for the net baryon and lepton numbers of the universe to be generated at the electroweak scale? This requires three ingredients: baryon- and lepton-changing processes;

CP violation; and a system out of equilibrium. The first has already been demonstrated in the standard model. CP violation also exists in the standard model, as evidenced by neutral kaon oscillations. The requirement that the universe be out of equilibrium is certainly possible if the standard model has a first-order electroweak phase transition. Much work has been done in this context, but the consensus is that there is no first-order electroweak phase transition in the minimal standard model; see Section 15.4. An extension of the minimal standard model to include extra Higgs bosons generally does allow for a first-order phase transition. There also seems to be a consensus that a second-order phase transition is not sufficient to generate baryon and lepton numbers anywhere near their observed values. What happens beyond the minimal standard model is a topic of much current research.

16.6 Decay of a heavy particle

Presumably there is physics beyond the standard model. This may include grand unified theories (GUT), supersymmetry (SUSY), and string theory. A feature common to all of these is the existence of new particles that have masses well above the electroweak scale of 100 GeV. These particles could have been in thermal and chemical equilibrium in the very early universe when the temperature was comparable with or greater than their masses. Since these particles are not observed today they must have been unstable and have decayed to lighter particles. The methods developed in previous chapters are perfectly adapted to describe the physics of these decays at finite temperature.

Following Weldon [30], consider a very heavy scalar field Φ with mass M that decays into a pair of lighter scalar fields ϕ_a and ϕ_b with masses m_a and m_b ($M > m_a + m_b$). The interaction responsible for the decay is taken to be $\mathcal{L}_{\text{int}} = g_s \Phi \phi_a \phi_b$. The self-energy of the Φ can be computed in the one-loop approximation in the usual way:

$$\Pi(k_0 = i\omega_n, \mathbf{k})$$
$$= -g_s^2 T \sum_{j=-\infty}^{\infty} \int \frac{d^3 p}{(2\pi)^3} \frac{1}{\omega_j^2 + \mathbf{p}^2 + m_a^2} \frac{1}{(\omega_j - \omega_n)^2 + (\mathbf{p} - \mathbf{k})^2 + m_b^2}$$
(16.78)

Here ω_n and ω_j are the Matsubara frequencies. After performing the sum the self-energy may be expressed as

$$\Pi(k_0 = i\omega_n, \mathbf{k}) = g_s^2 \int \frac{d^3 p}{(2\pi)^3} \frac{1}{2E_a 2E_b} \left(\frac{1 + n_a + n_b}{k_0 - E_a - E_b} + \frac{n_a - n_b}{k_0 + E_a - E_b} \right.$$
$$\left. + \frac{n_b - n_a}{k_0 - E_a + E_b} - \frac{1 + n_a + n_b}{k_0 + E_a + E_b} \right)$$
(16.79)

The energies are $E_a = \sqrt{\mathbf{p}^2 + m_a^2}$ and $E_b = \sqrt{(\mathbf{p} - \mathbf{k})^2 + m_b^2}$, and the n_a and n_b are the Bose–Einstein occupation numbers.

Since the Φ is unstable its self-energy has both real and imaginary parts. The imaginary part is what concerns us most here. As in Section 6.6, we write $k_0 = \omega - i\gamma$ and assume weak damping, $\gamma \ll \omega$. Then it is easy to see that

$$
\begin{aligned}
\operatorname{Im}\Pi(\omega, \mathbf{k}) = -\pi g_s^2 \int & \frac{d^3 p}{(2\pi)^3} \frac{1}{2E_a 2E_b} \\
\times & \{[(1 + n_a)(1 + n_b) - n_a n_b] \\
\times & [\delta(\omega - E_a - E_b) - \delta(\omega + E_a + E_b)] \\
+ & [n_a(1 + n_b) - n_b(1 + n_a)] \\
\times & [\delta(\omega + E_a - E_b) - \delta(\omega - E_a + E_b)]\}
\end{aligned}
\tag{16.80}
$$

The product $n_a n_b$ has been added and subtracted in each of the terms to provide a transparent physical interpretation. Under the conditions stated above, the kinematically allowed processes are the decay $\Phi \to \phi_a + \phi_b$ and the formation $\phi_a + \phi_b \to \Phi$. The former involves the factor $(1 + n_a)(1 + n_b)$, that is a Bose enhancement of the final state. The latter involves the factor $n_a n_b$ and a relative minus sign as is appropriate for a formation reaction. The overall normalization is governed by the decay amplitude g_s times kinematic factors. At zero temperature all the Bose–Einstein occupation numbers go to zero and $\gamma = -\operatorname{Im}\Pi/2\omega$ just represents the in-vacuum decay. The other terms represent processes that are kinematically forbidden in the present situation but could occur under different ones. They include $\Phi + \phi_a \to \phi_b$, $\Phi + \phi_b \to \phi_a$, $\Phi + \phi_a + \phi_b \to 0$, $\phi_a \to \Phi + \phi_b$, $\phi_b \to \Phi + \phi_a$, $0 \to \Phi + \phi_a + \phi_b$.

It may also be possible for the Φ to decay into a fermion–antifermion pair. This could happen via the interaction $\mathcal{L}_{\text{int}} = g_f \bar{\psi} \psi \Phi$. In that case the imaginary part would be

$$
\begin{aligned}
\operatorname{Im}\Pi(\omega, \mathbf{k}) = -2\pi g_f^2 \int & \frac{d^3 p}{(2\pi)^3} \frac{s - 4m_f^2}{2E_a 2E_b} \\
\times & \{[(1 + n_a)(1 + n_b) - n_a n_b] \\
\times & [\delta(\omega - E_a - E_b) - \delta(\omega + E_a + E_b)] \\
+ & [n_a(1 + n_b) - n_b(1 + n_a)] \\
\times & [\delta(\omega + E_a - E_b) - \delta(\omega - E_a + E_b)]\}
\end{aligned}
\tag{16.81}
$$

The physical interpretation of these terms is exactly analogous to those for the decay of the Φ into bosons.

The imaginary part due to the coupling to either bosons or fermions can be written in a universal format:

$$
\begin{aligned}
\operatorname{Im} \Pi(\omega, \mathbf{k}) = -\frac{1}{2} \int & \frac{d^3 p_a}{2 E_a (2\pi)^3} \frac{d^3 p_b}{2 E_b (2\pi)^3} (2\pi)^4 \\
& \times \big\{ \delta^4(k - p_a - p_b) |\mathcal{M}(\Phi \to a + b)|^2 \\
& \qquad \times [(1 - n_a)(1 - n_b) - n_a n_b] \\
& \quad + \delta^4(k + p_a - p_b) |\mathcal{M}(\Phi + a \to b)|^2 \\
& \qquad \times [n_a(1 - n_b) - n_b(1 - n_a)] \\
& \quad + \delta^4(k - p_a + p_b) |\mathcal{M}(\Phi + b \to a)|^2 \\
& \qquad \times [n_b(1 - n_a) - n_a(1 - n_b)] \\
& \quad + \delta^4(k + p_a + p_b) |\mathcal{M}(\Phi + a + b \to 0)|^2 \\
& \qquad \times [n_a n_b - (1 - n_a)(1 - n_b)] \big\}
\end{aligned}
\tag{16.82}
$$

Here $\mathcal{M}$ is the corresponding amplitude for a given process, whether for bosons or fermions.

This result is of wide application. It applies to final states involving more than two particles also. It easily generalizes to the decay of vector mesons and to the decay of a heavy fermion in an obvious way.

16.7 Exercises

16.1   Derive the formulas for the asymptotic mass and radius of a white dwarf star given in Section 16.1.

16.2   Derive the expression for the charge symmetry coefficient (16.22) given in Section 16.2.

16.3   Using the numbers given in the text, calculate the mean field potential at nuclear saturation density for nucleons and for the Λ, Σ, and Ξ hyperons.

16.4   Calculate the neutrino emissivity for an ultrarelativistic degenerate electron gas ($\mu_e \gg T \gg m_e$).

16.5   Show that the formulas for Z_{T} and Z_{L}, (16.35), (16.36), follow from the previous formulae.

16.6   Look up the relevant matrix element and use it to calculate (16.44).

16.7   Derive formulae for and plot the temperature $T(R)$ and hadronic volume fraction $h(R)$ assuming an idealized Maxwell construction for a QCD phase transition in the early universe.

16.8   Derive the equations of motion for f and h that start from the sphaleron ansatz (16.63).

16.9   Show that the baryon number of a sphaleron is $N_{\mathrm{fam}}/2$ by using (16.67).

16.10 Derive the formulae (16.74) for the baryon and lepton densities.

16.11 Suppose that the baryon-changing rate is given by $Ag^{10}\ln(1/g^2)\,T^4$. If the baryon-to-photon ratio η has the value 10^{-9} at $T = 100$ GeV, what would it have been at earlier times and temperatures? What is your estimate for the temperature T_{**} discussed in the text?

16.12 Consider a very heavy boson of mass M that decays into a massless fermion–antifermion pair. Write down the rate equation for the abundance of these heavy bosons. Solve this equation in the temperature range $M \gg T_0 > T > 100$ GeV in terms of the initial density $n_M(T_0)$.

References

1. Chandrasekhar, S., *Astrophys. J.* **74**, 81 (1931); *Mon. Not. Roy. Astron. Soc.* **95**, 207 (1935).
2. Rufa, M., Stöcker, H., Maruhn, J. A., Greiner, W., and Reinhard, P. G., *J. Phys. G* **13**, L143 (1987).
3. Keil, C. M., Hofmann, F., and Lenske, H., *Phys. Rev. C* **61**, 064309 (2000).
4. Maessen, P. M. M., Rijken, Th. A., and de Swart, J. J., *Phys. Rev. C* **40**, 2226 (1989).
5. Braaten, E., and Segel, D., *Phys. Rev. D* **48**, 1478 (1993).
6. Gamow, G., and Schoenberg, M., *Phys. Rev.* **59**, 539 (1941).
7. Yakovlev, D. G., Kaminker, A. D., Gnedin, O. Y., and Haensel, P., *Phys. Rep.* **354**, 1 (2001).
8. Iwamoto, N., *Phys. Rev. Lett.* **44**, 1637 (1980); *Phys. Rev. D* **28**, 2353 (1983).
9. Burrows, A., *Phys. Rev. D* **20**, 1816 (1979).
10. Witten, E., *Phys. Rev. D* **30**, 272 (1984).
11. Applegate, J. H., Hogan, C. J., and Scherrer, R. J., *Phys. Rev. D* **35**, 1151 (1987).
12. Alcock, C., Fuller, G. M., and Mathews, G. J., *Astrophys. J.* **320**, 439 (1987).
13. Kapusta, J. I., and Olive, K. A., *Phys. Lett.* **B209**, 295 (1988).
14. Kurki-Suonio, H., Matzner, R. A., Olive, K. A., and Schramm, D. N., *Astrophys. J.* **353**, 406 (1990).
15. Kainulainen, K., Kurki-Suonio, H., and Sihvola, E., *Phys. Rev. D* **59**, 083505 (1999).
16. Baym, G., Monien, H., Pethick, C. J., and Ravenhall, D. G., *Phys. Rev. Lett.* **64**, 1867 (1990).
17. Csernai, L. P., and Kapusta, J. I., *Phys. Rev. D* **46**, 1379 (1992).
18. Miller, J. C., and Pantano, O., *Phys. Rev. D* **40**, 1789 (1989); **42**, 3334 (1990).
19. Csernai, L. P., Kapusta, J. I., Kluge, Gy., and Zabrodin, E. E., *Z. Phys. C* **58**, 453 (1993).

20. Bell, J. S., and Jackiw, R., *Nuovo Cimento* **60A**, 47 (1969).
21. Adler, S. L., *Phys. Rev.* **117**, 2426 (1969).
22. 't Hooft, G., *Phys. Rev. Lett.* **37**, 8 (1976); *Phys. Rev. D* **14**, 3432 (1976).
23. Kuzmin, V., Rubakov, V., and Shaposhnikov, M. E., *Phys. Lett.* **B155**, 36 (1985).
24. Klinkhamer, F. R., and Manton, N. S., *Phys. Rev. D* **30**, 2212 (1984).
25. Arnold, P., and McLerran, L., *Phys. Rev. D* **36**, 581 (1987).
26. Carson, L., Li, Xu, McLerran, L., and Wang, R. T., *Phys. Rev. D* **42**, 2127 (1990).
27. Baacke, J., and Junker, S., *Phys. Rev. D* **49**, 2055 (1994); **50**, 4227 (1994).
28. Arnold, P., Son, D., and Yaffe, L. G., *Phys. Rev. D* **55**, 6264 (1997).
29. Bödeker, D., *Phys. Lett.* **B426**, 351 (1998).
30. Weldon, H. A., *Phys. Rev. D* **28**, 2007 (1983).

Bibliography

General relativity

Weinberg, S. (1972). *Gravitation and Cosmology* (Wiley, New York).
Misner, C. W., Thorne, K. S., and Wheeler, J. A. (1973). *Gravitation* (Freeman, San Francisco).

Compact astrophysical objects

Shapiro, S. L., and Teukolsky, S. A. (1983). *Black Holes, White Dwarfs, and Neutron Stars* (Wiley, New York).
Glendenning, N. K. (2000). *Compact Stars*, 2nd edn (Springer, New York).

Newborn neutron stars

Prakash, M., Bombaci, I., Prakash, M., Ellis, P. J., Lattimer, J. M., and Knorren, R., *Phys. Rep.* **280**, 1 (1997).

Conclusion

In this book we have developed relativistic quantum field theory at finite temperature and density. We have studied extensively the theories of three of the four fundamental forces of nature: QED, QCD, and the Glashow–Weinberg–Salam theory of the weak interactions. In its nonrelativistic quantum mechanical guise, QED is responsible for the structure of atomic and molecular systems. Here we have focused on the properties of relativistic plasmas as realized in astrophysical environments. We have studied the screening of static electric charges, the propagation of collective excitations with the quantum numbers of the photon and the electron, shear and bulk viscosities, and thermal and electrical conductivities. We have also used the cold equation of state of dense electrons to calculate the masses and radii of white dwarf stars.

Spontaneous symmetry breaking is an important concept in both the strong and the electroweak interactions. When such symmetries are broken, the result is Goldstone bosons that reflect the underlying symmetry. In simple models illustrating this phenomenon, the spontaneously broken symmetry is restored at high enough temperatures, often via a second-order phase transition. An extension of these models to include gauge bosons reveals the Higgs mechanism, whereby one of the would-be Goldstone bosons combines with a gauge boson to produce a massive vector boson with three spin states. In simple enough models, this symmetry is restored at high temperatures.

QCD is the theory of quarks and gluons. We have studied it using perturbation theory and have found the limitations of the latter. The minimum extension is to sum the set of ring diagrams. This gives a contribution of order g^3 to the pressure at high temperature. Contributions of order g^4, $g^4 \ln g^2$, g^5, and $g^6 \ln g^2$ have all been computed at high temperature, with rather slow convergence. The ring diagrams spawned a more elaborate technique that goes under the title of hard thermal loops.

They are important for calculating various linear-response properties of quark–gluon plasma, such as the emission of electromagnetic radiation in the form of photons and lepton pairs. At asymptotically high temperatures asymptotic freedom forces $g^2(T)$ to go to zero, albeit only logarithmically. Since individual quarks and gluons are never observed at zero and low temperatures, due to confinement, only color-neutral objects, or hadrons, can exist there. Numerical calculations with lattice gauge theory show conclusively that for the physical three-color theory without quarks, there is a first-order phase transition separating the two phases. For two flavors of massless quarks it should be a second-order transition, and for three massless flavors it should be first order. The answer for two up and down quarks, which are light, and one slightly heavier strange quark is still not known with certainty. Cold dense quark matter has been shown to be color superconducting. Various ways of pairing quarks can occur, including two-flavor superconducting and color-flavor-locked superconducting.

At subcritical baryon densities, the most economical way to describe the system is in terms of nucleon and hyperon degrees of freedom. The simplest model that displays the main features of nuclear matter is the Walecka model, which is readily solved in the mean field approximation. Sophistications can include more interactions and more fields, and solving to a higher number of loops. Complications with the former occur at high densities when the baryons are densely packed and multiparticle interactions become important. Complications with the latter are due to the large, order of 10, coupling constants. In any case, the philosophy is to construct the most sophisticated Lagrangian possible, that reflects the symmetries of QCD and low-energy scattering properties, and then to calculate the partition function to the best of one's abilities. The goal is to extrapolate to high densities, such as those in a neutron star. In fact, dozens of such stars have been observed with masses measured to be twice that of a star composed of neutrons alone, thereby showing the crucial importance of including interactions and/or other degrees of freedom.

Hot hadronic matter occurs at subcritical energy densities and with small or zero baryon density. The symmetries of QCD, particularly chiral symmetry, again restrict the form of effective Lagrangians used to describe the properties of this matter. The equation of state at small temperatures is quite well determined. As the temperature rises, more and more of the hundreds of hadrons observed in particle physics experiments are created, and the interactions among them are complicated and generally unknown. Still, it is important to understand this type of matter for it is the ultimate fate of quark–gluon plasma created in high-energy heavy ion collisions, as explored at accelerators at Brookhaven National Laboratory and at CERN. Signatures of the formation of quark–gluon plasma include the thermal emission of photons and lepton pairs, J/ψ production, strangeness production, and the relative abundances of numerous species of mesons and baryons.

The early universe provides an ideal setting to study matter at extraordinarily high temperatures. If QCD, for example, does undergo a first-order phase transition with its physical parameters then one may study the nucleation of the low-density hadronic phase from the high-density quark–gluon phase and the subsequent evolution of the bubbles and drops. The resulting inhomogeneities in energy density, baryon density, and isospin density may even influence nucleosynthesis at later times. At an even earlier epoch it was suspected that the spontaneously broken symmetry of the combined electroweak interactions might have been restored. A mean field approximation yields a second-order phase transition, but this becomes a very weak first-order transition when a resummation of the ring diagrams is done. This might have bided well for baryogensis occurring at this time via nonperturbative field configurations or sphalerons. However, it turns out that the order and even existence of a transition depends on the value of the quartic coupling in the Higgs sector, or rather on the Higgs mass. Lattice calculations in the three-dimensional sector show that present limits on the as yet undiscovered Higgs boson preclude a phase transition.

The reader should now be in a position to read the current literature on finite-temperature field theory and to make original contributions. There are a large number and variety of topics that require investigation. Neutron stars are being discovered all the time. Refined calculations of dense nuclear matter are still needed. Comparing their computed mass, radius, glitch characteristics, and cooling rates with observation should be invaluable for learning about the matter inside the densest objects in the universe. Since this is likely to be the only environment where superconducting quark matter may exist, it is necessary to understand it thoroughly. It has been suggested that quark matter at modest densities is actually in a color-superconducting crystalline state; this need to be worked out. The matter formed in high-energy nuclear collisions at RHIC seems to be behaving as a near perfect fluid. What is the nature of quark–gluon matter just above the critical, or crossover, temperature? What are the correlations between quarks and gluons there and how strong are they? Lattice calculations may be the best approach for studying the strongly coupled region in this vicinity. Much has been accomplished, but more work needs to be done even though the first lattice calculations at finite temperature were made twenty-five years ago. Analytical results are always appealing and welcome; the order g^6 and g^7 contributions to QCD should be available in the near future. How far can one go? A topic that has not been covered in this text is the absorption of high-energy jets at RHIC. This may well provide important information on the nature of the matter the jets traverse.

The full equation of state of electroweak theory has not been computed to the same level as it has for QCD. The importance of this theory for the early universe, and the possibility that it affects baryogenesis, strongly suggests that more work ought to be done. The same is true of grand unified theories (GUTs), which attempt to unify the strong, weak, and electromagnetic forces. Supersymmetry and supersymmetric extensions of the standard model have been studied to some extent in the literature but not at the level that QCD has. Hawking radiation has been discussed briefly in this book. It is unique in the sense that so far it is the only concrete connection we have between quantum theory and gravity. How was it manifested in the early universe, and where might it possibly be manifested today? More generally, how can one use thermal field theory in a possible theory-of-everything, namely, string theory? What about dark matter and dark energy?

We hope that, in some way, this book stimulates people to make further progress. There is much to be done. There is work for all!

Appendix

A1.1 Thermodynamic relations

Following is a list of the most commonly encountered thermodynamic functions. They are expressed in terms of their natural variables. This means that if a variational parameter, such as a condensate field, is introduced, the given function is an extremum with respect to variations in the parameter with all natural variables held fixed. To obtain an intensive function from an extensive function in the large-volume, thermodynamic, limit either divide by the volume or differentiate with respect to it. Only one chemical potential is indicated; the generalization to an arbitrary number of conserved charges is obvious. For a general reference, see Landau and Lifshitz [1] and Reif [2].

Grand canonical partition function:

$$Z(\mu, T, V) = \text{Tr} \exp[-\beta(H - \mu\hat{N})] \tag{A1.1}$$

Thermodynamic potential density:

$$\Omega(\mu, T) = -\frac{T \ln Z}{V} = -P(\mu, T)$$

$$V d\Omega = -SdT - PdV - Nd\mu$$

$$\frac{S}{V} = \left(\frac{\partial P}{\partial T}\right)_\mu \tag{A1.2}$$

$$\frac{N}{V} = \left(\frac{\partial P}{\partial \mu}\right)_T$$

Energy:

$$E = E(N, S, V)$$
$$dE = TdS - PdV + \mu dN$$
$$T = \left(\frac{\partial E}{\partial S}\right)_{N,V}$$
$$P = -\left(\frac{\partial E}{\partial V}\right)_{N,S} \tag{A1.3}$$
$$\mu = \left(\frac{\partial E}{\partial N}\right)_{S,V}$$

Helmholtz free energy:

$$F = F(N, T, V) = E - TS$$
$$dF = -SdT + PdV + \mu dN$$
$$S = -\left(\frac{\partial F}{\partial T}\right)_{N,V}$$
$$P = -\left(\frac{\partial F}{\partial V}\right)_{N,T} \tag{A1.4}$$
$$\mu = \left(\frac{\partial F}{\partial N}\right)_{T,V}$$

Gibbs free energy:

$$G = G(N, P, T) = E - TS + PV$$
$$dG = -SdT + VdP + \mu dN$$
$$S = -\left(\frac{\partial G}{\partial T}\right)_{N,P}$$
$$V = \left(\frac{\partial G}{\partial P}\right)_{N,T} \tag{A1.5}$$
$$\mu = \left(\frac{\partial G}{\partial N}\right)_{P,T}$$

A1.2 Microcanonical and canonical ensembles

The level density is defined as

$$\sigma(E) = \sum_{\text{states } s} \delta(E - E_s) \tag{A1.6}$$

The number of states with energies between E and $E + \Delta E$ is the integral

$$\mathcal{N}(E, \Delta E) = \int_E^{E+\Delta E} dE' \sigma(E') \qquad (A1.7)$$

This will be a choppy discontinuous function for low energies but will approach a smooth continuous function at high energies when many states are contained within the energy window ΔE. If there are conserved charges, such as baryon number or electric charge, the sum over states should be restricted to those that have the specified values. For one conserved charge with fixed value N,

$$\sigma(E, N) = \sum_s \delta(E - E_s) \delta_{N, N_s} \qquad (A1.8)$$

The conserved charge involves a Kronecker rather than a Dirac delta function because charge is always discrete. Specifying the exact energy and charge numbers of a system leads to the microcanonical ensemble. This is the situation for an isolated system.

The level density can always be expressed as the Laplace transform of the grand canonical partition function. For example, for a system with no conserved charges,

$$\sigma(E) = \frac{1}{2\pi i} \int_{-i\infty + \epsilon}^{i\infty + \epsilon} d\beta \, e^{\beta E} Z(\beta) \qquad (A1.9)$$

where

$$Z(\beta) = \text{Tr} \, e^{-\beta H}$$

This may be illustrated by applying it to the massless, self-interacting, scalar field theory discussed in Chapter 3. From (3.56) we know that

$$\ln Z = V \left(\frac{\pi^2}{90\beta^3} \right) c(\lambda) \qquad (A1.10)$$

where

$$c(\lambda) = 1 - \frac{5}{24} \left(\frac{9\lambda}{\pi^2} \right) + \frac{5}{18} \left(\frac{9\lambda}{\pi^2} \right)^{3/2} + \cdots$$

Hence

$$\sigma(E) = \frac{1}{2\pi i} \int_{-i\infty + \epsilon}^{i\infty + \epsilon} d\beta \, e^{f(\beta)} \qquad (A1.11)$$

where

$$f(\beta) = \beta E + \ln Z \qquad (A1.12)$$

Asymptotically, when $V \to \infty$ and $E \to \infty$ with E/V fixed, we can evaluate the level density using the saddle-point approximation. The location of the saddle point is determined by $df/d\beta = 0$. This occurs when

$$\beta^4 = \frac{\pi^2 V}{30 E} c(\lambda) \qquad (A1.13)$$

(It is legitimate to neglect the β-dependence of λ induced by the renormalization group to the order $\lambda^{3/2}$ at that we are working.) Then

$$\sigma(E) \approx \left[\frac{e^f}{\sqrt{2\pi d^2 f/d\beta^2}} \right]_{\text{saddle point}}$$
$$= aV^{1/8} E^{-5/8} \exp\left(bV^{1/4} E^{3/4} \right) \qquad (A1.14)$$

where

$$a = \frac{1}{2} \left(\frac{c(\lambda)}{480\pi^2} \right)^{1/8} \qquad b = \frac{4}{3} \left(\frac{\pi^2 c(\lambda)}{30} \right)^{1/4} \qquad (A1.15)$$

The saddle point value of β is therefore just the inverse temperature. Notice that the saddle point condition (A1.13) can also be written as

$$\frac{E}{V} = \frac{\pi^2}{30} T^4 c(\lambda) \qquad (A1.16)$$

that agrees with the energy density obtained via $-P + TdP/dT$ from (3.56). Furthermore, the level density (A1.14) agrees with that derived on the basis of single-particle phase space [3] when we set $\lambda = 0$.

The canonical ensemble refers to a system in a box of volume V, maintained at temperature T by thermal contact with a heat reservoir but with a fixed number of conserved charges. For a system with just one conserved charge, say baryon number, the canonical partition function is

$$Z_c(N, T, V) = \frac{1}{2\pi} \int_{-\pi}^{\pi} d\theta \, e^{-i\theta N} Z(\theta) \qquad (A1.17)$$

where

$$Z(\theta) = \text{Tr} \, e^{-\beta H + i\theta \hat{N}}$$

Notice the integral representation of the Kronecker delta on account of the discreteness of baryon number. Make the change of variable $\theta = -i\beta\mu$. Then

$$Z = \text{Tr} \, e^{-\beta(H - \mu\hat{N})} \qquad (A1.18)$$

that is the familiar form, albeit with an imaginary chemical potential.

As an illustration, recall the partition function for a massless noninteracting gas of fermions:

$$\ln Z = \frac{V}{12\pi^2\beta^3}\left(\beta^4\mu^4 + 2\pi^2\beta^2\mu^2 + \frac{7}{15}\pi^4\right) \qquad (A1.19)$$

Then

$$Z_{\rm c} = \frac{\beta}{2\pi i}\int d\mu\, {\rm e}^{f(\mu)} \qquad (A1.20)$$

where

$$f = -\beta\mu N + \ln Z$$

The saddle point is determined by the condition

$$\frac{N}{V} = \frac{\mu}{3\pi^2}\left(\mu^2 + \pi^2 T^2\right) \qquad (A1.21)$$

which is just the expression for the baryon density in the grand canonical ensemble, namely, $\partial P(\mu, T)/\partial\mu$. In the large-volume limit with fixed intensive quantities,

$$Z_{\rm c}(N, T, V) \approx V^{-1/2}\left(\frac{2T\mu^2}{\pi} + \frac{2\pi T^3}{3}\right)^{-1/2}$$

$$\times \exp\left[\frac{V}{12\pi^2}\left(-\frac{3\mu^4}{T} - 2\pi^2 T\mu^2 + \frac{7\pi^4 T^3}{15}\right)\right] \quad (A1.22)$$

In this equation, μ is given by (A1.21) as a function of N/V and T. Up to corrections of relative order $(\ln V)/V$ the canonical partition function is

$$T\ln Z_{\rm c} = T\ln Z - \mu N = PV - \mu N = -F \qquad (A1.23)$$

It is also possible to fix the total three-momentum of the system [4] and to pick out the singlet states of SU(N) gauge theories [5]. Different boundary conditions on the surface, such as periodic, Dirichlet, Neumann, and Cauchy, result in contributions to the free energies that scale as the surface area but with differing coefficients. Compared with the volume contributions they are of no importance in the large-volume, thermodynamic, limit and so we do not discuss them further.

A1.3 High-temperature expansions

Frequently a high-temperature ($T \gg m$) expansion of an integral like

$$h_n(y) = \frac{1}{\Gamma(n)}\int_0^\infty \frac{dx\, x^{n-1}}{\sqrt{x^2 + y^2}}\frac{1}{{\rm e}^{\sqrt{x^2+y^2}} - 1} \qquad (A1.24)$$

is desired, where $y = m/T$. These integrals satisfy the differential equation

$$\frac{dh_{n+1}}{dy} = -\frac{yh_{n-1}}{n} \tag{A1.25}$$

The high-temperature expansion is obtained by using the identity

$$\frac{1}{e^z - 1} = \frac{1}{z} - \frac{1}{2} + 2\sum_{l=1}^{\infty} \frac{z}{z^2 + (2\pi l)^2} \tag{A1.26}$$

multiplying the integrand by $x^{-\epsilon}$, integrating term by term, and letting $\epsilon \to 0$ at the end. One obtains

$$h_1(y) = \frac{\pi}{2y} + \frac{1}{2}\ln\left(\frac{y}{4\pi}\right) + \frac{1}{2}\gamma_E - \frac{1}{4}\zeta(3)\left(\frac{y}{2\pi}\right)^2 + \frac{3}{16}\zeta(5)\left(\frac{y}{2\pi}\right)^4 + \cdots \tag{A1.27}$$

where $\gamma_E = 0.5772\ldots$ is Euler's constant and $\zeta(3) = 1.202\ldots$, $\zeta(5) = 1.037\ldots$ are specific values of the Riemann zeta function $\zeta(n)$. Also

$$h_2(y) = -\ln\left(1 - e^{-y}\right) \tag{A1.28}$$

For example, the pressure of a noninteracting spinless boson field is

$$P = \frac{4T^4}{\pi^2} h_5\left(\frac{m}{T}\right) = \frac{\pi^2}{90}T^4 - \frac{m^2 T^2}{24} + \frac{m^3 T}{12\pi} \\ - \frac{m^4}{32\pi^2}\left[\ln\left(\frac{4\pi T}{m}\right) - \gamma_E + \frac{3}{4}\right] + \mathcal{O}\left(\frac{m^6}{T^2}\right) \tag{A1.29}$$

The analysis for a noninteracting charged spinless boson field is only slightly more complicated. See Haber and Weldon [6] for details. In the limit $T \gg m > |\mu|$ the pressure is

$$P = \frac{\pi^2}{45}T^4 - \frac{(m^2 - 2\mu^2)T^2}{12} + \frac{(m^2 - \mu^2)^{3/2}T}{6\pi} + \frac{(3m^2 - \mu^2)\mu^2}{24\pi^2} \\ - \frac{m^4}{16\pi^2}\left[\ln\left(\frac{4\pi T}{m}\right) - \gamma_E + \frac{3}{4}\right] + \mathcal{O}\left(\frac{m^6}{T^2}, \frac{m^4\mu^2}{T^2}\right) \tag{A1.30}$$

For fermions with zero chemical potential the integral of interest is

$$f_n(y) = \frac{1}{\Gamma(n)}\int_0^\infty \frac{dx\; x^{n-1}}{\sqrt{x^2 + y^2}} \frac{1}{e^{\sqrt{x^2+y^2}} + 1} \tag{A1.31}$$

The f_n satisfy the same differential equation as the h_n,

$$\frac{df_{n+1}}{dy} = -\frac{yf_{n-1}}{n} \tag{A1.32}$$

To evaluate the fermion integral, insert the factor $x^{-\epsilon}$, integrate term by term using the expansion

$$\frac{1}{e^z + 1} = \frac{1}{2} - \sum_{l=-\infty}^{\infty} \frac{z}{z^2 + (2l+1)^2\pi^2} \tag{A1.33}$$

and let $\epsilon \to 0$ at the end. One obtains [7]

$$f_1(y) = -\frac{1}{2}\ln\left(\frac{y}{\pi}\right) - \frac{1}{2}\gamma_E + \cdots$$
$$f_2(y) = \ln(1 + e^{-y}) \tag{A1.34}$$

For a noninteracting gas of fermions with $\mu = 0$ the pressure is

$$P = \frac{16T^4}{\pi^2}f_5\left(\frac{m}{T}\right) = \frac{7\pi^2}{180}T^4 - \frac{m^2T^2}{12}$$
$$+ \frac{m^4}{8\pi^2}\left[\ln\left(\frac{\pi T}{m}\right) - \gamma_E + \frac{3}{4}\right] + \mathcal{O}\left(\frac{m^6}{T^2}\right) \tag{A1.35}$$

Notice the absence of an m^3T term, that is present for bosons. For small mass and small chemical potential the high-temperature expansion begins as

$$P = \frac{7\pi^2}{180}T^4 + \frac{(2\mu^2 - m^2)T^2}{12} + \cdots \tag{A1.36}$$

A1.4 Expansion in the degeneracy

The pressure of a noninteracting gas may be expressed as

$$P = (2s + 1)T \int \frac{d^3p}{(2\pi)^3} \ln\left(1 \pm e^{-\beta(\omega-\mu)}\right)^{\pm 1} \tag{A1.37}$$

Here s is the spin, while the upper sign refers to fermions and the lower sign to bosons. The logarithm may be expanded in powers of the exponential and then integrated term by term:

$$P = \frac{(2s+1)m^2T^2}{2\pi^2} \sum_{l=1}^{\infty} \frac{(\mp)^{l+1}}{l^2} e^{l\mu\beta} K_2(lm\beta) \tag{A1.38}$$

Here K_2 is a modified Bessel function of the second kind. This is an expansion in powers of the quantum degeneracy.

The number density, entropy density, and energy density may be calculated using the thermodynamic identities:

$$n = \frac{(2s+1)m^2 T}{2\pi^2} \sum_{l=1}^{\infty} \frac{(\mp)^{l+1}}{l}\, e^{l\beta\mu} K_2(l\beta m)$$

$$s = \frac{(2s+1)m^2 T^2}{2\pi^2} \sum_{l=1}^{\infty} \frac{(\mp)^{l+1}}{l^2}\, e^{l\beta\mu} \Big[(2 - l\beta\mu) K_2(l\beta m)$$

$$+ \tfrac{1}{2}\beta m \left(K_1(l\beta m) + K_3(l\beta m) \right) \Big]$$

$$\epsilon = \frac{(2s+1)m^3 T}{2\pi^2} \sum_{l=1}^{\infty} \frac{(\mp)^{l+1}}{l}\, e^{l\beta\mu} \left[K_1(l\beta m) + \frac{3}{l\beta m} K_3(l\beta m) \right] \quad \text{(A1.39)}$$

These expressions do not include contributions from the antiparticles, if they exist; they may be obtained by the substitution $\mu \to -\mu$. The nonrelativistic limit may be obtained by using the expansions of the Bessel functions $K_n(x)$ when $x \gg 1$:

$$K_n(x) = \sqrt{\frac{\pi}{2x}}\, e^{-x} \left[1 + \frac{4n^2 - 1}{8x} + \frac{(4n^2 - 1)(4n^2 - 9)}{2!(8x)^2} + \cdots \right] \quad \text{(A1.40)}$$

Numerical approximations for both bosons and fermions have been worked out for arbitrary values of m, T, μ by Johns, Ellis, and Lattimer [8].

References

1. Landau, L. D., and Lifshitz, E. M. (1959). *Statistical Physics* (Pergamon Press, Oxford).
2. Reif, F. (1965). *Fundamentals of Statistical and Thermal Physics* (McGraw-Hill, New York).
3. Magalinski, V. B., and Terletskii, Ia. P., *ZhETF (USSR)* **32**, 584 (1957) (*JETP (Sov. Phys.)* **5**, 483 (1957)).
4. Kapusta, J., *Nucl. Phys.* **B196**, 1 (1982).
5. Redlich, K., and Turko, L. Z., *Z. Phys.* **C5**, 201 (1980).
6. Haber, H. E., and Weldon, H. A., *Phys. Rev. Lett.* **46**, 1487 (1981); *Phys. Rev. D* **25**, 502 (1982); *J. Math. Phys.* **23**, 1852 (1982).
7. Dolan, L., and Jackiw, R., *Phys. Rev. D* **9**, 3320 (1974).
8. Johns, S. M., Ellis, P. J., and Lattimer, J. M., *Astrophys. J.* **473**, 1020 (1996).

Index

action, 17, 33, 117, 129, 157–158, 203, 205–206, 208, 244, 271–272, 291–293, 375
Adler–Bell–Jackiw anomaly, 242, 402
analytic continuation, 41–43, 50, 74–76, 89, 153, 177, 215
anomalous dimension, 59, 142
antiparticle, 10, 19, 29, 50, 167
asymptotic freedom, 135–136, 139–145
axial anomaly, 213–214, 242
axial gauge, 65–69, 92, 101, 103, 143, 147, 152–156
axial symmetry, 213
axial-vector current, 254, 277–278, 280, 282

bag constant, 164–165, 321, 396–397
Bardeen–Cooper–Schrieffer (BCS) theory, 123, 166
baryogenesis, 402–408
beta function, 58–59, 81, 139, 142–145
Bjorken model, 318–324, 355
blackbody radiation, 1, 6, 68–70
Boltzmann equation, *see also* Vlasov equation, 190–192, 324
Bose–Einstein condensation, 19–23, 31–32, 50, 118
Bose–Einstein distribution, 4, 18–19, 22, 42, 45, 75–76
bounce solution, 291–294, 299
boundary condition
  antiperiodic (fermions), 28–29, 375
  periodic (bosons), 15, 17, 27, 160, 375
  spatial, 7, 207, 291, 404
Brillouin zone, 206

Cabibbo–Kobayashi–Maskawa matrix, 364
Chandrasekhar limit, 380–382

charge symmetry energy, 385
charmonium, 345–349
chemical equilibrium, 164, 326–327, 356, 384
chiral perturbation theory, 240–247
chiral symmetry, 196, 213–216, 237, 241–242, 254, 256, 261–264
coarse graining, 296, 299–303
collective excitations, 7, 8, 101–107, 156, 193, 390–392
color–flavor locking, 172
color superconductivity, 166–174
color symmetry, 136–137
commutation relations
  bosons, 4, 25, 90–91
  charges, 254
  fermions, 6, 25, 204
completeness, 3, 30, 45
compressibility, 224, 232, 237
condensate, 50, 221–222, 266–267, 271, 273–274, 284–286, 362, 366, 383, 385
conductivity
  electrical, 113–115
  thermal, 109–115, 298, 309, 313
confinement, 135–138, 157, 160, 201–202, 321, 345
connected diagram, 37–38, 40–41, 49
conserved current, 19–20, 24–25, 108–112, 124, 137
contour integral, 41–42, 50, 75–76
correlation length, 128, 294, 300, 304–306, 309, 396–397
correlations, 46, 154, 224
Coulomb field, 94, 123, 201, 219, 236
Coulomb gauge, 92, 101, 103, 143, 147, 156, 168, 182
covariant gauge, 69–73, 92, 101, 103–104, 138–145, 147, 154

425

Printed in the United States
by Baker & Taylor Publisher Services